Cellular and Tissue Engineering: Concepts and Applications

Cellular and Tissue Engineering: Concepts and Applications

Edited by **Shay Fisher**

Syrawood Publishing House
New York

Published by Syrawood Publishing House,
750 Third Avenue, 9th Floor,
New York, NY 10017, USA
www.syrawoodpublishinghouse.com

Cellular and Tissue Engineering: Concepts and Applications
Edited by Shay Fisher

International Standard Book Number: 978-1-68286-243-8 (Hardback)

Printed in the United States of America.

Contents

Preface

Tissue engineering combines biological science with engineering applications. This book consists of contributions made by international experts on complex topics such as types of cells, assembly methods, tissue culture, bioreactors, etc. Also included in this book are detailed elaborations of the applications of cellular and tissue engineering like tissue replacement, repair and regeneration, etc. This book attempts to assist those with a goal of delving into the field of tissue engineering.

After months of intensive research and writing, this book is the end result of all who devoted their time and efforts in the initiation and progress of this book. It will surely be a source of reference in enhancing the required knowledge of the new developments in the area. During the course of developing this book, certain measures such as accuracy, authenticity and research focused analytical studies were given preference in order to produce a comprehensive book in the area of study.

This book would not have been possible without the efforts of the authors and the publisher. I extend my sincere thanks to them. Secondly, I express my gratitude to my family and well-wishers. And most importantly, I thank my students for constantly expressing their willingness and curiosity in enhancing their knowledge in the field, which encourages me to take up further research projects for the advancement of the area.

Editor

1

Stable nuclear transformation of *Pandorina morum*

Kai Lerche and Armin Hallmann*

Abstract

Background: Volvocine green algae like *Pandorina morum* represent one of the most recent inventions of multicellularity diverged from their unicellular relatives. The 8–16 celled *P. morum* alga and its close multicellular relatives constitute a model lineage for research into cellular differentiation, morphogenesis and epithelial folding, sexual reproduction and evolution of multicellularity. *Pandorina* is the largest and most complex organism in the volvocine lineage that still exhibits isogamous sexual reproduction. So far, molecular-biological investigations in *P. morum* were constricted due to the absence of methods for transformation of this species, which is a prerequisite for introduction of reporter genes and (modified) genes of interest.

Results: Stable nuclear transformation of *P. morum* was achieved using chimeric constructs with a selectable marker, a reporter gene, promoters and upstream and downstream flanking sequences from heterologous sources. DNA was introduced into the cells by particle bombardment with plasmid-coated gold particles. The aminoglycoside 3′-phosphotransferase VIII (*aph*VIII) gene of *Streptomyces rimosus* under control of an artificial, heterologous promoter was used as the selectable marker. The artificial promoter contained a tandem arrangement of the promoter of both the heat shock protein 70A (*hsp*70A) and the ribulose-1,5-bisphosphat-carboxylase/-oxygenase S3 (*rbc*S3) gene of *Volvox carteri*. Due to the expression of *aph*VIII, transformants gained up to 333-fold higher resistance to paromomycin in comparison to the parent wild-type strain.
The heterologous luciferase (*gluc*) gene of *Gaussia princeps*, which was previously genetically engineered to match the nuclear codon usage of *Chlamydomonas reinhardtii*, was used as a co-transformed, unselectable reporter gene. The expression of the co-bombarded *gluc* gene in transformants and the induction of *gluc* by heat shock were demonstrated through bioluminescence assays.

Conclusion: Stable nuclear transformation of *P. morum* using the particle bombardment technique is now feasible. Functional expression of heterologous genes is achieved using heterologous flanking sequences from *Volvox carteri* and *Chlamydomonas reinhardtii*. The *aph*VIII gene of the actinobacterium *S. rimosus* can be used as a selectable marker for transformation experiments in the green alga *P. morum*. The *gluc* gene of the marine copepod *G. princeps*, expressed under control of heterologous promoter elements, represents a suitable reporter gene for monitoring gene expression or for other applications in *P. morum*.

Keywords: Co-transformation, *Gaussia princeps* luciferase gene, Genetic engineering, Green algae, Heterologous expression, Reporter genes, Selectable markers, *Streptomyces rimosus aph*VIII gene, Volvocaceae, Volvocine algae

* Correspondence: armin.hallmann@gmx.de
Department of Cellular and Developmental Biology of Plants, University of Bielefeld, Universitätsstr. 25, D-33615 Bielefeld, Germany

Background

The volvocine green algae, a group of ~50 species involving the families Chlamydomonaceae, Goniaceae and Volvocaceae [1], serve as a model lineage for research into cellular differentiation, morphogenesis, epithelial folding and evolution of multicellularity [2-4]. Furthermore, this group is used to study the evolution of sexual reproduction, including the evolution of anisogamy [2,5]. In many lineages, like higher plants and animals, the evolution of multicellularity and sexual reproduction occurred way back in evolutionary history and the molecular origins therefore are obscured by genetic drift. In the volvocine algae the development from unicellular species, which were similar to the present species *Chlamydomonas reinhardtii* [5], to multicellular species and its relatives occurred only about 200 million years ago [1]. Well-known multicellular species in this group are *Volvox*, *Pleodorina*, *Eudorina*, *Pandorina* and *Gonium* [2,4]. The 8–16 celled *Pandorina morum* (Figure 1) [6,7] is intermediate in organizational complexity between the unicellular alga *C. reinhardtii* and the multicellular alga *Volvox carteri* (2,000–4,000 cells) and it has not been subject to molecular analyses and genetic engineering due to a lack of biotechnological possibilities.

The cells of *P. morum* form an ellipsoid spheroid with a distinct anterior-posterior polarity. The way of vegetative reproduction and mode of daughter colony formation where the young colonies break out of the old parental layers reminded Bory of the legend of Pandora's box; thus, he applied the French diminutive, *Pandorina*, to the genus he named [7]. There are several distinctive features of *Pandorina* that warrant its inclusion in the series of volvocine algae to be analyzed molecularly in the future. *Pandorina* has a special morphology among the volvocine algae: the cells fill nearly the entire volume of the colony, as opposed to other, larger volvocine algae in which the cells are arranged at the periphery of the colony, with extracellular matrix filling the space between cells. The compact morphology characteristic of *Pandorina* (Figure 1) appears to be derived from ancestors with expanded ECM [8]. Furthermore, *Pandorina* is the smallest volvocine alga to exhibit a 'colonial boundary,' the tripartite boundary that surrounds the entire colony. This 'colonial boundary' is the most conserved morphological feature of the Volvocalean extracellular matrix [9]. *Pandorina* is moreover the largest and most complex organism in the volvocine lineage that still exhibits isogamous sexual reproduction [2,4]. Thus, the gametes of *P. morum* are almost identical in morphology, differing genetically only in allele expression in the mating-type locus (mating type plus or minus) [10]. The molecular-biological investigation and the comparison of molecular characteristics between the isogamous sexual reproduction in *P. morum* and anisogamous species like *Eudorina* could reveal insights into the evolution of anisogamy, but such studies are hampered by the absence of methods for genetic engineering of *P. morum*. In contrast, the anisogamous species *Eudorina* is accessible for genetic manipulations [11]. The technical feasibility of genetically manipulating a species of interest is a key factor for any detailed molecular analysis of gene or protein functions.

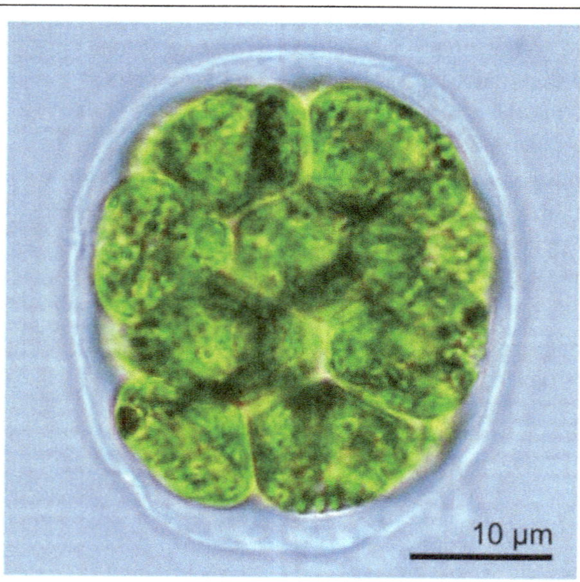

Figure 1 The wild-type phenotype of *P. morum*. A vegetatively grown, ellipsoid spheroid of *P. morum* with ~16 cells. The cells are close together and even slightly narrowed at the inner end to appear keystone- or pear-shaped. Each cell has two flagella with two contractile vacuoles at their base, a single massive, cup-shaped chloroplast with at least one pyrenoid, and an eyespot. The cells are surrounded by a complex but transparent glycoprotein-rich extracellular matrix that holds all of the cells in place.

The nuclear genome of *P. morum* has not been sequenced, only its mitochondrial genome has been revealed [12]. Actually, only the genomes of *C. reinhardtii* [13] and *V. carteri* [14] have been sequenced in the group of volvocine algae. For both species effective transformation protocols exist and they are accessible to molecular manipulations [15-19]. Nuclear transformation protocols also exist for the volvocine algae *Gonium pectorale* [20] and *Eudorina elegans* [11]. There is a single transformation technique that worked in all of the previously transformed volvocine species, namely particle bombardment.

Aside from the nuclear transformation protocol, selectable markers and reporter genes are required for genetic manipulations. In the volvocine algae, selectable markers and reporter genes have mainly been established for *C. reinhardtii* [5,21-28] and, to a lesser extent, for *V. carteri* [28-31], *G. pectorale* [20] and *E. elegans* [11]. In the present study, we wanted to check whether the vectors and transformation methods that were developed earlier for volvocine algae — in particular those for *G. pectorale* [20] and *E. elegans* [11] — can be applied with equal success to *Pandorina*.

The most promising candidate for a selectable marker gene in *Pandorina* was the *aph*VIII gene of *Streptomyces rimosus* that has been used as a selectable marker in *C. reinhardtii* [25], *V. carteri* [30,31], *G. pectorale* [20] and *E. elegans* [11]. The *aph*VIII gene codes for an aminoglycoside 3′-phosphotransferase enzyme that catalyzes the transfer of the γ-phosphate of ATP to the 3′-position of an aminoglycoside antibiotic like paromomycin; this transfer chemically modifies and inactivates the antibiotic molecule. Paromomycin otherwise inhibits cell growth by blocking the protein synthesis of ribosomes at the translocation stage [32].

The best candidate for a useful reporter gene in *P. morum* seemed to be the luciferase gene (*gluc*) of the deep sea copepod *Gaussia princeps* [33,34], which was previously adapted to the nuclear codon usage of *C. reinhardtii* [27]. The Gluc protein catalyzes the oxidation of the substrate coelenterazine into the excited form of coelenteramide that releases a photon with a wavelength of 480 nm upon electron transition to the ground state [34]. The blue light produced by this reaction can be detected by various methods like luminometer measurements and assays on light sensitive films [35]. The *gluc* reporter gene has successfully been introduced into *C. reinhardtii* [27], *G. pectorale* [20] and *E. elegans* [11] and it was functionally expressed in all three organisms.

The selectable markers and reporter genes also require appropriate promoters. Based on previous results, promising candidates were the *C. reinhardtii* promoter of the abundant protein of photosystem I complex gene (*PSAD*) [36], the *C. reinhardtii* promoter of the heat shock protein 70A (*HSP70A*) [37], and the *V. carteri* tandem promoter of the *hsp*70A and *rbc*S3 genes [31]. The *C. reinhardtii HSP70A* promoter and the *V. carteri hsp*70A/*rbc*S3 tandem promoter have already proved useful for expression of the *aph*VIII gene in *V. carteri, C. reinhardtii, G. pectorale* and *E. elegans* [11,20,30].

Here we demonstrate the stable nuclear transformation of *P. morum* using chimeric constructs containing heterologous *aph*VIII and *gluc* genes under the control of heterologous 5′ and 3′ flanking sequences from *V. carteri* and *C. reinhardtii*. The enzyme activity of the heterologously expressed *Gaussia* luciferase was detectable by luminometer measurements and bioluminescence assays. The expression of the heterologous *gluc* gene was further increased in transformants containing the *gluc* gene under control of the heterologous tandem promoter including the heat shock promoter, if the transformants were subject to a temporary shift to higher temperatures.

Results

Antibiotic tolerance of wild-type *Pandorina* algae

To allow for selection of transformants with even weak transgene-mediated resistance, the lowest concentration of the antibiotic paromomycin that kills all wild-type *Pandorina* cells was determined. To achieve this, a series of aliquots with the same number of wild-type *P. morum* cells was exposed to increasing concentrations of paromomycin, incubated for 10 days, and screened for living (green) or dead (white) cells. The alga *P. morum* tolerated concentrations of up to 0.15 μg/ml paromomycin. A concentration of 0.20 μg paromomycin/ml or higher led to 100% cell death (Figure 2A). To allow for an objective and rapid discrimination between aliquots containing living cells and aliquots containing dead cells even in large-scale screenings of culture plates, red-shifted false-color images were created from standard photographs of the plates (Figure 2A, right panel). Red-shifting makes living cells much easier to distinguish from dead cells.

Transformation experiments using *aph*VIII as a selectable marker

The establishment and optimization of the transformation protocol was done by changing several parameters systematically (Additional file 1). In the most successful protocol, 150 ml of a logarithmically growing *P. morum* culture with a density of 7×10^4 cells/ml was harvested by centrifugation. The concentrated algae were spread and immobilized on a cellulose acetate membrane filter and subjected to particle bombardment. The total number of cells used for each transformation experiment was about 9×10^6. Different settings of the Biolistic PDS-1000/He (Bio-Rad) particle delivery system were tested to obtain the maximum number of paromomycin resistant transformants (Additional file 1). The most effective protocol for the biolistic transformation of *P. morum* using DNA-coated gold microprojectiles is described in the Materials and Methods section. The parameters for the most effective transformation protocol are summarized in Table 1.

For all transformation experiments, a chimeric, *aph*VIII-based plasmid (pPmr3) [31] was used as a selectable marker that confers resistance to paromomycin. The pPmr3 plasmid contains the coding sequence of the *Streptomyces rimosus aph*VIII gene, a 5′-flanking sequence that includes an artificial tandem promoter from the *hsp*70A and *rbc*S3 genes of *V. carteri*, and a 3′-flanking sequence derived from the *rbc*S3 gene of *V. carteri* (Figure 3A) [11,20].

Unselectable plasmids were co-transformed with the selectable marker plasmid pPmr3. For use as a reporter gene, the *gluc* gene was brought under control of different heterologous 5′ and 3′ flanking sequences from *V. carteri* and *C. reinhardtii*. The utilized plasmids were pPsaD-GLuc, pHsp70A-GLuc or pHRLucP (Figure 3B-D) [11,20]. After the transformation procedure, the algae were incubated in liquid medium and allowed to recover for 24 h before paromomycin was added. Within 48 h all non-transformed algae died, which resulted in a temporary

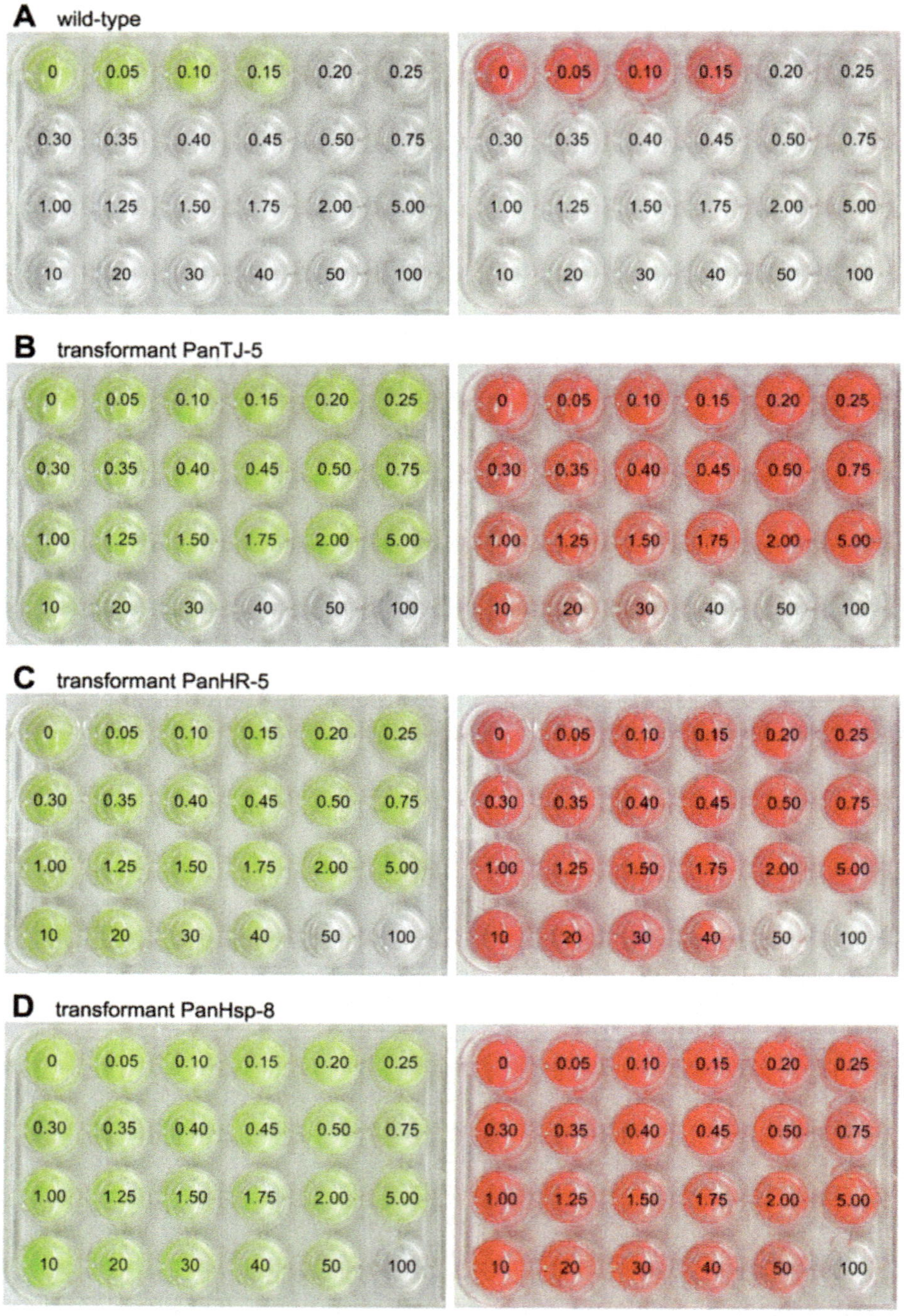

Figure 2 Analysis of paromomycin resistance in wild-type and transgenic *P. morum* strains. For detailed analysis of resistance, identical quantities of *P. morum* cells were exposed to increasing concentrations of paromomycin, and incubated for 10 days. Numbers refer to the concentration of paromomycin [μg/ml] utilized. Natural color (left) and red-shifted, false-color images (right) are shown. **(A)** The wild-type *P. morum* strain. **(B-D)** Transgenic *P. morum* strains co-transformed with pPmr3, the selectable marker plasmid, in addition to a second, non-selectable reporter gene plasmid. **(B)** Transformant PanTJ-5 was co-transformed with the plasmids pPmr3 and pPsaD-GLuc. **(C)** Transformant PanHR-5 was co-transformed with the plasmids pPmr3 and pHRLucP. **(D)** Transformant PanHsp-8 was co-transformed with the plasmids pPmr3 and pHsp70A-GLuc.

Table 1 Most successful combination of parameters for *P. morum* transformation using a PDS-1000/He biolistic device

Parameter	Parameter specification
Material of microprojectiles	gold
Size of microprojectiles	0.6 μm in diameter
Selectable marker plasmid	pPmr3
Coating of microprojectiles	plasmid-DNA/microcarrier/ $CaCl_2$/spermidine/ ethanol -precipitation
Target cells	immobilized on cellulose acetate membrane filter; almost free of liquid
Burst pressure of rupture disk	1350 psi
Rupture disk-macrocarrier distance	7 mm
Macrocarrier-stopping screen distance	8 mm
Stopping screen-target cell distance	6 cm
Chamber evacuation	27 inch Hg
Cultivation after particle bombardment	in liquid medium

clarification of the medium. After 10–14 days of incubation, re-greening of the incubation flasks indicated the initial presence and reproduction of at least one paromomycin-resistant transformant per flask. Using this protocol, a total of 38 paromomycin resistant clones were obtained in 4 independent transformation experiments.

Paromomycin resistance of transformants

The antibiotic resistance of the 38 transformants was justified and concretized by incubation for 10 days in 1 ml JM medium with different concentrations of paromomycin ranging from 0.05 to 100 μg/ml (for concentrations see Figure 2). Transformants with the most robust resistance tolerated concentrations up to 50 μg/ml paromomycin (Figure 2B-D, Table 2). Thus, transformants of *P. morum* generated with the selectable marker plasmid pPmr3 exhibit an up to 333-fold higher tolerance to paromomycin than the wild-type strains. The variation in resistance between different clones is quite common in transformation experiments [11,20]. This could be the result of gene-dosage effects caused by variable number of copies integrated into the genome and/or it can reflect position effects on expression of the transgene due to its random integration into the genome.

Stable genomic integration and expression of the artificial *aph*VIII gene

PCR was used to verify the stable integration of the pPmr3 vector with the artificial *aph*VIII gene into the genome of *P. morum* transformants. The artificial *aph*VIII gene was driven by the *V. carteri hsp*70A/*rbc*S3 tandem promoter and it confers resistance to paromomycin. Genomic DNA was isolated from the 38 transformants about 100 generations after the transformants had been generated. The isolated DNA was used as a template for PCR and a 324 bp chimeric fragment of genomic DNA, containing 107 bp of the *rbc*S3 promoter region and 217 bp of the *aph*VIII coding region, was expected to be amplified with the corresponding specific primers. PCR fragments of the expected size were obtained from many clones (Figure 4A-C); the fragments were cloned and sequenced (Figure 4E). In Figures 4B and 4C the lanes for the transformants PanHR-4, PanHR-10, PanHsp-2, PanHsp-9, PanHsp-10 and PanHsp-11 show no or only weak bands, which might reflect disturbance of the PCR reaction caused by too small DNA amounts, impurities of the genomic DNA or other reasons. Anyway, the repetition of the PCRs with fresh DNA isolates from these six clones yielded fragments of the expected size (data not shown). Moreover, growth of all 38 transformants in the presence of antibiotic already demonstrated the functional expression of the *aph*VIII gene (Figure 2, Table 2).

In addition to the provided indirect evidence of *aph*-VIII expression through the mediated antibiotic resistance in transformants, the transcription of the *aph*VIII gene was also examined directly by RT-PCR analysis. For it, RNA from paromomycin-resistant transformants was reverse transcribed, and products were amplified by PCR using *aph*VIII specific primers. A cDNA fragment with 324 bp was expected. Actually, a PCR fragment of the expected size was obtained from all investigated paromomycin-resistant transformants, but not from the wild-type (Figure 4D). The amplified fragments were cloned and sequenced (Figure 4E).

Calculation of transformation frequency

The transformation frequency was calculated as the ratio of transformants that exhibited both resistance against paromomycin and the correct *aph*VIII fragment following PCR and the total number of cells used for each transformation experiment (9×10^6 cells). Based on the results of 4 independent transformation experiments, a transformation frequency of 1.2×10^{-6} per cell was calculated. This corresponds to a transformation frequency of about 1.9×10^{-5} per organism.

Plasmids for co-transformation with non-selectable reporter gene constructs

To investigate the efficiency of co-transformation in *P. morum,* three different plasmids were used containing the *gluc* gene of *G. princeps*, which was previously engineered to match the codon usage of *C. reinhardtii* [27].

Plasmid pPsaD-GLuc contains the *gluc* gene under control of 5′ and 3′ control elements of the *PSAD* gene of *C. reinhardtii* [27,36] (Figure 3B). In three volvocine species, i.e., *C. reinhardtii, G. pectorale* and *E. elegans*, the

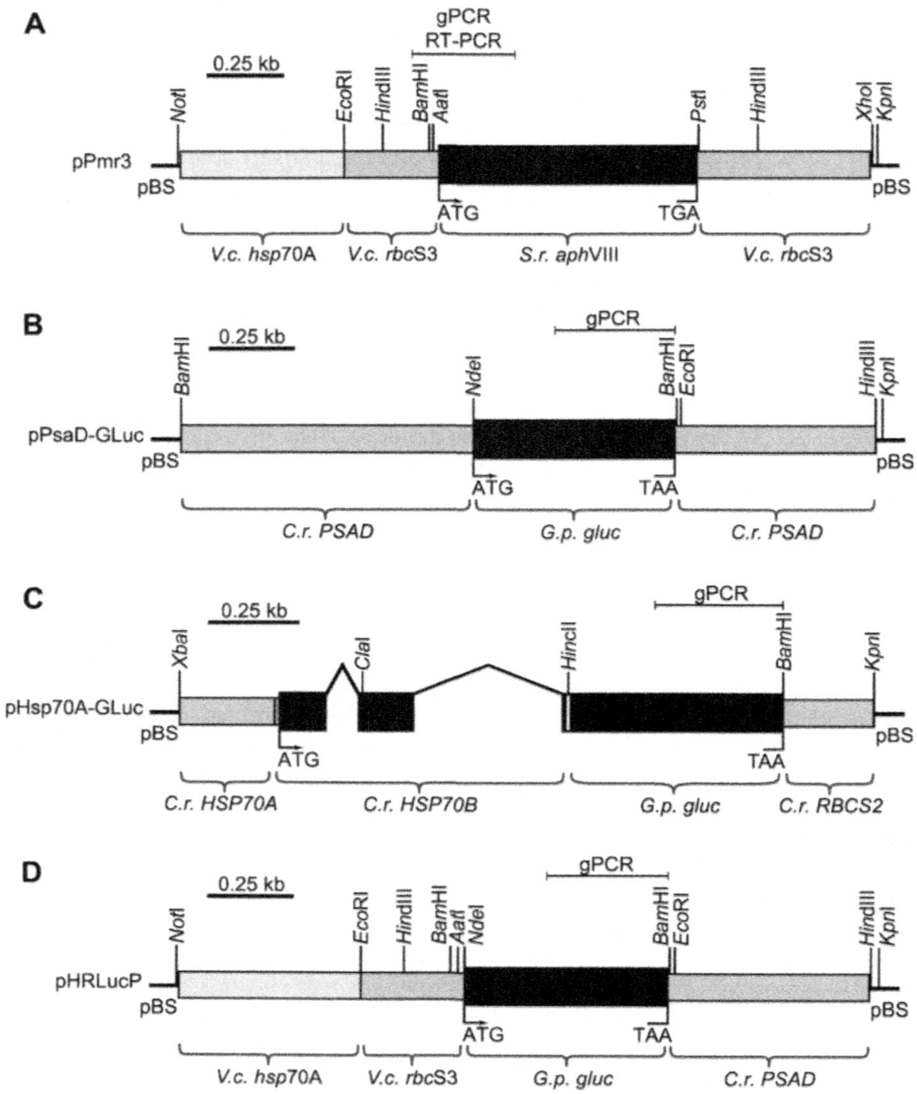

Figure 3 Schematic diagram of the selectable marker plasmid and co-transformed, non-selectable plasmids. (A) The chimeric selectable marker plasmid pPmr3. **(B-D)** Co-transformed, non-selectable plasmids containing reporter genes: **(B)** pPsaD-GLuc, **(C)** pHsp70A-GLuc, and **(D)** pHRLucP. The angled lines in (C) indicate introns. (A-D) Amplified genomic (gPCR) or RT-PCR fragments are indicated. *V.c., Volvox carteri; C.r., Chlamydomonas reinhardtii; S.r., Streptomyces rimosus; G.p., Gaussia princeps; gluc*, luciferase gene; pBS, pBluescript II vector.

PSAD promoter was shown to mediate strong expression [11,20,27].

Plasmid pHsp70A-GLuc [11,20,27] (Figure 3C) contains the *gluc* gene flanked by the *C. reinhardtii HSP70A* 5′UTR and the *RBCS2* 3′UTR of *C. reinhardtii*. The *HSP70A* promoter was shown to mediate not only strong but also heat-inducible expression. The plasmid also contains the first three exons and two introns of the *C. reinhardtii HSP70B* gene that codes for the chloroplast transit peptide of the HSP70B protein [27,38,39].

Except for flanking sequences of *C. reinhardtii*, we also tested flanking sequences of *V. carteri*. Plasmid pHRLucP contains the coding region of the *gluc* gene fused to a *V. carteri hsp*70A/*rbc*S3 tandem promoter and a *C. reinhardtii PSAD* 3′UTR (Figure 3D). Like in plasmid pHsp70A-GLuc, the tandem promoter should mediate not only strong but also heat-inducible expression.

Stable integration of non-selectable reporter gene constructs into the genome

Transformants were generated by co-transformation with the selectable pPmr3 plasmid and one of the three *gluc*-derived, non-selectable reporter gene plasmids at a time. The obtained transformants were screened for stable

Table 2 Results of paromomycin resistance assays*

Transformant	Resistance up to x µg/ml paromomycin	Transformant	Resistance up to x µg/ml paromomycin	Transformant	Resistance up to x µg/ml paromomycin
PanTJ-1	50	PanHR-1	50	PanHsp-1	40
PanTJ-2	50	PanHR-2	50	PanHsp-2	50
PanTJ-3	50	PanHR-3	40	PanHsp-3	50
PanTJ-4	50	PanHR-4	50	PanHsp-4	50
PanTJ-5	30	PanHR-5	40	PanHsp-5	50
PanTJ-6	40	PanHR-6	30	PanHsp-6	40
PanTJ-7	50	PanHR-7	40	PanHsp-7	40
PanTJ-8	40	PanHR-8	40	PanHsp-8	50
PanTJ-9	40	PanHR-9	40	PanHsp-9	50
PanTJ-10	40	PanHR-10	40	PanHsp-10	50
		PanHR-11	50	PanHsp-11	50
		PanHR-12	50	PanHsp-12	40
		PanHR-13	50		
		PanHR-15	50		
		PanHR-21	40		
		PanHR-22	40		

*tolerance of the wild-type strain *P. morum* SAG 32.96: 0.15 µg/ml paromomycin.

integration of the reporter gene constructs by PCR. For it, a PCR reaction was performed using genomic DNA of transformants as a template and *gluc*-specific oligonucleotides. The expected fragment was 342 bp in size. Such a fragment was obtained from transformants produced by co-transformation with the plasmids pPsaD-GLuc (7 positive transformants), pHRLucP (16 positive transformants) and pHsp70A-GLuc (10 positive transformants) (Figure 5A-C). As expected, PCRs with genomic DNA of the wild-type yielded no such product. The identity of the amplified fragments was verified by cloning and sequencing (Figure 5D). Based on the results of the PCR experiments, the co-transformation rate in *P. morum* was determined to be 84% (±15%). The co-transformation rate was defined as the rate of transformants containing the *gluc* gene among the transformants containing the *aph*VIII gene.

Functional expression of co-transformed reporter genes

Transformants generated by co-transformation with the selectable marker plasmid pPmr3 and the *gluc*-derived plasmids pPsaD-GLuc, pHRLucP and pHsp70A-GLuc were assayed for functional expression of luciferase. For it, cell extracts of all transformants were checked for luciferase activity using a luminometer (Figure 6A-C).

For some reason, no significant luciferase activity was observed in transformants generated with pPsaD-GLuc (Figure 6A). The relative light units (rlu) detected in these transformants are about the same as in wild-type cells and correspond to the background light, which is at about 1,000 rlu.

In contrast, luciferase activity was detected in several transformants generated by co-transformation with the plasmid pHRLucP (Figure 6B). In this DNA construct the *gluc* gene is under control of the artificial tandem promoter from the *hsp*70A and *rbc*S3 genes of *V. carteri*. Extracts of ten out of sixteen transformants clearly showed emission of light and, thus, functional expression of luciferase. The amount of detected relative light units (rlu) is significantly higher than in the wild-type strain (853 ± 11 rlu). The transformants PanHR-9 and PanHR-11 exhibited the highest Gluc activity with 407,500 ± 45,000 rlu and 378,500 ± 57,000 rlu, respectively.

Extracts of transformants generated by co-transformation with the plasmid pHsp70A-GLuc showed a significant but mostly moderate luciferase activity when cultivated under standard conditions at 29°C (Figure 6C). In this DNA construct the *gluc* gene is under the control of the *C. reinhardtii HSP70A* promoter; there is also the genomic sequence of the *C. reinhardtii HSP70B* chloroplast transit peptide containing two introns that precedes the *gluc* coding sequence. Only transformant PanHsp-4 exhibited a high Gluc activity of 786,000 ± 20,000 rlu under these conditions, while the wild-type strain showed only 966 ± 50 rlu.

The DNA constructs in pHRLucP and pHsp70A-GLuc are driven by heat-shock promoters that show inducibility by heat in their donor organisms *V. carteri* and *C. reinhardtii*, respectively.

Analysis for inducibility of these promoter elements in *P. morum* was done by subjecting the algae transformed with the plasmids pHRLucP and pHsp70A-GLuc to a

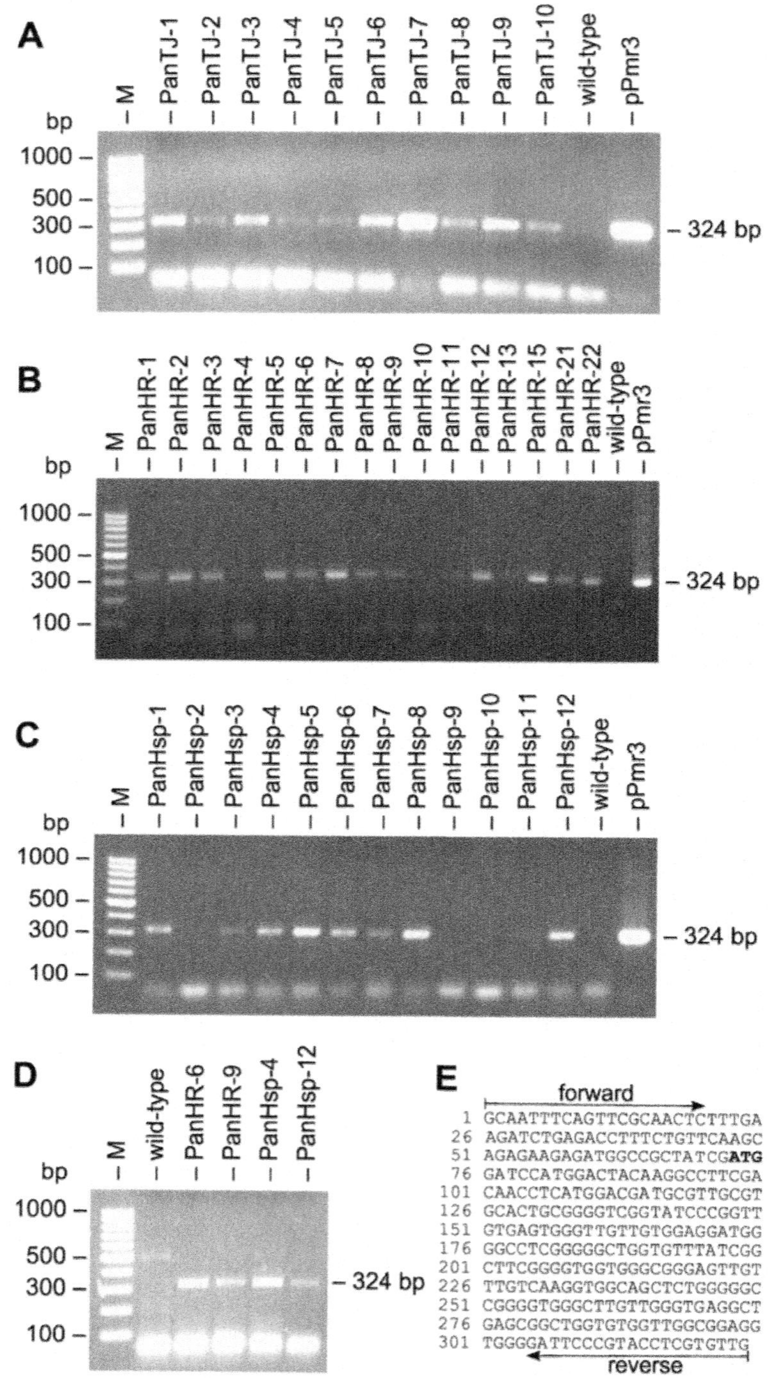

Figure 4 Detection of the *aph*VIII gene in transformants and demonstration of its transcription. (A-C) Paromomycin resistant transformants and the parental wild-type strain were analyzed for presence of the *aph*VIII gene in the genome by genomic PCR. The expected size of the PCR fragment produced in transformants was 324 bp (Figure 3A). The rightmost lane (pPmr3) shows a positive control using pPmr3 plasmid DNA as the template. Lane M refers to the molecular weight marker. **(D)** RT-PCR analysis of *aph*VIII expression. RNA from paromomycin-resistant transformants was reverse transcribed and products were amplified by PCR using primers specific for the *aph*VIII gene construct. Transformants with *aph*VIII expression were expected to yield a 324-bp cDNA fragment (Figure 3A). The parent wild-type strain was used as a control. Lane M, molecular weight marker. **(E)** Sequence obtained from the amplified and cloned genomic and cDNA fragments. The 324-bp fragment contains part of the *aph*VIII gene and part of the 5′ untranslated region from *V. carteri rbc*S3. The positions of the primers and the start codon (bold) are indicated.

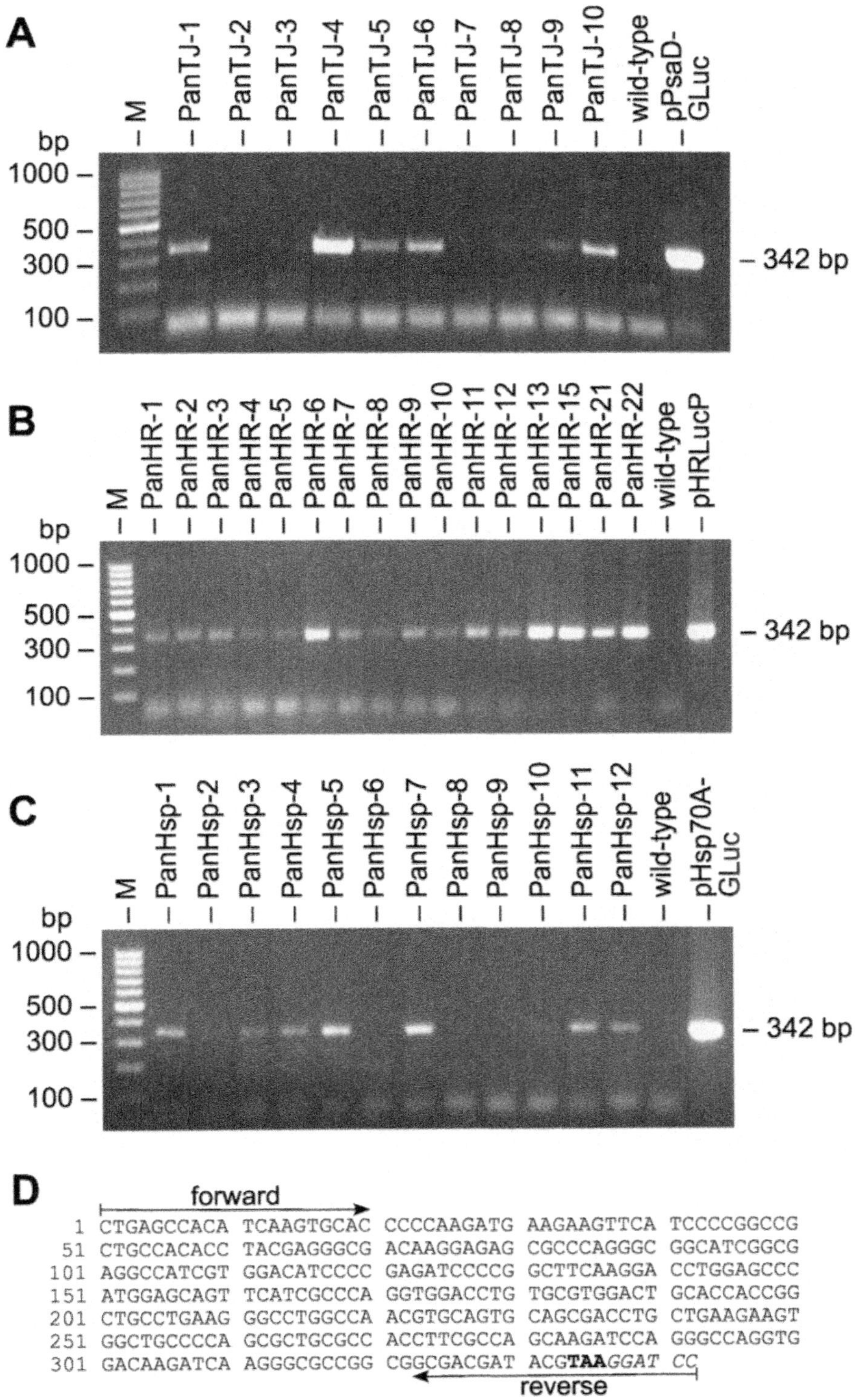

Figure 5 Demonstration of co-transformation of heterologous genes. Paromomycin-resistant transformants were analyzed for the presence of the co-transformed, non-selectable DNA in the genome. PCRs were conducted using genomic DNA from transformants co-transformed with the pPsaD-GLuc **(A)**, pHRLucP **(B)** or pHsp70A-GLuc **(C)** plasmids as template. The parental wild-type strain was analyzed as a control. Primers were specific for the heterologous *gluc* gene, and a 342-bp PCR fragment (Figure 3B-D) was expected in co-transformants. The rightmost lane shows a positive control using DNA of the co-transformed plasmid as the template. Lane M refers to the molecular weight marker. **(D)** Sequence obtained from the amplified and cloned *gluc* fragments. Primer positions, the stop codon (bold), and a *Bam*HI restriction site (italics) are indicated.

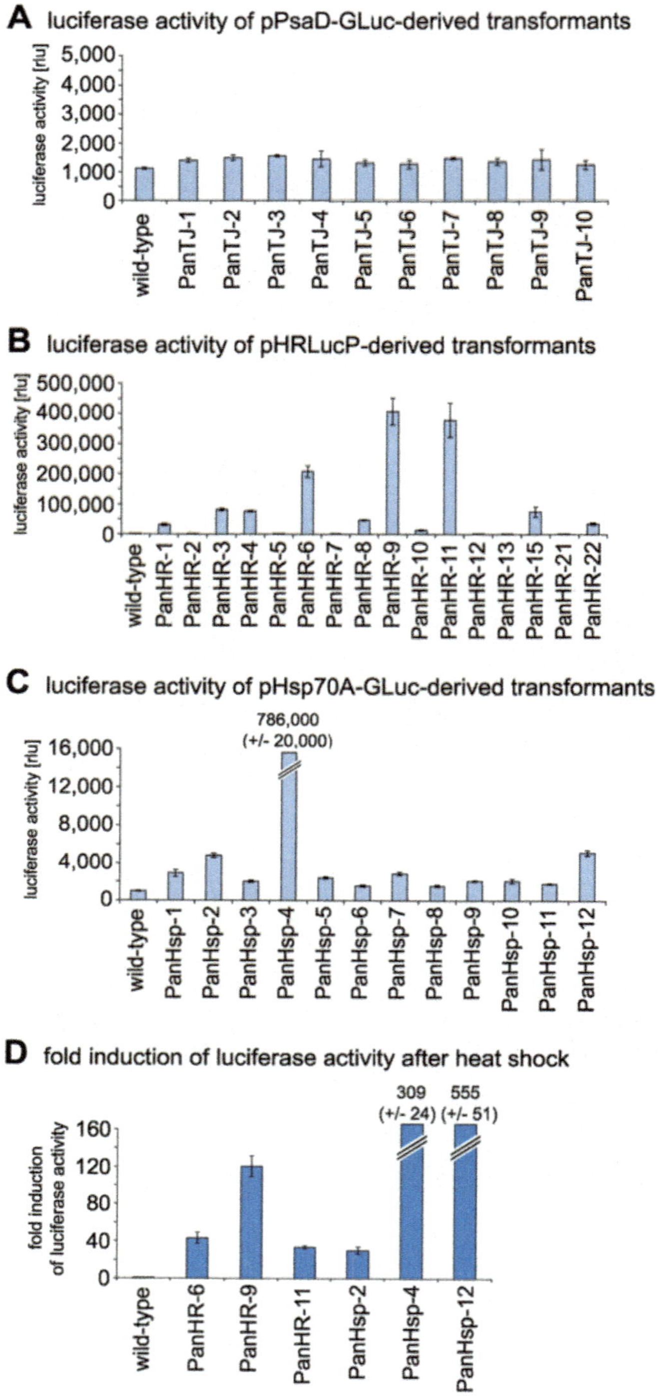

Figure 6 Quantification and inducibility of luciferase activity in transformants expressing the luciferase gene. Luciferase activity in transformants was assayed using a luminometer. The bars represent the mean of three independent experiments. The standard deviation is indicated. **(A)** Luciferase activity of pPsaD-GLuc derived transformants (PanTJ…) compared to the wild-type strain. **(B)** Luciferase activity of pHRLucP derived transformants (PanHR…) compared to the wild-type strain. **(C)** Luciferase activity of pHsp70A-GLuc derived transformants (PanHsp…) compared to the wild-type strain. **(D)** Fold induction of luciferase activity in heat-shocked transformants (39°C) compared to non-heat-shocked transformants. In pHRLucP-derived transformants (PanHR…) the luciferase is driven by the *V. carteri hsp70A/rbcS3* tandem promoter. For pHsp70A-GLuc-derived transformants (PanHsp…) the luciferase gene is driven by the *C. reinhardtii HSP70A* promoter. The parental wild-type strain was analyzed as a control. Transformants and wild-type colonies were subjected to a temperature shift from 29°C to 39°C for 1 h. After a 15 min recovery phase at 29°C, cells were lysed, and luciferase activity was assayed. As a reference control, non-heat-shocked transformants were analyzed in an identical fashion.

temperature shift. The temperature was changed from 29°C to 9 different temperatures between 27°C and 51°C. The strongest increase in luciferase activity was achieved by a temperature shift to 39°C for 1 h. In extracts of transformants generated by co-transformation with plasmid pHRLucP the luciferase activity increased between 33- and 120-fold compared to algae that were kept at 29°C. Extracts of transformants generated by co-transformation with plasmid pHsp70A-GLuc showed a 30- to 555-fold increase in luciferase activity (Figure 6D).

The heat-inducibility of luciferase activity also was investigated by an assay using light sensitive films and intact, undisrupted cells. For it, an aliquot of the transformant cells produced with plasmid pHRLucP was subject to a temperature shift from 29°C to 39°C for 1 h; as a reference, another aliquot of the transformant cells was kept at 29°C. The aliquots were pipetted into separate wells of a 24-well microtiter plate, substrate solution was added and the plates were set down on a light sensitive film for 2 h in the dark. In heat-treated transformants, the light emission caused by luciferase activity was strong enough to turn the light sensitive film totally black (Figure 7). No signal was detectable in transformants that were not subject to the increase in temperature. As expected, both untreated and heat-treated wild-type cells also show no signal.

The molecular weight of the mature Gluc protein in *P. morum* was determined by comparison with a molecular weight marker in an in-gel activity assay. For it, cell extracts of heat induced transformants and controls were separated on a standard SDS-PAGE without thiol reagents. After electrophoresis, SDS was removed from the gel to allow for refolding of the luciferase enzyme into its active conformation. After addition of the substrate solution, the gel was set down on a light sensitive film for 2 h in the dark.

In cell extracts from heat-treated transformants, generated by co-transformation with plasmid pHRLucP (Figure 8, transformant PanHR-11), light emission produced a band with a size of about 20 kDa that reflects the molecular weight of the Gluc protein (Figure 8). For transformants generated by co-transformation with plasmid pHsp70A-GLuc, a size of 31 kDa was calculated for the intact, unprocessed HSP70B/Gluc fusion protein. The in-gel activity assay loaded with cell extracts from heat-treated transformants (Figure 8, transformant PanHsp-4) shows a band of only about 20 kDa. This size reveals that the HSP70B chloroplast transit peptide has been cleaved off as expected.

The results from the in-gel activity assays substantiate the heat-inducibility of heat-shock promoters from heterologous sources in *P. morum*.

Long term stability of DNA integration and gene expression

For any biotechnological approach with transgenic organisms, the stability of the transgenes is of fundamental importance. Therefore, the stability and expression of the *aph*VIII and *gluc* genes introduced into *P. morum* transformants was investigated after twelve months of cultivation or, in other words, after about 300 generations. Cultivation was without selective pressure during the twelve months. The presence of the heterologous genes was rechecked by PCR and the functionality of the expression products was reexamined by paromomycin resistance assays and luminometer assays. In all transformants, PCR fragments of the expected sizes were amplified and, thus, the genomic integration of the artificial gene constructs could be confirmed by PCR. All transformants exhibited about the same antibiotic resistance compared to the values determined right after transformation; the range of fluctuations was only +/−6%. Likewise, luciferase activity in the transformants was stable. The comparison of the values after twelve months of cultivation with the values determined right after transformation gave a fluctuation range of only +/−7%. Thus, both integration and expression of the heterologous *aph*VIII and *gluc* genes are stable in *P. morum*.

Discussion

Here we demonstrate the stable nuclear transformation of the green alga *P. morum* by particle bombardment with plasmid-coated gold particles. Our approach only uses plasmids containing chimeric constructs from heterologous sources. Particle bombardment also was efficient for transformation of *C. reinhardtii* [40], *V. carteri* [15], *G. pectorale* [20] and *E. elegans* [11]. The capability for stable genetical modification of an organism of interest is an important requirement in modern molecular biology, because this technique facilitates the identification of gene functions and allows for the introduction of additional characteristics and abilities.

Antibiotics resistance

For stable nuclear transformation of *P. morum* the *aph*VIII gene of *S. rimosus* driven by a *V. carteri hsp*70A/*rbc*S3 tandem promoter proved to be a suitable selectable marker, even though all parts of the DNA construct came from heterologous sources. Transformants generated with the *aph*VIII construct show an up to 333-fold higher tolerance to the aminoglycoside antibiotic paromomycin compared to the parent wild-type strain. When the *aph*VIII gene was used as a selectable marker in other volvocine algae, the factor for the increase in antibiotic tolerance in transformants was similar or lower. For *V. carteri* the increase in antibiotic tolerance was about 100-fold [30,31], it was 100 to 200 fold in *C. reinhardtii* [25,30], about 330-fold in *E. elegans* [11] and 15 to 40 fold in *G. pectorale* [20]. Even if the antibiotic tolerance of transformants varies between the investigated volvocine species,

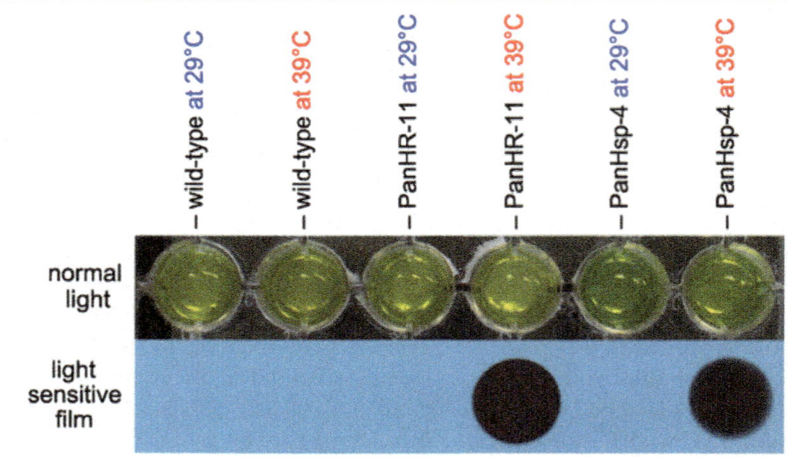

Figure 7 Visualization of inducible luciferase activity. Luciferase assay results for parental wild-type colonies and for several transformants with or without heat shock. The transformants PanHR-11 and PanHsp-4 were generated with the co-bombarded plasmids pHRLucP and pHsp70A-GLuc, respectively. Algae cultures were divided into two aliquots; one aliquot was subjected to a heat shock at 39°C for 1 h, and the other aliquot was maintained at 29°C. Upper row: standard photo showing the assay setup (normal light). Lower row: exposure to a light-sensitive film.

the *aph*VIII gene allowed for selection of transformants in all these species.

Transformation frequency

The frequency of transformation was calculated to be 1.2 $\times$ 10^{-6} per cell in *P. morum*. The transformation frequency in *V. carteri* was estimated to be 2.5 $\times$ 10^{-5} [15]. In *C. reinhardtii* the frequency of transformation was 1.3 - 1.9 $\times$ 10^{-7} [41]; it was 1.1 – 6.6 $\times$ 10^{-7} in *G. pectorale* [20] and 3.7 $\times$ 10^{-7} in *E. elegans* [11]. Thus, the transformation frequency in *P. morum* is about 20 times lower than in *V. carteri* but about 2 to 10 times higher than in *C. reinhardtii*, *G. pectorale* and *E. elegans*. On an overall basis, the transformation frequencies in volvocine algae vary by about two orders of magnitude (1.1 $\times$ 10^{-7} to 2.5 $\times$ 10^{-5}). There is limited information about transformation frequencies in other algae with the exception of some diatom species. The reported frequencies are 1 $\times$ 10^{-7} - 1 $\times$ 10^{-6}

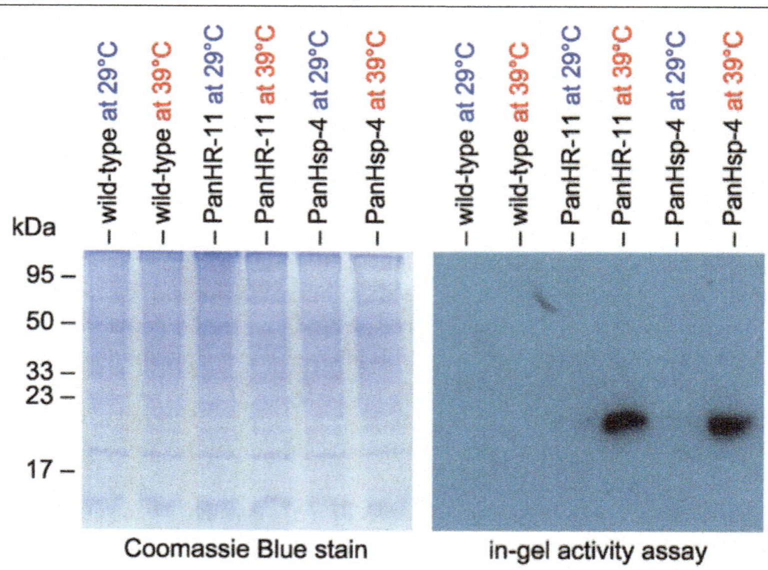

Figure 8 Detection of luciferase activity by in-gel activity assays. The transformants PanHR-11 and PanHsp-4 were generated with the co-bombarded plasmids pHRLucP and pHsp70A-GLuc, respectively. The wild-type strain was used as a control. Algae cultures were divided into two aliquots; one aliquot was subjected to a heat shock at 39°C for 1 h, and the other aliquot was maintained at 29°C. Cell extracts with equal amounts of protein were subjected to SDS-PAGE. Subsequently, in-gel renaturation was performed using β-cyclodextrin to remove SDS from the gel, the coelenterazine-substrate was added, and the gel was exposed to a light-sensitive film (right panel). As a loading control, the same extracts were also stained with Coomassie Blue following SDS-PAGE (left panel).

per cell in *Phaeodactylum tricornutum* [42] and 3×10^{-8} - 3×10^{-7} per cell in *Cyclotella cryptica* and *Navicula saprophila* [43]. Therefore, in diatoms the frequency of transformation is similar or somewhat lower compared to that in volvocine algae.

Co-transformation rate

Co-transformation of a selectable marker plasmid and the gene of interest on a separate plasmid is a common strategy especially when large genes and complex gene constructs have to be integrated into the genome [16,20,24,26,29,31,44-46]. Another advantage of co-transformation is that the same selectable marker plasmid can be used for all transformation experiments. In *P. morum* the co-transformation rate was calculated to be at about 80%. For *V. carteri* co-transformation rates of 10% - 60% [29] or 40% - 80% [15] have been reported. The co-transformation rate was about 80% in *C. reinhardtii* [26], 30% - 50% in *G. pectorale* [20] and 50% - 100% in *E. elegans* [11]. The co-transformation rates in volvocine algae are high in all volvocine species and particularly in *P. morum*. Therefore, transformation of *P. morum* does not require construction of vectors that contain all required genes and regulatory elements on a single plasmid.

Expression of heterologous genes using heterologous promoters

Some promoters from *C. reinhardtii* have been shown to work in *V. carteri* and vice versa [30]. Likewise, some promoter elements from *C. reinhardtii* and *V. carteri* also work in *G. pectorale* and *E. elegans* [11,20]. Here we show for *P. morum* that both the heterologous *hsp*70A/*rbc*S3 tandem promoter of *V. carteri* and the heterologous *HSP70A* promoter of *C. reinhardtii* work in this species. Three out of the four constructs clearly lead to functional expression of the heterologous Gluc luciferase and the heterologous aminoglycoside 3′-phosphotransferase VIII (AphVIII) enzymes. It is unclear why transformants produced with pPsaD-GLuc (PanTJ...) didn't produce significant luciferase activity.

The heterologous heat shock promoters retained their inducibility by heat when expressed in *P. morum*. The expression was up to 555 times higher in heat-shocked versus non-heat-shocked cultures.

Conclusion

The introduction of foreign genes into the nuclear genome of *P. morum* by particle bombardment is now feasible. Both the heterologous *hsp*70A/*rbc*S3 tandem promoter of *V. carteri* and the heterologous *HSP70A* promoter of *C. reinhardtii* drive the expression of heterologous genes in *P. morum*. Both promoters retained their heat-inducibility in *P. morum*. Furthermore, the heterologous *aph*VIII gene of *S. rimosus* was established as a selectable marker and the *gluc* luciferase gene was found to be a useful reporter gene in *P. morum*. It is thus to be emphasized that genetic engineering of a species is possible even without the availability of any endogenous genes and promoters.

Methods

Strains and culture conditions

The wild-type *P. morum* strain SAG 32.96 was obtained from the Culture Collection of Algae at Goettingen University (SAG), Germany. Cultures were maintained in Jaworski medium (JM) [47] at 29°C in an 8 h dark/16 h light (~10,000 lux) cycle [11,20]. Cultures were grown in 10 ml glass tubes, 50 or 300 ml Erlenmeyer flasks, or 1,000 ml Fernbach flasks. The glass tubes had caps that allow for gas exchange, and Erlenmeyer and Fernbach flasks were aerated via Pasteur pipettes with approximately 50 cm^3 sterile air/min [11,20]. Transgenic strains that express the *aph*VIII gene were grown in JM in the presence of 1 µg paromomycin/ml (paromomycin sulfate, Sigma-Aldrich, St. Louis, MO).

Transformation vectors

The pPmr3 plasmid (Figure 3A) contains the 0.8 kb coding region of the *S. rimosus aphVIII* selectable marker gene, a *V. carteri hsp*70A-*rbc*S3 tandem promoter (0.5 kb of *V. carteri hsp*70A and 0.27 kb of *V. carteri rbc*S3 sequences), the 3′ UTR from the *V. carteri rbc*S3 gene (0.53 kb of downstream sequence), and the pBluescript II vector backbone [GenBank: AY429514] [31]. The total size of the pPmr3 plasmid is 5.1 kb.

The pPsaD-GLuc plasmid (Figure 3B) contains the 0.57 kb coding region of the luciferase (*gluc*) gene from *G. princeps*, which was engineered to match the codon usage in *C. reinhardtii*, 0.8 kb of upstream sequence, which includes the *C. reinhardtii PSAD* promoter, the 3′-UTR of the *C. reinhardtii psaD* gene (0.56 kb of downstream sequence), and the pBluescript vector backbone [27,36] [GenBank: EU372000, AF335592]. The pPsaD-GLuc plasmid is 5.0 kb in size.

The pHsp70A-GLuc plasmid (Figure 3C) contains the 0.57 kb coding region of the luciferase (*gluc*) gene from *G. princeps* (codon-optimized for *C. reinhardtii*) [27] fused downstream of a 0.8 kb genomic DNA fragment that contains the first three exons and two introns of the *HSP70B* gene of *C. reinhardtii*; this fragment of the *HSP70B* gene contains the sequence coding for the chloroplast transit peptide of HSP70B. The hybrid *HSP70B*/*gluc* gene is driven by the *C. reinhardtii HSP70A* promoter (0.26 kb of upstream sequence) and the 3′ UTR comes from the *C. reinhardtii RBCS2* gene (0.22 kb of downstream sequence) [27]. The total size of the pHsp70A-GLuc plasmid is 4.9 kb.

The pHRLucP plasmid (Figure 3D) contains the 0.57 kb coding region of the luciferase (*gluc*) gene of *G. princeps*, which was engineered to match the codon usage in *C. reinhardtii*, the *V. carteri hsp*70A-*rbc*S3 tandem promoter (0.5 kb of *V. carteri hsp*70A and 0.27 kb of *V. carteri rbc*S3 sequences), the *C. reinhardtii PSAD* 3′ UTR (0.55 kb of downstream sequence), and the pBluescript II vector backbone [11,27,36]. The pHRLucP plasmid is 5.0 kb in size.

Preparation of plasmid DNA

Plasmid DNA was purified using the Nucleospin® Plasmid Kit according to the manufacturers' instructions (Macherey-Nagel, Düren, Germany).

Coating of microprojectiles

Gold microprojectiles (0.6 μm in diameter, Bio-Rad, Hercules, CA) were coated with the required plasmids for biolistic transformation as previously described [11,20]. The DNA-coated microprojectiles were resuspended in 60 μl of ethanol and kept at 4°C; these microprojectiles were used within 3 h of preparation.

Determination of cell concentration

In *P. morum* the number of cells per colony varies. Therefore, we refer to "cells/ml" rather than "colonies/ml". The concentration of cells was determined using a hemacytometer with Neubauer ruling.

Stable nuclear transformation by particle bombardment

The stable transformation of *P. morum* was performed using a Biolistic® PDS-1000/He (Bio-Rad) particle gun. To this end, a logarithmically growing *P. morum* culture (150 ml) at a cell concentration of ~7×10^4 cells/ml was harvested by centrifugation (1,000 g, 5 min, swing-out rotor) and resuspended in 6 ml of JM. One milliliter of the suspension was evenly spread on a cellulose acetate membrane filter and excess liquid was removed as previously described [11,20]. One-sixth of the DNA-coated microprojectiles were evenly spread on a macrocarrier (Bio-Rad) that was placed in a macrocarrier holder (Bio-Rad). The transformation procedure was as previously described [11,20] and the most successful combination of parameters is shown in Table 1. After the bombardment step, the algae were washed off from the membrane filter with JM. The procedure was repeated five times and the algae from six rounds of bombardment were pooled and then evenly distributed among twelve 50 ml Erlenmeyer flasks containing a volume of ~40 ml of JM. The bombarded colonies were incubated under standard conditions for 24 h to allow for regeneration and expression of the transgenes. Then, 2 μg of paromomycin/ml was added. The large fraction of non-transformed cells died within 48 h and the medium clarified. After 10–14 days of incubation in the presence of antibiotic, re-greening of flasks indicated the initial presence and reproduction of at least one paromomycin-resistant *P. morum* transformant that led to a population of transformants. No more than one transformant per flask was analyzed [11,20].

Re-isolation of transformants

Transformants were re-isolated to ensure uniform genetic conditions for all further analyses. For this purpose, a serial dilution of an exponentially growing *P. morum* culture was performed in a Terasaki plate (Nunc™ MicroWell™ MiniTrays; Thermo Fisher Scientific, Langenselbold, Germany) [11,20]. Each well of the Terasaki plate was filled with 10 μl of JM. A single *P. morum* colony was finally transferred into a standard glass tube with JM, containing 1 μg of paromomycin/ml. Further incubation was under standard conditions.

Paromomycin-resistance assay

Cells of transformed or wild-type *P. morum* strains were transferred into the wells of a 24-well culture plate (Sarstedt, Nümbrecht, Germany) with a wide range of paromomycin concentrations from 0 to 100 μg/ml in 1 ml JM (concentrations are given in Figure 2). At the beginning of the assay, each well contained approximately 4,000 cells, which corresponds to 250–300 colonies. After incubation under standard conditions for 12 days, the wells were analyzed for viable green cells or white remains of lysed cells.

Primer design

Oligonucleotide primers were designed using the primer analysis software Oligo 6 (Molecular Biology Insights, Cascade, CO), DNASIS™ (version 7.00; Hitachi Software Engineering, San Francisco, CA), and Primer Express® (Applied Biosystems, Foster City, CA).

Isolation of genomic DNA

Ten milliliters of a logarithmically growing *P. morum* culture with a density of 6×10^6 cells/ml was harvested by centrifugation (3,500 g for 10 min). The resulting pellet, which had a wet weight of ~80 mg, was washed with water and resuspended in 40 μl water. Isolation of genomic DNA was as previously described [11,20].

Genomic PCR

PCR reactions with genomic DNA as a template were carried out in a total volume of 50 μl that contained ~100 ng of genomic DNA, 300 nM of each primer, 0.2 mM dNTP, 1.5 mM $MgCl_2$, and 2.6 units of Expand High Fidelity enzyme mix in 1× Expand High Fidelity buffer (Roche Applied Science, Basel, Switzerland). The PCR reactions were performed on a T3 Thermocycler

PCR system (Biometra, Göttingen, Germany) using the following conditions: 40 cycles of 94°C for 20 s, 55°C for 30 s, and 72°C for 45 s and a final extension was at 72°C for 10 min. The PCR products were cloned and sequenced (Eurofins, Ebersberg, Germany).

Isolation of total RNA

Total RNA was extracted from ~100 mg of concentrated, frozen *P. morum* algae using 1 ml of phenol-based TRI Reagent (Sigma-Aldrich, St. Louis, MO) and 300 µl trichloromethane. RNA precipitation and RNA purification was as previously described [11,20].

Reverse Transcription (RT)-PCR

First strand cDNA synthesis was carried out using 1 µg of total RNA and Moloney murine leukemia virus (MMLV) reverse transcriptase lacking ribonuclease H activity (H minus) according to the manufacturer's instructions (Promega). The subsequent PCR step was performed using the Expand High Fidelity PCR system (Roche Applied Science, Basel, Switzerland), a T3 Thermocycler PCR system (Biometra), and the following cycling conditions: 40 cycles of 94°C for 20 s, 55°C for 30 s, and 72°C for 45 s and a final extension at 72°C for 10 min. The RT-PCR products were cloned and sequenced.

Induction of luciferase activity

Preliminary tests showed that the shift of a *P. morum* culture (3–6 × 10^5 cells/ml) from 29°C to 39°C for 1 h, followed by a recovery phase at 29°C for 15 min, leads to the strongest induction of luciferase activity in the temperature range investigated (27°C to 51°C).

Luciferase assays

For assays on light-sensitive films, *P. morum* cultures with a cell density of 3–6 × 10^5 cells/ml were divided into aliquots of 50 ml. One aliquot was incubated at the optimal temperature for heat stress-induced expression of luciferase (39°C) for 1 h, and another aliquot was incubated at 29°C as a reference control. Further treatment and execution of the luciferase enzyme assay was as previously described [11,20]. The final exposure to a chemiluminescence sensitive film (Retina XBA; Fotochemische Werke) was for 2 h at 20°C [11,20,35].

The quantification of luciferase activity was conducted as previously described [20,27] using a MiniLumat LB9506 luminometer (Berthold, Bad Wildbad, Germany). For this purpose, aliquots of a *P. morum* culture (3–6 × 10^5 cells/ml) were heat-stressed at 39°C or kept at 29°C (control), respectively. Further treatment was as previously described [20]. Luminescence was recorded as relative light units (rlu). Induction factors were calculated by comparison of samples from heat-shocked versus non-heat-shocked cultures.

In-gel activity assay

One liter of logarithmically growing, heat-stressed or untreated cultures of transformed or wild-type *P. morum* cultures with cell densities of 4–7 × 10^5 cells/ml were harvested by filtration on a 10-µm mesh nylon screen. Cells passing the screen were harvested by centrifugation (4,000 g, 5 min) and added to cells collected on the screen. Cells lysates were produced as previously described [11]. The protein concentration of the cell extracts was determined photometrically using the Bio-Rad Protein Assay Dye Reagent (Bio-Rad). For gel electrophoresis, 50 µg of total protein was loaded onto a standard SDS-polyacrylamide gel. The gel electrophoresis, the in-gel renaturation, the execution of the luciferase enzyme assay and the light detection on a chemiluminescence sensitive film was performed as previously described [11].

Additional file

Additional file 1: Influence of different parameters on transformation efficiency.

Abbreviations

*aph*VIII: *S. rimosus* aminoglycoside 3'-phosphotransferase VIII gene; GFP: Green fluorescent protein; *gluc*: *G. princeps* luciferase gene (adapted to *C. reinhardtii* nuclear codon usage); *hsp*70A: Heat shock protein 70A gene; JM: Jaworski medium; kDa: Kilodalton; MMLV: Moloney murine leukemia virus; PCR: Polymerase chain reaction; *PSAD*: Abundant protein of photosystem I complex gene; *rbc*S3: Ribulose bisphosphate carboxylase small chain gene 3; rlu: Relative light units; RT: Reverse transcription; SDS: Sodium dodecyl sulfate; UTR: Untranslated region.

Competing interests

The authors declare that they have no competing interests.

Authors' contributions

KL conducted the experiments, analyzed the data, and wrote the first version of the manuscript with advice and guidance from AH. AH (corresponding author) conceived and coordinated the study, critically evaluated the data, and finalized the manuscript. Both authors read and approved the final manuscript.

Acknowledgements

We are grateful to T. Jakobiak and R. Schmitt (University of Regensburg, Germany) for providing the *aph*VIII gene, to N. Shao and R. Bock (Max-Planck-Institut für Molekulare Pflanzenphysiologie, Potsdam-Golm, Germany) for providing the *gluc* gene, and to K. Puls for technical assistance. We acknowledge the support of the Deutsche Forschungsgemeinschaft and the Open Access Publication Funds of Bielefeld University, which funded the article processing charges.

References

1. Herron MD, Hackett JD, Aylward FO, Michod RE: **Triassic origin and early radiation of multicellular volvocine algae.** *Proc Natl Acad Sci U S A* 2009, **106**(9):3254–3258.
2. Kirk DL: *Volvox: Molecular-Genetic Origins of Multicellularity and Cellular Differentiation.* 1st edition. Cambridge: Cambridge University Press; 1998.
3. Höhn S, Hallmann A: **There is more than one way to turn a spherical cellular monolayer inside out: type B embryo inversion in *Volvox globator*.** *BMC Biol* 2011, **9**(1):89.

4. Hallmann A: **Evolution of reproductive development in the volvocine algae.** *Sex Plant Reprod* 2011, **24:**97–112.
5. Harris EH, Stern DB, Witman GB: *The Chlamydomonas Sourcebook (2nd Edition).* 2nd edition. San Diego, CA: Academic Press; 2009.
6. Fulton AB: **Colonial development in *Pandorina morum*. II. Colony morphogenesis and formation of the extracellular matrix.** *Dev Biol* 1978, **64**(2):236–251.
7. Lamouroux JV, de Saint-Vincent JBGM B, Deslongschamps E: *Encyclopédie méthodique ou par ordre de matières. Histoire naturelle des zoophytes, ou animaux rayonnés, faisant suite à l'histoire naturelle des vers de Bruguière.* Paris: Chez Mme Veuve Agasse; 1824.
8. Herron MD, Michod RE: **Evolution of complexity in the volvocine algae: transitions in individuality through Darwin's eye.** *Evolution* 2008, **62**(2):436–451.
9. Nozaki H, Kuroiwa T: **Ultrastructure of the extracellular matrix and taxonomy of *Eudorina*, *Pleodorina* and *Yamagishiella* gen. nov. (Volvocaceae, Chlorophyta).** *Phycologia* 1992, **31**(6):529–541.
10. Nozaki H, Kazaki H: **The sexual process of Japanese *Pandorina morum* Bory (Chlorophyta).** *J Jpn Bot* 1979, **54**(12):363–370.
11. Lerche K, Hallmann A: **Stable nuclear transformation of *Eudorina elegans*.** *BMC Biotechnol* 2013, **13:**11.
12. Moore LJ, Coleman AW: **The linear 20 kb mitochondrial genome of *Pandorina morum* (Volvocaceae, Chlorophyta).** *Plant Mol Biol* 1989, **13**(4):459–465.
13. Merchant SS, Prochnik SE, Vallon O, Harris EH, Karpowicz SJ, Witman GB, Terry A, Salamov A, Fritz-Laylin LK, Marechal-Drouard L, Marshall WF, Qu LH, Nelson DR, Sanderfoot AA, Spalding MH, Kapitonov VV, Ren Q, Ferris P, Lindquist E, Shapiro H, Lucas SM, Grimwood J, Schmutz J, Cardol P, Cerutti H, Chanfreau G, Chen CL, Cognat V, Croft MT, Dent R, *et al*: **The *Chlamydomonas* genome reveals the evolution of key animal and plant functions.** *Science* 2007, **318**(5848):245–250.
14. Prochnik SE, Umen J, Nedelcu AM, Hallmann A, Miller SM, Nishii I, Ferris P, Kuo A, Mitros T, Fritz-Laylin LK, Hellsten U, Chapman J, Simakov O, Rensing SA, Terry A, Pangilinan J, Kapitonov V, Jurka J, Salamov A, Shapiro H, Schmutz J, Grimwood J, Lindquist E, Lucas S, Grigoriev IV, Schmitt R, Kirk D, Rokhsar DS, Hellsten U, Chapman J, *et al*: **Genomic analysis of organismal complexity in the multicellular green alga *Volvox carteri*.** *Science* 2010, **329**(5988):223–226.
15. Schiedlmeier B, Schmitt R, Müller W, Kirk MM, Gruber H, Mages W, Kirk DL: **Nuclear transformation of *Volvox carteri*.** *Proc Natl Acad Sci U S A* 1994, **91**(11):5080–5084.
16. Kindle KL: **High-frequency nuclear transformation of *Chlamydomonas reinhardtii*.** *Proc Natl Acad Sci U S A* 1990, **87**(3):1228–1232.
17. Klein RM, Wolf ED, Wu R, Sanford JC: **High-velocity microprojectiles for delivering nucleic acids into living cells. 1987.** *Biotechnology* 1992, **24:**384–386.
18. Brown LE, Sprecher SL, Keller LR: **Introduction of exogenous DNA into *Chlamydomonas reinhardtii* by electroporation.** *Mol Cell Biol* 1991, **11**(4):2328–2332.
19. Kumar SV, Misquitta RW, Reddy VS, Rao BJ, Rajam MV: **Genetic transformation of the green alga *Chlamydomonas reinhardtii* by *Agrobacterium tumefaciens*.** *Plant Sci* 2004, **166:**731–738.
20. Lerche K, Hallmann A: **Stable nuclear transformation of *Gonium pectorale*.** *BMC Biotechnol* 2009, **9**(1):64.
21. Nelson JA, Savereide PB, Lefebvre PA: **The *CRY1* gene in *Chlamydomonas reinhardtii*: structure and use as a dominant selectable marker for nuclear transformation.** *Mol Cell Biol* 1994, **14**(6):4011–4019.
22. Randolph-Anderson BL, Sato R, Johnson AM, Harris EH, Hauser CR, Oeda K, Ishige F, Nishio S, Gillham NW, Boynton JE: **Isolation and characterization of a mutant protoporphyrinogen oxidase gene from *Chlamydomonas reinhardtii* conferring resistance to porphyric herbicides.** *Plant Mol Biol* 1998, **38**(5):839–859.
23. Kovar JL, Zhang J, Funke RP, Weeks DP: **Molecular analysis of the acetolactate synthase gene of *Chlamydomonas reinhardtii* and development of a genetically engineered gene as a dominant selectable marker for genetic transformation.** *Plant J* 2002, **29**(1):109–117.
24. Berthold P, Schmitt R, Mages W: **An engineered *Streptomyces hygroscopicus aph 7"* gene mediates dominant resistance against hygromycin B in *Chlamydomonas reinhardtii*.** *Protist* 2002, **153**(4):401–412.
25. Sizova I, Fuhrmann M, Hegemann P: **A *Streptomyces rimosus aphVIII* gene coding for a new type phosphotransferase provides stable antibiotic resistance to *Chlamydomonas reinhardtii*.** *Gene* 2001, **277**(1–2):221–229.
26. Stevens DR, Rochaix JD, Purton S: **The bacterial phleomycin resistance gene *ble* as a dominant selectable marker in *Chlamydomonas*.** *Mol Gen Genet* 1996, **251**(1):23–30.
27. Shao N, Bock R: **A codon-optimized luciferase from *Gaussia princeps* facilitates the in vivo monitoring of gene expression in the model alga *Chlamydomonas reinhardtii*.** *Curr Genet* 2008, **53**(6):381–388.
28. Hallmann A: **Algal transgenics and biotechnology.** *Transgenic Plant J* 2007, **1:**81–98.
29. Hallmann A, Rappel A: **Genetic engineering of the multicellular green alga *Volvox*: a modified and multiplied bacterial antibiotic resistance gene as a dominant selectable marker.** *Plant J* 1999, **17**(1):99–109.
30. Hallmann A, Wodniok S: **Swapped green algal promoters: *aphVIII*-based gene constructs with *Chlamydomonas* flanking sequences work as dominant selectable markers in *Volvox* and vice versa.** *Plant Cell Rep* 2006, **25**(6):582–591.
31. Jakobiak T, Mages W, Scharf B, Babinger P, Stark K, Schmitt R: **The bacterial paromomycin resistance gene, *aphH*, as a dominant selectable marker in *Volvox carteri*.** *Protist* 2004, **155**(4):381–393.
32. Davies J, Wright GD: **Bacterial resistance to aminoglycoside antibiotics.** *Trends Microbiol* 1997, **5**(6):234–240.
33. Barnes AT, Case JF: **Bioluminescence in the mesopelagic copepod, *Gaussia princeps* (T. Scott).** *J Exp Mar Biol Ecol* 1972, **8**(1):53–71.
34. Szent-Gyorgyi C, Ballou BT, Dagmal E, Bryan B: **Cloning and characterization of new bioluminescent proteins.** *Proc SPIE* 1999, **3600:**4–11.
35. Jacobsen I, Hennig-Pauka I, Baltes N, Trost M, Gerlach GF: **Enzymes involved in anaerobic respiration appear to play a role in *Actinobacillus pleuropneumoniae* virulence.** *Infect Immun* 2005, **73**(1):226–234.
36. Fischer N, Rochaix JD: **The flanking regions of *PsaD* drive efficient gene expression in the nucleus of the green alga *Chlamydomonas reinhardtii*.** *Mol Genet Genomics* 2001, **265**(5):888–894.
37. Schroda M, Blöcker D, Beck CF: **The *HSP70A* promoter as a tool for the improved expression of transgenes in *Chlamydomonas*.** *Plant J* 2000, **21**(2):121–131.
38. Shao N, Krieger-Liszkay A, Schroda M, Beck CF: **A reporter system for the individual detection of hydrogen peroxide and singlet oxygen: its use for the assay of reactive oxygen species produced in vivo.** *Plant J* 2007, **50**(3):475–487.
39. Drzymalla C, Schroda M, Beck CF: **Light-inducible gene *HSP70B* encodes a chloroplast-localized heat shock protein in *Chlamydomonas reinhardtii*.** *Plant Mol Biol* 1996, **31**(6):1185–1194.
40. Kindle KL, Schnell RA, Fernandez E, Lefebvre PA: **Stable nuclear transformation of *Chlamydomonas* using the *Chlamydomonas* gene for nitrate reductase.** *J Cell Biol* 1989, **109**(6 Pt 1):2589–2601.
41. Sizova IA, Lapina TV, Frolova ON, Alexandrova NN, Akopiants KE, Danilenko VN: **Stable nuclear transformation of *Chlamydomonas reinhardtii* with a *Streptomyces rimosus* gene as the selective marker.** *Gene* 1996, **181**(1–2):13–18.
42. Apt KE, Kroth-Pancic PG, Grossman AR: **Stable nuclear transformation of the diatom *Phaeodactylum tricornutum*.** *Mol Gen Genet* 1996, **252**(5):572–579.
43. Dunahay TG, Jarvis EE, Roessler PG: **Genetic transformation of the diatoms *Cyclotella cryptica* and *Navicula saprophila*.** *J Phycol* 1995, **31:**1004–1012.
44. Hallmann A, Sumper M: **Reporter genes and highly regulated promoters as tools for transformation experiments in *Volvox carteri*.** *Proc Natl Acad Sci U S A* 1994, **91**(24):11562–11566.
45. Hallmann A, Sumper M: **The *Chlorella* hexose/H^+ symporter is a useful selectable marker and biochemical reagent when expressed in *Volvox*.** *Proc Natl Acad Sci U S A* 1996, **93**(2):669–673.
46. Hallmann A, Rappel A, Sumper M: **Gene replacement by homologous recombination in the multicellular green alga *Volvox carteri*.** *Proc Natl Acad Sci U S A* 1997, **94**(14):7469–7474.
47. Thompson AS, Rhodes JC, Pettman I: *Culture Collection of Algae and Protozoa - Catalogue of Strains.* 5th edition. Cumbria, UK: CCAP; 1988.

2

Screening and genetic characterization of thermo-tolerant *Synechocystis sp.* PCC6803 strains created by adaptive evolution

Ulrich M Tillich[1,2], Nick Wolter[1], Philipp Franke[1], Ulf Dühring[3] and Marcus Frohme[1*]

Abstract

Background: Temperature tolerance is an important aspect for commercial scale outdoor cultivation of microalgae and cyanobacteria. While various genes are known to be related to *Synechocystis* sp. PCC6803's heat shock response, there is very limited published data concerning the specific genes involved in long term thermal tolerance. We have previously used random mutagenesis and adaptive evolution to generate a mixture of strains of *Synechocystis* sp. PCC6803 with significantly increased thermal tolerance. The genetic modifications leading to the phenotypes of the newly generated strains are the focus of this work.

Results: We used a custom screening platform, based on 96-deepwell microplate culturing in an in house designed cultivation chamber integrated in a liquid handling robot for screening and selection; in addition we also used a more conventional system. The increased thermal tolerances of the isolated monoclonal strains were validated in larger bioreactors and their whole genomes sequenced. Comparison of the sequence information to the parental wild type identified various mutations responsible for the enhanced phenotypes. Among the affected genes identified are *clpC, pnp, pyk2, sigF, nlpD, pyrR, pilJ* and *cya1*.

Conclusions: The applied methods (random mutagenesis, in vivo selection, screening, validation, whole genome sequencing) were successfully applied to identify various mutations, some of which are very unlikely to have been identified by other approaches. Several of the identified mutations are found in various strains and (due to their distribution) are likely to have occurred independently. This, coupled with the relatively low number of affected genes underscores the significance of these specific mutations to convey thermal tolerance in *Synechocystis*.

Keywords: Cyanobacteria, *Synechocystis*, Thermal tolerance, HTS, NGS, Adaptive evolution

Background

Microalgae and cyanobacteria could be used for large scale outdoor cultivation, e.g. for the production of biofuels. This has led to a revamped interest in biotechnology of cyanobacteria in recent years [1-4]. Wild-type strains however are not ideal for such artificial growth environments. One possibility to create better production strains is classical strain improvement by adaptive evolution, which is defined as the propagation of advantageous mutations through positive selection. Genome wide random mutagenesis can be used to increase the rate of mutations above its natural level, while *in vivo* selection is used to apply the selective pressure. Herein, with only little knowledge about the underlying molecular biology, significant alterations in the phenotype may be promoted. It is also possible to achieve effects which require simultaneous modifications of many seemingly unrelated genes [5].

One important target for optimization is the growth temperature range, in particular when considering desert areas for outdoor cultivation. In a previous work we demonstrated the use of adaptive evolution to produce a mix of strains of *Synechocystis* sp. PCC6803 – a model organism for cyanobacteria and phototrophs as a whole - with a significantly increased thermal tolerance compared to the parental wild type [6]. However this work was limited to the description of the mutagenesis method itself, as well as the phenotype of the generated mix; no monoclonal strains were isolated and no genetic characterization

* Correspondence: mfrohme@th-wildau.de
[1]Molecular Biotechnology and Functional Genomics, Technical University of Applied Sciences Wildau, Bahnhofstraße 1, 16-2001, D-15745 Wildau, Germany
Full list of author information is available at the end of the article

was performed. The genetic modifications leading to the phenotypes of the newly generated strains are the focus of this work.

With the combination of agents used for the generation of the mutants (Ultraviolet Light (UV) and Methylmethanesulphonate (MMS)), all kinds of mutations (transitions, transversions, frameshifts, and deletions) are possible, though the mutagens used have varied tendencies towards the induction of different kinds of mutations. UV induces mainly transitions from Guanine/Cytosine to Adenosine/Thymine, while deletions mainly occur in A/T rich regions [7], whereas MMS induces all kinds of base substitutions (transitions and transversions), though mainly transversions from GC to TA and vice versa [8,9]. Introduced mutations can be repaired by specific cellular mechanisms, though the mutagenesis protocols used should minimize their activity [6]. Additionally unspecific, error-prone cellular repair mechanisms (such as the SOS response) occur, which are able to induce any type of indirect mutations [7,10,9,11].

Synechocystis is polyploid with the number of genome copies per cell reported to vary from 58 to 218 [12]. Numerous genes are already known to be related to *Synechocystis* heat shock response, such as hrcA [13] and hik34 [14] which regulate the expression of heat shock proteins. The heat shock response however, while related, is distinct from long term temperature tolerance. The knockout of hik34 for example, while strongly increasing expression of heat shock proteins in *Synechocystis*, barely affected its survival at the elevated temperature of 47°C (3 h survived, compared to 2 h for the wild type strain) [14]. Many genes (such as chaperonins and proteases) have been shown to be differentially expressed after short term acclimation to higher temperatures [15,16], however, there is very limited published data concerning the specific genes involved in long term thermal tolerance in *Synechocystis*.

Inoue et al. analyzed the growth of *Synechocystis* at various temperatures and determined a very similar growth rate for temperatures ranging from 25 to 40°C. Above (and below) this temperature the growth rate drops rapidly, with no growth at 45°C [17]. They speculated that cell death at higher temperatures does not occur directly because of photosynthesis inhibition, as is often assumed because of Photosystem II sensitivity to higher temperatures [18,15], but rather because of damage to other cell mechanisms, such as the functions of the plasma membrane [17].

Before the genotype of the various previously generated thermo-tolerant strains can be analyzed to determine which genes were affected by the mutagenesis, they have to be isolated and cultured as individual strains. A screening method to determine the individual temperature tolerances of many clones would allow narrowing the focus towards the best performing strains. However, there are currently no commercial systems available for mid- or high-throughput screening of phototrophic microorganisms such as cyanobacteria. We have previously developed a prototype for such a system with the capability for precise temperature control. Furthermore this prototype has recently been greatly expanded and improved with subsequent adaption to a more modern and larger pipetting platform (unpublished results).

The availability of an annotated genome for *Synechocystis* [19], which can be used as a mapping reference greatly reduces the amount of sequencing-reads needed at any given position for a sufficient coverage. Together with increasingly inexpensive next generation sequencing (NGS) techniques it became feasible to sequence various strains with enhanced temperature tolerance (as well as the parental wild type), to identify genes of interest.

We utilized the 454 GS Junior (Roche) and the Ion Torrent PGM (Life Technologies) bench-top sequencers, both of which have enough throughput for the re-sequencing of bacterial genomes. While their technical implementations differ, in both platforms, insertions or deletions (Indels) in homopolymer regions are the main source of errors [20,21]. This leads to an overall higher error rate compared to the competing MySeq platform (Illumina); however both have higher accuracy when detecting base substitutions at the same coverage [22,23]. Though it has been noted for the IonTorrent, that while the percentage of correctly called true SNPs is higher compared to Illumina's MySeq, the false positive rate is also increased [24].

Here we use our established screening platform, as well as a more conventional system based on a shaking incubator, to screen for the best strains out of the previously generated mix, i.e. those with the highest thermotolerance. The phenotypes of these strains are then characterized and their genetic makeup analyzed with NGS and Sanger sequencing. Various mutations conferring the increased thermal tolerance to the strains are identified and their possible effects discussed.

Results

Wild type reference

To identify new mutations generated after mutagenesis and selection under thermal stress, the parental wild-type strain was sequenced as a baseline for further analysis. The combined sequencing coverage was 19-fold with an average raw accuracy across each individual base position in a read of 99.57% and a median read length of 524 bp. Several changes were found compared to the originally published Kazusa-wild-type:

There were four large deletions detected in the genome; three on the chromosome (1.200.305 – 1.201.486; 2.048.411 – 2.049.595; 3.400.331 – 3.401.515) and one on plasmid pSYSM (117.276 – 118.460). Also various SNPs (Additional file 1: Table S1) were identified. All alterations in the wild type were also present in the sequenced thermo-tolerant strains. Additionally, the plasmids pCC5.2 and pCB2.4 could not be identified; no single read properly mapped unto them, even with separate blast alignment.

Screening and characterization of temperature tolerant strains

A mix of thermo-tolerant strains (mutIV-mix) was previously generated through random mutagenesis and adaptive evolution [6]. To obtain monoclonal strains dilutions were plated on agar-media and single colonies then picked. It is important to note however that while strains obtained via this method originate from a single ancestral cell and are thus by definition monoclonal, due to the polyploidic character of *Synechocystis* each monoclonal strain could possibly contain various differing variants of genes within the many genome copies per cell.

Initial attempts at isolating random single clones from plate without screening for performance yielded strains without notably increased thermal tolerance compared to the wild type (data not shown). Therefore two distinct screening methods were applied.

The first is based on culturing isolated monoclonal strains in small cell culture flasks within a commercially available shaking incubator (incubator screening). In this screening, out of 27 strains analyzed, 6 demonstrated higher thermal tolerance than the wild type (wt) controls. The second approach is based on an automated high throughput screening system for phototrophic microorganisms within Deepwell plates (robot screening), which was developed in house. This system identified 8 (out of 81) strains with significantly higher thermal tolerance than the parental wild type. None of the control wells containing only media showed any increase in OD, which would have indicated cross contamination.

For both approaches the growth rates were determined over a range of temperatures and data then fitted using a modified Ratkowsky-Equation. The data-fitting was not used to determine the maximum tolerated temperature of the strains but merely to allow an easier visualization and comparison of data. In the incubator screening, the strain C5K was identified as having the highest temperature tolerance (Figure 1) and was therefore validated with bioreactor cultivation as well as sequenced. Figure 2 shows the results of the robot screening for four strains (A1, A6, H2, and H12), which were selected for further validation in a 1 l bioreactor and sequencing, since their temperature tolerances were the most promising. In the bioreactor validation the maximum temperature was determined by stepwise increase in temperature until stable growth was no longer possible.

The results for the determined maximally tolerated temperature for the various strains (or mixes) as identified either in the screening systems or the 1 l photobioreactos, as well as the number of mutations identified are summarized in Table 1.

The goal of the screening was to identify strains with notably increased thermal tolerance over the wild-type. This was achieved in both screening setups for the selected strains, though the wild-type control did outperform many other strains included in the screening (data not shown). Meanwhile the previously generated thermo-tolerant strain-mix, out of which all strains were isolated (mutIV-mix), was also included in the screening as a positive control and consistently demonstrated the highest thermal tolerance. The thermal tolerances of the strains selected for bioreactor validation and subsequent sequencing (C5K, A1, A6, H2, and H12) while considerably higher than the wild type's, but did not quite match mutIV-mix. The mix also appears to have a somewhat more robust growth rate over a range of temperatures while the single clones have a sharper optimal range. This effect is especially noticeable in the incubator screening (Figure 1).

The characterization of the selected strains in the target bioreactors yielded maximally tolerated temperatures between 44.9 and 45.1°C. The determined tolerances are very close to the 45°C the strain mix (mutIV-mix) was growing stably at, at the point they were isolated.

Mutations in selected thermo tolerant stains

The total DNA from the five individual strains selected by the screening process, as well as a wild-type control, were each sequenced with NGS technology. The sequencing coverage for all sequencings was sufficiently high (wt-control: 41-fold, C5K: 28-fold, A1: 24-fold, A6: 20-fold, H2: 18-fold, H12: 32-fold) and the average raw accuracy across each individual base position in a read was well above 99% with a median read lenght between 141 bp and 226 bp (wild-type control: 203 bp, C5K 218 bp, A1 216 bp, A6 141 bp, H2 226 bp, H12 221 bp). As expected the wild-type control did not show any new mutations compared to the parental wild type, even after the screening process.

For the thermo-tolerant strains various mutations were identified. To exclude false positives introduced by the sequencing technology or the mapping algorithm used, results were confirmed with Sanger sequencing. Table 2 shows a matrix of the true positive mutations which were identified in the thermotolerant strains by NGS and validated with Sanger sequencing. As suspected, all tested indels in homopolymer regions identified with NGS were artifacts of the sequencing technologies (discussed in background) and were confirmed as false positives through Sanger sequencing.

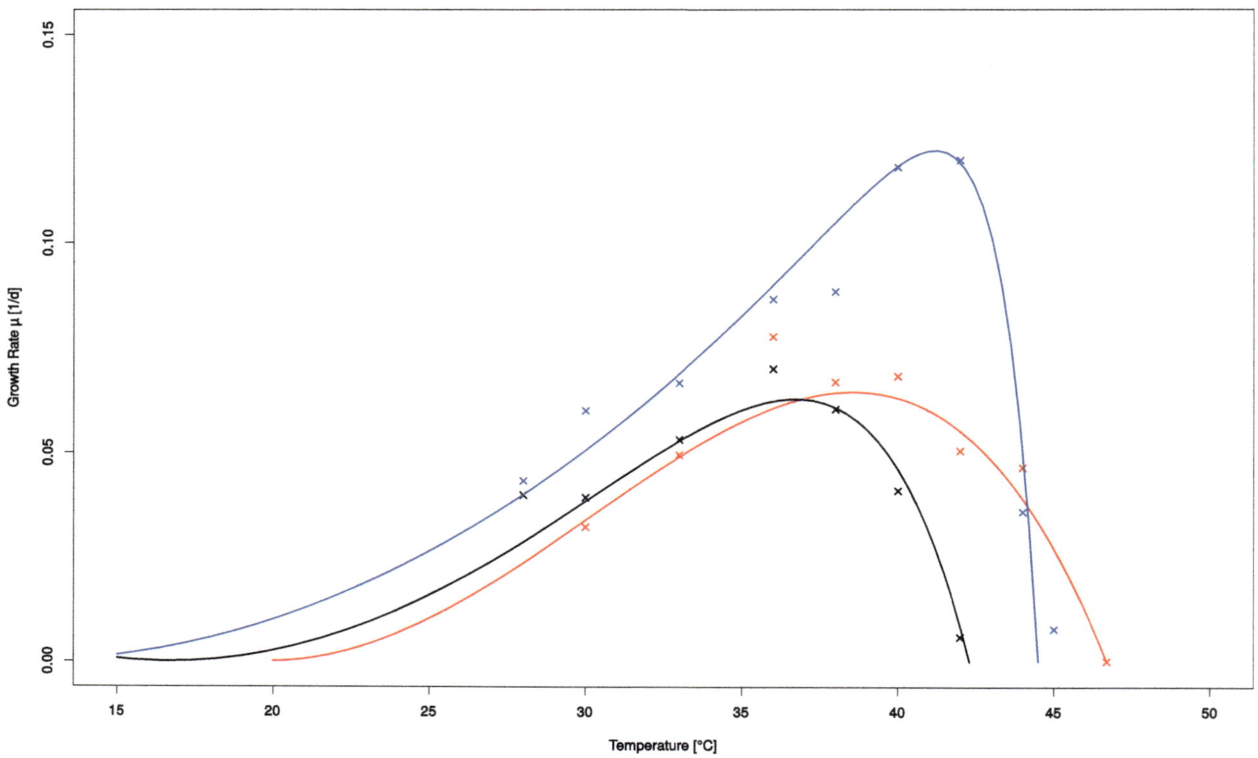

Figure 1 Growth rates over temperature as determined by incubator screening for wt (black), mutIV-mix (red), and C5K (blue).

Prolonged In vivo selection within the strain mix

The various strains used for screening were isolated out of mutIV-mix, a mix of many strains generated through random mutagenesis and *in vivo* selection growing stably at 45°C. Continuing selection on mutIV-mix for one year after isolation of the monoclonal strains (by keeping the selective pressure high and raising the temperature whenever stable growth was achieved) further raised the maximally tolerated temperature with stable growth up to 45.8°C (see Table 1). This new strain-mix with higher thermal tolerance was dubbed mutIV-mix-2. It was sequenced with NGS to validate mutations identified in the monoclonal strains and to gauge the distribution of mutations after prolonged selection. To asses the distribution in of the mutations within the mix a high coverage was achieved (123.5-fold); the average raw accuracy across each individual base position in a read was at 99% with a median read length of 188 bp. The results are listed in Tables 3 and 4.

Identified genes

The gene that stands out the most from those identified (Tables 2 and 3 and 4) is ***clpC***, an ATP-dependent Clp protease ATPase subunit. In total five distinct mutations where identified in this gene. Four within the monoclonal strains, and an additional one in the strain mix. No single strain appears to contain all of the mutations however. C5K contains the most, with three distinct mutations, one of which A1 also carries, while H12 has another unique mutation (Table 2). Proteases are responsible for digesting misfolded proteins, and are typically up-regulated as part of the heat shock response. In *Synechocystis clpC* has been linked to various stress responses, its expression has been shown to be increased after heat shock [16], high-salt acclimation [25] and UV or high-light irradiation [26]. Four of the mutations found in *clpC* (positions 2466798, 2466808, 2466882 & 2467723 - present in H12/mutIV-mix-2, A1/C5K/mutIV-mix-2, mutIV-mix-2, and C5K/mutIV-mix-2 respectively) are located between the known domains of the protein. Of these, the Mutation at position 2466798 is by far the most common in mutIV-mix-2, being represented in 45% of all reads. The last mutation (position 2468031; in C5K/mutIV-mix-2) is located in the first ATPase (ATP binding) site of the protein and is the second most common in mutIV-mix-2.

A second gene presenting more than one mutation is ***pnp***, a polyribonucleotide nucleotidyltransferase. It has one amino acid (AA) change in A1, H12, C5K & mutIV-mix-2 (position 98129) and an additional one in C5K (position 96437). Though in C5K both mutations again seem to be not fully segregated. In mutIV-mix-2 the first

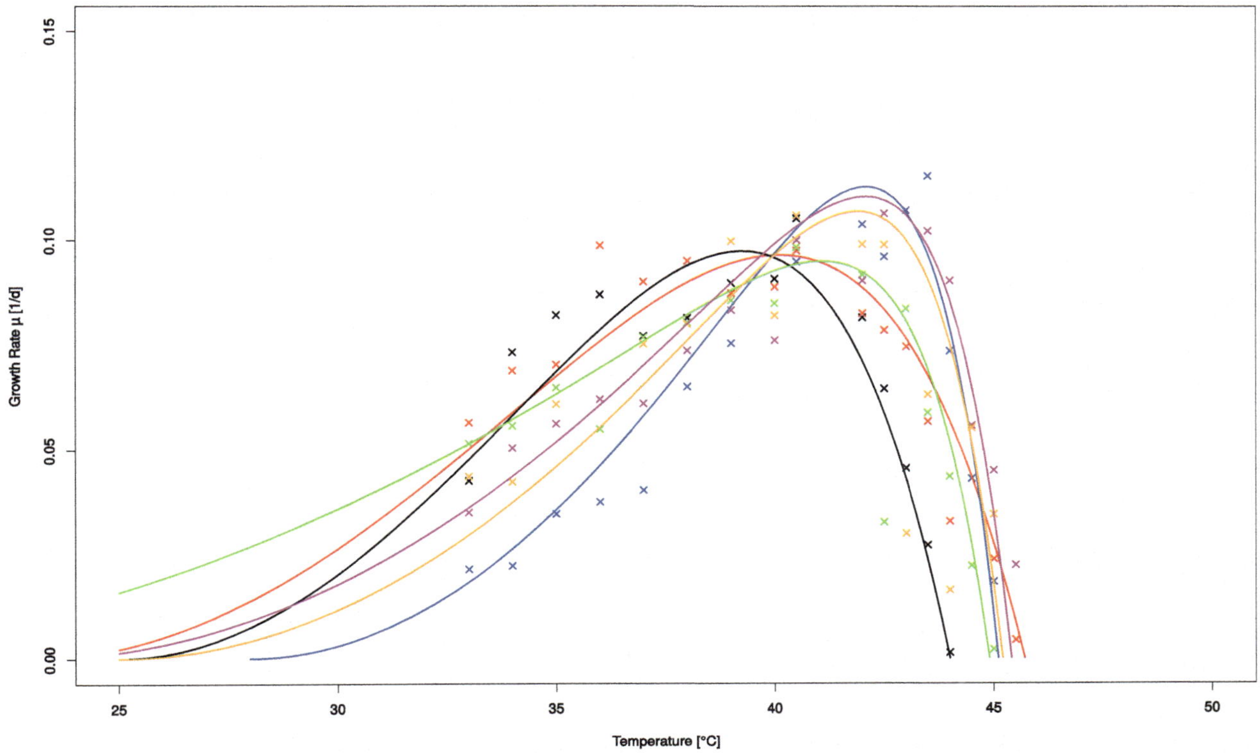

Figure 2 Growth rates over temperature as determined by robot screening: wt (black), mutIV-mix (red), A1 (green), A6 (blue), H2 (magenta), and H12 (orange).

mutation was found in all reads, while the second was not detectable. *Pnp* is part of the degradosome, which is involved in the processing of ribosomal RNA and the degradation of messenger RNA. The expression of *pnp* has previously been reported to be up-regulated in *Synechocystis* in acid stressed [27] and cold acclimated [28] cultures. The first identified mutation (position 96437; C5K) lies in the S1-like superfamily region, which contains the RNA binding sites, while the second (position 98129; C5K, A1, H12, C5K & mutIV-mix-2) is located within the RNAse domain of the protein.

Another very interesting mutation was identified is the amino acid exchange in ***sigF***, a group 3 RNA polymerase sigma factor. *SigF* has been shown to affect various processes in *Synechocystis*. It is needed for high-salt and high-light tolerances [29] and for the expression of thick pili [30,31] and motility [29,31,32]. The identified mutation occurred in the region-4 domain, which is the region of Sigma-70 proteins containing the DNA binding sites.

The same three strains (A6, H12 & H2) with a mutation in *SigF* also showed an amino acid exchange in ***pilJ***, a chemotaxis protein required for thick pili and motility

Table 1 Overview of the different maximally tolerated temperatures of the various strains in both screening platforms, and in the 1 l bioreactor cultivation

Strain	Max. temp robot screening	Max. temp incubator screening	Max. temp bioreactor	Confirmed mutations
wt	44°C	42.3°C	<43°C	None
MutIV-mix	45.7°C	48°C	45.0°C	Various
MutIV-mix-2			45.8°C	Various
MutIV-C5K		44.5°C	44.9°C	10
MutIV-A1	44.9°C		45.0°C	7
MutIV-A6	45.1°C		45.0°C	2
MutIV-H2	45.4°C		45.2°C	3
MutIV-H12	45.2°C		45.0°C	7

Table 2 Matrix of true-positive mutations (marked with X) identified in the various thermo tolerant strains

Position	Mutation type	Gene	WT	Mutant	Strains				
					C5K	A1	A6	H2	H12
96437	AA change	pnp	gAg	gGg	X*				
98129	AA change	pnp	aTg	aCg	X*	X			X*
195523	AA change	slr1098	gGc	gTc	X*				
494155	AA change	PilJ	Cgg	Tgg			X	X*	X*
1103495	AA change	pyk2	ttC	ttA				X*	
1103586	AA change	pyk2	gGa	gAa	X	X			X*
1133812	Intergenic	Intergenic	G	T		X			
1436521	Frameshift	cya1	-	GCAA		X			
2260225	Intergenic	Intergenic	-	A	X	X			X*
2466798	AA change	clpC	gTg	gCg					X*
2466808	AA change	clpC	Cgt	Tgt	X*	X			
2467723	AA change	clpC	Gac	Tac	X*				
2468031	AA change	clpC	gTt	gCt	X*				
2579829	AA change	sll0064	Ttg	Atg	X*				X*
2717197	Frameshift	pyrR	-	TAATTAACTCCAC	X	X			
3371838	AA change	SigF	Cgt	Tgt			X	X*	X

Mutations marked with *indicate mixed reads in Sanger (& NGS), where the wt variant can also still be detected.

[33]. The identified mutation lies within the signaling (receptor) domain of the protein.

The final gene showing more than one mutation (positions 1103495 & 1103586) is ***pyk2***, the gene coding for pyruvate kinase 2, which is essential for glycolysis [34]. No strain contains both mutations; with the exception of A6, all contain either one or the other. In mutIV-mix-2 the first mutation is in 94.4% of the reads, while the second is barely measurable. Both mutations are located within the conserved domain of the protein.

Additionally an intergenic mutation at position **2260225** was found in C5K, A1, H12 as well as mutIV-mix-2. It is upstream of ***nlpD***, a putative peptidase, and thus might affect its expression. In *E. coli nlpD* has functions in cell

Table 3 Mutations identified in mutIV-mix-2 in genes also identified in the monoclonal strains (Table 2)

Position	Mutation type	Gene	WT	Mutant	Percentage of reads
98129*	AA change	pnp	aTg	aCg	100
194955	Frameshift	slr1098	-	T	11.70
195111	AA change	slr1098	Agt	Cgt	23.7
1103495*	AA change	pyk2	ttC	ttA	0.68
1103586*	AA change	pyk2	gGa	gAa	94.4
1133812*	Intergenic	Intergenic	G	T	51.4
1435867	Frameshift	cya1	ATGGTCGACG	-	41.3
2260225*	Intergenic	Intergenic	-	A	100
2466798*	AA change	clpC	gTg	gCg	45.5
2466808*	AA change	clpC	Cgt	Tgt	8.16
2466882	AA change	clpC	gTa	gCa	10
2467723*	AA change	clpC	Gac	Tac	4.65
2468031*	AA change	clpC	gTt	gCt	20.5
2579430	Frameshift	sll0064	-	CCCTG	61
2717197	Frameshift	pyrR	AAGGTTAA	GTGGCTTTAA...	10.9
2717187	Frameshift	pyrR	-	TTAACTCCACTAA	70.4

Positions marked with a *correspond to identical mutation positions for monoclonal strains.

Table 4 Mutations in genes only identified in mutIV-mix-2 and not previously observed in a monoclonal strain

Position	Mutation type	Gene	WT	Mutant	Percentage of reads
57479	AA change	glgA	gCc	gTc	22
853615	AA change	rpoC2	gCc	gTc	38
2499596	AA change	recG	gCc	gTc	70
2850065	AA change	hemH	Gat	Tat	22
3169828	AA change	cmpA	Ggt	Agt	54
3553027	Intergenic	Intergenic	A	G	16

wall formation and maintenance [35], and has been shown to be required for cell separation (peptidoglycan degradation) [36,37].

Both the A1 and C5K strains, presented a mutation in ***pyrR***, an uracil phosphoribosyltransferase. In mutIV-mix-2 two separate frameshift mutations were detected in *pyrR*, one insertion at position 2717183 (in 10.9% of reads) and a second one at position 271787 (in 70.4% of reads). *PyrR* catalyzes conversion of uracil back to uridine 5'-monophosphate, and thereby allows its use for new RNA synthesis without de novo synthesis. The mutation in A1 and C5K inserted a repeat sequence (TAATTAA CTCCA) into the end of the gene, and thus changed the end of its amino acid sequence from VELIKG to WS, shortening the 178 AA long protein by four amino acids. The mutations detected in mutIV-mix-2 also changed the terminal part of the protein. The more common mutation from KG to NSTN, the other removes a stop-codon changing the last 13 AA and adding 40 new AA (STQNAEFFCHWFSILSAAFWCVLDFWRLSDPTGG A GIPLFRPGHRGDRLKKTG). The *pyrR* multi-domain and its active sites are not affected by any of the mutation.

Two distinct frameshift mutations introducing a stop codon were detected in the gene ***cya1***, the major class III adenylate cyclase in *Synechocystis*: An insertion in the the strain A1, and a deletion in 41.3% of mutIV-mix-2 reads. The mutation in A1 truncates the 337 AA long protein by 37 AA in length and changes the last eight AA (to SNCSAAHH). The mutation in mutIV-mix-2 reduces the length by 262 AA and changes the last AA.

In the strains C5K and H12 a not fully segregated mutation causing an AA exchange, was found in the active domain of ***sll0064***, a putative polar amino acid transport system substrate-binding protein located in the periplasm. In mutIV-mix-2 a different mutation causing a frameshift leading to a truncation of 10 AA outside of the active domain was detected.

Three distinct mutations where detected in the hypothetical protein ***slr1098***. Two AA exchanges (C5K & 11.7% of mutIV-mix-2) and one frameshift (23.7% of mutIV-mix-2). The frameshift detected in mutIV-mix-2 inserts a stop codon early in the sequence, reducing the protein to just 38AA.

Additionally a base substitution was also identified on position 1133812 in the strain A1 and in 51.9% of mutIV-mix-2 reads. The mutation upstream of the hypothetical proteins ***sll0875 & slr0876***.

A few mutations were only fond in mutIV-mix-2 but not in any of the monoclonal strains (Table 4). These include mutations the genes ***recG***, ***cmpA***, ***rpoC2***, ***glgA*** and ***hemH*** as well as a mutation at position 3553027 (upstream of ***fur & slr0596***).

The most prevalent of these are the mutations in ***recG*** (70% of reads) and in ***cmpA*** (54% of reads). *RecG* is involved in DNA repair over homolog recombination [38]. The mutation leads to a AA change in the nucleotide binding region of the HELICc-Domain. *CmpA* is part of the cyanobacterial carbon concentrating mechanism (CCM) and specifically binds bicarbonate [39]. The mutation leads to an AA exchange in the active domain (NMT1_2; TauA multidomain).

Discussion

Temperature tolerances

We were able to identify strains with superior characteristics than the wild type by two different screening approaches. The modified Ratkowsky-Equation used for data fitting, allowed for good visualization and easier comparison of the data. The maximally tolerated temperatures determined during the screening (see Table 1) however varied somewhat from those seen in 1 l photobioreactor cultivation. The actually tolerated temperature is expected to vary to some degree depending on other cultivation conditions, such as, light, aeration, and cultivation vessel shape. Therefore selection and validation should be performed in the target cultivation system if possible. In our case the whole approach is based on the conditions in the 1 l bioreactor, (used for selection and validation) which are a good approximation for outdoor bioreactor conditions.

The variation between cultivation systems is especially high for the incubator screening: The maximally tolerated temperature of the C5K strain determined in this system for example, is quite a bit lower (0.4°C) than the one determined in the photobioreactor validation. Meanwhile for the strain mix (mutIV-mix) a far higher thermal tolerance was determined in the incubator screening compared to the photobioreactor validation (0.7°C; see Table 1). In comparison the maximally tolerated temperatures determined in the high throughput robot system were much more in line with bioreactor cultivation data, demonstrating the superiority of this system, not only in throughput and handling, but also in the quality of generated data.

During screening, the strain mix (mutIV-mix-1) also showed a more robust growth over a range of temperatures, which is to be expected as the measured curve results from the average of a number of strains with differing optima. Overall validation of the selected strains in 1 l photobioreactors showed them to be as thermo tolerant (~45°C) as the strain mix they were isolated out of, underpinning the validity of the approach.

Prolonged cultivation of the starting mix (mutIV-mix) over one year under selective conditions in the bioreactor further increased temperature tolerance by an additional 0.8°C (mutIV-mix-2). We assume that no new mutations took place during the cultivation of mutIV-mix-2, after the isolation of the monoclonal strains described above. This is likely to be the case due to the relatively short time frame, lack of additional rounds of mutagenesis and low natural mutation rate. Thus this further increase in thermo-tolerance can probably be attributed to further selection and enrichment of higher performing strains already present within mutIV-mix. Those better strains were not identified with the screening because of the relatively low throughput of the systems used with only 27 candidate strains in the incubator screening and 81 strains in the robot screening. Unlike the incubator screening, the robot screening platform however is highly scalable and could be expanded to create a true high throughput culturing and screening system for phototrophic organisms.

It is also of interest that so few of the strains isolated from the mix outperformed the wild type at all. One would expect all strains contained within the mix to have a thermal tolerance to at least the temperature they were cultured at within the mix. Otherwise they should have been out-competed by other strains, however this does not seem to have been the case. There are several possible explanations for this behavior. It could be, that many strains within the mix (perhaps even those with the highest thermal tolerance) are deficient in other aspects. They could be unable to grow on agar plates, or they could lack the ability to grow monoclonally in any medium (for example because they no longer produce certain vitamins which are readily available in the mix culture or other synergistic effects). They could also have become dependent on specific factors within the photobioreactor cultivation system for fast growth, such as the controlled pH-range, the strong mixing, or the relatively high cell density, where other cells provide shading.

Genetic mutations

The reference wild type appears to have amassed various mutations over time compared to the genome published in 1996. Most are SNPs which may or may not affect the phenotype, while the larger genomic rearrangements appear to be due to mobile transposon elements (ISY203b, ISY203e, ISY203g and ISY203j respectively). Additionally, it should be noted, that the plasmids pCC5.2 and pCB2.4 could not be detected with NGS technology. This has previously been reported, and appears to be an artifact of the technologies used, as both plasmids are detectable by PCR [40]. The genetic drift of the parental wild-type strain during extended laboratory cultivation is not a main topic for this work however, and was not examined in detail. Variation found in our parental wild-type strain compared to the Kazusa strain sequenced in 1996 [19] were not validated with Sanger sequencing. However a re-sequencing of a wt clone after screening (as well as the sequenced enhanced strains) presented the same identified variations.

The lack of new mutations in the wild type, even after selective screening, shows a relative stability in the genome, and demonstrates the need for a mutagen to obtain advantageous mutations in an acceptable time frame.

In the enhanced strains various (though overall surprisingly few) mutations were found. As expected some mutations identified by NGS in homopolymer regions turned out to be false positives upon Sanger validation and were excluded from the analysis. However, the remaining mutations identified are likely to offer some form of increased fitness at higher temperatures, as disadvantageous or neutral mutations are not expected to propagate through the many genomes in *Synechocystis* and are thus likely to be lost after a couple of generations or at the very least not be detectable by NGS or Sanger sequencing.

Advantageous mutations in contrast are expected to multiply within *Synechocystis* genomes during the long selection process, while separating from neutral or disadvantageous mutations which might have generated at the same time. This, together with the relatively low mutagen dosages used (which were specifically selected for optimal point mutation rate), explains the low number of specific hits, and lack of noise from other mutations within the sequencing data.

Overall, the mutation distribution identified in the monoclonal strains selected via the screening procedure (see Table 2) is not easily explained trough linear ancestry between strains. At first glance it appears that within the 5 strains analyzed there are two distinct lineages, one containing the strains C5K & A1 and the other containing A6, H2 & H12, (possibly with A6 as a common ancestor). A6, H2 & H12 also showed very similar growth rates during the screening (see Figure 2). However some of the strains share mutations across these two putative lineages. For example distinct mutations in *pyk2* and *pnp* are found in C5K, A1 & H12 and another in *sll0064* is found in C5K and H12. This indicates that at least some of the mutations must have occurred independently at the same position in various strains. This validates the significance and importance of these specific mutations to achieve a higher thermal tolerance in *Synechocystis.*

All mutations identified are assumed to have at least some degree of effect on the strains temperature tolerances. Speculation on the possible mechanism for these mutations to facilitate a higher temperature tolerance is presented below.

Five distinct mutations were identified in the gene ***clpC***. It is very likely that these affect the activity of the protein, either making it better able to cope with the increased demand or helping it to keep its normal activity up at increased temperature. It is also interesting to note how the mutations are distributed. The monoclonal strain containing more than one mutation in *clpC* (C5K) always had mixed sequencing signals, indicating that they are not fully segregated through all of the genomes. The sequencing of mutIV-mix-2 also show various mutations, with differing percentages. These results indicate that the wild-type copy might still represented in some of the genomes, and that there are various differing mutant copies of the gene, each with one or more of the base substitutions present. Overall the results point towards each point-mutation excluding others within the same gene copy, rather than a combination of various mutations within a single copy. For the first three mutation positions, which have less than 100 bp distance from each other and can thus be covered by a single sequencing-read, no read showed more than one of the mutations simultaneously. This implies that the various mutations are not complementary but instead reach a similar phenothypic outcome through different means. The most common mutation in mutIV-mix-2 (at position 2466798) might provide the highest fitness advantage at increased temperatures, however this effect does not appear to be high enough to fully out-compete all other mutations even after one year of stringent selection.

Both mutations detected in ***pnp*** again might increase its activity at higher temperatures. The position of the first mutation within the binding site could also indicate a change the enzymes specificity towards certain RNAs, thus changing overall protein expression.

In mutIV-mix-2 however only the second mutation which lies within the RNAse domain was found, indicating that this mutation is superior to the other and thus propagated across all strains during the prolonged cultivation. Together with the intergenic mutation at position 2260225, it is the only mutation found in all reads of mutIV-mix-2, and thus appears to confer a large fitness advantage at increased temperature. A propagation of this mutation across all the strains merely through early formation can be excluded, as at the time of single clone isolation the monoclonal strains A6 and H2 did not contain it.

The identified mutation in ***sigF*** occurred within the DNA binding sites and thus probably affects its binding characteristics. The effect of altered binding fro a sigma factor can potentially be very far reaching (and unfortunately unpredictable) by affecting the expression of many proteins. This characteristic has previously been exploited to significantly increase ethanol tolerance in yeast, in a method called global transcription machinery engineering (gTME), where a library of random mutagenized transcription factors is screened for their phenotypic effects [41]. Two out of three monoclonal strains identified carrying this mutation (A6, H2) show only one or two additional mutations (both in *pilJ*, and H2 also one in *pyk2*). These strains do however have a strongly increased thermal tolerance; comparable to that of strains with far more mutations (see Table 1), indicating the strong phenotypic effects of this single mutation.

SigF was one of only two genes for which, while they presented a mutation in some monoclonal strains, none could be detected in mutIV-mix-2 (the other being *pilJ*). This implies that there is a limit for the maximally tolerated temperature reachable with this mutation, as no strain form the mix growing at higher temperatures contains it. Or it might also lead to negative consequences due to deregulation of other genes, giving other mutations conferring increased thermal tolerance an advantage in selection.

The postion of the ***pilJ*** *m*utation indicates it might affect the genes function. It could also have a synergistic effect with the mutation in identified *SigF* as these mutations were only ever identified in conjunction in the same three strains (A6, H12 & H2). Either way, as discussed above, both mutations were not detectable in any reads of mutIV-mix-2, demonstrating that they might only lead to a local maximum concerning temperature tolerance in *Synechocystis.*

Two mutations were detected in ***pyk2***. Since pyruvate kinase is essential for glycolysis, and thus needed for the survival of the organism, a knockout or strong inhibition of *pyk2* can be excluded. Both mutations are within the conserved domain of *pyk2* and could indicate that in *Synechocystis* the reduced activity of the pyruvate kinase at higher temperatures could be one of the bottlenecks inhibiting growth, which is alleviated by either of the identified mutations. The distribution in mutIV-mix-2 indicates that the second mutation seems to be less effective at increasing temperature tolerance, as it has been practically completely displaced by the first.

The mutation upstream of ***nlpD*** could lead to increased or reduced expression of the protein, thus it might affect cell wall composition or aggregation behavior of the cells, making them more likely to stick together, thus protecting the inner cells from the increased temperature. Within the screening platform the strains carrying this mutation where among those where a higher adhesion towards the well walls was observed (data not shown). Together with the previously discussed mutation in *pnp* this is the only mutation identified in all reads of mutIV-mix-2. This indicates that all (or at least almost all) strains contain this mutation in all genome copies indicating a high fitness advantage at increased temperature. A large distribution of this mutation due to early formation can again be excluded, as not all monoclonal strains contained it.

Even though no mutation in ***pyrR*** was within the conserved domain, they could still affect protein activity. Especially when considered in conjunction with the previously discussed mutations in *pnp*, this could indicate an increased activity for for both Enzymes because of a higher turnover of mRNA. In mutIV-mix-2 in total 81.3% of the reads showed a mutation in pyrR, which demonstrates a high importance for increased thermal tolerance.

For ***cya1*** even the more conservative mutation identified, already removes some of the catalytic sites in the conserved CyaA domain of the protein, and thus probably inhibit its function. Adenylate cyclases are important regulators for various cell processes. They generate cAMP, which is a second messenger that participates in a wide variety of signal transduction systems. In *Synechocystis* a *cya1* knockout has been shown to be impaired in motility [42,43]. The effects of a frameshift mutation in *cya1* could be very far reaching however and affect cellular regulation in many ways beyond its motility. In mutIV-mix-2 the mutation from A6 was no longer detectable, instead a mutation was detected which is even more likely to inhibit protein function. Its spread across 41.3% of the reads indicates that this inhibition has a notable fitness advantage at higher temperatures.

For ***sll0064, slr1098*** as well the mutation upstream of the hypothetical proteins ***sll0875 & slr0876*** it is difficult to speculate upon the possible effects, as the full function of the proteins in *Synechocystis* is still unknown.

While sll0064 is classified as a putative amino acid transporter, *sll0875 & slr0876* have unknown functions and have either very low identity to known proteins, or are only similar to other proteins of unknown function, making it impossible to speculate of the discrete effects. *For slr1098* ortholog search results however seem to indicate it contains a HEAT domain, and thus might be involved in intracellular transport. [44,45]. The early insertion of a Stop-codon in *slr1098* definitely constitutes a knock out.

For *sll0064 and sll0875/slr0876* the incidence of the mutations seams to have increased after the prolonged section process (from 2/5 and 1/5 of monoclonal strains, to 61% and 51.9% of all reads in mutIV-mix-2 respectively), indicating an advantage at higher temperatures.

Meanwhile the mutations in *slr1098* do not appear to be as important for high temperature tolerance as only one of the monoclonal strains, and only a low percentage of mutIV-mix-2 reads, included one. The effects should also not be fully excluded however as three distinct mutations propagated across the stains/genomes.

For the mutations only identified in mutIV-mix-2 with very few reads (**rpoC2, glgA, hemH, fur & slr0596**) it is impossible to determine if these are very rare true positives, or if it is a sequencing error (some of which are statistically expected to be identical in more than one read). A validation via PCR and Sanger-sequencing can also not be performed as the competitive PCR process would confer an advantage the more common variant.

For the mutations identified in ***recG*** and ***cmpA*** however, the large number of reads identifies them as true positives. As no further mutagenesis was performed after the isolation of the monoclonal strains it is very likely that these two mutations were already present in the mix before the prolonged selection, albeit probably in lower numbers (less strains).

The mutations in *recG* and *cmpA* are both within the active (/binding) sites and thus might affects their binding characteristics (to nucleotides or bicarbonate respectively). Again these mutations could increase the activity or maintaining the activity at higher temperatures. The mutation at CmpA is especially interesting as phototrophic organisms typically require a more specific RuBisCo at higher temperatures to compensate for the different concentration ratios of dissolved O_2/CO_2 [46]. Cyanobacteria instead increase the local CO_2 concentration via the CCM. Thus this mutation might increase the overall efficacy of the CCM.

Conclusions

Overall the approach of using random mutagenesis, followed by *in vivo* selection, selective screening, and then genetic sequencing was found to be very promising. Many interesting mutations could be identified which increase the thermal tolerance in *Synechocystis.* With only two exceptions all genes containing mutations in the isolated monoclonal strains were confirmed by sequencing the strain mix after prolonged selection, further validating the results. Some mutations in genes (such as *pyk2* – coding for pyruvate kinase 2) are not obvious candidates for temperature tolerance increases, and are thus very unlikely to have been identified by other approaches.

With quite a few genes now identified as being involved in the temperature tolerance of the strains, further strains could be selected at random and then analyzed at low cost with a PCR panel and Sanger sequencing. Thereby providing data to analyze the distribution of the mutations in the strain mix.

Further studies beyond the scope of this work should isolate the individual effect on the temperature tolerance of each mutation by transforming it into a wt strain and characterizing it.

Additionally gene expression profiling of some of the strains, or of a wild type with specific mutations introduced, would probably also yield very interesting results. This is of special interest for mutations such as the one found in sigma factor F (*sigF*) which could affect the expression of many different genes. Sigma factor F is now also an obvious candidate (over the other nine sigma factors present in *Synechocystis*) for global transcription machinery engineering (gTME).

The underlying functional differences resulting from the mutated genes could now also be characterized (e.g. *pyk2* activity over temperature) and be compared to the wild-type protein.

The function of the hypothetical proteins found to be altered should also be investigated, by creating respective knockout strains in *Synechocystis*.

Introducing various identified mutations into a single strain and analyzing the resulting phenotype would also be of considerable interest. It is quite likely that some combinations could lead to strains with even higher temperature tolerances, which could then be used as a starting point for further enhancements, either with targeted methods on known genes, or with new rounds of random mutagenesis.

Methods

Organisms

- *Synechocystis* sp. PCC 6803 salt adapted wild type, provided by Algenol Biofuels Germany GmbH
- *Synechocystis* sp. PCC 6803 - mutIV, mix of various thermo-tolerant strains [6]

Media

All strains were cultivated in sterile mBG11 Media (63 g/l Instant Ocean®, 17.65 mM NaNO3, 0.18 mM K_2HPO_4, 0.03 mM Citric acid, 0.003 mM EDTA (disodium magnesium), 0.19 mM Na_2CO_3, 0.03 mM Ferric ammonium citrate and trace metals).

Generation of thermo-tolerant mutant pool

The starting pool of mutants (mutIV-mix) was generated through four rounds of mutagenesis (2 X UV at 50 J/m^2 and 2 X MMS at 0.1v%/1v% for 1 min). After each mutagenic treatment cultures were kept under highly selective conditions, thereby performing an *in vivo* selection for cells with increased temperature tolerance. The generated pool was able to stably grow at 45°C while the wild type would not tolerate 43°C. For details please see [6].

Monoclonal strain isolation

Monoclonal strains were isolated from the previously generated mix of thermo-tolerant strains (cultivated at 45°C at the time of isolation), by plating 500 cells on BG11-1%-Agar plates. Plates were cultivated at a constant temperature of 40°C to keep selection pressure on thermal tolerance. After about 2 weeks single clones were picked with a sterile loop and transferred into 1.5 ml mBG11 in 96-Deepwell plates (DWPs) or 15 ml mBG11 in T25 cell culture flasks for temperature tolerance screening.

Temperature tolerance screening

Isolated monoclonal strains were grown competitively in either T25 cell culture flasks in an Infors Minitron Incubator (27 strains) or in a 96-Deepwell plate (81 strains) in a custom build automated cultivation chamber integrated into a liquid handling robot.

In both setups, the starting thermo-tolerant strain-mix (4 wells for robot screening), as well as a wild-type clone (6 wells for robot screening) were also cultivated as positive and negative controls respectively. In the DWPs 4 wells contained sterile media, and were used as a control for the detection of potential cross contaminations.

In the incubator for the T25 cell culture flasks night and day cycles were set to 12 h each. During the day phase, illumination was set to 19 μE/m^2, 55 rpm orbital shaking, aeration with 0.5% CO2 in the chamber, and temperature starting at 28°C up to 47°C at the end of the screening. Night temperature was set to 25°C and 25 rpm shaking. OD750 was measured daily with a Shimadzu UV-1800 photometer.

The cultivation chamber is integrated into a Tecan RSP150 robot and can be fully handled by the robots object manipulation arm. Cultures were kept in suspension by orbital shaking at 750 rpm. To avoid cell aggregates and cell adhesion to walls of the DWP, three glass beads (1 mm ø) were added to each well. The atmosphere was set to 2% CO_2. The setup allows automated detection of OD (with Tecan Genios Pro Plate reader). Cultures were temperated through a heated metal plate on which the DWP rests. Temperature (measured in a reference well) is PID controlled and has a maximum measured deviation of 0.2°C from the set value. Day temperature and light were increased stepwise from 30°C and 3 μE/(m^{2*}s) (first 3 days after inoculation) up to 47°C and 85 μE/(m^{2*}s). Night temperature was set to 26°C.

For both systems day-temperatures were raised in 0.5°C increments until 35°C (approx. optimal growth rate of the wt). Depending on behavior of the mix of thermotolerant strains (mutIV-mix), which was cultivated as control, the temperature was raised in maximum increments of 0.2 or 0.1°C per day. Each temperature for which a growth rate was determined was held at least until a minimum of three OD values could be measured (3–7 days).

Analysis of screening data

The growth rate (μ) was calculated for each strain at any given temperature by comparing the OD_{750} at the time point 0 (X_0) with OD_{750} after time t (X_t):

$$\mu = \ln\frac{X_t}{X_0}\cdot\frac{1}{t}$$

To allow an easier comparison of data, and a faster identification of performing strains, the behavior of the strains over temperature was fitted using a modified Ratkowsky-Equation [47,48]:

$$\mu = (b(T-T_{\min}))^2 \cdot\left(1-e^{c(T-T_{\max})}\right)$$

where T_{min} (°C) is the minimum, while T_{max} (°C) is the maximum temperature allowing growth.

The Ratkowsky parameters b and c were calculated according to the following formulas:

$$b = \frac{\Sigma(X_i-\overline{X})\ (y_i-\overline{y})}{\Sigma(X_i-\overline{X})^2}$$

$$c(T-T_{\max}) = log1-\ \frac{\sqrt{\mu}}{b(T-T_{\max})}$$

where x_i are the calculated growth rates with their mean $\overline{X}$, while y_i and $\overline{y}$ represent the respective temperatures and mean temperatures.

Minimum growth temperature was not measured, but instead determined using the first Ratkowsky-Equation:

$$\sqrt{\mu = b\cdot(T-T_{\min})}$$

Only data in the linear range, below the optimal growth rate, was used for this calculation.

Validation of thermal tolerance in bioreactors

The five strains identified in the screening as the most thermal-tolerant were validated through cultivation in a 1 l photobioreactor before being sequenced. The pH value was controlled through the periodic addition of 10% CO2/air mixture. During a 12 h daylight phase, pH was kept at 7.3+/−0.05, light at about 125 μE/m^{2}*s (170 from one side and 80 μE/(m^{2}*s) from the other) and the culture was actively temperated. During the night phase cultures were bubbled with 10 ml/min air, and (passively) cooled down to 23–26°C. Cultures were diluted with fresh media so that they stayed in the logarithmic growth phase, within an OD_{750} between 1 and 2 (measured with a Shimadzu UV-1800 photometer). Cultivation temperature was increased stepwise until the maximum tolerated temperature was found. A culture was considered to be growing stably at a given temperature, if it grew consistently at that temperature for at least two weeks. The maximum tolerated temperature was determined by raising the temperature in in 0.1°C increments until stable growth (as defined above) was no longer possible. The previous lower temperature still allowing stable growth is the maximum tolerated temperature.

Prolonged in vivo selection within thermo-tolerant mix

Over the time span of one year, the strain-mix (mutIV), was continually cultivated under selective pressure in a bioreactor with the conditions described above. The temperature was raised in 0.1°C increments whenever stable growth was achieved, and lowered (allowing for recovery) if a lack of growth showed the maximum tolerated temperature was exceeded.

DNA extraction

DNA was extracted with a novel two step protocol minimizing the amount of foreign (non *Synechocystis*) DNA based on the protocol of Franche and Damerval [49].

All samples were prepared by centrifugation at low g to reduce the content of potential heterotrophic contaminants. 1 ml of culture (OD_{750} = 1) was centrifuged three times at 500 g for 10 min, the supernatant discarded and the resulting pellet resuspended in sterile media.

For the first lysis step the sample was centrifuged at 13,000 g for 10 min. The pellet was washed twice in TE buffer (10 mM TRIS, 1 mM EDTA, pH 8) and then resuspended in 500 μl TES buffer (50 mM Tris, 100 mM EDTA, 25% Sucrose, pH 8) for an osmotic shock and incubated 60 min at 4°C. 500 μg Lysozyme was added before 240 min incubation at 37°C. The sample was then centrifuged at 13000 g for 10 min. The pellet was washed once with Tris buffer and then resuspended in Tris with two units of DNaseI and incubated for 30 min at 37°C. The sample was then centrifuged, washed and resuspended in TE buffer.

A second, stronger, lysis was then performed by repeating the above steps up to and including the addition of Lysozyme, incubating for 30 min at 37°C before adding 50 μl 20% SDS and 8.4 μl ProteaseK (600 mAU/ml) and then incubating overnight (12-16 h). DNA was extracted by adding one volume of phenol/chloroform, vortexing and phase separation by centrifugation at 13000 g for 10 min using PhaseLockGel tubes. The aqueous phase was transferred to a new tube and one volume of chloroform/isoamylalcohol (24:1) was added. After vortexing and centrifugation with the parameters above the aqueous phase was transferred and containing DNA was precipitated using 0.7 volumes of isopropanol and centrifugation at 13000 g for 30 min and 4°C. The pellet was washed twice with 70% ethanol, then dried and resuspendend in 30 µl DEPC water.

RNA was then digested by adding 1 µl RNAseI (10 U/µl) and incubation for 30 min. Afterwards RNAseI was deactivated by heat shock (75°C, 5 min). The DNA content was measured using a Nanodrop Spectrophotometer.

Next-geneneration-sequencing

To identify the mutations in the various strains with thermal tolerance, each of the selected five strains, as well as a wt-control (which also went through the thermal tolerance screening process), and the mutIV-mix-2 strain mixture were sequenced with the Ion Torrent PGM. As a baseline reference, an axenic wt clone, without any thermal tolerance selection or mutagenesis, was sequenced with the Roche 454 Junior Sequencer.

For the reference wt two sequencing runs were performed with DNA fragmented via nitrogen nebulization followed by end repair and library preparation. Library quality was controlled with a capillary gel electrophoresis (Agilent Bioanalyzer), and quantified via the fluorescent beads (TBS 380 Fluorometer) before performing an emulsion PCR amplification and then the shotgun sequencing. All steps were performed according to manufacturer protocols (Rapid Library Preparation Method Manual rev. March 2011; emPCR Amplification Method Manual – Lib-L re. April 2011; Sequencing Method Manual rev. June 2010).

For each screening candidate selected for sequencing and mutIV-mix-2, isolated genomic DNA was fragmented with the Biorupter followed by construction of a shotgun library using the Ion Plus Fragment Library Kit. The quality of the fragments was checked with the 2100 Bioanalyzer (DNA High Sensitivity Kit) and the quantity was assessed on the Roche LightCycler 480 (Ion Library Quantitation Kit). Template preparation and enrichment were performed automatically on the Ion OneTouch devices by means of Ion OneTouch 200 Template Kit v2 DL. Quality of the template preparation and enrichment was checked on the Qubit 2.0 Fluorometer (Ion Sphere Quality Control Kit). The enriched template was loaded to a 314 or 316 (for mutIV-mix-2) chip and sequenced on the IonTorrent PGM using the Ion PGM 200 Sequencing Kit (wt-control; two runs) or Ion PGM 300 Sequencing Kit (C5K, A1, A6, H2, H12 & mutIV-mix-2).

Identification of mutations

Data from the NGS sequencing runs (in SFF format) was mapped against the previously sequenced reference wt [GenBank:NC_000911.1; NC_005229.1; NC_005232.1; NC_005230.1; NC_005231.1; CP003272.1; CP003270.1; L25424.1] using the software Roche gsMapper [50]. Data was then further analyzed with snpEFF [51], and various scripts (bash or Perl) especially written for the analysis; all scripts are attached as additional files, as well as available online at github [52].

First relevant lines were extracted from the generated output file for high confidence differences generated by gsMapper (454HCDiffs.txt) and the chromosome/plasmid names changed with the bash script Additional file 2.

The newly generated file was then converted with the Perl script Additional file 3 to a VCF file which is accepted as input for snpEff. For all enhanced strains, this output file was then compared to the mutations found in the starting wt strain, subtracting all mutations already present in the wild type using the Perl script Additional file 4.

SnpEff, a program to determine the effect of SNPs [51], was then run on the resulting VCF file against the available *Synechocystis* database in snpEff [snpEff:Synechocystis_PCC_6803_uid57659].

The third Perl script Additional file 5 was then used to extend the output of snpEff by adding information on the function, category, subcategory as listed in the category.csv flatfile provided by Cyanobase [53] (Additional file 6) and a link to the Cyanobase entry for mutations within specific genes. The output from this script is in a human readable format and can be opened with a spreadsheet program.

Finally mutations contained in this list were manually verified using Tablet Genome viewer [54] by looking up the mutation positions in the BAM files generated by gsMapper.

Validation of identified mutations and effect determination

Mutations identified by the analysis of next-gen-sequencing data, were validated with Sanger sequencing (except those which were obviously false positives - such as missing bases in some reads within homopolymer regions). Selected regions were amplified by PCR (melting 30 s, 95°C; annealing 30 s, 56°C; extension 30 s, 72°C; 30 Cycles) with specific primers, purified with a PCRapace Kit (Stratec GmbH) according to manufacturer specifications, and then send to Starseq GmbH for sequencing.

The Sanger-sequencing data was visualized with finchtv [55] and the validity of the putative mutation verified.

Additionally five Indel mutations identified within homopolymer regions with the most reads showing missing bases, thus being the most likely true positives, where also verified as described above to validate the assumption of classifying them as false positives.

To better speculate on the possible effects of the identified mutations, the amino acid sequence was compared against the NCBI Conserved Domain Database (CDD).

It was checked if the changes in the protein affect known domains.

Primer design and sequences

Primers were designed with Perl Primer [56] to be 100-150 bp (fw primer) or 350-400 bp (rv primer) from the target base (or first target base for primecr pairs covering multiple mutations). Melting temperature was set to 58-62°C with a maximum difference of 2°C between primer pairs.

All primers were ordered from Thermo Fisher Scientific and are RP-HPLC purified. The full list of primers can be found in Additional file 7: Table S2.

Additional files

Additional file 1: Table S1. List of SNP mutations identified with NGS in the starting wt compared to the Kazusa-wt used as a reference for mapping.

Additional file 2: Bash script to extract relevant lines from output generated by gsMapper (454HCDiffs.txt) and to the chromosome/ plasmid names changed. Also available online at https://github.com/utillich/NGS_Scripts.

Additional file 3: Converts output from "Extract_454HCDiffs.sh" to a VCF file which is accepted as input for snpEff. Also available online at https://github.com/utillich/NGS_Scripts.

Additional file 4: Description of data: Substracts output from one VCF file from another. Also available online at https://github.com/utillich/NGS_Scripts.

Additional file 5: Parses output of snpEff (adding information on the function, category, subcategory, and a link to the Cyanobase entry) and puts out a human readable spreadsheet. Also available online at https://github.com/utillich/NGS_Scripts.

Additional file 6: Slightly edited flatfile containing gene category data, provided by Cyanobase [50]; used as second imput for "vcf_Syn6803_identify.pl". Also available online at https://github.com/utillich/NGS_Scripts.

Additional file 7: Table S2. List of primers used for PCR and Sanger validation of identified mutations.

Competing interests

The submitted work was performed in cooperation with Algenol Biofuels Germany GmbH. Algenol Biofuels actively researches various systems to produce biofuels with cyanobacteria and has a vested interest in its success.

Authors' contributions

UMT conceived and designed the experiments, performed the *in vivo* selection on the mutIV-mix, the bioreactor validation of the selected strains, the PCRs for Sanger Sequencing, the 454 Sequencing runs (together with Markus Grohme), analyzed NGS and Sanger results, co-designed and established the screening platform, and wrote the manuscript. NW co-designed and established the screening platform, performed the screening of the strains, analyzed screening data, extracted DNA from target strains, and reviewed drafts of the manuscript. PF performed all Ion torrent sequencing runs, and reviewed drafts of the manuscript. UD assisted with experimental design, helped with cyanobacteria biotechnology know-how and reviewed drafts of the manuscript. MF supervised the research, provided laboratory facilities and contributed to the writing of the manuscript. All authors read and approved the final manuscript.

Acknowledgements

The authors would also like to acknowledge Markus Grohme for assistance with the two 454 Sequencing runs on the *Synechocystis* wt. Additionally we would like to thank Stefanie Grüber for establishing work on the automated cultivation system. We also wish to express our gratitude to Christian Rockmann for initial assistance in the mapping of the sequencing data. Furthermore we would like to thank Prof. Wolfgang Lockau from the Plant Biochemistry Workgroup at the Humboldt-University Berlin for general guidance and fruitful discussions.

Author details

[1]Molecular Biotechnology and Functional Genomics, Technical University of Applied Sciences Wildau, Bahnhofstraße 1, 16-2001, D-15745 Wildau, Germany. [2]Institute of Biology, Humboldt-University Berlin, Berlin, Germany. [3]Algenol Biofuels Germany GmbH, Berlin, Germany.

References

1. Mascarelli AL: **Gold rush for algae.** *Nature* 2009, **461**:460–461.
2. Savage PE, Hestekin JA: **A perspective on algae, the environment, and energy.** *Environ Progr Sustain Energ* 2013, **32**:877–883.
3. Haag AL: **Algae bloom again.** *Nature* 2007, **447**:520–521.
4. Waltz E: **Biotech's green gold?** *Nat Biotechnol* 2009, **27**:15–18.
5. Jia K, Zhang Y: **Systematic engineering of microorganisms to improve alcohol tolerance.** *Engineering in Life Sci* 2010, **10**:422–429.
6. Miller J: **Mutagenic specificity of ultraviolet light.** *J Mol Biol* 1985, **182**:45–65.
7. Todd P, Monti-Bragadin C: **MMS mutagenesis in strains of Escherichia coli carrying the R46 mutagenic enhancing plasmid: phenotypic analysis of Arg + revertants.** *Mech Mutagen* 1979, **62**:227–237.
8. Tillich UM, Lehmann S, Schulze K, Dühring U, Frohme M: **The optimal mutagen dosage to induce point-mutations in synechocystis sp. PCC6803 and its application to promote temperature tolerance.** *PLoS One* 2012, **7**:e49467.
9. Foster PL: **In vivo mutagenesis.** *Methods Enzymol* 1991, **204**:114–125.
10. Witkin EM: **Ultraviolet mutagenesis and inducible DNA repair in Escherichia coli.** *Bacteriol Rev* 1976, **40**:869–907.
11. Shevell D, Friedman B: **Resistance to alkylation damage in Escherichia coli: role of the Ada protein in induction of the adaptive response.** *Mutat Res* 1990, **233**:53–72.
12. Griese M, Lange C, Soppa J: **Ploidy in cyanobacteria.** *FEMS Microbiol Lett* 2011, **323**:124–131.
13. Nakamoto H, Suzuki M, Kojima K: **Targeted inactivation of the hrcA repressor gene in cyanobacteria.** *FEBS Lett* 2003, **549**:57–62.
14. Suzuki I, Kanesaki Y, Hayashi H, Hall JJ, Simon WJ, Slabas AR, Murata N: **The histidine kinase Hik34 is involved in thermotolerance by regulating the expression of heat shock genes in synechocystis.** *Plant Physiol* 2005, **138**:1409–1421.
15. Rowland JG, Pang X, Suzuki I, Murata N, Simon WJ, Slabas AR: **Identification of components associated with thermal acclimation of photosystem II in synechocystis sp. PCC6803.** *PLoS One* 2010, **5**:e10511.
16. Suzuki I, Simon WJ, Slabas AR: **The heat shock response of Synechocystis sp. PCC 6803 analysed by transcriptomics and proteomics.** *J Exp Bot* 2006, **57**:1573–1578.
17. Inoue N, Taira Y, Emi T, Yamane Y, Kashino Y, Koike H, Satoh K: **Acclimation to the growth temperature and the high-temperature effects on photosystem II and plasma membranes in a mesophilic cyanobacterium, synechocystis sp. PCC6803.** *Plant Cell Physiol* 2001, **42**:1140–1148.
18. Mamedov M, Hayashi H, Murata N: **Effects of glycinebetaine and unsaturation of membrane lipids on heat stability of photosynthetic electron-transport and phosphorylation reactions in Synechocystis PCC6803.** *Biochim Biophys Acta Bioenerg* 1993, **1142**:1–5.
19. Kaneko T, Sato S, Kotani H, Tanaka A, Asamizu E, Nakamura Y, Miyajima N, Hirosawa M, Sugiura M, Sasamoto S, Kimura T, Hosouchi T, Matsuno A, Muraki A, Nakazaki N, Naruo K, Okumura S, Shimpo S, Takeuchi C, Wada T, Watanabe A, Yamada M, Yasuda M, Tabata S: **Sequence analysis of the genome of the unicellular cyanobacterium synechocystis sp. Strain PCC6803. II. Sequence determination of the entire genome and assignment of potential protein-coding regions (supplement).** *DNA Res: Int J for rapid Pub of reports on genes and Genom* 1996, **3**:185–209.
20. Glenn TC: **Field guide to next-generation DNA sequencers: FIELD GUIDE TO NEXT-GEN SEQUENCERS.** *Mol Ecol Resour* 2011, **11**:759–769.

21. Loman NJ, Constantinidou C, Chan JZ, Halachev M, Sergeant M, Penn CW, Robinson ER, Pallen MJ: **High-throughput bacterial genome sequencing: an embarrassment of choice, a world of opportunity.** *Nat Rev Microbiol* 2012, **10**:599–606.
22. Jünemann S, Sedlazeck FJ, Prior K, Albersmeier A, John U, Kalinowski J, Mellmann A, Goesmann A, von Haeseler A, Stoye J: **Updating benchtop sequencing performance comparison.** *Nat Biotechnol* 2013, **31**:294–296.
23. Suzuki S, Ono N, Furusawa C, Ying B-W, Yomo T: **Comparison of sequence reads obtained from three next-generation sequencing platforms.** *PLoS One* 2011, **6**:e19534.
24. Quail MA, Smith M, Coupland P, Otto TD, Harris SR, Connor TR, Bertoni A, Swerdlow HP, Gu Y: **A tale of three next generation sequencing platforms: comparison of Ion torrent, pacific biosciences and illumina MiSeq sequencers.** *BMC Genomics* 2012, **13**:341.
25. Marin K, Kanesaki Y, Los D, Murata N: **Gene expression profiling reflects physiological processes in salt acclimation of Synechocystis sp. strain PCC 6803.** *Plant \ldots* 2004, **136**:3290–3300.
26. Huang L, McCluskey M: **Global gene expression profiles of the cyanobacterium Synechocystis sp. strain PCC 6803 in response to irradiation with UV-B and white light.** *J of \ldots* 2002, **184**:6845–6858.
27. Kurian D, Phadwal K, Mäenpää P: **Proteomic characterization of acid stress response in Synechocystis sp. PCC 6803.** *Proteomics* 2006, **6**:3614–3624.
28. Rowland JG, Simon WJ, Prakash JSS, Slabas AR: **Proteomics reveals a role for the RNA helicase crhR in the modulation of multiple metabolic pathways during cold acclimation of Synechocystis sp. PCC6803.** *J Proteome Res* 2011, **10**:3674–3689.
29. Huckauf J, Nomura C, Forchhammer K, Hagemann M: **Stress responses of Synechocystis sp. strain PCC 6803 mutants impaired in genes encoding putative alternative sigma factors.** *Microbiol (Reading, England)* 2000, **146**(Pt 1):2877–2889.
30. Barker M, de Vries R, Nield J, Komenda J, Nixon PJ: **The deg proteases protect Synechocystis sp. PCC 6803 during heat and light stresses but are not essential for removal of damaged D1 protein during the photosystem two repair cycle.** *The J of Biol Chem* 2006, **281**:30347–30355.
31. Bhaya D, Watanabe N, Ogawa T, Grossman AR: **The role of an alternative sigma factor in motility and pilus formation in the cyanobacterium Synechocystis sp. strain PCC6803.** *Proc Natl Acad Sci U S A* 1999, **96**:3188–3193.
32. Asayama M, Imamura S: **Stringent promoter recognition and autoregulation by the group 3 sigma-factor SigF in the cyanobacterium Synechocystis sp. strain PCC 6803.** *Nucleic Acids Res* 2008, **36**:5297–5305.
33. Chung YH, Cho MS, Moon YJ, Choi JS, Yoo YC, Park YI, Lee KM, Kang KW, Park YM: **ctr1, a gene involved in a signal transduction pathway of the gliding motility in the cyanobacterium Synechocystis sp. PCC 6803.** *FEBS Lett* 2001, **492**:33–38.
34. Knowles VL, Plaxton WC: **From genome to enzyme: analysis of Key glycolytic and oxidative pentose-phosphate pathway enzymes in the cyanobacterium synechocystis sp. PCC 6803.** *Plant Cell Physiol* 2003, **44**:758–763.
35. Lange R, Hengge-Aronis R: **The nlpD gene is located in an operon with rpoS on the Escherichia coli chromosome and encodes a novel lipoprotein with a potential function in ceil wall formation.** *Mol Microbiol* 1994, **13**:733–743.
36. Uehara T, Parzych KR, Dinh T, Bernhardt TG: **Daughter cell separation is controlled by cytokinetic ring-activated cell wall hydrolysis.** *EMBO J* 2010, **29**:1412–1422.
37. Uehara T, Dinh T, Bernhardt TG: **LytM-domain factors Are required for daughter cell separation and rapid ampicillin-induced lysis in Escherichia coli.** *J Bacteriol* 2009, **191**:5094–5107.
38. Domain F, Houot L, Chauvat F, Cassier-Chauvat C: **Function and regulation of the cyanobacterial genes lexA, recA and ruvB: LexA is critical to the survival of cells facing inorganic carbon starvation.** *Mol Microbiol* 2004, **53**:65–80.
39. Koropatkin NM, Koppenaal DW, Pakrasi HB, Smith TJ: **The structure of a cyanobacterial bicarbonate transport protein, CmpA.** *J Biol Chem* 2007, **282**:2606–2614.
40. Trautmann D, Voss B, Wilde A, Al-Babili S, Hess WR: **Microevolution in cyanobacteria: Re-sequencing a motile substrain of synechocystis sp. PCC 6803.** *DNA Res* 2012, **19**:435–448.
41. Alper H, Fischer C, Nevoigt E, Stephanopoulos G: **Tuning genetic control through promoter engineering.** *PNAS* 2005, **102**:12678–12683.
42. Bhaya D, Nakasugi K, Fazeli F, Burriesci MS: **Phototaxis and impaired motility in adenylyl cyclase and cyclase receptor protein mutants of Synechocystis sp. strain PCC 6803.** *J Bacteriol* 2006, **188**:7306–7310.
43. Terauchi K, Ohmori M: **An adenylate cyclase, Cya1, regulates cell motility in the cyanobacterium Synechocystis sp. PCC 6803.** *Plant Cell Physiol* 1999, **40**:248–251.
44. Andrade MA, Bork P: **HEAT repeats in the Huntington's disease protein.** *Nat Genet* 1995, **11**:115–116.
45. Groves MR, Hanlon N, Turowski P, Hemmings BA, Barford D: **The structure of the protein phosphatase 2A PR65/a subunit reveals the conformation of its 15 tandemly repeated HEAT motifs.** *Cell* 1999, **96**:99–110.
46. Tcherkez GGB, Farquhar GD, Andrews TJ: **Despite slow catalysis and confused substrate specificity, all ribulose bisphosphate carboxylases may be nearly perfectly optimized.** *Proc Natl Acad Sci U S A* 2006, **103**:7246–7251.
47. Ratkowsky DA, Olley J, McMeekin TA, Ball A: **Relationship between temperature and growth rate of bacterial cultures.** *J Bacteriol* 1982, **149**:1–5.
48. Zwietering MH, de Koos JT, Hasenack BE, de Witt JC, van't Riet K: **Modeling of bacterial growth as a function of temperature.** *Appl Environ Microbiol* 1991, **57**:1094–1101.
49. Franche C, Damerval T: **Tests on nif probes and DNA hybridizations.** In *Methods in Enzymology*. Edited by Lester Packer ANG Academic Press; 1988:803–808. Volume 167.
50. **Roche GS Reference Mapper.** http://www.454.com/products/analysis-software/.
51. Cingolani P, Platts A, Coon M, Nguyen T, Wang L, Land SJ, Lu X, Ruden DM: **A program for annotating and predicting the effects of single nucleotide polymorphisms, SnpEff: SNPs in the genome of Drosophila melanogaster strain w1118; iso-2; iso-3.** *Fly* 2012, **6**:80–92.
52. 115fx: **UT's Github Repository for NGS Skripts.** https://github.com/utillich/NGS_Scripts.
53. **CyanoBase *Synechocystis* gene categories.** http://genome.microbedb.jp/cyanobase/Synechocystis/genes/category.csv.
54. Milne I, Bayer M, Cardle L, Shaw P, Stephen G, Wright F, Marshall D: **Tablet–next generation sequence assembly visualization.** *Bioinformatics* 2010, **26**:401–402.
55. **FinchTV 1.31.** http://www.geospiza.com.
56. Marshall OJ: **PerlPrimer: cross-platform, graphical primer design for standard, bisulphite and real-time PCR.** *Bioinformatics* 2004, **20**:2471–2472.

3

The transcription factor Ap-1 regulates monkey 20α-hydroxysteroid dehydrogenase promoter activity in CHO cells

Tsevelmaa Nanjidsuren and Kwan-Sik Min*

Abstract

Background: Monkey 20α-hydroxysteroid dehydrogenase (20α-HSD) is a catabolic enzyme responsible for converting progesterone into biologically inactive 20α-hydroxyprogesterone, thereby playing a key role in the estrous cycle or pregnancy and allowing ovulation and parturition to occur in most mammalian animals. Monkey 20α-HSD was highly abundant in ovarian and placental tissues during the pre-ovulation and pre-parturition phase and was primarily localized in the syncytiotrophoblast of the placenta. In this study, we focused on the molecular characterization of the monkey *20α-HSD* promoter region by conducting reporter assays in Chinese hamster ovary (CHO) K1 cells.

Results: A reporter assay using constructs of various lengths of the 5′-flanking region (−890-Luc, −513-Luc, −306-Luc, −273-Luc, and −70-Luc) revealed that a region corresponding to the activator protein 1 (Ap-1) located between −281 and −274 bp was essential for the transcriptional activity. Absence of the Ap-1 site in −273-Luc dramatically decreased the transcription levels to the control levels. When the reporter constructs were co-transfected with Ap-1 (Jun) and specificity protein (Sp-1) genes, the transcription activities of the constructs increased with the exception of −273 and −70, while that of the double construct was reduced compared to that of Ap-1 alone. Furthermore, mutational analysis demonstrated that a putative *Ap-1* site played an important role in the expression of the reporter gene. These findings were confirmed by EMSA examining the interactions of the protein Ap-1 in a nuclear extract from CHO-K1 cells and the expression levels of the Ap-1 transcription factor in pre-parturition placenta and CHO-K1 cells. Although mut-1 and mut-2 of Ap-1 bound with nuclear extracts from CHO-K1 cells, the transcriptional activity of mut-3 was almost completely suppressed.

Conclusions: Our results indicate that the Ap-1 site (−281 → −274) (5′-TGTCTCAT-3′) plays a crucial role in the activation of the monkey *20α-HSD* gene. Thus, we demonstrated that monkey *20α-HSD* promoter activity is regulated by the transcription factor Ap-1 in CHO-K1 cells.

Keywords: Macaque monkey, 20α-HSD, Transcription factor, Ap-1, Sp-1, CHO-K1 cells

Background

The steroid-metabolizing enzyme 20α-hydroxysteroid dehydrogenase (20α-HSD; EC.1.1.1.149) catalyzes the conversion of progesterone, a potent progestin, to the metabolically inactive form 20α-dihydroprogesterone (20α-OHP) [1]. This enzyme belongs to the aldo-keto reductase (AKR) superfamily of nicotinamide adenine dinucleotide phosphate (NADPH)-dependent reductases that act on a wide range of substrates, including steroid hormones and endogenous prostaglandins [2,3]. 20α-HSD plays a crucial role in the modulation and regulation of steroid hormones, such as androgen, estrogen, and progestins [4].

The activity of 20α-HSD is known to be down-regulated by prolactin (PRL) [5], allowing for the maintenance of progesterone secretion during the first half of pregnancy. However, $PGF_2\alpha$ stimulates the expression of 20α-HSD at the end of the pregnancy [6,7]. Administration of $PGF_2\alpha$ to pregnant rats increases luteal 20α-HSD activity [8]. $PGF_2\alpha$ is also known to induce abortion in

* Correspondence: ksmin@hknu.ac.kr
Animal Biotechnology, Graduate School of Bio and Information Technology, Institute of Genetic Engineering, Hankyong National University, Ansung 456-749, Republic of Korea

many species, including rodents [9]. $PGF_2\alpha$ receptor-deficient mice do not show the normal pre-partum drop in progesterone [10,11]. Placental lactogen (PL) is another suppressor of *20α-HSD* gene expression. The luteotrophic effects of both PRL and PL on the corpus luteum (CL) are mediated via the PRL receptor [12]. Progesterone production in the rodent CL is regulated by hormones, including PRL and $PGF_2\alpha$, which have luteotropic and luteolytic functions, respectively [7]. There is additional evidence that $PGF_2\alpha$ may act as an inhibitor of the lactogenic suppression of 20α-HSD expression by increasing the expression of a member of the suppressor of cytokine signaling (SOCS) family, SOCS-3, which has been demonstrated to interfere with cytokine signaling through the Jak2/Stat5 pathway [13].

20α-HSD-deficient mice sustain high progesterone levels and present a delay in parturition of several days, demonstrating that 20α-HSD regulates parturition downstream of the $PGF_2\alpha$ receptor in an essential and nonredundant manner. 20α-HSD deficiency partially corrected the abortion of pregnancies associated with Stab5b deficiency [14]. The duration of the estrous cycle, pseudopregnancy, and pregnancy was significantly prolonged in 20α-HSD $^{-}/^{-}$ mice [15]. Studies on the 5′-flanking region of the mouse 20α-HSD gene demonstrated that the putative Sp-1 site was critical for the expression of the reporter gene on rat luteinized granulose cells [16]. In rats, the regulatory elements of 20α-HSD (Nur77, Ap-1, CRE, PRLRE, and PRE) are present within a 2.5-kb 5′-flanking region of the gene. In addition, it was shown that the deletion of this promoter at the −289 bp positions significantly decreased the basal promoter activity [17]. $PGF_2\alpha$ induces the expression of the nuclear orphan receptor and transcription factor Nur77 in the CL, which in turn leads to the stimulation of 20α-HSD, triggering a decrease in serum progesterone, which is essential for parturition [18]. Thus, Nur77 plays an important role in ovarian physiology by mediating the $PGF_2\alpha$-mediated induction of 20α-HSD in rats. In the human *20α-HSD* gene, the transcription factor-binding site for NF-Y/CEBP located between −128 and −88 is involved in controlling the basal transcription of 20α-HSD in ovarian, lung, and liver carcinoma cells [19]. The transcription factor Sp-1 regulates the overexpression of 20α-HSD in HT29 human colon cells, which are resistant to methotrexate [20]. The activity of the human *20α-HSD* promoter luciferase was increased by 1.7-fold upon treatment with $PGF_2\alpha$ and oxytocin on porcine ovarian follicle cells [21].

Among primates, 20α-HSD has been found in the placenta of humans [22], rhesus monkeys [23], and baboons [24]. In humans, the activity of 20α-HSD increased by 5-fold in the placenta from mid to late gestation [25]. The progesterone levels in baboon fetuses declined by almost 6-fold from mid to late gestation and this decrease was accompanied by a substantial increase in the concentration of 20α-OHP, a major metabolite of progesterone [24]. Fetal progesterone concentrations decreased during late pregnancy in the rhesus monkey [26]. In recent reports, the expression of human 20α-HSD was detected in the kidneys [27], adipose cells [28], skin [29], osteoblasts [30], and carcinoma cells from the lung [31], endometrium [32,33], ovaries [34], guts [35] and skin [36]. Over expression of 20α-HSD has been observed in non-small-cell lung cancer [33] and in esophageal and breast cancer [37,38].

We previously showed that 20α-HSD was highly detected in ovarian and placental tissues during the pre-ovulation and pre-parturition phases. It was mainly localized in the syncytiotrophoblast of the placenta and isthmus cells of the oviduct in macaque monkeys [23]. However, the function and regulation of this enzyme in primate reproductive physiology are not well known. Human *20α-HSD* gene is controlled by Sp-1 [19], but there is no report on the regulation of the monkey *20α-HSD* promoter activity. Therefore, we analyzed that the *20α-HSD* gene could be regulated by different transcriptional factors depending on the species.

In the present study, to gain detailed information on the transcriptional regulation factors, we investigated the 5′-flanking region of the monkey *20α-HSD* gene promoter. We demonstrated that the Ap-1 site of the monkey *20α-HSD* gene is essential for its promoter activity, is related to the interaction of the Ap-1 protein with the nuclear extract from Chinese hamster ovary (CHO) K1 cells, and influences the expression of the Ap-1 transcription factor in pre-parturition placenta and CHO-K1 cells.

Results

Analysis of putative transcription factor binding sites in the monkey *20α-HSD* gene promoter

The monkey *20α-HSD* gene was used to construct the 5′-2005 bp basal promoter region. The transcription factor binding sites of the monkey *20α-HSD* gene promoter was analyzed using the TFsearch software. The promoter region includes a TATA (−52 → −55) box and two putative CCAAT (−59 → −63; −105 → −109) boxes. Moreover, plural transcription factor binding sites were identified, such as Ap-1, Oct-1, Sp-1, NUR77, NF-Y (nuclear factor-Y), CRE-BP (cAMP-response element), SRY, and PRLRE (prolactin response element) (Figure 1A). Various lengths of the 5′-flanking region were constructed to identify the regulatory region crucial for the transcription activity of the *20α-HSD* gene. The deleted mutant promoters were designed as −890-Luc (HSF-2), −510-Luc (XFD), −273-Luc (Ap-1), and −70-Luc (Sp-1), and three mutants were cloned into the PGL-3 Basic vector (Figure 1B).

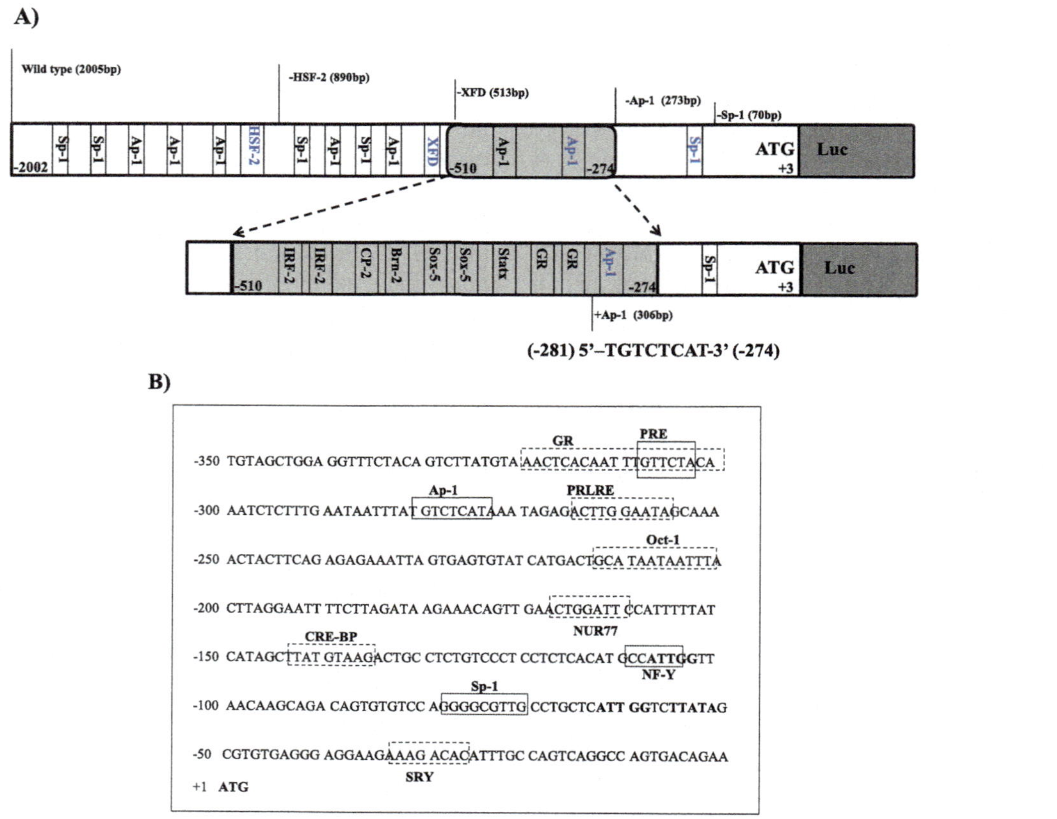

Figure 1 Putative transcription factor binding sites in the 5′-flanking region of the monkey *20α-HSD* gene. A) Sequences approximately 2.5 kb upstream from the translation start codon of the monkey 20α-HSD gene. Putative transcription factor binding sites were identified using the TFsearch software. The region includes GR, Ap-1, Oct-1, Sp-1, CRE-BP1 and others. The TATA (−52 → −55) box and two putative CCAAT (−59 → −63; −105 → −109) box sites are indicated in bold. **B)** The translation start codon is indicated as +1. The region between −350 and −53 has a cluster of transcription factor binding sites that are believed to be important for the basal expression of the monkey *20α-HSD* gene.

Activity of the deleted promoter regions of monkey 20α-HSD

The pGL3-basic vectors were constructed with promoter deletion mutants of different lengths and their activities were assayed in hamster ovarian cells in comparison with the 2005-bp promoter + pGL3 vector (positive control) and non-transfected cells (negative control). As shown in Figure 2, the series of deletion constructs had similar promoter activities. However, when the two constructs −273-Luc and −70-Luc were transfected, the luciferase activity dramatically decreased. In particular, -70-Luc activity was identical to that of the basic pGL3 expression plasmid. Thus, we postulated that the Ap-1 site (−281 to −274) plays an important role in the activation of the promoter of the monkey *20α-HSD* gene. The -306-Luc and −513-Luc constructs containing the Ap-1 site were generated, and their luciferase activities were almost the same (Figure 3). The activity of −273-Luc highly decreased compared to that of −513-Luc and −306-Luc. Next, to investigate whether these factors affected the transcription of monkey *20α-HSD* gene, we transiently

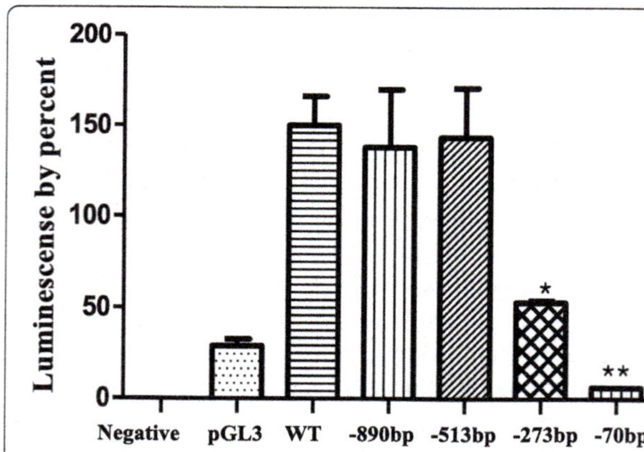

Figure 2 Promoter activity of the serially deleted 5′-flanking region of the monkey *20α-HSD* gene. The −890-Luc, −513-luc, −273-Luc, and −70-Luc fragments were generated by PCR amplification and inserted into the pGL3-basic vector. CHO-K1 cells were transfected with the constructs. The results are presented as a percentage of the relative luciferase activity compared with the wild-type. Graphs show the mean ± SEM of three independent experiments.

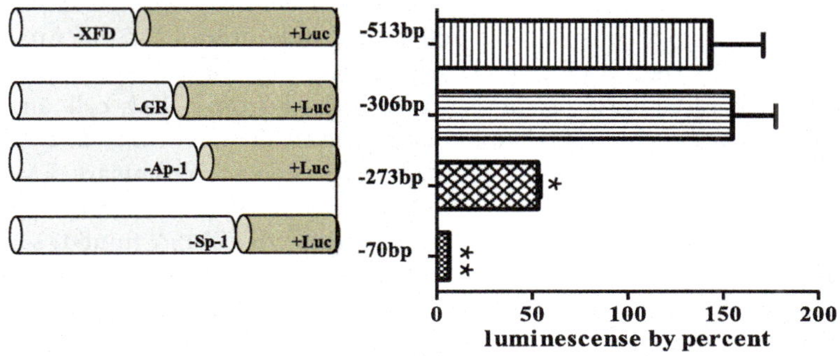

Figure 3 Promoter activity of the deletion regions with/without the Ap-1 site. The deletion constructs with/without the Ap-1 site were amplified and then inserted into the pGL3-basic vector. The promoter activities of the constructs were analyzed by transfection into the CHO-K1 cells. The −306-Luc fragment has an Ap-1 site (5′-TGTCTCAT-3′) from −281 to −274. The −273-Luc fragment does not have an Ap-1 site.

co-transfected the constructs into CHO cells in combination with the Ap-1 or Sp-1 expression plasmids individually and together. The luciferase activity of the three constructs, with the exception of −273-Luc and −70-Luc, co-transfected with the Ap-1 and Sp-1 expression plasmids was high compared to that when only one plasmid was transfected. However, the activity of the −273-Luc and −70-Luc constructs, which did not contain the Ap-1 binding site, did not increase when co-transfected with the Ap-1 and Sp-1 expression plasmids (Figure 4). Therefore, the promoter activity of the monkey *20α-HSD* gene may be related to the Ap-1 transcription factor located between −281 and −274. Moreover, the Sp-1 site between −79 and -70 bp is also dispensable to promoter activity (Figures 3 and 4). This region specifically contains IRF-1, 2, Brn-2, CP-2, Sox-5, and GR (Figure 1A). It is also interesting to note that the region between −273 and −70 contains critical regulation sites (e.g., NF-Y, NUR77 and Sp-1; Figure 1B).

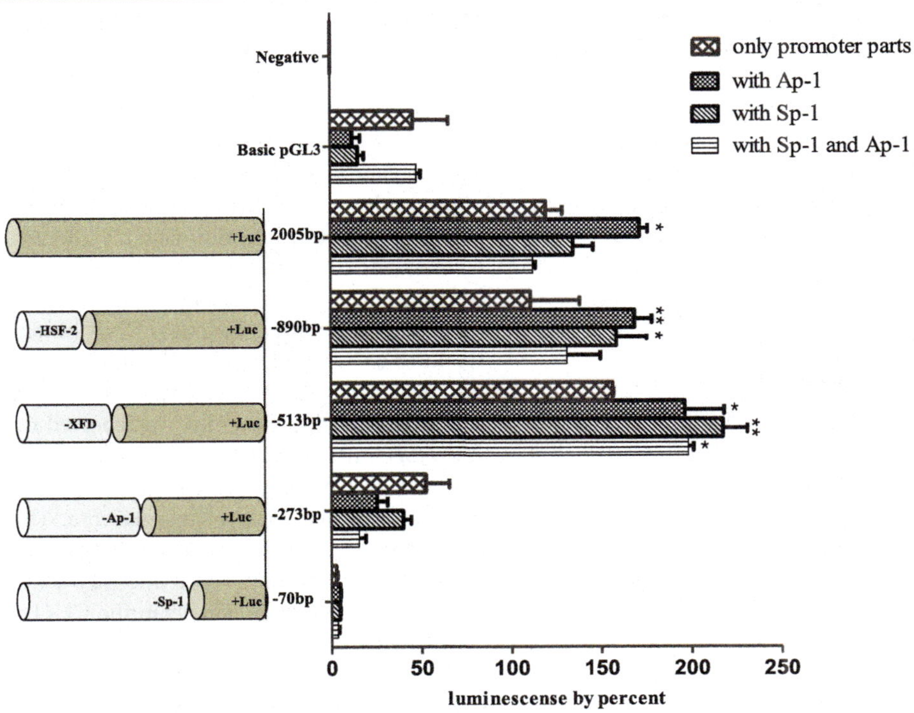

Figure 4 Transcription activity of the serially deleted promoter regions. The pGL3-basic vector constructs containing fragments of various lengths were transfected or co-transfected with the pCMV-Sp1 or pSG5-Ap1 expression vectors into CHO cells. For the co-transfection experiments, the luciferase reporter constructs and factor expression vectors were used at 1:4 concentration ratio. The relative luciferase activity (mean ± SEM for triplicate wells) is presented as the percentage of the activity resulting from the promoter transfection alone in each group.

Analyses of the putative Ap-1 binding site by site-directed mutagenesis

Based on the above results, to investigate whether preventing the binding of Ap-1 resulted in decreased the transcription of the *20α-HSD* gene, we constructed the −288-Luc deletion mutant including the Ap-1 binding site (5′-TGTCTCAT-3′). Additionally, we also constructed mutants within the Ap-1 binding site (Table 1), transfected them into CHO cells alone or with the Ap-1 transcription factor, and compared the results with the positive control (−288-Luc). As shown in Figure 5, the luciferase activities of mutant-1 (mut-1) and mutant-2 (mut-2), in which half of the Ap-1 site nucleotides were mutated, were not significantly affected. However, the transcription activity of mutant-3 (mut-3), which had all eight nucleotides of the Ap-1 region mutated, significantly decreased. Moreover, the activity of −288-Luc co-transfected with the Ap-1 expression plasmid increased more than that of mut-1. The activity of mut-2 and mut-3 co-transfected with the Ap-1 expression plasmid did not change. However, as shown in Figure 5, the activity of mut-3 co-transfected with Ap-1 decreased. These results demonstrate that Ap-1 transcription factor (5′-TGTCTCAT-3′) has a specific binding site for the monkey *20α-HSD* gene located in the −281 → −274 region.

Ap-1 expression in CHO cells and monkey placenta and analysis by EMSA

The Ap-1 expression level in the placenta from the pre-parturition period and the CHO-K1 cell line was determined by RT-PCR and Western blot (Figure 6A,B). Strong expression of Ap-1 mRNA was detected in both CHO-K1 cells and placenta at pre-parturition. We also found that the Ap-1 protein was coordinately expressed as a nuclear and cytoplasmic protein in CHO cells. However, the expression of the Ap-1 protein in the monkey placenta was weak. Consensus fragments of the oligonucleotide probes for the Ap-1 mutants were also synthesized. For the EMSA, a nuclear protein extract was prepared from CHO cell and used with 3′-end biotin-labeled 30 bp (from -292 nt to -262 nt) oligonucleotide probes. Preliminary EMSA analysis showed the presence of Ap-1 protein-DNA complexes. The mutants used for the EMSA (mut-1, −2, and −3) were similar to those employed in the luciferase assay. The binding of mut-1 and mut-2, which had half of the nucleotides mutated, did not compete with the DNA binding as well as the WT construct. However, mut-3, which had all the nucleotides mutated, did compete with the DNA binding (Figure 6C). Thus, partial mutations of the Ap-1 binding site did not affect the basal activity. Therefore, in the present study, we have functionally characterized the monkey *20α-HSD* gene basal proximal promoter and provided evidence of the role of the Ap-1 in CHO cells.

Table 1 Oligonucleotides used for the construction of the reporter vectors and EMSA

Constructs	Nucleotide sequence (5′ 3′)
pGL3 (−2005-Luc)	GAG ACG GGG TTT CTC CAT
pGL3 (−890-Luc)	TCT TAC AAG GCT AAT AAG AA
pGL3 (−513-Luc)	AGT AAA CTT TAA TTT TTA AT
pGL3 (−273-Luc)	AAA TAG AGA CTT GGA ATA GC
pGL3 (−306-Luc)	ACA AAT CTC TTT GAA TAA TT
pGL3 (−288-Luc)	TTT ATG TCT CAT AAA TAG AG
pGL3 (−70-Luc)	CTG CTC ATT GGT CTT ATA GC
pGL3 Reverse	CAT TTC TGT CAC TGG CCT
EMSA probe WT	GAA TAA TTT ATG TCT CAT AAA TAG AGA CTT
EMSA probe mt-1	GAA TAA TTT Aca caT CAT A AA TAG AGA CTT
EMSA probe mt-2	GAA TAA TTT ATG TCg agg AAA TAG AGA CTT
EMSA probe mt-3	GAA TAA TTT Aca cag agg AAA TAG AGA CTT

Discussion

In the present study, we characterized for the first time the transcription activity and regulation of the 5′-flanking region of the monkey *20α-HSD* gene located in the −281 → −274 region relative to the translational start. This study demonstrated that the Ap-1 site plays a crucial role in the transcription of the monkey 20α-HSD gene in CHO cells. Moreover, we clearly demonstrated that Ap-1 is indispensable for the regulation of the expression of the *20α-HSD* gene. The Ap-1 transcription factor is crucially involved in a multitude of cellular processes, including development and differentiation, cell proliferation, apoptosis, oncogenic transformation, and the response to genotoxic agents [39,40]. The activity of this transcription factor is modulated by growth factors, cytokines, and tumor promoters, such as the phorbol ester 12-O-tetradecanoylphorbol 13-acetate (TPA) [40]. Jun homodimers have less DNA binding activity than do Jun/Fos heterodimers [41]. Moreover, the function of the Ap-1 dimer is cell type-specific and cell cycle-dependent [42,43]. Recently, we reported that the monkey *20α-HSD* mRNA was highly expressed in pre-parturition placenta and pre-ovulation ovary, and it was localized in the syncytiotrophoblast of the placenta. The promoter region of the 5′-flanking region was cloned and the EGFP expression signal under the monkey *20α-HSD* 2005-bp promoter functioned normally in the CHO cell line [23].

In this study, we analyzed the expression of the Ap-1 factor in the pre-parturition placenta of monkey and in hamster ovarian cells. In rat ovarian cells, the Ap-1 factor exhibited a specific expression pattern and response to hormones. Rapid induction of JunB, c-Fos, and Fra-2 by FSH revealed a pattern of other immediate-early genes in granulose cells [44]. Ap-1 proteins are specific

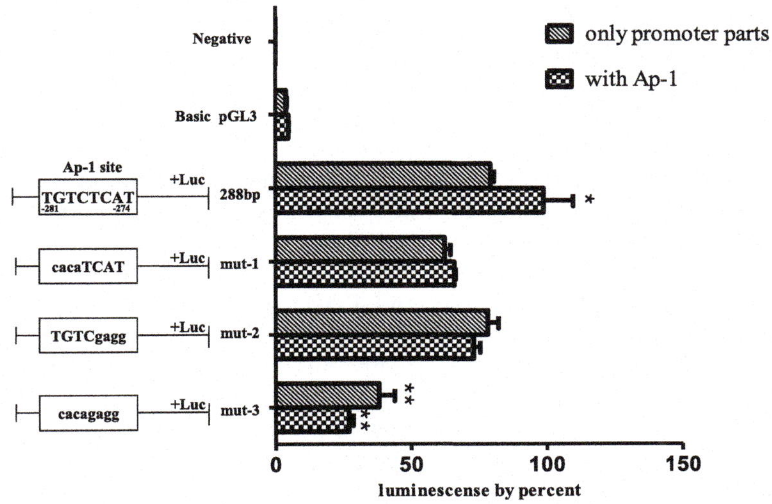

Figure 5 Effect of nucleotide conversion mutation in the putative Ap-1 site on the promoter region of the monkey *20α-HSD* gene on the transcription activity. The Ap-1 transcription factor binding sites are indicated in bold and the conversion-mutated nucleotides are indicated by a smaller character. The luciferase reporter constructs and Ap-1 transcription factor expression vectors were co-transfected at 1:4 concentration ratio. The relative luciferase activity (mean ± SEM for triplicate wells) is presented as the percentage of the activity of the 288 bp-pGL3 construct.

in the process of invasion of human maternal tissue by the trophoblast. Most of the Ap-1 family factors are also expressed in the intermediate trophoblast of the placenta during the first trimester and later during pregnancy, with the expression being stronger for JunD and Fra2, followed by c-Jun, c-Fos, and FosB. Furthermore, JunD was expressed in the syncytiotrophoblast [45]. Ap-1 is involved in placenta-specific expression of the mouse and human lactogen I (PL) gene [46,47], human chorionic somatomammotropin A and B [48], human gonadotropin-releasing hormone receptor gene [49], and the collagen digesting enzyme metalloproteinase (MMP)-1 gene that

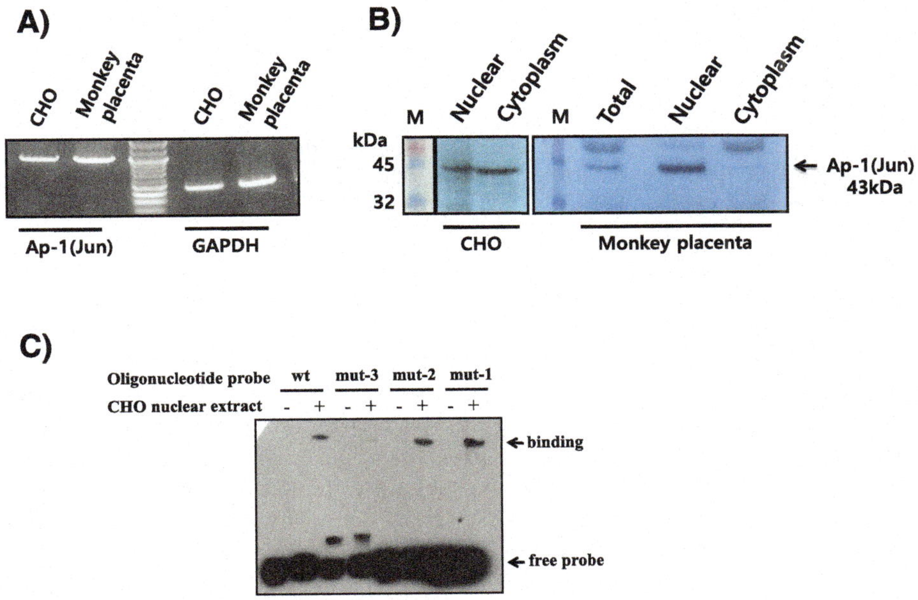

Figure 6 Expression of the Ap-1 (Jun) factor and EMSA analysis with nuclear extract from CHO-K1 cells. A) Ap-1 mRNA was detected in CHO cells and monkey placenta by RT-PCR. **B)** Ap-1 protein was analyzed in the total, nuclear, and cytoplasmic proteins extracted from CHO cells by western blot. **C)** EMSA was performed for the wild-type (−288-Luc) and nucleotide-converted mutants (mut-1, mut-2, mut-3) with nuclear extract (3 μg) from CHO cells using 20 fmol of biotin-labeled double-stranded oligonucleotide probe. The protein-DNA complex was transferred to a Zeta probe membrane. Detection of biotin-labeled DNA was performed using the LightShift chemiluminescent EMSA kit and exposed to X-ray film.

contains an AP-1 response element in its promoter region [50]. In humans, the expression of Ap-1 and NFkB increased in the cervical stroma at the time of parturition. In particular, c-Jun significantly increased in the end of pregnancy and c-Fos immediately increased after parturition, suggesting MAPK activation during cervical ripening before labor [51,52].

In the present study, a reporter assay using constructs of various lengths of the 5′-flanking region revealed that the Ap-1 region between −281 and −274 bp was essential for transcriptional activity. According to the gel shifting assay, the mutant mut-3, which had all the nucleotides mutated, did not have DNA binding activity compared to the wild type, mut-1, or mut-2 Ap-1 factor binding sites. Recent studies demonstrated that human 20α-HSD was regulated by the nuclear factor-Y (NF-Y) in human ovarian, liver, and lung carcinoma cells [19]. However, gel shift assays revealed an increase in Sp1 and Sp2 binding compared to sensitive cells in methotrexate-resistant HT29 human colon cancer cells, without differences in the Sp1 protein levels [20]. A reporter assay of the 5′-flanking region of the mouse *20α-HSD* gene revealed that the region between −83 and 60 bp upstream of the transcription start site was essential for the transcriptional activity in rat luteinizing granulose cells [16]. A putative Sp1 site in this region was critical for the expression of the reporter gene. Finally, the promoter activity was enhanced by the co-transfection of Sp1 expression vector. Thus, the Sp-1 site between −79 and -70 bp needs to be mutated and the activity of the monkey *20α-HSD* mutant promoter tested. However, our results indicate that the monkey *20α-HSD* gene was regulated by Ap-1. Thus, we suggest that the *20α-HSD* gene could be regulated by different transcriptional factors depending on the species.

PRL signals through the Jak2/Stat5 pathway downregulate rat 20α-HSD expression in the decidua. Thus, decidual PRL plays an important role in pregnancy by repressing the expression of interleukin-6 (IL-6) and 20α-HSD in the decidua [53]. The different types of human 20α-HSD inhibitors were reported in an update on the design of potent and active inhibitors based on the crystal structure of the enzyme-inhibitor complex [52]. $PGF_2\alpha$ induces the expression in the CL of the nuclear orphan receptor and transcription factor NUR77, which in turn leads to the transcriptional stimulation of rat 20α-HSD, triggering the decrease in serum progesterone, which is essential for parturition [7]. The activation of Jun D through a calmodulin-dependent activation of extracellular signal regulated kinases (ERk) 1/2 mediates NUR77 induction by $PGF_2\alpha$ [18]. When forskolin was used to treat luteal cells transfected with the 2.5-kb rat *20α-HSD* promoter reporter gene, it caused a dose-dependent inhibition of luciferase activity within 24 h [17]. We also collected data on the acceleration and downregulation of the monkey *20α-HSD* gene promoter by $PGF_2\alpha$ and forskolin (unpublished data). In humans, during cervical ripening and dilation, the presence of the 20α-HSD enzyme results in progesterone inactivation, attenuation of progesterone action, and cervical ripening [54]. In mouse, knockout experiments have revealed non-redundant functions of most of the Ap-1 family members. With regard to the utero-placental system, these experiments pointed to a specific role for JunB and Fra-1. Inactivation of each of the corresponding genes led to severe growth retardation and early fetal death due to impaired vascularization of the decidua [55,56]. Although the NUR77, NF-Y, and Sp1 sites were contained inside the Ap-1 site of the monkey *20α-HSD* promoter, we did not perform analyses to detect the complete suppression of the transcriptional activity of the deletion mutants of the Ap-1 site. However, the *20α-HSD* gene was regulated by these factors in humans, mice, and rats. Thus, further studies aiming to clarify the regulation of the *20α-HSD* gene promoter are needed.

Conclusions

Taken together, these findings indicate that the Ap-1 transcription factor might play a dynamic role in monkey and mammalian reproductive physiology. Our previous study determined that the 20α-HSD enzyme is important in the monkey estrous cycle and parturition. In this study, we suggest that *20α-HSD* gene expression in the ovary and placenta is regulated by Ap-1 in the monkey reproductive system during pregnancy and pre-parturition. Furthermore, we suggest that the 20α-HSD activity may be important for protecting the fetus from high progesterone levels during parturition because Ap-1 regulates *20α-HSD* gene expression in the placenta and ovaries.

Methods

Animal tissues and cell lines

Tissue samples were obtained from the Korea National Primate Research Center (Ohchang, Korea). Placental tissues from an 8-year-old female rhesus monkey of Chinese origins were collected by cesarean section in the pre-parturition period as described previously [23]. The tissues were immediately frozen in liquid nitrogen and stored at −80°C until use. CHO-K1 cells were obtained from the Japanese Cancer Research Resources Bank (Tokyo, Japan). CHO-K1 cells were cultured in Ham's F12 medium (Gibco, MD) containing penicillin (50 U/mL), streptomycin (50 mg/mL), glutamine (2 mM), and 10% fetal bovine serum (FBS). The cells were incubated at 37°C under 5% CO_2 atmosphere. All animal housing and experiments were performed in accordance with the

Korea Research Institute of Bioscience and Biotechnology (KRIBB) Institutional Animal Care and Use Committee Guidance (Accepted No. KRIBB-AFC-09017).

Construction of luciferase reporter vectors and Ap-1 site mutants

The monkey *20α-HSD* gene promoter (2005-bp) was cloned using the long amplification polymerase chain reaction (LA-PCR) cloning method as described previously [23] and the potential transcription factor binding sites within the *20α-HSD* gene promoter were screened using the TFsearch program (Kyoto University, Japan). To identify the critical trans-activation region of the *20α-HSD* promoter, a series of deletion fragments [−890-Luc, −513-Luc, −306-Luc, −273-Luc, and −70-Luc relative to the translational start codon (+1)] were generated by PCR amplification using the template cloning vector containing the 2005-bp *20α-HSD* promoter region and the primers shown in Table 1. The primers contained the *Kpn*I site at the 5′-end and the *Xho*I site at the 3′-end of each fragment. The PCR fragments were subcloned into the pCR2.1 vector and verified by sequencing to rule out the possibility of any PCR error. The correct plasmids were digested with *Kpn*I and *Xho*I, and then inserted into the same digested sites of the pGL3-Basic luciferase reporter vector (Promega, Madison, WI). Nucleotide inversion mutants (mut-1, mut-2, and mut-3) of the putative Ap-1 site were constructed *according to* the two-step overlapping PCR method [57] using the primers shown in Table 1. PCR was performed with Ex taq polymerase (TaKaRa, Osaka, Japan), and the mutant fragments were subcloned into the pCR 2.1. Finally, the mutants were sequenced and inserted into the pGL3 luciferase vector as described above.

Transient transfection and luciferase assay

For the luciferase assay, CHO cells were plated at density of 0.5×10^5 cells per well in 500 μL of growth medium without antibiotics 24 h before transfection. On the next day, the cells were washed twice, and 150 μL of Opti-MEM (Invitrogen, Carlsbad, CA) was added to each well. Each vector (0.8 μg) was mixed with 1 μL of Lipofectamine™2000 (Invitrogen, Carlsbad, CA) reagent in 100 μL of Opti-MEM and incubated at room temperature for 20 min. Afterwards, the Lipofectamine-DNA complex was added to each well. After 4 h, 250 μL of CHO growth medium containing 20% (v/v) FBS was added. After incubation for 24 h, the cells were washed twice, 500 μL of CHO-S-SFM-II was added to each well, and the cells were incubated for additional 24 h. Finally, the luciferase activity was measured using a Dual-Glo™ luciferase assay system kit and a 20/20 Glomax Luminometer (Promega, Madison, WI) according to the manufacturer's instructions. Ap-1 (c-Jun: AF069514) and Sp-1 expression vectors were provided by Dr. K. Imakawa (The University of Tokyo) and Dr. YC. Chang (Catholic University of Daegu School of Medicine). Rabbit polyclonal antibody against mouse Jun was bought from Santa Cruz Biotechnology (Santa Cruz, CA, USA).

In co-transfection experiments, pCMV-Sp1 or pSG5-Ap1 expression vector was mixed with the luciferase vectors at 1:4 concentration ratio. All experiments were repeated at least three times, and representative results were shown. The results of the luciferase assays are expressed as the mean ± standard error of the mean (SEM).

cDNA synthesis and polymerase chain reaction

Total RNA was extracted using TRIzol reagent (Invitrogen, Carlsbad, CA). Monkey placenta and CHO cells (100 mg) were added to 1 mL of TRIzol reagent and homogenized. Then, 0.2 mL of chloroform was added per 1 mL of TRIzol reagent, and the samples were vigorously vortexed for 15 s. Samples were centrifuged at $12{,}000 \times g$ for 15 min at 4°C. After centrifugation, the upper aqueous phase was transferred into a fresh tube and 0.5 mL of isopropyl alcohol was added. The RNA pellet was washed twice with 75% DEPC water. Finally, the concentration and purity of the RNA was determined by spectrophotometry at 260 nm and 280 nm. cDNA was synthesized as described previously [23]. Two microliters of cDNA was used in each PCR. The Ap-1 (Jun) region was amplified by using forward (5′-ATG ACT GCA AAG ATG GAA ACG-3′) and reverse (5′-TCA AAA TGT TTG CAA CTG CTG CTG-3′) primers. The PCR was performed under the following conditions: 94°C for 1 min; 30 cycles of 94°C for 30 s, 58°C for 30 s, and 72°C for 50 s; final extension at 72°C for 10 min. Primers for glyceraldehyde 3-phosphate dehydrogenase (GAPDH) were used for the normalization of the expression of Ap-1. The sequences of the forward and reverse primers were 5′-ACC ACA GTC CAT GCC ATC AC-3′ and 5′-TCC ACC ACC CTG TTG CTG TA-3′, respectively. The expected length of the PCR fragment was 452 bp. The PCR conditions were 26 cycles of 98°C for 10 s, 55°C for 20 s, and 72°C for 20 s. Gel electrophoresis was used to analyze 5 μL of the PCR products.

Preparation of nuclear extracts and Western blot analysis

Nuclear extracts were prepared according to the protocol of the NE-PER Nuclear and Cytoplasmic Extraction Reagents Kit (Thermo Scientific, Marietta, OH). For nuclear extraction, the CHO cells were seeded at a density of 1.5×10^6 cells in 100-mm culture dishes. After 100% of the dish area was covered by cells, the cells were trypsinized and washed with ice-cold 1× PBS and centrifuged at $500 \times g$ for 5 min. Then, the cell pellet was dried. The cells were re-suspended in 200 μL ice-cold

CERI buffer containing 2 μL of a protease inhibitor cocktail, vortexed for 15 s, and incubated on ice for 10 min. Then, 11 μL of ice-cold CERII buffer was added to the tube, the samples were vortexed for 5 s, and incubated on ice for 1 min. The tube was centrifuged for 5 min at 4°C, and the supernatant (cytoplasmic extract) was transferred to a pre-chilled tube. Finally, the insoluble fraction containing the nuclei was suspended in 100 μL of ice-cold NER buffer containing a protease inhibitor cocktail, vortexed for 15 s, and placed on ice for 40 min. The tube was centrifuged at 4°C, 16,000 × g for 10 min, the supernatant (nuclear extract) was recovered to a pre-chilled tube, and the extracts were stored at −80°C until use. The protein concentration was determined by the Bradford's colorimetric assay (Bio-Rad, Hercules, CA) using bovine serum albumin (BSA) as the standard [58].

Monkey placenta and CHO cell total proteins were extracted using PRO-PREP (Intron Biotech, Seoul) protein extraction solution. Approximately 15 mg of tissues were minced, and the confluent CHO cells in 100-mm dishes were washed in 1× PBS. The samples were then homogenized in 600 μL of PRO-PREP solution. The cell lysis reaction was induced by incubating the cells for 30 min on ice. Then, the samples were centrifuged at 13,000 rpm at 4°C for 5 min and the supernatant was transferred to a fresh tube. Finally, the protein concentration was measured using the Bradford's method [40]. The protein samples (10 μg) were subjected to SDS-PAGE and transferred to a PVDF membrane using a semidry electro-blotter apparatus for 3 h. Then, the membrane was washed with 1× PBST solution and blocked with 5× skim-milk blocking solution for 1 h. The membrane was probed with a c-Jun rabbit monoclonal antibody (Cell Signaling Technology, Danvers, MA) washed to remove the unbound antibody, and incubated with the secondary antibody goat anti-rabbit horseradish peroxidase (HRP; Abcam, Cambridge, MA), for 30 min. Finally, the membrane was incubated for 5 min with 3 mL of Lumi-Light substrate solution and exposed on X-ray film for 30 s–5 min.

Electrophoretic mobility shifting assay (EMSA)

Wild-type and mutant probes were amplified as double-stranded oligonucleotides (Table 1) from the −291 to −262 region of the *20α-HSD* promoter. Mutation of the binding consensus site was checked by sequencing, and the fragment was purified by the silica-based Gene Clean II kit (MP Biomedicals, Cleveland, OH) and ammonium acetate DNA precipitation method. All the probes were labeled with biotin using the Biotin 3′-End DNA Labeling kit (Thermo Scientific, Marietta, OH) according to the manufacturer's instructions. For EMSA, 3 μg of the nuclear extract was utilized for the binding reactions with 20 fmol of biotin-labeled double stranded oligonucleotide probe. The EMSA binding reactions were performed at room temperature for 30 min. in a reaction buffer containing 2.5% glycerol, 5 mM $MgCl_2$, 0.05% NP-40, and 50 ng of poly-(dI-dC) for each reaction. The reaction mixtures were loaded on a 6% non-denaturation polyacrylamide gel in pre-chilled 0.5% Tris-EDTA buffer and run at 100 V. The protein-DNA complexes were then transferred to a Zeta probe membrane using the Trans-Blot semi-dry method (Bio Rad, Hercules, CA) and cross-linked using the BLX-254 UV cross-linker. Detection of biotin-labeled DNA was performed using the LightShift chemiluminescent EMSA kit (Thermo Scientific, Marietta, OH), and the membrane was exposed to X-ray film.

Data and statistical analysis

One-way ANOVA Newman–Keuls Multiple Comparison tests were used to compare the results between samples using GraphPad Prism 5 (GraphPad Software, San Diego, CA). Asterisks indicate significant differences from the control group. (*$P < 0.05$, **$P < 0.01$, ***$P < 0.001$).

Competing interests

The authors declare that they have no competing interests.

Authors' contributions

TN performed the experiments. MHK and DJK performed data analysis. KSM wrote the manuscript. Both authors read and approved the final manuscript.

Acknowledgements

The authors thank Dr. HW Seong (Institute of Animal Science) for his helpful discussion and to Mrs, Kang for her technical assistance. T. Nanjidsuren was supported by a scholarship from the BK21 Plus Program (31Z20130012928), the Ministry of Education, Science and Technology, Korea.

References

1. Naidansuren P, Park CW, Kim SH, Nanjidsuren T, Park JJ, Yun SJ, Sim BW, Hwang S, Kang MH, Ryu BY, Hwang SY, Yoon JT, Yamanouchi K, Min KS: **Molecular characterization of bovine placental and ovarian 20α-hydroxysteroid dehydrogenase.** *Rerpoduction* 2011, **142:**723–731.
2. Jez JM, Bennett MJ, Schlegel BP, Lewis M, Penning TM: **Comparative anatomy of the aldoketo reductase superfamily.** *Biochem J* 1997, **326:**625–636.
3. Penning TM: **Molecular endocrinology of hydroxysteroid dehydrogenase.** *Endocr Rev* 1997, **18:**281–305.
4. Penning TM, Jin Y, Steckelbroeck S, Linzer LT, Lewis M: **Structure function of human 3 alpha-hydroxysteroid dehydrogenase genes and proteins.** *Mol Cell Endocrinol* 2004, **215:**63–72.
5. Albarracin CT, Gibori G: **Prolactin action on luteal protein expression in the corpus luteum.** *Endocrinology* 1991, **129:**1821–1830.
6. Albarracin CT, Parmer TG, Duan WR, Nelson SE, Gibori G: **Identification of a major prolactin-regulated protein as 20α-hydroxysteroid dehydrogenase: coordinate regulation of its activity, protein content, and messenger ribonucleic acid expression.** *Endocrinology* 1994, **134:**2453–2460.
7. Stocco CO, Zhong L, Sugimoto Y, Ichikawa A, Lau LF, Gibori G: **Prostaglandin $F_2α$ induced expression of 20α-hydroxysteroid dehydrogenase involves the transcription factor NUR77.** *J Biol Chem* 2000, **275:**37202–37211.
8. Telleria CM, Stocco CO, Stati AO, Deis RP: **Progesterone receptor is not required for progesterone action in the rat corpus luteum of pregnancy.** *Steroids* 1999, **64:**760–766.
9. Deis RP: **Induction of lactogenesis and abortion by prostaglanding F2-alpha in pregnant rats.** *Nature* 1971, **229:**568.

10. Sugimoto Y, Yamasaki A, Segi E, Tsuboi K, Aze Y, Nishimura T, Oida H, Yhosida N, Tanaka T, Katsuyama M, Hasumoto K, Murata T, Hirata M, Ushikubi F, Negishi M, Ichikawa A, Narumiya S: **Failure of parturition in mice lacking the prostaglandin F receptor.** *Science* 1997, **277:**681–683.
11. Narumiya S, Sugimoto Y, Ushikubi F: **Prostanoid receptors: structures, properties, and functions.** *Physiol Rev* 1999, **79:**1193–1225.
12. Clevenger CV, Kline JB: **Prolactin receptor signal transduction.** *Lupus* 2001, **10:**706–718.
13. Stephen TA, Naajia NMI, Johanna LB, Michael JW, Jon DC: **Maximal expression of suppressors of cytokine signaling in the rat ovary occurs in late pregnancy.** *Reproduction* 2009, **138:**537–544.
14. Piekorz RP, Gingras S, Hoffmeyer A, Ihle JN, Weinstein Y: **Regulation of progesterone levels during pregnancy and parturition by signal transducer and activation of transcription 5 and 20α-hydroxysteroid dehydrogenase.** *Mole Endocrinol* 2005, **19:**431–440.
15. Ishida M, Choi JH, Hirabayashi K, Matsuwaki T, Suzuki M, Yamanouchi K, Horai R, Sudo K, Iwakura Y, Nishihara M: **Reproductive phenotype in mice with targeted disruption of the 20α-hydroxysteroid dehydrogenase gene.** *J Reprod Dev* 2007, **53:**499–508.
16. Hirabayashi K, Ishida M, Suzuki M, Yamanouchi K, Nishihara M: **Characterization and functional analysis of the 5'-flanking region of the mouse 20α-hydroxysteroid dehydrogenase.** *Biochem J* 2004, **382:**975–980.
17. Zhong L, Ou J, Barkai U, Mao JF, Frasor J, Gibori G: **Molecular cloning and characterization of the rat ovarian 20α-hydroxysteroid dehydrogenase gene.** *Biochem Biophys Res Commun* 1998, **249:**797–803.
18. Stocco CO, Lau LF, Gibori G: **A calcium/calmodulin-dependent activation of ERK1/2 mediates JunD phosphorylation and induction of Nur77 and 20alpha-HSD genes by prostaglandin F2alpha in ovarian cells.** *J Biol Chem* 2002, **277:**3293–3302.
19. Pallai R, Simpkins H, Chen J, Parekh HK: **The CCAAT box binding transcription factor, nuclear factor-Y (NF-Y) regulates transcription of human aldo-keto reductase 1C1 (AKR1C1) gene.** *Gene* 2010, **459:**11–23.
20. Selga E, Noé V, Ciudad CJ: **Transcriptional regulation of aldo-keto reductase 1C1 in HT29 human colon cancer cells resistant to methotrexate: Role in the cell cycle and apoptosis.** *Biochem Pharmacol* 2008, **75:**414–426.
21. Nishizawa M, Nakajima T, Yasuda K, Kanzaki H, Sasakuri Y, Watanabe K, Ito S: **Close kinship of human 20α-hydroxysteroid dehydrogenase gene with three aldo-keto reductase genes.** *Genes Cells* 2000, **5:**111–125.
22. Fukuda T, Hirato K, Yanaihara T, Nakayama T: **Microsomal 20α-hydroxysteroid dehydrogenase activity for progesterone in human placenta.** *Endocrinol Jpn* 1986, **33:**361–368.
23. Nanjidsuren T, Naidansuren P, Park CW, Park JJ, Yun SJ, Yun BW, Kang MH, Lee SR, Chang KT, Min KS: **Expression and localization of the 20α-hydroxysteroid dehydrogenase (HSD) enzyme in the reproductive tissue of the cynomolgus monkey *Macaca fascicularis*.** *J Steroid Biochem Mol Biol* 2011, **127:**337–344.
24. Waddel BJ, Pepe GJ, Albrecht ED: **Progesterone and 20 alpha-hydroxypregn-4-en-3-one (20alpha-OHP) in the pregnant baboon: Selective placental secretion of 20 alpha-OHP into fetal compartment.** *Biol Reprod* 1996, **55:**854–859.
25. Diaz-Zagoya JC, Wiest WG, Arias F: **Metabolism of progesterone by placentas from several mammalian species in vitro.** *Am J Obstet Gynecol* 1979, **135:**809–813.
26. Thau R, Lanman JT, Brinson A: **Declining plasma progesterone concentration with advancing gestation in blood from ambilical and uterine veins and fetal heart in monkeys.** *Biol Reprod* 1976, **14:**507–509.
27. Quinkler M, Bumke-Vogt C, Meyer V, Bahr V, Olekers W, Diederich S: **The human kidney is a progesterone-metabolizing and androgen-producing organ.** *J Clin Endocrinol Metab* 2003, **88:**2803–2809.
28. Zhang Y, Nadeau M, Faucher F, Leschlleur O, Biron S, Daris M, Rheaume C, Luu-The V, Tchernor A: **Progesterone metabolism in adipose cells.** *Mol Cell Endocrinol* 2009, **298:**76–83.
29. Marin YE, Seiberg M, Lin CB: **Aldo-keto reductase 1C subfamily genes in skin are UV-inducible: possible role in keratinocytes survival.** *Exp Dermatol* 2009, **18:**611–618.
30. Quinkler M, Kaur K, Hewison M, Stewart PM, Cooper MS: **Progesterone is extensively metabolized in osteoblasts: implications for progesterone-action on bone.** *Horm Metab Res* 2008, **40:**679–684.
31. Hsu NY, Ho HC, Chow KC, Lin TY, Shih CS, Wang LS, Tsai CM: **Overexpression of dihydrodiol dehydrogenae as a prognostic marker of non-small cell lung cancer.** *Cancer Res* 2001, **61:**2727–2731.
32. Rizner TL, Smuc T, Rupreht R, Sinkovec J, Penning TM: **AKR1C1 and AKR1C3 may determine progesterone and estrogen ratios in endometrial cancer.** *Mol Cell Endocrinol* 2006, **248:**126–135.
33. Smuc T, Rinzer TL: **Aberrant pre-receptor regulation of estrogen and progesterone action in endometrial cancer.** *Mol Cell Endocrinol* 2009, **301:**74–82.
34. Chen YJ, Yuan CC, Chow KC, Wang PH, Lai CR, Yen MS, Wang LS: **Overexpression of dihydrodiol dehydrogenase is associated with cisplatin-based chemotherapy resistance in ovarian cnacer patients.** *Gynecol Oncol* 2005, **97:**110–117.
35. Chang HC, Chen YL, Chan CP, Yeh KT, Kuo CJ, Fang HY: **Overexpression of dihydordiol dehydrogenase as a prognostic marker in resected gastric cancer patients.** *Dig Dis Sci* 2009, **54:**342–347.
36. Chow KC, Lu MP, Wu MT: **Expression of dihydrodiol dehydrogenase plays important roles in apoptosis- and drug-resistance of A431 squamous cell carcinoma.** *J Dermatol Sci* 2006, **41:**205–212.
37. Wang LS, Chow KC, Wu YC, Lin TY, Li WY: **Inverse expression of dihydrodiol dehydrogenase and glutathione-s-transferase in patients with esophageal squamous cell carcinoma.** *Int J Cancer* 2004, **111:**246–251.
38. Zhang D, Tai LK, Wong L, Chiu LL, Sethi SK, Koay ES: **Proteomic study reveals that proteins involved in metabolic and detoxification pathways are highly expressed in HER-2/neu-positive breast cancer.** *Mol Cell Proteomic* 2005, **4:**1686–1696.
39. Schreiber M, Baumann B, Cotton M, Angel P, Wagner EF: **Fos is an essential component of the mammalian UV response.** *EMBO J* 1995, **14:**5338–5349.
40. Karin M, Liu Z, Zandi E: **AP-1 function and regulation.** *Curr Opin Cell Biol* 1997, **9:**240–246.
41. Ryseck RP, Bravo R: **c-Jun, JunB, and JunD differ in their binding affinities to AP-1 and CRE consensus sequences: effect of Fos proteins.** *Oncogene* 1991, **6:**533–542.
42. Watson A, Eilers A, Lallemand D, Kyriakis J, Rubin LL, Ham J: **Phosphorylation of c-Jun is necessary for apoptosis induced by survival signal withdrawal in cerebellar granule neurons.** *J Neurosci* 1998, **18:**751–762.
43. Lallemand D, Spyrou G, Yaniv M, Pfarr CM: **Variations in Jun and Fos protein expression and AP-1 activity in cycling, resting and stimulated fibroblasts.** *Oncogene* 1997, **14:**819–830.
44. Sharma SC, Richards JS: **Regulation of AP1 (Jun/Fos) factor expression and activation in ovarian granulosa cells: relation of JunD and Fra2 to terminal differentiation.** *J Biol Chem* 2000, **275:**33718–33728.
45. Bamberger AM, Bamberger CB, Aupers S, Milde-Langosch K, LoEning T, Makrigiannakis A: **Expression pattern of the activating protein-1 family of transcription factors in the human placenta.** *Mol Hum Rep* 2004, **10:**223–228.
46. Shida MM, Ng YK, Soares MJ, Linzer DL: **Trophoblast-specific transcription from the mouse placental lactogen-I gene promoter.** *Mol Endocrinol* 1993, **7:**181–188.
47. Peters TJ, Chapman BM, Wolf MW, Soares MJ: **Placental lactogen- I gene activation in differentiating trophoblast cells: extrinsic and intrinsic regulation involving mitogen-activated protein kinase signaling pathways.** *J Endocrinol* 2000, **165:**443–456.
48. Oury C, Alsat E, Jacquemin P, Evain-Brion D, Martial JA, Muller M: **A one-nucleotide difference in a cAMP and phorbol ester response element leads to differential regulation of the human chorionic somatomammotropin A and B gene transcription.** *Mol Endocrinol* 1997, **18:**87–99.
49. Kwai WC, Billy KC, Peter CK: **Functional mapping of a placenta-specific upstream promoter for human gonadotropin-releasing hormone receptor gene.** *Endocrinology* 2001, **142:**1506–1516.
50. Jonat C, Stein B, Ponta H, Herrlich P, Rahmsdorff H: **Positive and negative regulation of collagenase gene expression.** *Matrix Supple* 1992, **1:**145–155.
51. Wang H, Stjernholm Y: **Plasma membrane receptor mediated MAPK signaling pathways are activated in human uterine cervix at parturition.** *Reprod Biol Endocrinol* 2007, **5:**3.
52. Bao L, Tessier C, Prigent-Tessier A, Li F, Buzzio OL, Callegari EA, Horseman ND, Gibori G: **Decidual prolactin silence the expression of genes detrimental to pregnancy.** *Endocrinology* 2007, **148:**2326–2334.
53. El-Kabbani O, Dhagat U, Hara A: **Inhibitors of human 20α-hydroxysteroid dehydrogenase (AKR1C1).** *J Steroid Biochem Mole Biol* 2011, **125:**105–111.
54. Anderson S, Minjarez D, Yost NP, Word RA: **Estrogen and Progesterone Metabolism in the Cervix during Pregnancy and Parturition.** *J Clin Endocrinol Metab* 2008, **93:**2366–2374.

55. Schorpp-Kistner M, Wang ZQ, Angel P, Wagner EF: **JunB is essential for mammalian placentation.** *EMBO J* 1999, **18**:934–948.
56. Schreiber M, Wang ZQ, Jochum W, Fetka I, Elliott C, Wagner EF: **Placental vascularization requires the AP-1 component Fra1.** *Development* 2000, **127**:4939–4948.
57. Min KS, Hiyama T, Seong HH, Hattori N, Tanaka S, Shiota K: **Biological activities of tethered equine chorionic gonadotropin (eCG) and its deglycosylated mutants.** *J Reprod Dev* 2004, **50**:297–304.
58. Bradford MM: **A rapid and sensitive method for the quantitation of microgram quantities of protein utilizing the principle of protein-dye binding.** *Anal Biochem* 1976, **72**:248–254.

4

Transgene autoexcision in switchgrass pollen mediated by the Bxb1 recombinase

Maria N Somleva[1*], Chang Ai Xu[1], Kieran P Ryan[1], Roger Thilmony[2], Oliver Peoples[1], Kristi D Snell[1] and James Thomson[2*]

Abstract

Background: Switchgrass (*Panicum virgatum* L.) has a great potential as a platform for the production of biobased plastics, chemicals and energy mainly because of its high biomass yield on marginal land and low agricultural inputs. During the last decade, there has been increased interest in the genetic improvement of this crop through transgenic approaches. Since switchgrass, like most perennial grasses, is exclusively cross pollinating and poorly domesticated, preventing the dispersal of transgenic pollen into the environment is a critical requisite for the commercial deployment of this important biomass crop. In this study, the feasibility of controlling pollen-mediated gene flow in transgenic switchgrass using the large serine site-specific recombinase Bxb1 has been investigated.

Results: A novel approach utilizing co-transformation of two separate vectors was used to test the functionality of the Bxb1/*att* recombination system in switchgrass. In addition, two promoters with high pollen-specific activity were identified and thoroughly characterized prior to their introduction into a test vector explicitly designed for both autoexcision and quantitative analyses of recombination events. Our strategy for developmentally programmed precise excision of the recombinase and marker genes in switchgrass pollen resulted in the generation of transgene-excised progeny. The autoexcision efficiencies were in the range of 22-42% depending on the transformation event and assay used.

Conclusion: The results presented here mark an important milestone towards the establishment of a reliable biocontainment system for switchgrass which will facilitate the development of this crop as a biorefinery feedstock through advanced biotechnological approaches.

Keywords: Autoexcision, Bxb1 site-specific recombinase, Developmentally programmed transgene excision, Pollen-mediated gene flow, *Panicum virgatum* L., Switchgrass

Background

Biomass from perennial grasses is considered a "low-input high-diversity" feedstock [1] for the production of second generation biofuels in the US and Europe [2]. Switchgrass (*Panicum virgatum* L.), a warm-season perennial grass native to North America, is recognized as a premium candidate and has been identified by the U.S. Department of Energy as a model herbaceous energy crop due to its potential for high biomass production on marginal land, low agricultural inputs and positive environmental impacts [3]. Traditional breeding efforts have been focused on the improvement of switchgrass as a forage crop [4], while the current bioenergy research is targeting high cellulose and starch content for bioethanol production and low ash content for combustion systems [5] combined with significantly increased biomass yield. However, conventional breeding alone may not be able to meet the milestones for the commercial implementation of renewable biofuels as defined by the US government in the 2005 Energy Policy Act, the 2007 Energy Independence and Security Act and the 2008 Food, Conservation and Energy Act. Thus, a biotech approach to engineering bioenergy-related traits in switchgrass has been explored by many research groups [6]. Moreover, the recently demonstrated potential of switchgrass as a production platform for PHA bioplastics and biochemicals [6,7] makes this crop a promising candidate for the

* Correspondence: somleva@metabolix.com; James.Thomson@ARS.USDA.GOV
[1]Metabolix, Inc., 21 Erie St., Cambridge, MA 02139, USA
[2]USDA-ARS-CIU, 800 Buchanan St., Albany, CA 94710, USA

development of a value added biorefinery feedstock for the production of biobased coproducts and energy [8].

A major challenge for the commercialization of transgenic switchgrass and other outcrossing perennial grasses is to prevent the pollen-mediated transfer of transgenes to non-transgenic crops and wild relatives [9,10]. Although several biocontainment methods have been developed to reduce transgenic pollen dispersal, they have been tested mainly in model plants [11]. Among these methods, site-specific recombination is recognized as a promising strategy for transgene removal and other manipulations of the plant genome [12].

In this study, we have investigated the feasibility of engineering site-specific autoexcision of transgenes in switchgrass pollen mediated by the large serine recombinase Bxb1. Bxb1 is a 500 amino acid enzyme that binds minimal recognition sites 48 bp *attP* and 38 bp *attB* without the need for cofactors [13] to execute unidirectional site-specific recombination [14]. Essentially all reactions are fixed in place, which is advantageous for the stability of genomic engineering. The Bxb1/*att* system has previously been tested in tobacco [15], *Arabidopsis* [16] and wheat [17]. This is the first report demonstrating the functionality of a site-specific recombination system in switchgrass, an exclusively self-incompatible and highly heterozygous species.

Results

Bxb1/*att* recombination system is functional in switchgrass

A gain-of-function strategy was used to test the Bxb1/*att* site-specific recombination system in switchgrass. Two vectors, the recombinase vector pMBXS638 (Figure 1a) and the target vector pMBXS640 (Figure 1b), were created and used for co-transformation. Their design allows for the formation of a functional expression cassette of the β-glucuronidase-enhanced Green Fluorescent Protein (*gusAeGFP*) bifunctional reporter gene and $P^{OsActin1}$ after Bxb1-mediated site specific recombination removes the stuffer DNA (Figure 1c).

The vectors were co-introduced into switchgrass callus cultures using a two-strain co-transformation method. Five co-transformants obtained from two independent callus lines were identified in tissue culture by PCR using primers specific for the coding regions of the transgenes. The co-transformed plants were transferred to soil and grown in a greenhouse for further analyses. Transgenic plants transformed with the vector pMBXS640 and wild-type plants (regenerated from untransformed callus cultures) grown under the same conditions were used as controls.

Southern blot analysis confirmed that the plants obtained from different callus lines were independent co-transformation events. The results also revealed the presence of different copy numbers of the pMBXS638 and pMBXS640 T-DNAs (Figure 1d).

Bxb1-mediated excision of the stuffer in co-transformed plants was analyzed by PCR with the primers *s1&s2* specific for the stuffer sequence and *e1&e2* annealing to sequences outside the region flanked by the Bxb1 recombination sites (Figure 1c). This analysis revealed the lack of the stuffer in the co-transformed plants (Figure 1e). The size of the amplified fragment suggested precise excision of the DNA sequence between *attB* and *attP* and the formation of the predicted footprint *attL* was confirmed by sequencing of the PCR product (Figure 1f). RT-PCR analysis demonstrated the expression of the recombinase gene and the transcriptional activation of the reporter gene (Figure 1g). As expected, *gusA* and *bxb1* transcripts were not detected in the plant transformed with the stuffer vector pMBXS640 alone (lane 6 in Figure 1g).

Characterization of the rice pollen-specific promoters *PS1* and *PS3* in switchgrass

In an effort to identify promoters suitable for transgene biocontainment in switchgrass, the spatiotemporal activity of two pollen-specific promoters from rice was evaluated in plants transformed with the vectors pGPro8-PS1 and pGPro8-PS3 which contain *PS1* and *PS3* promoter-GUSPlus reporter gene fusions, respectively.

To test the activity of these promoters in vegetative tissues, 27 plants representing 6 pGPro8-PS1 events and 36 plants representing 12 independent pGPro8-PS3 events were analyzed in tissue culture by histochemical staining for GUS activity and by RT-PCR. No expression of the reporter gene was detected in leaves and roots (data not shown). Randomly selected transgenic plants from 6 independent transformation events for each promoter construct were grown in soil. At the reproductive stage, reporter gene expression was monitored in the perfect terminal and vestigial lower florets of spikelets before, at, and after anthesis by histochemical staining. GUS activity was observed at different stages of pollen development – uninuclear microspores and bicellular pollen grains as well as in mature tricellular pollen grains in the *PS3* transgenic lines, while strong expression of the reporter gene driven by the rice *PS1* promoter was detected only in mature pollen (Figure 2).

Bxb1-mediated transgene autoexcision in switchgrass pollen

A gain-of-function strategy based on pollen-specific autoexcision was also used to obtain switchgrass plants with transgene-excised pollen. The test vector pMBXS824, harboring the expression cassettes of the optimized recombinase gene *bxbNom* and the marker gene *bar* was created for this purpose (Figure 3a). To

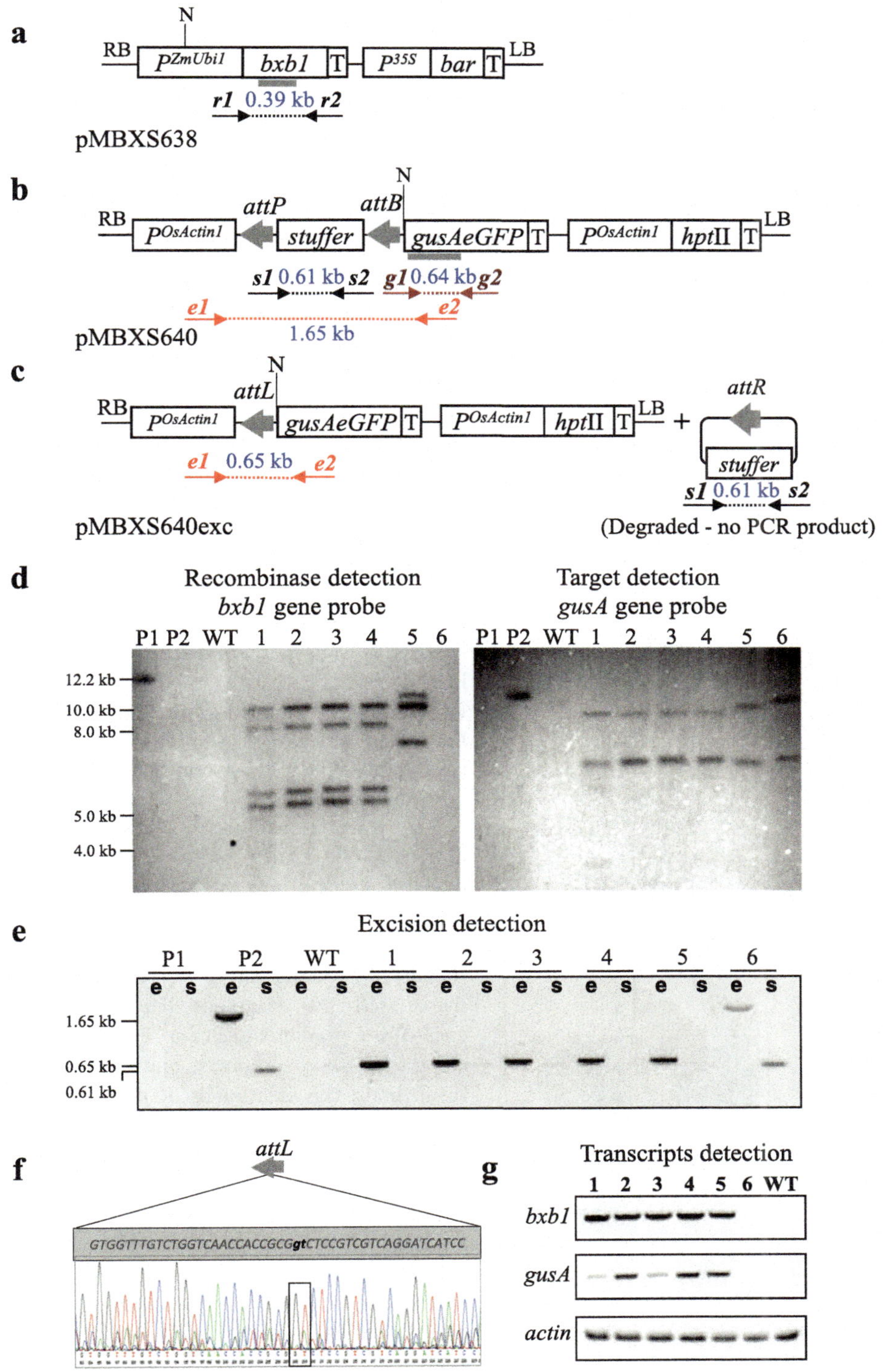

Figure 1 (See legend on next page.)

(See figure on previous page.)

Figure 1 Evaluation of Bxb1 functionality in switchgrass. Schematic of **(a)** the pMBXS638 T-DNA containing the *bxb1* recombinase gene, **(b)** the pMBXS640 T-DNA carrying a non-coding DNA fragment (stuffer), and **(c)** the pMBXS640 predicted T-DNA structures after stuffer excision. PCR primers are shown as *r1&r2, g1&g2, e1&e2* and *s1&s2, att* sites as grey arrows and hybridization probes as grey rectangles. **(d)** Southern blot with probes for the *bxb1* gene (left panel) and the *gusA* gene (right panel). *Lanes*: *P1*, pMBXS638 DNA; *P2*, pMBXS640 DNA; *WT*, wild-type plant; *1–4*, individual co-transformed plants from the same callus line; *5*, co-transformed plant regenerated from another callus line; *6*, plant transformed with the vector pMBXS640. **(e)** PCR for detection of Bxb1-mediated excision of the stuffer: *e*, PCR reactions for the stuffer construct using primers annealing to sequences outside the region flanked by the *att* sites (primers *e1&e2*); *s*, PCR reactions for the stuffer fragment (primers *s1&s2*); *Lanes*: *P1*, pMBXS638 DNA; *P2*, pMBXS640 DNA; the plant samples are in the same order as in **(d)**. **(f)** Sequence of a PCR product (obtained with primers *e1&e2*) representing the *attL* footprint. **(g)** RT-PCR detection of *bxb1* (with primers *r1&r2* in 5 ng of total RNA), *gusA* (*g1&g2*, 200 ng RNA), and actin (*ActA* and *ActB*, 0.5 ng RNA) as a loading control. *Lanes*: *1–6*, transgenic plants in the same order as in **(d)** and **(e)**; *Abbreviations*: P^{ZmUbi1}, promoter and first intron of the maize *ubiquitin1* gene; *bxb1*, Bxb1 recombinase gene; P^{35S}, *CaMV35S* promoter; *bar*, marker gene conferring resistance to bialaphos; $P^{OsActin1}$, promoter and first intron of the rice *actin1* gene; *gusAeGFP*, β-glucuronidase-enhanced Green Fluorescent Protein fusion; *attB* and *attP*, Bxb1 recombinase recognition sites; *attL* and *attR*, sequences formed after excision; *T*, transcription terminator; *LB*, left border sequence; *RB*, right border sequence; *N*, *Nco*I site.

facilitate the identification of autoexcision events, the two expression cassettes, flanked by the Bxb1 recombination sites, separate the *PS1* promoter (proximal to *attP50* site) from the reporter gene *gusAeGFP*, which is silent prior to the removal of the transgenes between the *att* sites. The events leading to the pollen-specific elimination of the recombinase and marker genes from the plant genome (autoexcision) are illustrated in Figure 3b.

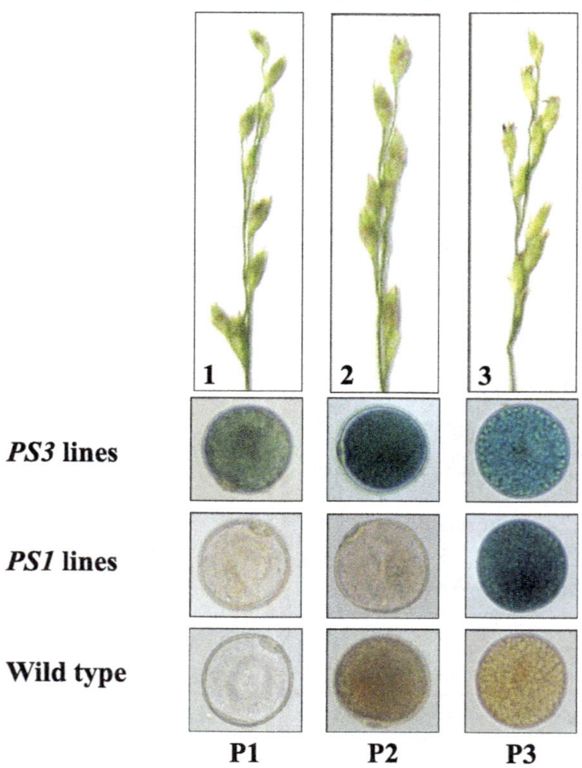

Figure 2 Activity of the rice pollen-specific promoters *PS1* and *PS3* in switchgrass pollen. *1–3*, Images of switchgrass spikelets showing their morphology at representative developmental stages before anthesis (*1–2*) and at anthesis (*3*) corresponding to the stages of pollen development (P1 and P2) and maturation (P3) according to [18]. The stained pollen grains were examined at magnification of 63×.

Switchgrass transformation with pMBXS824 yielded 26 bialaphos-resistant callus lines and 78 primary transformants regenerated from them were identified by PCR. In total, 38 transgenic plants from 22 independent transformation events were grown in a greenhouse and analyzed by molecular and histochemical techniques to identify plants with Bxb1-mediated autoexcision in pollen. For this purpose, pollen grains were collected from perfect florets at anthesis and used for DNA and RNA isolation as well as for GUS activity staining.

Seven switchgrass lines with transgene excision in pollen were identified by PCR using primers *e3&e4* (Figure 3). A single copy of the inserted T-DNA was detected in four of these lines by Southern blot hybridization (Figure 4a). A more detailed PCR analysis was conducted with four transgenic lines using different sets of primers (Figure 3) for detection of autoexcision, *bxbNom*, *bar*, and the excised DNA fragment. The results revealed the presence of the predicted 0.51 kb band obtained after excision along with the recombinase and marker genes (Figure 4b). In all lines, the excision PCR product contained the expected *attL* footprint as verified by sequencing (data not shown). Since no strong GFP fluorescence was detected in switchgrass pollen, all reporter gene analyses consisted of evaluating transgenic plants for *gusA* gene expression. Although GUS activity was detected in some mature pollen grains (Figure 4c), the observed staining patterns were not conclusive. A possible explanation could be reduced expression of the reporter gene due to the presence of the *attL* sequence between the promoter and the coding region in the newly-formed expression cassette as suggested previously [19]. Pollen-specific autoexcision resulting in expression of the *gusA* gene was demonstrated by RT-PCR with total RNA isolated from pollen (Figure 4d). The observed accumulation of the *bxbNom* transcripts confirmed the presence of the recombinase gene detected by PCR.

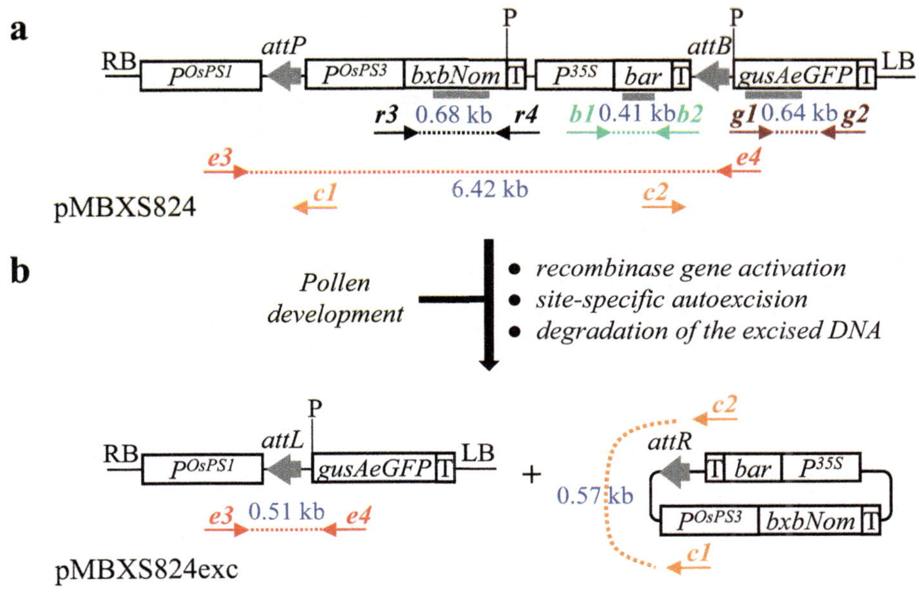

Figure 3 Schematic illustration of Bxb1 recombinase-mediated transgene autoexcision in pollen. (a) T-DNA of the vector pMBXS824. PCR primers are indicated as *r3&r4*, *b1&b2*, *g1&g2*, *e3&e4*, and *c1&c2* (Table 3), *att* sites as grey arrows and hybridization probes as grey rectangles. **(b)** Predicted pollen-specific autoexcision event and T-DNA fragments. P^{OsPS3} drives the *bxbNom* gene expression during pollen development and P^{OsPS1} drives the *gusAeGFP* gene expression after autoexcision. *Abbreviations:* P^{OsPS1} and P^{OsPS3}, pollen specific promoters; *bxbNom*, optimized Bxb1 recombinase gene; *P*, *Pst*I restriction site; other abbreviations are the same as in Figure 1.

T_1 progeny analyses

Since switchgrass is an exclusively cross-pollinating species, T_1 seeds were obtained from controlled crosses between transgenic plants (MB5, MB17, MB18 and MB24) and wild type plants representing different Alamo genotypes. In these reciprocal crosses, the T_0 plants were used as both male (WT × MB) and female (MB × WT) parents.

In one set of experiments, PCR was used to distinguish between transgenic and non-transgenic T_1 plants. Since the Bxb1-mediated autoexcision occurs in the male gametes, transgene-excised progeny are only expected to be obtained from crosses with T_0 plants used as the pollen donor. In the offspring from the reciprocal crosses (MB × WT), the transgenes are expected to be transmitted through maternal inheritance and to be present in the genome of the transgenic T_1 plants.

In total, 556 T_1 plants obtained from crosses of three lines with pollen-specific autoexcision were analyzed by PCR. The results from the molecular analyses of MB17 progeny are shown in Figure 5. Fifteen transgene-excised WT × MB17 seedlings were identified based on the presence of the 0.51 kb excision positive amplicon (see Figure 3) combined with the lack of the *bxbNom* and *bar* genes (some shown in Figure 5a). The offspring from the reciprocal crosses (87 T_1 plants) were also screened to confirm the segregation ratio (Figure 5b). The complete elimination of the *att*-flanked target DNA from the genome of transgene-excised progeny was confirmed by DNA blot hybridization. As shown in Figure 5c for randomly selected WT × MB17 progeny, the recombinase and marker genes were not detected in T_1 plants identified as transgene-excised by PCR. The reporter gene located on the pMBXS824 T-DNA but not included in the *att*-flanked fragment was detected in the transgenic progeny with the *gusA* gene probe (Figure 5c).

For all lines, the obtained segregation ratios of transgenic-to-non-transgenic T_1 plants were analyzed by χ^2-test (Table 1). Complete excision of the marker and recombinase genes was detected via genomic PCR in 23 of the transgenic WT × MB18 plants resulting in an autoexcision efficiency of 23.5% (23 of 98). No transgene elimination was observed in the progeny of MB5 possibly due to the insufficient number of T_1 plants analyzed. The approximate 3:1 segregation ratio in the offspring of the multi-copy line MB17 when used as the male parent indicated that the transgenes were integrated in two loci as confirmed by the analysis of the progeny from the reciprocal cross. Fifteen transgene-excised plants representing 22.4% of the transgenic WT × MB17 T_1 plants were identified (Table 1).

For functional analyses of autoexcision events, 344 T_1 seedlings obtained from reciprocal controlled crosses of two transgenic lines were treated with the herbicide Basta (Table 2). Since both non-transgenic and transgene-excised plants would be Basta sensitive, the excision

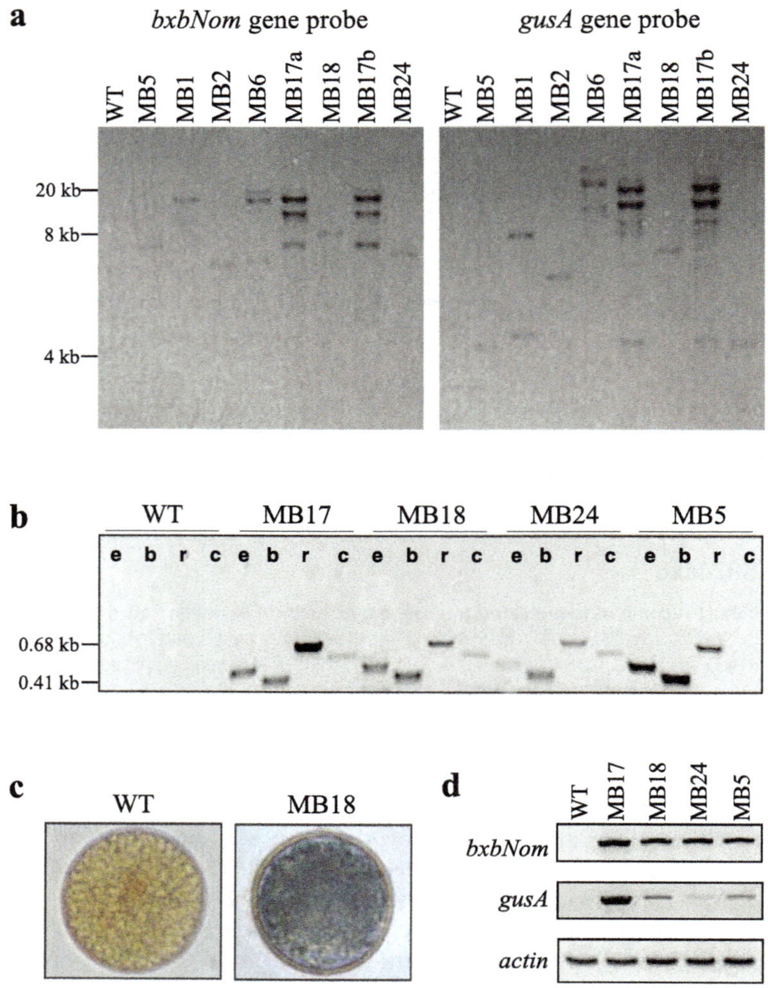

Figure 4 Analyses of T_0 plants with transgene autoexcision in pollen. (a) Determination of the copy number of the pMBXS824 T-DNA (Figure 3**a**) by Southern blot hybridization of genomic DNA isolated from leaves of transgenic and control plants. *Lanes*: *WT*, a non-transgenic plant; *MB*, individual transgenic plants from lines MB1, 2, 5, 6, 17a, 17b, 18, and 24. **(b)** PCR analysis for detection of Bxb1-mediated autoexcision in pollen from select transgenic lines. Genomic DNA was isolated from mature pollen grains collected from perfect florets at anthesis. *PCR reactions*: designated lane *e*, detection of autoexcision (primers *e3&e4*, PCR product size 0.51 kb); *b*, detection of the marker gene *bar* (primers *b1&b2*, 0.41 kb); *r*, detection of the recombinase gene *bxbNom* (primers *r3&r4*, 0.68 kb); *c*, detection of the excised DNA fragment (primers *c1&c2*, 0.57 kb). **(c)** Detection of GUS activity in mature pollen (magnification 63×). **(d)** RT-PCR for detection of the *bxbNom* and *gusA* transcripts in pollen. The following amounts of total RNA per reaction were used: 10 ng for the *bxbNom* gene, 25 ng for the *gusA* gene, and 2 ng for actin as a loading control. The plant samples are in the same order as in **(b)**.

efficiency was estimated based on the expected segregation ratio. In the offspring of the single-copy line MB24, 50% of the plants (38 of 76) should be non-transgenic and 50% - transgenic (1:1 segregation ratio). In WT × MB24 progeny, 25 plants were tolerant to the herbicide due to expression of the *bar* gene. Assuming that 38 of the 51 Basta sensitive plants were non-transgenic, 13 plants were presumed to be transgene-excised. They represented 34.2% of the expected number of transgenic plants. The autoexcision efficiency of 41.9% in the offspring of the multi-copy line MB17 (when used as the pollen donor) was calculated considering the ratio of 3:1.4 obtained by PCR analysis (Table 1, Table 2).

Discussion

Plant genetic engineering provides a powerful tool for crop improvement and the development of crop-based production systems. However, the potential dispersal of the introduced transgenes via different biological vectors (pollen, seeds, and/or vegetative propagules) is a key issue for the commercial use of transgenic crops. Since most forage and turf grasses, including many promising

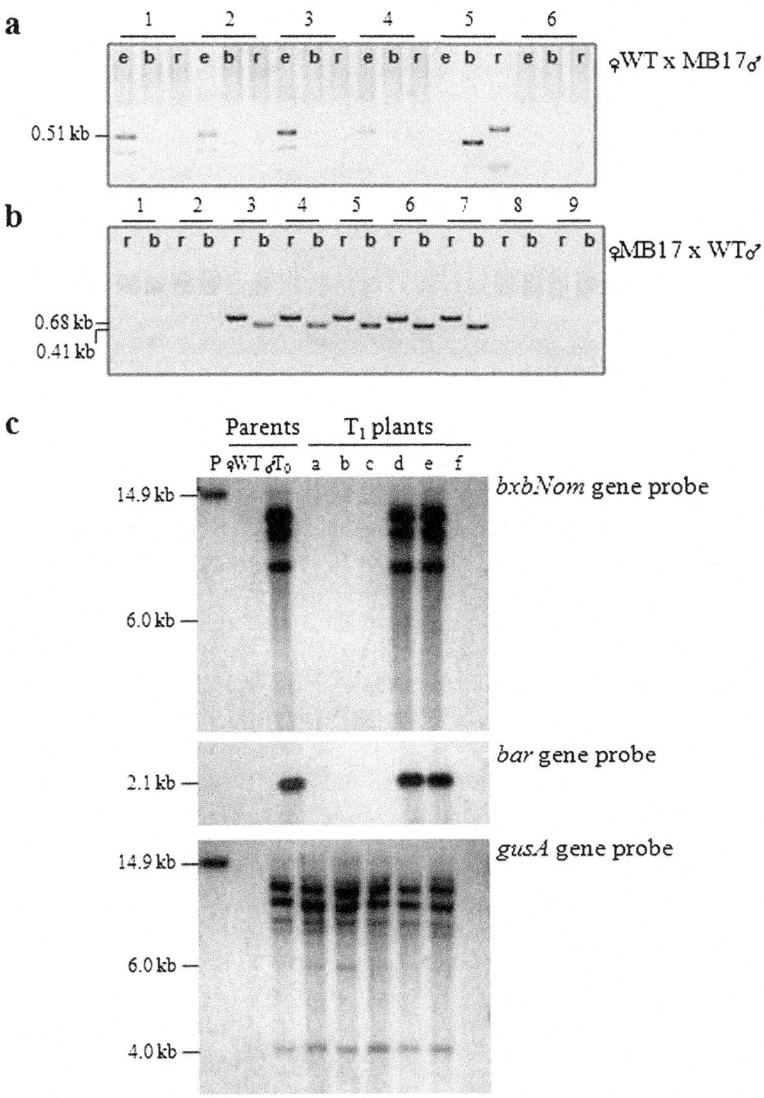

Figure 5 Molecular analyses of T_1 plants from controlled crosses between MB17 and non-transgenic plants. (a) PCR for identification of transgene-excised progeny when MB17 was used as the male parent (WT × MB17 cross). *1–6*, individual T_1 plants. *Lanes*: *e*, PCR reactions for detection of autoexcision (primers *e3&e4*); *b*, reactions for the marker gene *bar* (primers *b1&b2*); *r*, reactions for the recombinase gene *bxbNom* (primers *r3&r4*). **(b)** PCR for determination of the segregation ratio in the T_1 generation when MB17 was used as the female parent (MB17 × WT cross). *1–9*, individual T_1 plants. *PCR reactions*: same as reactions *r* and *b* in **(a)**. **(c)** Southern blot hybridization for detection of the pMBXS824 T-DNA (Figure 3) in the genome of WT × MB17 progeny. *Lanes*: *P*, pMBXS824 DNA; *WT*, wild-type plant used as the female parent; *T_0*, MB17 plant used as the pollen donor; *a-f*, individual T_1 plants. Based on the results from PCR screening (Table 1), the analyzed T_1 plants were categorized as follows: *a-c*, transgene-excised; *d-e*, transgenic; *f*, a non-transgenic T_1 plant (a null segregant).

bioenergy crops, are exclusively outcrossing and less domesticated, the risk of transgene transfer from these species to their wild relatives is very high, which raises commercial and environmental concerns [9,10,20,21]. Although different biological containment strategies have been proposed and developed, most of them only have a proof of concept in model plants. Some of these strategies utilize site specific recombination systems for the removal of superfluous DNA from the genome of genetically engineered plants [12,22,23].

In this study, we have demonstrated the possibility to reduce pollen-mediated transgene flow in switchgrass using developmentally programmed site-specific recombination for transgene autoexcision from the plant genome. In general, this strategy requires a combination of pollen- or microspore-specific promoters and highly active recombinases suitable for the targeted crop. Due to the lack of such information for switchgrass, our research started with the identification and functional characterization of potential candidates. Two recombination systems - Bxb1/*att*

Table 1 Segregation analysis of the T_1 progeny from reciprocal controlled crosses

Cross	Number of T_1 plants				Ratio TR:NT	χ^2
	Total	Transgenic (TR)	Transgene-excised	Non-transgenic (NT)		
WT × MB18	183	98	23	85	1:0.9	0.92
MB18 × WT	112	61	NA	51	1:0.8	0.89
WT × MB5	42	22	0	20	1:0.9	0.10
MB5 × WT	33	16	NA	17	1:1.1	0.03
WT × MB17	99	67	15	32	3:1.4	2.83
MB17 × WT	85	58	NA	27	3:1.4	2.07

The T_1 plants were analyzed by genomic PCR using primers *e3&e4*, *b1&b2*, *r3&r4*, and *g1&g2* (see Table 3). *MB*, switchgrass plants engineered for pollen-specific transgene autoexcision; *NA* not applicable, *WT* wild-type plants.

and ParA/*MRS* [12] - and a set of four promoters active in monocot floral and reproductive organs – the barley *Lem1* promoter [24,25] and the rice promoters *LP2*, *PS1* [26,27], and *PS3* - have been tested in stably transformed switchgrass plants in tissue culture and soil. Based on this initial screening (data not shown), we have chosen the well characterized large serine recombinase Bxb1 [14,28] with demonstrated activity in different plant species [15-17], and the rice pollen-specific promoters *PS1* and *PS3*.

A novel approach was used to test the Bxb1/*att* recombination system in switchgrass. A recombinase activity detection system based on two transformation vectors was designed for this purpose. The target and recombinase vectors were co-introduced into callus cultures resulting in the production of stably co-transformed plants with precise (without the gain or loss of nucleotides) excision of the *att*-flanked DNA fragment (Figure 1). When the recombinase gene and the target DNA (a marker and/or a gene of interest) are located on different vectors, they can be combined in a transgenic plant by re-transformation or genetic crosses or transient expression of the recombinase gene [29]. Our results demonstrate the possibility of achieving the complete removal of undesired DNA from the switchgrass genome in the same time frame as required for our standard transformation protocol. Moreover, no crosses between transgenic lines were necessary, which is advantageous for genetic engineering of switchgrass and other self-incompatible crops with a long reproductive cycle and overall complex genetics due to high levels of ploidy and heterozygosity. To our knowledge, this is the first report on engineering site specific recombinase-mediated excision in a monocot species by co-transformation.

After successfully validating the functionality of the Bxb1/*att* recombination system in switchgrass, we evaluated the spatiotemporal activities of candidate promoters prior to their use for pollen-specific autoexcision. The rice *PS1* and *PS3* promoters did not show any background activity in somatic tissues detectable by molecular and histochemical techniques. The observed differences in the functionality profile of these promoters in switchgrass pollen (Figure 2) led us to choose the *PS3* promoter for achieving high expression of the Bxb1 recombinase during pollen development and maturation.

To evaluate whether transgene flow via pollen in switchgrass can effectively be controlled by Bxb1-mediated site specific recombination, the autoexcision test vector pMBXS824 was created with the expression cassettes of the recombinase and marker genes positioned between the *att* target sequences (Figure 3). For more effective performance, an optimized form of the *bxb1* coding sequence (*bxbNom*) was used that contained a C-terminal nuclear localization signal [17]. The developmentally programmed pollen-specific *bxbNom* expression was engineered by placing the *PS3* promoter upstream of the recombinase gene. For screening purposes, pMBXS824 also contains the rice pollen-specific promoter *PS1* and the reporter *gusAeGFP* gene outside of the *att*-flanked target DNA fragment. Activation of *bxbNom* expression by the *PS3* promoter during pollen development led to Bxb1-mediated autoexcision of the genetic material between the *att* recombination sites.

Table 2 Functional analyses of autoexcision events in the T_1 generation

Cross	T_1 plants assayed for Basta tolerance			Transgene-excised T_1 plants	Estimated autoexcision efficiency [%]
	Total	Tolerant	Sensitive		
WT × MB24	76	25	51	13	34.2
MB24 × WT	91	45	46	NA	NA
WT × MB17	126	50	76	36	41.9
MB17 × WT	51	31	20	NA	NA

The number of transgene excised plants was estimated based on the segregation ratios. *NA* not applicable, *NT* non-transgenic plants, *TR* transgenic plants.

As a result, a new expression cassette is formed where the *PS1* promoter and the *attL* footprint are fused to *gusAeGFP* reporter gene within the heritable pollen genome.

In total, 22 independent transgenic switchgrass lines were used to examine the efficiency of the autoexcision strategy. Transgene-excised pollen was detected in 7 of these lines using PCR and GUS activity staining. The results from DNA blot analysis demonstrated that four of these T_0 lines contained a single copy of pMBXS824 T-DNA, while the other lines had 2–5 copies (Figure 4a). Three of the single-copy lines (MB5, MB18 and MB24) and the multi-copy line MB17 were used for more detailed molecular and genetic analyses. PCR analysis of genomic DNA isolated from mature pollen grains revealed the presence of the *bxbNom* and *bar* genes in all of the transgenic lines along with the 0.51 kb amplicon that indicates site-specific autoexcision. The minicircle excision PCR product (Figure 3b) was detected in three of the lines, MB17, MB18 and MB24 (PCR reactions *c* in Figure 4b). These circular DNA molecules formed upon site-specific recombination are expected to be lost via degradation and cell division [30] but they may be stable in non-proliferating cells, such as mature pollen grains. The excised DNA fragments and/or the genomic DNA of pollen without autoexcision could serve as template for the amplification of the *bar* and *bxbNom* transgenes observed in our experiments. The presence of the recombinase gene in the pollen genome was also confirmed by the detection of its transcripts by RT-PCR (Figure 4d). Taken together, the results suggest that either complete autoexcision has not occurred in all of the transgenic pollen grains, or that there is retention of the excised DNA. An optimized recombinase-promoter combination designed specifically for switchgrass could ensure complete removal of the transgenes in the germline or microspores (see Conclusions), while the minicircles would be degraded in the actively proliferating cells of the developing embryo and growing seedling.

The efficiency of Bxb1-mediated transgene autoexcision in switchgrass pollen or transmission of the resulting recombination event to the next generation was assessed by molecular and functional analyses. T_1 progeny was obtained from reciprocal crosses between transgenic lines and wild-type plants from different genotypes to mimic cross pollination in natural switchgrass populations [7]. To date, pollen-specific autoexcision systems have only been demonstrated in diploid self-pollinating model plants (see below), and only the offspring of single-copy events have been analyzed in detail. Since all transgenic and wild-type switchgrass plants used in our study are from the tetraploid cultivar Alamo, we have also included a multi-copy line along with the single-copy ones to illustrate the complexity of progeny analyses and the potential of Bxb1-mediated autoexcision in a polyploid heterozygous crop. The MB17 line contains at least 5 copies of the pMBXS824 T-DNA and some of them are probably arranged as tandem repeats based on the DNA blot hybridization analysis (Figure 4a) and the transgene segregation observed in the T_1 generation. The progeny segregation ratio (Table 1) indicates that the T-DNAs are integrated in two unlinked loci, which can potentially result in the deletion or rearrangement of genomic DNA fragments between different transgenic loci upon Bxb1-mediated recombination in the pollen. Based on morphological observations, if genomic DNA has been removed or rearranged, it has not caused detectable changes in the phenotype and development of the analyzed WT × MB17 progeny. On the other hand, the unaffected alleles present in the polyploid switchgrass genome might compensate for any DNA loss.

Overall, the results from progeny analyses revealed that the offspring of all lines segregated in the expected Mendelian ratios for transgenic:non-transgenic plants suggesting that the Bxb1-mediated site specific recombination did not affect pollen viability and germination. The identification of transgene-excised and non-excised transgenic T_1 plants in the offspring of each line (Table 1) is in agreement with the detected presence of both autoexcision products and transgenes in T_0 pollen. Consistent with the transmission of the recombined allele(s) through the male gametes, transgene-excised progeny were only obtained from controlled crosses when the primary transformants were used as pollen donors. The identified transgene-excised T_1 plants represented approximately 22-42% of the offspring depending on the transformation event and the assay used. Interestingly, similar autoexcision efficiencies were observed in the progeny of single- and multi-copy lines (Tables 1 and 2).

Pollen-specific autoexcision strategies have previously been reported in tobacco and *Arabidopsis* [21,29,31,32]. Germline-, microspore- and pollen-specific promoters have been utilized for developmentally regulated expression of CRE/*loxP*, CinH/*RS2* or FLP/*FRT* recombination systems resulting in high frequency (up to ~100%) of production of transgene-excised plants in the progeny (T_1-T_3 generations) obtained through selfing or backcrossing of primary transformants. The applicability of these approaches in other plant species and crops, however, is yet to be evaluated.

Conclusions

This is the first study demonstrating a functional and relatively simple strategy for the reduction of pollen-mediated gene flow in transgenic switchgrass. Our results indicate that the Bxb1/*att* site specific recombination system can be used for the precise removal of the recombinase and marker genes from switchgrass pollen resulting in the generation of transgene-excised progeny.

The genetic elements included in our test vector for screening purposes can be eliminated from an optimized autoexcision vector to obtain plants with transgene-free pollen. Although the Bxb1 recombinase has been shown to have activity in other plant species, no Bxb1-mediated autoexcision has previously been reported. In addition, we also provide experimental evidence that our Bxb1-based autoexcision vector can be used for the transcriptional activation of a gene of interest at a defined developmental stage. A similar strategy utilizing an *attP* and *attB* flanked selectable marker combined with tissue-specific Bxb1 expression could be used to generate marker-free transgenic switchgrass that contain only the introduced genes of interest.

Further studies are necessary to improve the recombination efficiency in switchgrass pollen. Consistent with results from research with model plants, the results presented here clearly indicate that high levels of recombinase activity conferred by promoters with crop-specific functionality is crucial for the successful elimination of transgenes in gametes or somatic tissues. Our future work on the improvement of the Bxb1/*att* system will include codon optimization of the recombinase gene sequence for increased expression specifically in switchgrass. Intron-mediated enhancement of the *bxb1* gene expression [33] is also considered. Another approach to improving the Bxb1 activity would be mutagenesis by a combination of error prone PCR and alanine scanning which could both increase the recombination rate and/or modify the native binding preferences of the enzyme as reported previously for the related recombinase phiC31 [34]. The outcome of these strategies would also depend on the activity and tissue specificity of the promoter controlling the recombinase gene expression. As described above, the highest autoexcision efficiencies have been achieved in dicotyledonous plants using promoters specifically active in germline or during microsporogenesis. However, monocot promoters with this type of spatiotemporal activity are not currently available, which combined with the lack of thorough analyses of major aspects of switchgrass gene flow hinder the development of both risk assessment strategies and versatile gene biocontainment systems for this valuable bioenergy crop.

Methods

Construction of plant transformation vectors

All gene constructs were made using standard molecular biology techniques. Five binary vectors were used for switchgrass transformation and co-transformation.

The vectors pMBXS638 and pMBXS640 were co-introduced in callus cultures to test the functionality of the Bxb1/*att* recombination system in switchgrass. The vector pMBXS638 contains the *bxb1* recombinase gene driven by P^{ZmUbi1} (the promoter and first intron of the maize *ubiquitin1* gene) and the *bar* gene (conferring resistance to bialaphos) as a marker for plant selection. The target vector pMBXS640 was constructed using the vector pGPro1 [35] containing the *hptII* gene (conferring resistance to hygromycin) as a plant selectable maker. The vector also harbors a 0.69 kb non-coding DNA fragment (stuffer) flanked by the Bxb1 recombination sites *attP50* and *attB42* in direct orientation, $P^{OsActin1}$ (the promoter and first intron of the rice *actin1* gene) proximal to *attP50*, and a distal β-glucuronidase-enhanced Green Fluorescent Protein (*gusAeGFP*) bifunctional reporter gene [35]. For complete sequences of the Bxb1 attachment sites (*att*), see [28].

The vectors pGPro8-PS3 and pGPro8-PS1 were used to evaluate the spatiotemporal activity patterns of the rice pollen specific promoters *PS3* and *PS1*, respectively. A 2.2 kb fragment representing the *PS3* promoter sequence [EMBL: JN593331] was fused to the reporter β-glucuronidase gene *GUSPlus* [36] to assemble the vector pGPro8-PS3. The vector pGPro8-PS1 harbors the 1.8 kb sequence of the *PS1* promoter [GenBank: JN593329] [26,27] driving the *GUSPlus* gene. Both vectors contain the *hptII* gene as a marker for callus and plant selection under the control of the promoter and first intron of the rice *ubiquitin2* gene (P^{OsUbi2}) [7,37]. The presence of the transgenes in hygromycin-resistant putative transformants was confirmed by PCR using primers *g3&g4* and *h3&h4* (Table 3).

The vector pMBXS824 was constructed for Bxb1-mediated transgene autoexcision in switchgrass pollen and identification of transgene-excised progeny. The vector harbors the expression cassettes of an optimized *bxb1* coding sequence (*bxbNom*) [17] driven by the *PS3* promoter and the marker gene *bar* under the control of the *CaMV 35S* promoter between the Bxb1 recombination sequences. The *PS1* promoter and the reporter gene *gusAeGFP* are positioned outside of the *att*-flanked DNA fragment.

Plant material, transformation and co-transformation

Switchgrass plants from Alamo genotype 56 [7] grown under greenhouse conditions were used for initiation of immature inflorescence-derived callus cultures. The top culm nodes of elongating tillers with 3–4 visible nodes were used for development of inflorescences in tissue culture following a previously published procedure [38]. Callus cultures were initiated from individual spikelets from *in vitro* developed panicles and propagated by transferring on to a fresh medium for callus growth every four weeks. These highly embryogenic immature inflorescence-derived callus cultures were transformed with *Agrobacterium tumefaciens*, strain AGL1 and selected

Table 3 Primers used in this study

Primer ID	Gene/element	Sequence (5′- 3′)	Reference
b1	*bar*	GCACCATCGTCAACCACTACATCG	[7]
b2		TCATGCCAGTTCCCGTGCTTG	[7]
h1	*hptII (pMBXS640)*	TGATCGAAAAGTTCGACAGCGTCTC	This study
h2		CTCCAGTCAATGACCGCTGTTATGC	This study
h3	*hptII (pGPro8)*	CTCCCGATTCCGGAAGTGCT	This study
h4		CATCGCCTCGCTCCAGTCAA	This study
r1	*bxb1*	CTGGTAGTCATCCGCCTGT	This study
r2		CTTCTAATTCCATCTGCGCCAC	This study
r3	*bxbNom*	CAACTGGAGAGCTGCCAACAAC	This study
r4		ATATCCCAGCATGGCTTCGG	This study
g1	*gusAeGFP*	AGCGCGTTACAAGAAAGC	This study
g2		AGAGATAACCTTCACCCGG	This study
g3	*GUSPlus*	CGCAACCATATCGGATATGTCT	This study
g4		TGACATTCGGAATCTCCACG	This study
s1	Stuffer	CTCGATCGGTTAGCATAGGCAG	This study
s2		GACTTAGCGTAGCAATGGCAACTG	This study
e1	Stuffer construct	TGTAGTCTAGAGTGCTCCATTC	This study
e2		CGGCTTCAAATGGCGTATAGC	This study
e3	Autoexcision	TCTAGAGTGCTCCATTCTCTCTCCTCG	This study
e4		GTTAAAACTGCCTGGCACAGCAATT	This study
c1	Excised DNA fragment	GGGTTCCTATAGGGTTTCGCTCATG	This study
c2		TGATTTACCAGAAGCGGAGGAAGAA	This study
ActA	*actin*	CACTGGAATGGTCAAGGATG	[7]
ActB		CTCCATGTCATCCCAGTTG	[7]

with 10 mg l^{-1} bialaphos following the previously published protocols [7,39,40].

For *Agrobacterium*-mediated co-introduction of transgenes on two independent T-DNAs located on two binary vectors in different bacterial cells, callus cultures were incubated in a mixture of equal volumes of *A. tumefaciens* (strain AGL1) cultures (OD_{600} 0.4) under the conditions described previously [7,39,40]. Co-transformed callus cultures and plants were subjected to double selection with 10 mg l^{-1} bialaphos and 200 mg l^{-1} hygromycin for 2–4 months with transfers to a fresh selection medium every two weeks.

The growth conditions in tissue culture and soil were as described previously [7,40]. Wild-type plants obtained from untransformed callus cultures and grown under the same conditions were used as controls.

Molecular analyses

Primary switchgrass transformants and co-transformants were identified by PCR with total nucleic acid extracts from leaves of plants in tissue culture (REDExtract Plant PCR Kit, Sigma-Aldrich, St. Louis, MO, USA) using primers specific for the coding regions of the transgenes (Table 3) and the amplification conditions described previously [7]. The same procedure was used for PCR screening of T_1 plants. Excision and autoexcision were detected by PCR with genomic DNA isolated from leaves and pollen, respectively, of soil-grown plants using a Wizard Genomic DNA Extraction kit (Promega, Madison, WI, USA) according to the manufacturer's instructions. Primer pairs are specified in the text.

Southern blot hybridizations were performed with 15 μg of genomic DNA digested with *Nco*I for analysis of co-transformed plants and with *Pst*I for analysis of primary transformants and T_1 plants following a previously described procedure [7,40].

Expression analysis

For detection of the reporter and recombinase gene transcripts by reverse transcription-polymerase chain reaction (RT-PCR), total RNA was isolated from leaf tissues or pollen grains using the RNeasy Plant Mini Kit (Qiagen, Valencia, CA, USA) according to the manufacturer's instructions. After DNase treatment and column purification, different amounts of RNA (specified in figure legends) were subjected to reverse transcription and

DNA amplification in a one-step RT-PCR assay (Qiagen, Valencia, CA, USA) as described previously [7] using the primers listed in Table 3.

Histochemical GUS staining

For detection of β-glucuronidase (GUS) activity, transverse leaf and root sections from plants in tissue culture or pollen grains from soil-grown plants were incubated in X-Gluc (5-bromo-4-chloro-3-indolyl-β-D-glucuronic acid; Gold Biotechnology, Inc., St. Louis, MO, USA) solution [41] at 37°C for 24 or 48 hours.

Progeny analyses

T_1 seeds were obtained from controlled crosses between transgenic and wild-type plants as reported previously [7,40] and germinated after stratification at 4°C for two weeks. The resulting seedlings were transferred to soil and analyzed by PCR and Southern blot hybridization as described above. Preliminary PCR experiments showed no amplification of the excision product in T_1 plants obtained from crosses when the primary transformants with transgene-excision in pollen were used as the female parents.

T_1 seedlings were treated with the herbicide Basta as described previously [7,40].

The segregation ratios were analyzed by the χ^2-test at P = 0.05.

Competing interests

The authors declare that they have no competing interests.

Authors' contributions

MNS designed the study, performed molecular analyses, analyzed data and drafted the manuscript. CAX carried on the design and construction of the autoexcision vector, switchgrass transformation and co-transformation, genetic and molecular analyses, characterized monocot promoters, analyzed data and participated in manuscript editing. KPR performed expression analyses, participated in construction of vectors and manuscript editing. RT identified, cloned and characterized the pollen-specific promoters, built the promoter-reporter gene pGPro8 binary vector constructs and edited drafts of the manuscript. OP and KDS contributed to the overall research concept of the project and edited drafts of the manuscript. JT participated in manuscript drafting, editing, study design, optimizing and testing of the recombinases. All authors read and approved the final version of this manuscript.

Acknowledgements

The authors thank Dawit Abreham and Milly Nabalende for their technical assistance and R. Skadsen, USDA-ARS, for providing the *Lem1* promoter. This material is based upon work supported by the Department of Energy under Award Number DE-EE0004943. This report was prepared as an account of work sponsored by an agency of the United States Government. Neither the United States Government nor any agency thereof, nor any of their employees, makes any warranty, express or implied, or assumes any legal liability or responsibility for the accuracy, completeness, or usefulness of any information, apparatus, product, or process disclosed, or represents that its use would not infringe privately owned rights. Reference herein to any specific commercial product, process, or service by trade name, trademark, manufacturer, or otherwise does not necessarily constitute or imply its endorsement, recommendation, or favoring by the United States Government or any agency thereof. The views and opinions of authors expressed herein do not necessarily state or reflect those of the United States Government or any agency thereof. This research was also supported by USDA Agricultural Research Service CRIS projects 5325-21000-018 and 5325-21000-020 and by the Biotechnology Risk Assessment Program competitive grant number 2010-33522-21773 from the USDA - National Institute of Food and Agriculture. Mention of trade names or commercial products is solely for the purpose of providing specific information and does not imply recommendation or endorsement by the US Department of Agriculture. USDA is an equal opportunity provider and employer.

References

1. Tilman D, Hill J, Lehman C: **Carbon-negative biofuels from low-input high-diversity grassland biomass.** *Science* 2006, **314:**1598–1600.
2. Lewandowski I, Scurlock JMO, Lindvall E, Christou M: **The development and current status of perennial rhizomatous grasses as energy crops in the US and Europe.** *Biomass Bioenergy* 2003, **25:**335–361.
3. Sanderson MA, Adler PR, Boateng AA, Casler MD, Sarath G: **Switchgrass as a biofuels feedstock in the USA.** *Can J Plant Sci* 2006, **86:**1315–1325.
4. Schmer MR, Vogel KP, Mitchell RB, Perrin RK: **Net energy of cellulosic ethanol from switchgrass.** *Proc Natl Acad Sci U S A* 2008, **105:**464–469.
5. McLaughlin S, Bouton J, Bransby D, Conger B, Ocumpaugh W: **Developing Switchgrass as a Bioenergy Crop.** In *Perspectives on New Crops and New Uses.* Edited by Janick J. Alexandria, VA: ASHS Press; 1999:282–299.
6. Somleva MN, Peoples OP, Snell KD: **PHA bioplastics, biochemicals, and energy from crops.** *Plant Biotechnol J* 2013, **11:**233–252.
7. Somleva MN, Snell KD, Beaulieu JJ, Peoples OP, Garrison BR, Patterson NA: **Production of polyhydroxybutyrate in switchgrass, a value-added co-product in an important lignocellulosic biomass crop.** *Plant Biotechnol J* 2008, **6:**663–678.
8. Snell KD, Peoples OP: **PHA bioplastic: a value-added coproduct for biomass biorefineries.** *Biofuels Bioprod Biorefin* 2009, **3:**456–467.
9. Kausch AP, Hague J, Oliver M, Li Y, Daniell H, Mascia PN, Watrud L, Stewart CN Jr: **Transgenic perennial biofuel feedstocks and strategies for bioconfinement.** *Biogeosciences* 2010, **1:**163–175.
10. Wang Z-Y, Brummer EC: **Is genetic engineering ever going to take off in forage, turf and bioenergy crop breeding?** *Ann Bot* 2012, **110:**1317–1325.
11. Husken A, Prescher S, Schiemann J: **Evaluating biological containment strategies for pollen-mediated gene flow.** *Environm Biosafety Res* 2010, **9:**67–73.
12. Wang Y, Yau Y-Y, Perkins-Balding D, Thomson J: **Recombinase technology: applications and possibilities.** *Plant Cell Rep* 2011, **30:**267–285.
13. Kim AI, Ghosh P, Aaron MA, Bibb LA, Jain S, Hatfull GF: **Mycobacteriophage Bxb1 integrates into the *Mycobacterium smegmatis* groEL1 gene.** *Molec Microbiol* 2003, **50:**463–473.
14. Ghosh P, Kim AI, Hatfull GF: **The orientation of mycobacteriophage Bxb1 integration is solely dependent on the central dinucleotide of attP and attB.** *Mol Cell* 2003, **12:**1101–1111.
15. Yau Y, Wang Y, Thomson J, Ow D: **Method for Bxb1-mediated site-specific integration in planta.** *Methods Mol Biol* 2011, **701:**147–166.
16. Thomson J, Chan R, Smith J, Thilmony R, Yau Y-Y, Wang Y, Ow D: **The Bxb1 recombination system demonstrates heritable transmission of site-specific excision in *Arabidopsis*.** *BMC Biotechnol* 2012, **12:**9–19.
17. Blechl A, Lin J, Shao M, Thilmony R, Thomson J: **The Bxb1 recombinase mediates site-specific deletion in transgenic wheat.** *Plant Mol Biol Rep* 2012, **30:**1357–1366.
18. Fujita M, Horiuchi Y, Ueda Y, Mizuta Y, Kubo T, Yano K, Yamaki S, Tsuda K, Nagata T, Niihama M, Kato H, Kikuchi S, Hamada K, Mochizuki T, Ishimizu T, Iwai H, Tsutsumi N, Kurata N: **Rice expression atlas in reproductive development.** *Plant Cell Phisiol* 2010, **51:**2060–2081.
19. Thomson JG, Ow DW: **Site-specific recombination systems for the genetic manipulation of eucaryotic genomes.** *Genesis* 2006, **44:**465–476.
20. Kwit C, Moon HS, Warwick SI, Stewart CN Jr: **Transgene introgression in crop relatives: molecular evidence and mitigation strategies.** *Trends Biotechnol* 2011, **29:**284–293.
21. Moon H, Abercrombie J, Kausch A, Stewart CN Jr: **Sustainable use of biotechnology for bioenergy feedstocks.** *Environ Manag* 2010, **46:**531–538.
22. Gidoni D, Srivastava V, Carmi N: **Site-specific excisional recombination strategies for elimination of undesirable transgenes from crop plants.** *In Vitro Cell Dev Biol – Plant* 2008, **44:**457–467.
23. Lyznik LA, Gordon-Kamm W, Gao H, Scelonge C: **Application of site-specific recombination systems for targeted modification of plant genomes.** *Transgenic Plant J* 2007, **1:**1–9.

24. Skadsen RW, Sathish P, Federico ML, Abebe T, Fu J, Kaeppler HF: **Cloning of the promoter for a novel barley gene, Lem1, and its organ-specific promotion of Gfp expression in lemma and palea.** *Plant Mol Biol* 2002, **49**:545–555.
25. Somleva MN, Blechl AE: **The barley Lem1 gene promoter drives expression specifically in outer floret organs at anthesis in transgenic wheat.** *Cereal Res Comm* 2005, **33**:665–671.
26. Cook M, Thilmony R: **The *OsGEX2* gene promoter confers sperm cell expression in transgenic rice.** *Plant Mol Biol Rep* 2012, **30**:1138–1148.
27. Zou J-T, Zhan X-Y, Wu H-M, Wang H, Cheung AY: **Characterization of a rice pollen-specific gene and its expression.** *Am J Bot* 1994, **81**:552–561.
28. Singh S, Ghosh P, Hatfull GF: **Attachment site selection and identity in Bxb1 serine integrase-mediated site-specific recombination.** *PLoS Genet* 2013, **9**(5):1–14.
29. Verweire D, Verleyen K, De Buck S, Claeys M, Angenon G: **Marker-free transgenic plants through genetically programmed auto-excision.** *Plant Physiol* 2007, **145**:1220–1231.
30. Shao M, Kumar S, Thomson JG: **Precise excision of plastid DNA by the large serine recombinase Bxb1.** *Plant Biotechnol J* 2014, **12**:322–329.
31. Luo K, Duan H, Zhao D, Zheng X, Deng W, Chen Y, Stewart CN Jr, McAvoy R, Jiang X, Wu Y, He A, Pei Y, Li Y: **'GM-gene-deletor': fused loxP-FRT recognition sequences dramatically improve the efficiency of FLP or CRE recombinase on transgene excision from pollen and seed of tobacco plants.** *Plant Biotechnol J* 2007, **5**:263–274.
32. Mlynárová L, Conner AJ, Nap JP: **Directed microspore-specific recombination of transgenic alleles to prevent pollen-mediated transmission of transgenes.** *Plant Biotechnol J* 2006, **4**:445–452.
33. Rethmeier N, Seurinck J, Van Montagu M, Cornelissen M: **Intron-mediated enhancement of transgene expression in maize is a nuclear, gene-dependent process.** *Plant J* 1997, **12**:895–899.
34. Keravala A, Lee S, Thyagarajan B, Olivares EC, Gabrovsky VE, Woodard LE, Calos MP: **Mutational derivatives of phiC31 integrase with increased efficiency and specificity.** *Mol Ther* 2009, **17**:112–120.
35. Thilmony R, Guttman M, Chiniquy D, Blechl A: **pGPro1, a novel binary vector for monocot promoter characterization.** *Plant Mol Biol Rep* 2006, **24**:57–69.
36. Broothaerts W, Mitchell HJ, Weir B, Kaines S, Smith LMA, Yang W, Mayer JE, Rodriguez CR, Jefferson RA: **Gene transfer to plants by diverse bacteria.** *Nature* 2005, **433**:629–633.
37. Wang J, Jiang J, Oard J: **Structure, expression and promoter activity of two polyubiquitin genes from rice (*Oriza sativa* L.).** *Plant Sci* 2000, **156**:201–211.
38. Alexandrova KS, Denchev PD, Conger BV: ***In vitro* development of inflorescences from switchgrass nodal segments.** *Crop Sci* 1996, **36**:175–178.
39. Somleva M: **Switchgrass (*Panicum virgatum* L.).** In *Agrobacterium Protocols.* Volume 2nd edition. Edited by Wang K. New York: Humana Press; 2006:65–74.
40. Somleva MN, Tomaszewski Z, Conger BV: ***Agrobacterium*-mediated genetic transformation of switchgrass.** *Crop Sci* 2002, **42**:2080–2087.
41. Jefferson RA: **Assaying chimeric genes in plants: the GUS gene fusion system.** *Plant Mol Biol Rep* 1987, **5**:387–405.

5

Unmixing of fluorescence spectra to resolve quantitative time-series measurements of gene expression in plate readers

Catherine A Lichten[1,2], Rachel White[3], Ivan BN Clark[1] and Peter S Swain[1*]

Abstract

Background: To connect gene expression with cellular physiology, we need to follow levels of proteins over time. Experiments typically use variants of Green Fluorescent Protein (GFP), and time-series measurements require specialist expertise if single cells are to be followed. Fluorescence plate readers, however, a standard in many laboratories, can in principle provide similar data, albeit at a mean, population level. Nevertheless, extracting the average fluorescence per cell is challenging because autofluorescence can be substantial.

Results: Here we propose a general method for correcting plate reader measurements of fluorescent proteins that uses spectral unmixing and determines both the fluorescence per cell and the errors on that fluorescence. Combined with strain collections, such as the GFP fusion collection for budding yeast, our methodology allows quantitative measurements of protein levels of up to hundreds of genes and therefore provides complementary data to high throughput studies of transcription. We illustrate the method by following the induction of the *GAL* genes in *Saccharomyces cerevisiae* for over 20 hours in different sugars and argue that the order of appearance of the Leloir enzymes may be to reduce build-up of the toxic intermediate galactose-1-phosphate. Further, we quantify protein levels of over 40 genes, again over 20 hours, after cells experience a change in carbon source (from glycerol to glucose).

Conclusions: Our methodology is sensitive, scalable, and should be applicable to other organisms. By allowing quantitative measurements on a per cell basis over tens of hours and over hundreds of genes, it should increase our understanding of the dynamic changes that drive cellular behaviour.

Keywords: Gene expression, Fluorescence, Plate readers, Spectral unmixing, Budding yeast, High throughput measurements, Systems biology

Background

Most organisms live in changing environments, and gene and protein networks respond dynamically to extracellular change. Understanding these dynamic responses is one of the challenges of biology [1], but it is difficult to monitor changes within cells as they happen. Microscopy [2], particularly when combined with microfluidics [3], offers a way forward, but requires specialized expertise and equipment.

Most laboratories do, however, have access to a fluorescence plate reader. Fluorescence plate readers allow multiple experiments to be run in parallel on populations of cells in a temperature-controlled environment with optional agitation. Such experiments produce quantitative time-course data measuring absorbance and fluorescence. These data provide information about cell growth and, by using genetically encoded fluorescent reporters, gene expression. Plate readers report only on the mean response of the population of cells and information on the variation around that mean is lost. Nevertheless, such mean data is often suitable for many applications either in itself [4,5] or as additional data to support experiments at the single-cell level.

There are two factors that complicate the analysis of data from plate readers. First, when measuring absorbance, optical density (OD) does not always increase linearly with the density of cells, although this effect can

*Correspondence: peter.swain@ed.ac.uk
[1] SynthSys, University of Edinburgh, Mayfield Road, Edinburgh, UK
Full list of author information is available at the end of the article

be corrected with a calibration procedure [6,7]. Second, the measured fluorescence, in addition to containing emissions from the fluorescent proteins under observation, also includes contamination from other sources: autofluorescence of the plate and media and autofluorescence from the cells. For weakly expressed proteins, this autofluorescence can dominate the signal from the fluorescent reporter.

Developing an effective technique for separating out the signal of interest from the background autofluorescence would make it possible to collect quantitative data on the mean level of gene expression per cell over time. While various methods have been developed for *Escherichia coli* and applied to fusions of promoters and Green Fluorescent Protein (GFP) [4,8,9], the few published examples of measuring protein expression using a fluorescence reader in other organisms such as yeast have been limited to highly expressed proteins (for instance [10]).

In the following, we describe a technique for quantifying plate reader experiments using spectral unmixing and show that we can capture and quantify a wide range of levels of protein expression in the budding yeast *Saccharomyces cerevisiae*. Budding yeast, one of the world's most well-studied organisms, has long been a focus for high throughput experiments. We demonstrate that our method is robust enough to allow medium throughput measurements of protein levels in yeast using, for example, the GFP fusion collection [11]. It therefore provides an additional type of data to complement other high throughput characterizations of gene expression, which typically consider only transcription. We illustrate the methodology with two examples: the expression of genes for galactose import and metabolism (Figure 1), a classic example of a eukaryotic response to an environmental change [12], and the response to a shift from a poor to a rich carbon source, where we study expression of 44 genes from the GFP fusion collection (Figure 2).

Results and discussion

Correcting autofluorescence following a procedure for bacteria is not appropriate for budding yeast

For experiments with *Escherchia coli*, the fluorescence of interest is often corrected using the autofluorescence of wild-type cells as an estimate for the autofluorescence of cells expressing fluorescently tagged proteins [4,8,9]. For example, to correct for autofluorescence at a particular time point, either the fluorescence per cell for the wild-type cells (the fluorescence at that time point divided by the OD at the same time point) was subtracted from the fluorescence per cell for tagged strains [8] or the fluorescence levels from the wild-type strain interpolated to the OD of the tagged strain was subtracted from the fluorescence of the tagged strains [9].

Such correction procedures, however, require that both the tagged and wild-type strains have identical levels of autofluorescence. Typically, though, there are differences in growth curves between strains because of biological reproducibility, perturbations generated by the fluorescent reporter, or both. Such differences are important because autofluorescence need not only depend on the size and stage of growth of the population, but also on the contents of the growth media, which changes over time as nutrients are depleted by the cells and products excreted. Indeed, examining the autofluorescence of wild-type cells revealed that the autofluorescence does vary with the type and amount of sugar available, at least for budding yeast. Further, for low levels of gene expression, strains with fluorescently tagged proteins could have lower fluorescence values than wild-type cells, leading to predictions, with the correction procedure of [8], of negative levels of fluorescence per cell (Methods: Figure 3a). Ideally then, we would like to correct each cell by its own autofluorescence rather than by an average value calculated from other cells in a different well in the plate reader.

Spectral unmixing corrects autofluorescence

By measuring emissions at two wavelengths, rather than the usual one, we can infer the mean autofluorescence of a particular population of cells in a given well at a given time and so partly meet this goal. We use a single excitation wavelength, but measure emissions both at the wavelength appropriate to the fluorescent reporter and at a higher wavelength where most emissions come from autofluorescence. For example, the emission spectrum of autofluorescence is broader than that of enhanced GFP (Methods: Figure 3b), and we can infer the extent of autofluorescence emissions in the wavelength used for the fluorescent reporter from the autofluorescence emissions measured at the high wavelength.

We use spectral or linear unmixing to remove autofluorescence [13,14]. In linear unmixing, signals measured at each wavelength are assumed to be a linear combination of the different components in the sample. For the plate reader data, we assume that each measured fluorescence is a linear sum of autofluorescence and the fluorescence from any fluorescent reporter present in the cells. At each time point, we make two fluorescence measurements: excitation at 485 nm (near the excitation peak of GFP) and emission at both 525 nm (near the emissions peak of GFP) and at 585 nm (where GFP emissions are much reduced but autofluorescence is still substantial: Figure 3b). For all fluorescence measurements, we first remove the autofluorescence of the media and plas-

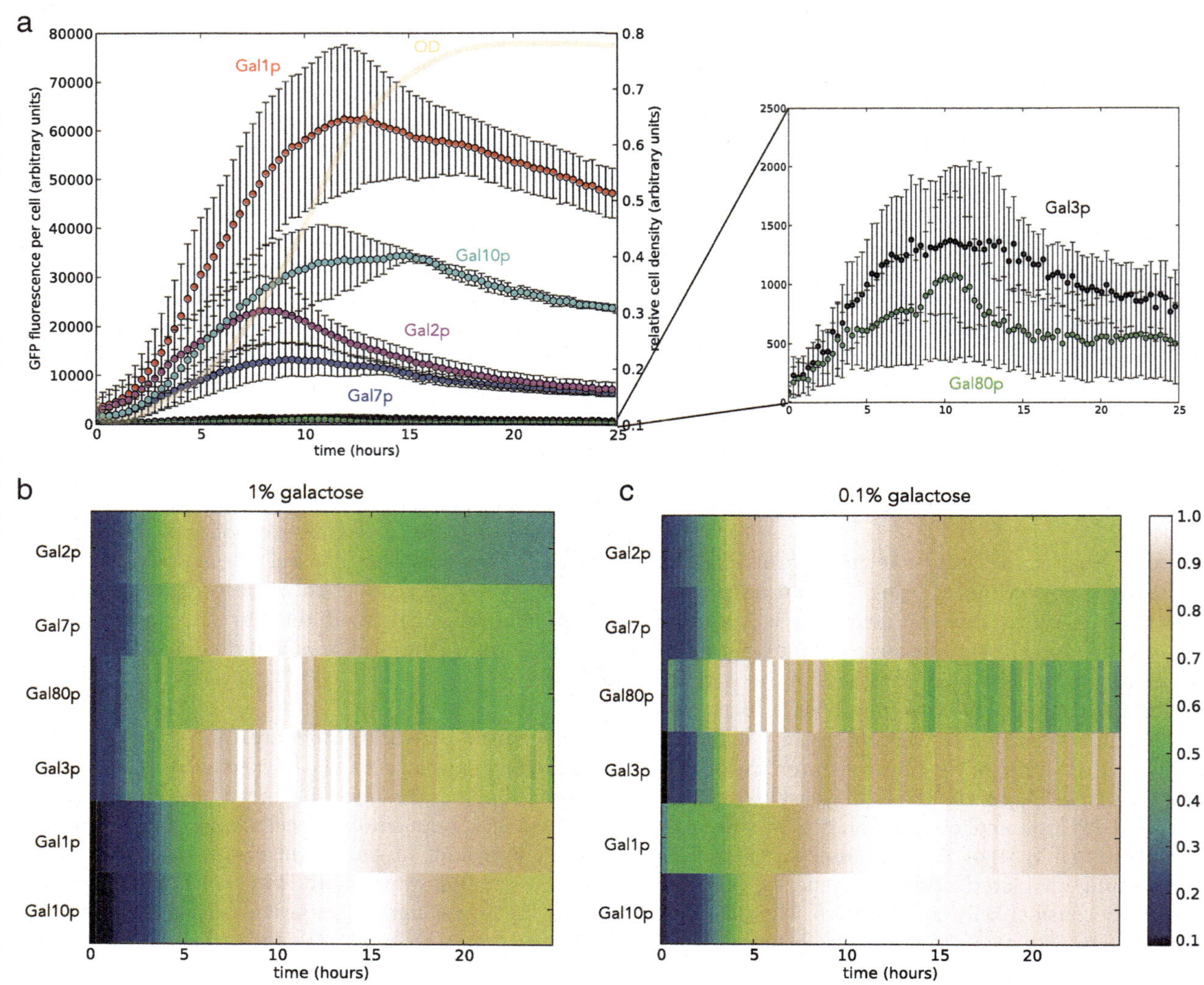

Figure 1 Spectral unmixing reveals the timing of expression of the GAL genes. Cells were grown in 2% raffinose, transferred into galactose and followed in the plate reader. **a)** Left-hand axis: Time-series data for mean fluorescence per cell for the Leloir enzymes Gal1, Gal7, and Gal10 and for the galactose permease, Gal2, and the regulators Gal3 (an intracellular sensor of galactose) and Gal80 (a repressor of GAL expression) in 1% galactose. Right-hand axis: Relative cell density (orange). **b)** The same data in **a**, but with the time-series of expression for each gene normalized by its maximum value. The normalized level of expression is indicated by the colour bar. **c)** A similar plot to **b** but for GAL expression in 0.1% galactose. Derivation of the error bars shown is given in Methods.

ticware by subtracting the fluorescence of wells containing only media.

For our analysis, we assume that: (i) the ratio of emissions of autofluorescence at 585 nm to 525 nm after excitation at 485 nm is the same for the wild-type and tagged strains at a given OD, and (ii) that the autofluorescence and the fluorescence of the tagged strain combine linearly. Our assumption (i) is weaker than assumptions of equal growth curves and equal levels of autofluorescence per cell required by the bacterial correction method. Assumption (ii), as mentioned, is standard (see, for example, [15]).

For each time point, we have four measurements of fluorescence: two for fluorescently tagged strains and two for wild-type strains. The two measurements for the tagged cells are:

$$\begin{aligned} f_{525} &= g + a \\ f_{585} &= r_g g + r_a a \end{aligned} \qquad (1)$$

where f_{525} is the fluorescence measured at 525 nm and f_{585} is the fluorescence measured at 585 nm. The symbol g denotes fluorescence from the protein tags with

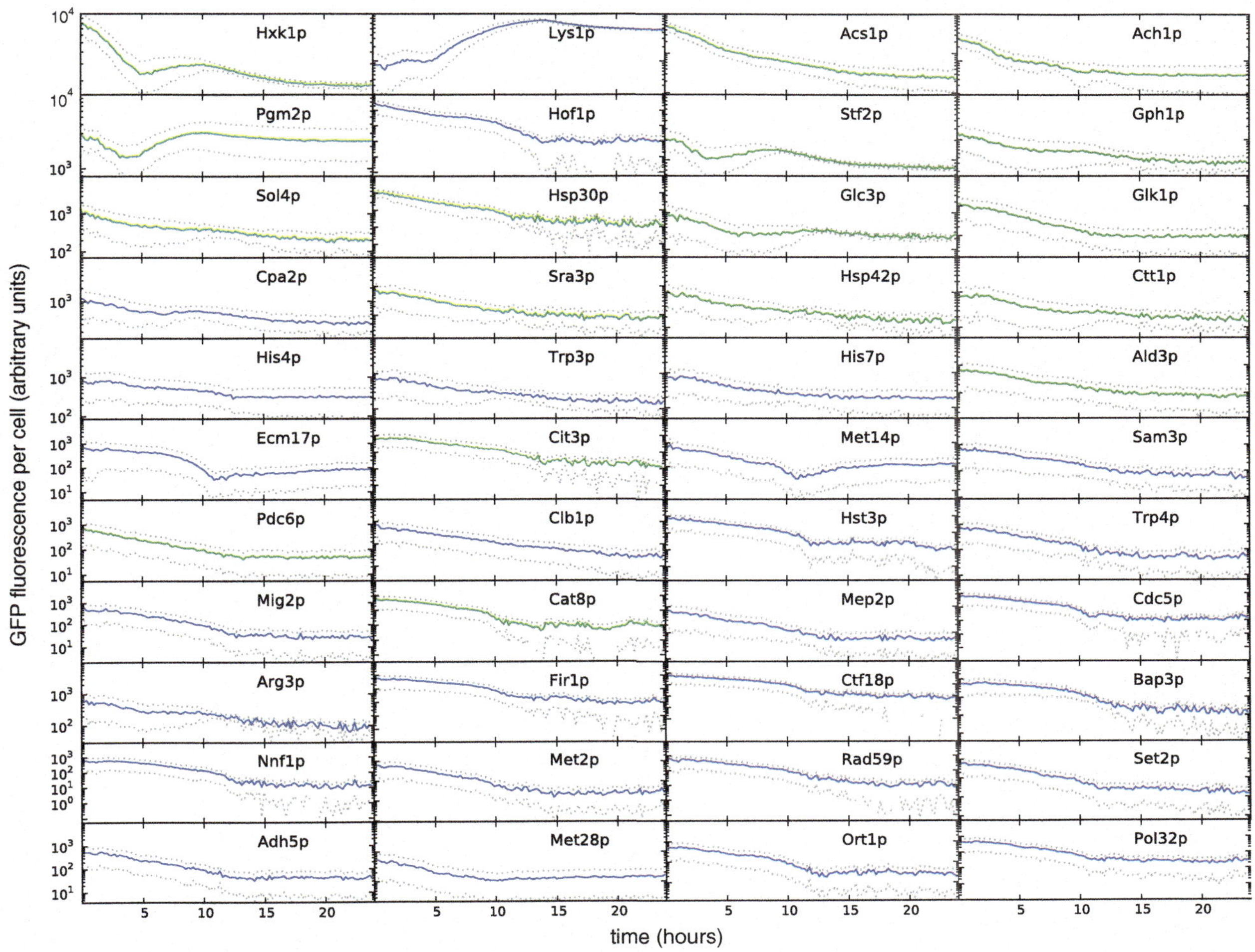

Figure 2 Spectral unmixing allows dynamic, medium throughput quantification of levels of proteins. Cells were grown in 3% glycerol, transferred into media with 3% glycerol and 2% glucose, and followed in the plate reader with data taken every 12 minutes. Genes are ordered by maximum level of expression with this maximum decreasing across the rows and down each column (Hxk1p has the highest maximum expression; Pol32p has the lowest maximum expression). Fluorescence values are in arbitrary units, with long tick marks showing a change in fluorescence of a factor of 10. The *y*-axes in the first two rows ranges from 10^3 to 10^4 units. The blue curves are protein levels per cell for genes with increased mRNA expression from 0-1 hours, using the data from Wang *et al.* [19]; the green curves are protein levels per cell for genes with decreased mRNA expression from 0-1 hours. Cell density initially increases and then plateaus at approximately 12 hours. Error bars are shown with a dashed line (Methods).

excitation at 485 nm and measurement at 525 nm, and a denotes autofluorescence also from excitation at 485 nm and measurement at 525 nm. We wish to determine g. For emissions at 585 nm, we write the fluorescence from the tagged proteins as a product of a constant r_g and g and the autofluorescence as the product of r_a and a. The constant r_g is the ratio of enhanced GFP emissions at 585 nm to that at 525 nm. Normalized emission of enhanced GFP at 525 nm is 0.570 and normalized emission at 585 nm is 0.065 implying that $r_g \simeq 0.114$ (data from R. Tsien laboratory at U.C. San Diego). This number is consistent with the same fluorescence ratio measured from cells strongly expressing an EGFP-tagged protein (Methods: Figure 3b). The two wild-type measurements only have contributions from autofluorescence:

$$\begin{aligned} f_{525}^{\mathrm{WT}} &= a^{\mathrm{WT}} \\ f_{585}^{\mathrm{WT}} &= r_a a^{\mathrm{WT}} \end{aligned} \qquad (2)$$

where r_a, the ratio of the autofluorescence at 585 nm to that at 525 nm, is assumed to be the same for both wild-type and tagged cells (assumption (i) above), but the autofluorescence of the wild-type cells, a^{WT}, can be different from the autofluorescence of the tagged cells, a.

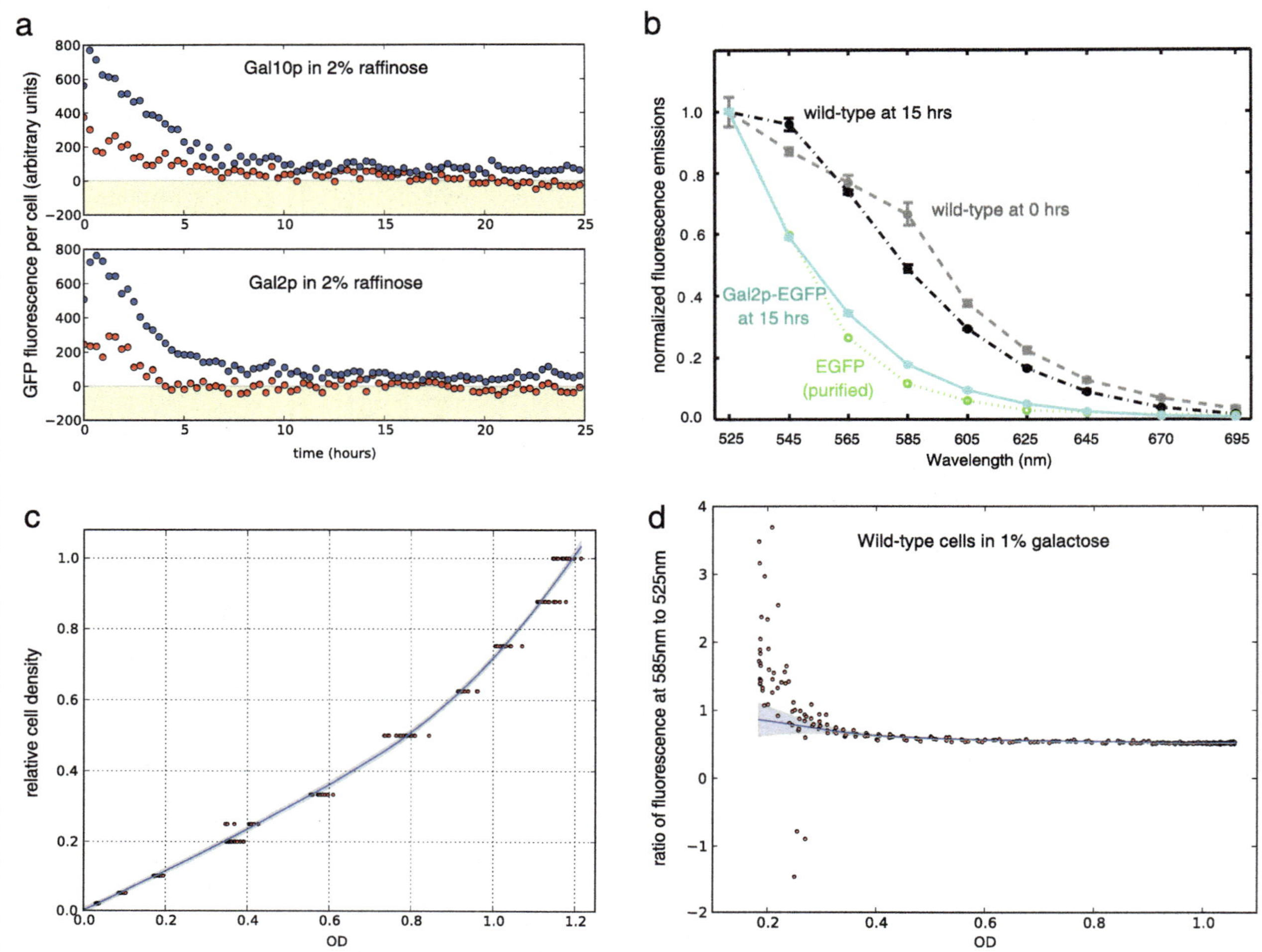

Figure 3 Estimating the mean fluorescence per cell requires several preliminary steps. a) Correcting autofluorescence using the wild-type fluorescence per cell, following the method of de Jong *et al.* [8] developed for bacteria, can lead to negative estimates in yeast. Correction using the bacterial method is shown in red; correction using spectral unmixing is shown in blue. Data are for cells grown in 2% raffinose and then grown further in 2% raffinose in the plate reader. Raffinose does not induce *GAL* expression. **b)** Emissions spectra for autofluorescence and enhanced GFP (EGFP) following excitation at 485 nm. Cells were grown in 2% raffinose and then in 2% galactose for the time indicated. Data for purified EGFP were obtained from the website of the R. Tsien laboratory at U.C. San Diego. **c)** The OD correction curve: relative cell density obtained from diluting a culture by known factors plotted against the measured OD of the diluted cultures. The blue line shows a regression and the shading gives 95% confidence limits (plus and minus twice the standard deviation of the posterior distribution found using a Gaussian process). **d)** Estimating r_a: the ratio of fluorescence emitted at 585 nm to 525 nm plotted versus OD. The blue line shows a regression and the shading gives 95% confidence limits (found using a Gaussian process). Cells were grown in 2% raffinose and transferred to 1% galactose in the plate reader.

In essence, our method is then to estimate r_a from the wild-type data, Eqs. 2, as

$$r_a = \frac{f_{585}^{\mathrm{WT}}}{f_{525}^{\mathrm{WT}}}, \tag{3}$$

and then solve Eqs.1 for g, the signal of interest:

$$g = \frac{r_a f_{525} - f_{585}}{r_a - r_g} \tag{4}$$

where the r_a is calculated for wild-type data at the same OD as the OD of the tagged strain. Dividing g by OD then gives an estimate of the fluorescence per cell.

In practice, the data are often noisy (Figure 3d) and we allow r_a to change smoothly with OD. Further, we often wish to calculate error bars on our estimates of fluorescence, g, and so need to determine how errors in estimating r_a propagate through to the estimation of g (Methods). Finally, we present the fluorescence per cell by dividing g by the appropriate relative cell density, which we find by correcting OD levels for any non-linear dependence on absolute cell density (Methods).

Time-series measurements for expression of *GAL* genes

Applying spectral unmixing to measurements of six proteins in the *GAL* pathway tagged with enhanced GFP reveals the timing of *GAL* gene expression (Figure 1). Cells were grown in 2% raffinose and then transferred to galactose and followed in the plate reader. The three Leloir enzymes, Gal1p, Gal7p, and Gal10p, are strongly expressed in response to galactose and convert galactose into a form of glucose [12]. Other *GAL* proteins regulate the response: Gal2p is a galactose permease; Gal3p is a sensor of intracellular galactose and its activation promotes induction of the *GAL* genes; and Gal80p is a transcriptional repressor and (indirectly) reduces *GAL* expression. Both *GAL3* and *GAL80* are only weakly induced, but their expression can still be detected (Figure 1a Inset).

We see (Figure 1b and c) that levels of Gal7p reach their maximum before levels of Gal1p and Gal10p, the other Leloir enzymes, and the ability of Gal7p to remove a toxic intermediate may explain this early expression. Galactose is converted to galactose-1-phosphate, glucose-1-phosphate, and then glucose-6-phosphate before being able to enter glycolysis [12]. This process must be efficient because galactose-1-phosphate is toxic to cells [16]. Gal7p is a uridyl transferase that converts UDP(uridine diphosphate)-glucose and galactose-1-phosphate to UDP-galactose and glucose-1-phosphate. It thus removes toxic galactose-1-phosphate, which is created by the combined action of Gal1p and Gal10p. Having sufficient quantities of Gal7p ready early may therefore prevent a deleterious build up of toxicity.

Excluding the regulatory proteins, Gal3p and Gal80p, the permease Gal2p is expressed first, which may ensure that cells take advantage of the new carbon source as quickly as possible. Both negative feedback through expression of the repressor Gal80p and positive feedback through expression of the sensor Gal3p initiate at similar times and levels of both proteins peak approximately simultaneously. Naively, Gal3p expression might have been expected to precede expression of Gal80p to ensure that the system is initially dominated by positive feedback generating a fast response. The strength of feedback, however, is determined not just by levels of proteins but also by binding affinities, and therefore positive feedback through Gal3p could still dominate expression at the promoters of the Leloir enzymes. Interestingly, both regulatory proteins peak earlier for lower concentrations of galactose (although levels of measurement noise are high).

Probing gene expression using the collection of GFP fusion proteins

Our method is sufficiently robust that we can simultaneously assay at least tens of genes. A collection of over 4,000 strains of budding yeast has been created [11], each of which expresses a different GFP fusion protein and that potentially allows for high throughput studies of gene expression. Surprisingly the collection has been rarely used in this way, and then only for tens of genes (for instance [17]) or with custom-built microfluidics [18] and never apparently with plate readers. We now show that our methodology combined with robotics could straightforwardly allow expression of hundreds of genes to be followed over time.

After growing 45 strains from the GFP collection in 3% glycerol, we transferred each strain into 2% glucose and followed expression in the plate reader (Figure 2). An increased availability of glucose to starved cells of budding yeast is known to cause substantial transcriptional reprogramming [19]. We had two replicates of each strain, which with two further wells containing wild-type strains and four containing media, filled a 96-well plate. One strain (*SHE1-GFP*) did not grow, but all other genes were amenable to our analysis.

A similar experiment has been performed by Wang *et al.* [19], but with gene expression followed by microarrays up to one hour after the cells were placed in glucose. We used this data to select our strains: those with the largest positive or negative changes in mRNA levels relative to their median level and that were in the GFP collection. To quantify the change in mRNA levels, we used the difference between levels at 0 and 60 minutes after exposure to glucose normalized by the median level of mRNA for that gene (measurements were taken at 0, 20, 40, and 60 minutes).

In general, this measure of change in mRNA based on the microarray data was a poor predictor of the dynamics of the corresponding protein (Figure 2), in agreement with conclusions drawn from proteomics [20]. Although all 17 genes that had decreasing levels of mRNA also had decreasing levels of protein (except perhaps Pgm2p – row 2 & column 1), only one out of the 27 genes with increasing levels of mRNA had increasing levels of protein (Lys1p – row 1 & column 2): the vast majority of genes had falling levels of protein. We emphasize though that our experiment followed cells for almost 24 hours whereas the microarray experiment ran for 1 hour and that the cells stopped growing after about 12 hours when the cell density, as measured by OD, flattened. Reflecting this loss of growth, expression from many genes levels off around 12 hours (Figure 2).

Genes with decreasing levels of mRNA do typically have high levels of proteins, at least initially (they mostly appear in the top half of Figure 2), indicating that these proteins are potentially required for growth on glycerol but not for growth on glucose. Indeed, the hexokinases, Hxk1p (row 1 & column 1) and Glk1p (row 3 & column 4) are induced by non-fermentable carbon sources and repressed in glucose [21]. Further, Acs1p (row 1 & column 3), an acetyl-coenzyme A synthetase, Ach1p (row 1 &

column 4), an acetyl coenzyme A hydrolase, Ald3p (row 5 & column 4), an aldehyde dehydrogenase, and Pdc6p (row 7 & column 1), a pyruvate decarboxylase, have been reported to have their expression repressed in glucose [22-25], as has Cat8p (row 8 & column 2), an activator of a variety of genes under non-fermentative growth [26].

Where expression has been investigated, our data largely support previous studies. They are consistent with the expectations for levels of Pgm2p (row 2 & column 1), phosphoglucomutase, which should rise as glucose falls [27], for Hsp30p (row 3 & column 2), a heat shock protein present in stationary phase [28], and for Glc3p (row 3 & column 3), glycogen branching enzyme, whose transcription peaks when glucose is exhausted (at around 12 hours) [29]. Nevertheless, we do not see evidence of the expected induction during late exponential phase of *GPH1* (row 2 & column 4) [30], a glycogen phosphorylase.

Conclusions

We have introduced a new method to quantify protein expression that uses plate reader measurements of emissions at two wavelengths and spectral unmixing to correct for autofluorescence. Our approach uses linear unmixing and is similar in spirit to others that have been proposed to solve the problem of autofluorescence during flow cytometry [31,32]. While we have considered just one fluorophore, the methodology could readily be applied to samples with two or more [13].

Our method allows medium throughput measurements of gene expression through resources, such as the collection of GFP-protein fusions in budding yeast [11]. Protein levels from the expression of hundreds of genes can be followed over tens of hours. Although other approaches have been applied to bacteria [4,8,9], these methods require that the growth curves of a wild-type and all tagged strains are identical and consequently do not scale to hundreds of strains with different tagged genes, which will not all have the same growth curves. The work in bacteria, though, used GFP-promoter fusions rather than GFP-protein fusions, and promoter fusions may perturb growth rates less (although biological reproducibility will still be an issue).

Time-series data describing the mean behaviour of a typical cell in population of cells is ideal for testing mathematical models of cellular behaviour based on ordinary differential equations [5], and our analysis allows such data to be generated from plate readers. Further, and unlike other methods, we provide error bars on the fluorescence per cell that combine both measurement error and the errors generated by correcting for autofluorescence. Such error bars are invaluable when fitting models to data [33,34].

High throughput studies of gene expression typically focus on levels of mRNA, but levels of mRNA need not correlate with levels of protein [20]. Our methodology combined with strain collections such as the GFP fusion collection will allow complementary data at the level of proteins to be gathered using standard laboratory equipment. This data will increase our understanding of physiological change and cellular decision-making because it is largely proteins, not mRNA, that enact such responses.

Methods

Reagents

Low fluorescence synthetic complete media (SC) was made following Sheff *et al.* [35]. We determined that the primary source of media fluorescence was the ammonium salts solution and that the media's fluorescence increased significantly after both autoclaving and a few days of storage at room temperature. To reduce media fluorescence we therefore used filter sterilisation instead of autoclaving, stored the ammonium salts solution at -20°C, made smaller batches of media to minimise the storage time, and stored the media at 4°C. Sugar solutions were made up from D_(+)_Raffinose pentahydrate, D_(+)_Glucose, and D_(+)_Galactose (Sigma-Aldrich, Steinheim, Germany).

We used black optical bottom 96-well plates (cat. no. 265301, ThermoScientific/Nunc, Rochester, New York, USA), in which we found reduced autofluorescence compared with clear plates. For sterilisation, they were placed in a UV cross linker on high for 2 min (1 min face up, 1 min face down). Film covers were Polyolefin Sealing Tape (cat. no. 235307 Nunc), which contributed less autofluorescence than clear polystyrene lids and reduced evaporation.

Yeast strains

The strain used was haploid *a* BY4741 *S. cerevisiae* (obtained from EUROSCARF, Frankfurt). To construct strains expressing *GAL* protein-EGFP fusions, we followed the procedure described by Janke *et al.* [36]. The *pYM28* plasmid with an EGFP tag and marker HIS3MX6 (accession number P30240, EUROSCARF toolbox) [37] was amplified with primers containing 40-50 bp of the region just upstream of the target gene's stop codon, followed by CGTACGCTGCAGGTCGAC (forward primer), and 40-50bp of the reverse strand for the region just beyond the target gene's coding sequence, followed by ATCGATGAATTCGAGCTCG (reverse primer). PCR products were verified with gel electrophoresis. Then, following PCR product purification with a Qiaquick PCR purification kit (Qiagen), yeast transformation was carried out following a modified LiOAc protocol [37]. For growth and selection, yeast were spread on SC plates containing 2% glucose lacking histidine and incubated for 2-3 days at 30°C. Positive colonies were purified by streaking onto fresh plates. Cultures were stored at -80°C in media containing 10% glycerol.

Preparing galactose induction experiments

Cells were grown overnight (17-19 hours) in 5 mL SC with 2% raffinose in glass test tubes at 30°C with shaking at 200-230 rpm. In the morning, cells were diluted 10× into fresh SC with 2% raffinose and incubated further at 30°C with shaking at 200-230 rpm. After 6-8 hours, 180 μL of each sample was placed in wells of a 96 well plate to measure the OD. Then, cells were transferred to 15 mL plastic tubes and washed: they were spun (5 min at 3000 rpm = 1811 RCF) to form a pellet, supernatant was discarded, pellet was resuspended in 11 mL sterile water, and then cells were spun again (5 min at 3000 rpm).

While cells were being washed, the OD was measured and dilutions were calculated to obtain starting OD values for the experiment of 0.25 (wild-type) or 0.3 (cells with enhanced GFP tag). Wild-type cells started at a lower OD to ensure that autofluorescence corrections are possible from the beginning of the induction period. When washing was complete, cells were resuspended in the volume of low fluorescence SC media indicated by the dilution calculations. OD was measured again and adjustments made as necessary to achieve desired starting OD values.

Finally, 180 μL aliquots of each sample were then pipetted into wells of a 96-well plate. 20 μL of the appropriate sterile sugar solutions in water were then added to the wells. A polyolefin film cover was used to cover the plate, and plate reader measurements were begun immediately.

Measuring using the plate reader

Plate readers were from the Tecan Infinity M200 series (for combined OD and dual-wavelength fluorescence measurements) (Tecan Group Ltd., Switzerland). We set the temperature at 29.9°C (range 29.4°-30.4°C) with linear shaking (6 mm amplitude at 200-220 rpm). For OD measurements, the absorbance wavelength was 595 nm and the measurement bandwidth was 9 nm with 15 reads and 0 ms settle time. For fluorescence measurements, the excitation wavelength was 485 nm, the excitation bandwidth was 9 nm, the emission wavelength was 525 nm (or 585 nm), the emission bandwidth was 20 nm, and reading mode was top with 0 μs lag time, 20 μs integration time, 10 reads, and 0 ms settle time. The gain was set manually to 105. To minimise photobleaching and fluorescence-induced toxicity, measurements were made no more often than every 10 minutes. The plate reader was switched on and adjusted to the correct temperature around 1-2 hours prior to the experiment.

Medium throughput measurements of gene expression

Selected strains from the yeast GFP fusion collection (Life Technologies, California) were cultured in 100 μl SC with 3% glycerol (w/v) in a 96 well plate at 30°C for 16 hours with shaking. A Biomek FX liquid handling robot (Beckman Coulter) was used to dilute the cultures 1/10 into a fresh plate in SC with 3% glycerol. This plate was incubated for 7 hours at 30°C to obtain log phase cells (OD between 0.2 and 0.6 in all cases). An assay plate was then prepared by the Biomek FX in which each well contained 100 μl of log phase cell suspension diluted into a total of 200 μl SC with 3% glycerol (w/v) and 2% glucose (w/v). Fluorescence and optical density data were then recorded using the plate reader.

Data analysis

A direct approach can be used, which although straightforward does not allow estimates of error and can give negative values of the fluorescence per cell. We estimate the curve r_a, which is the ratio of the autofluorescence at 585 nm to that at 525 nm, as a function of OD (Eq. 3) by calculating the mean of $f_{585}^{\mathrm{WT}} / f_{525}^{\mathrm{WT}}$ over all replicates at each time point. To determine the OD at each time point, we average OD values over all replicates and then smooth the resulting average OD curve using local regression with a second degree polynomial (the `rloess` option in Matlab's `smooth` command (Mathworks, Natick, MA)). Each time point therefore corresponds to one average OD value of the wild-type strain. To find the fluorescence of the tagged strain, we find the value of r_a at the corresponding OD, with interpolation when necessary, and then use Eq. 4 to find g for each replicate. We average g over all replicates and smooth the resulting curve using local regression. Dividing this smoothed curve by the (corrected) OD measurements for the tagged strain gives the mean fluorescence per cell as a function of time.

Gaussian processes

We use Gaussian processes to more robustly analyze the data because we can then specify the degree of smoothness of the fits, can make interpolations between and beyond data points, and can straightforwardly propagate errors from one stage of the data analysis to the next.

A Gaussian process is an infinite collection of random variables with each variable associated with a point in some range of an input variable and with any finite number of these random variables having a joint Gaussian distribution [38]. Time is a typical input variable. A Gaussian process is entirely specified by its mean and covariance, both of which are functions of the input variable. When fitting data, we initially consider a Gaussian process with zero mean and with a particular covariance function. The choice of covariance function determines an *a priori* distribution over functions so that samples from the Gaussian process plotted over many input points have characteristic behaviours. For example, if the covariance function is of the neural network form then sampled functions have an *a priori* sigmoid-like shape. If the covariance function is squared exponential, then sampled functions smoothly fluctuate. Given data with measurement noise that is Gaussian, additive, and independently and identically distributed, we can analytically calculate the *a posteriori* mean and covariance function of the Gaussian process by conditioning on the data the joint distribution of both the observed data and the predicted values. By sampling from this *a posteriori* Gaussian process, we can generate functions consistent with the data and our choice of *a priori* covariance function. The variation in these functions is a measure of the error in the fitting. Covariance functions are specified by parameters, called hyperparameters, and we choose the most appropriate hyperparameters by maximizing the marginal likelihood [38] – the probability of the data given the input points – and so by maximizing the evidence for our choice of covariance function. We use a truncated Newton algorithm (from the Python module SciPy) for all optimizations.

Correcting the growth curve to be proportional to cell density

To ensure a linear relationship between OD and cell density (the number of cells per unit volume), we used dilution to have an accurate if relative measure of cell density [7]. We diluted a yeast culture (grown overnight on 2% glucose) at a known high OD in a series from 1× to 256× in SC media. The cell density of the last sample in the series is therefore 256 times smaller than the cell density of the original sample. By comparing the measured ODs of these samples to their known relative cell densities, we observed that the deviation from linearity begins around an OD of 0.5 (Figure 3c). To correct OD measurements, we would like to find a relative cell density from a measured OD and so performed a regression on the data of Figure 3c. We used a Gaussian process for the regression with a covariance function that is the sum of a squared exponential covariance function and a linear covariance function. Such a covariance function generates smooth functions that increase as the OD increases [38]. For absolute quantification, we used a Neubauer bright-line cytometer (depth: 0.100 mm, grid surface: 0.0025 mm^2) imaged on a Nikon Eclipse Ti microscope to estimate that a relative cell density of 1.0 for the calibration curve, corresponding to an OD of around 1.2, has an absolute density of 1.46×10^7 cells/mL.

Estimating the mean fluorescence per cell and its error

To find the mean fluorescence per cell and an estimate of its error, we use Gaussian processes. First, we correct the data for the OD of the media in which the cells grow and for its fluorescence. Second, we estimate the curve r_a as a function of OD (Eq. 3) using data from the wild-type strain, and generate sample r_a functions to determine how errors in measuring r_a affect estimates of mean fluorescence. Third, for each time point, we use the measurements of fluorescence at 525 nm and 585 nm for the fluorescently tagged strain, the samples of r_a, and samples of the OD correction curve to find a probability distribution of the estimates of the fluorescence per cell. The mean and standard deviation of this distribution are our best-fit value and its error.

Correcting for the media: Although we expect both the OD and fluorescence measured in wells containing only media to remain constant over time, both change slightly during the experiment reflecting perhaps small changes in the plate reader. We use a Gaussian process with a squared exponential covariance function to fit data from the media for all replicates. Regression with Gaussian processes typically assumes a constant measurement error but the measurement error in our case changes with time, at least for some of the data sets. We therefore first estimate the measurement errors using the standard deviation of the 10 nearest data points to each data point, and then run a regression with a Gaussian process with measurement errors fixed at these estimated values. If we fail to find hyperparameters by maximizing the marginal likelihood because the optimization algorithm fails to converge, we try an second fit where the measurement error is assumed constant and is itself fit as part of the optimization. Using this combination of methods, we could automatically fit all media data. The OD of the media as a function of time was subtracted from all OD measurements, and the fluorescence of the media as a function of time was subtracted from all fluorescence measurements.

Fitting wild-type data to find r_a: From Eq. 3, r_a at a particular level of OD is given by the ratio of fluorescence at 585 nm to that at 525 nm for wild-type cells. Figure 3d shows a typical example of this data. We use a Gaussian process with a neural network covariance function, which generates sigmoid-like functions, to find r_a. We therefore assume that the measurement error in measuring the ratio of fluorescences has a Gaussian distribution. From Figure 3d, the measurement error changes with OD, and we empirically estimate the measurement error at each OD as the standard deviation of the 20 nearest data points to the data point at that OD. We use this estimated measurement error in our regression (Figure 3d).

We used sampling to estimate how errors in fitting r_a affect the fluorescence per cell. We sampled tens of r_a functions that are consistent with the data and which differ only because of measurement errors, at least given our assumptions. We can therefore determine the error in the estimated fluorescence per cell by finding the standard deviation of the distribution generated by correcting the fluorescence of the tagged strain by each of these sample r_a functions. Although using Gaussian processes make this sampling straightforward, we should sample r_a at the

OD levels measured for the tagged strain, not those for the wild-type strain. We can interpolate r_a to these OD values using standard techniques [38], but we must also interpolate the measurement errors to the expected measurement error in r_a at the OD levels of the tagged strain. We use regression with another Gaussian process to fit and interpolate the empirical measurement errors, with a squared exponential covariance function to generate smooth functions. With the interpolated measurement errors and the best-fit hyperparameters of the neural network covariance function found from the wild-type data, we can sample r_a functions.

Fitting the data from the tagged strain to find g: To find the fluorescence of the tagged strain, g, we use a Bayesian approach with uniform prior probabilities so that the posterior probability of g is proportional to the likelihood of g given the data [39]. From Eq. 1, and, assuming Gaussian measurement errors for measuring each fluorescence value, then the likelihood of g at a particular time point is

$$P(f_{525}, f_{585} | r_a, r_g, g, \sigma_{525}, \sigma_{585})$$

$$\sim \int_0^\infty da\, (\sigma_{585}\sigma_{525})^{-1} \exp\left[-\frac{(f_{525} - g - a)^2}{2\sigma_{525}^2}\right]$$

$$\times \exp\left[-\frac{(f_{585} - r_g g - r_a a)^2}{2\sigma_{585}^2}\right]$$

$$\sim \left(\sigma_{585}^2 + r_a^2\sigma_{525}^2\right)^{-\frac{1}{2}} \exp\left[-\frac{(f_{585} - r_a f_{525} - (r_g - r_a)g)^2}{2(\sigma_{585}^2 + r_a^2\sigma_{525}^2)}\right] \quad (5)$$

where σ_{525} is the measurement error at 525 nm and σ_{585} is the measurement error at 585 nm. In Eq. 5, we have extended the range of integration of the autofluorescence, a, to $-\infty$ to ∞ to perform the integration, which should give little effect if the posterior distribution of a is sufficiently peaked at a positive value of a [39].

We wish to determine both the best-fit value of g and its error. We do so by generating a distribution of samples of g consistent with the data and take the mean value of g as the best estimate and its standard deviation as the error in this estimate. We sample g from Eq. 5. For each replicate, we generate 50 samples of r_a functions at the OD levels of the tagged strain. Using the one-to-one relationship between OD and time (Figure 1a), we map the r_a values to the values expected at particular time points. Given a sample of a r_a function, we use rejection sampling to draw 1000 samples of g from Eq. 5 [40]. The measurement errors σ_{525} and σ_{585} are found empirically by taking the standard deviation of the 20 fluorescence measurements closest to the fluorescent measurement at the time point of interest. We include the errors in estimating the relative cell density by sampling a corrected OD function for each sample of r_a and dividing the samples of g by this corrected OD. Finally, averaging over 50 (samples of r_a and of corrected OD levels) $\times$ 1000 (samples of g) $\times$ the number of replicates, we find the mean and standard deviation of g at each time point.

Access to data

We have written code to implement our correction method that uses Python and other open source software (available with our raw data at swainlab.bio.ed.ac.uk/software/platereader and as Additional file 1).

Additional file

Additional file 1: Python software that implements our correction methodology and the data from Figures 1 and 2.

Competing interests

The authors declare that they have no competing interests.

Authors' contributions

CAL and IBNC developed the experimental method; CAL and PSS wrote the analysis routines; IBNC and RW gathered the data for Figures 1 and 2; CAL and PSS wrote the paper. All authors read and approved the final manuscript.

Acknowledgements

We thank Guido Sanguinetti for advice on Gaussian processes, the Tyers lab for generous access to strains and equipment, and gratefully acknowledge financial support from the B.B.S.R.C. (U.K) – grant number BB/I00906X/1 – and from the Scottish Universities Life Sciences Alliance (SULSA).

Author details

[1]SynthSys, University of Edinburgh, Mayfield Road, Edinburgh, UK. [2]Department of Physiology, McGill University, Promenade Sir William Osler, Montreal, Canada. [3]Biological Sciences, University of Edinburgh, Mayfield Road, Edinburgh, UK.

References

1. Nurse P: **Life logic and information Nature.** 2008, **454**(7203):424–426.
2. Locke JCW, Elowitz MB: **Using movies to analyse gene circuit dynamics in single cells.** 2009, **7**(5):383–392.
3. Bennett MR, Hasty J: **Microfluidic devices for measuring gene network dynamics in single cells.** *Nat Rev Genet* 2009, **10**(9):628–638.
4. Kalir S, McClure J, Pabbaraju K, Southward C, Ronen M, Leibler S, Surette MG, Alon U: **Ordering genes in a flagella pathway by analysis of expression kinetics from living bacteria.** *Science* 2001, **292**(5524):2080–2083.
5. Chen WW, Niepel M, Sorger PK: **Classic and contemporary approaches to modeling biochemical reactions.** *Genes Dev* 2010, **24**(17):1861–1875.
6. Dalgaard P, Ross T, Kamperman L, Neumeyer K, McMeekin TA 1994, **23**:391–404.
7. Warringer J, Blomberg A: **Automated screening in environmental arrays allows analysis of quantitative phenotypic profiles in Saccharomyces cerevisiae.** *Yeast* 2003, **20**(1):53–67.
8. de Jong H, Ranquet C, Ropers D, Pinel C, Geiselmann J: **Experimental and computational validation of models of fluorescent and luminescent reporter genes in bacteria.** *BMC Syst Biol* 2010, **4**:55.

9. Berthoumieux S, de Jong H, Baptist G, Pinel C, Ranquet C, Ropers D, Geiselmann J: **Shared control of gene expression in bacteria by transcription factors and global physiology of the cell.** *Mol Syst Biol* 2013, **9:**634.
10. Stagoj MN, Comino A, Komel R: **Fluorescence based assay of GAL system in yeast Saccharomyces cerevisiae.** *FEMS Microbiol Lett* 2005, **244**(1):105–110.
11. Ghaemmaghami S, Huh W-K, Bower K, Howson RW, Belle A, Dephoure N, O'Shea EK, Weissman JS: **Global analysis of protein expression in yeast.** *Nature* 2003, **425**(6959):737–741.
12. Sellick CA, Campbell RN, Reece RJ: **Galactose metabolism in yeast-structure and regulation of the leloir pathway enzymes and the genes encoding them.** *Int Rev Cell Mol Biol* 2008, **269:**111–150.
13. Zimmermann T, Rietdorf J, Pepperkok R: **Spectral imaging and its applications in live cell microscopy.** *FEBS Lett* 2003, **546**(1):87–92.
14. Garini Y, Young IT, McNamara G: **Spectral imaging: principles and applications.** *Cytometry A* 2006, **69**(8):735–747.
15. Gordon A, Colman-Lerner A, Yu RC, Brent R: **Single-cell quantification of molecules and rates using open-source microscope-based cytometry.** *Nat Methods* 2007, **4**(2):175–181.
16. Douglas HC, Hawthorne DC: **Enzymatic expression and genetic linkage of genes controlling galactose utilization in Saccharomyces.** *Genetics* 1964, **49:**837–844.
17. Bar-Even A, Paulsson J, Maheshri N, Carmi M, O'Shea E, Pilpel Y, Barkai N: **Noise in protein expression scales with natural protein abundance.** *Nat Genet* 2006, **38**(6):636–643.
18. Dénervaud N, Becker J, Delgado-Gonzalo R, Damay P, Rajkumar AS, Unser M, Shore D, Naef F, Maerkl SJ: **A chemostat array enables the spatio-temporal analysis of the yeast proteome.** *Proc Nat Acad Sci USA* 2013, **110**(39):15842–15847.
19. Wang Y, Pierce M, Schneper L, Güldal CG, Zhang X, Tavazoie S, Broach JR: **Ras and Gpa2 mediate one branch of a redundant glucose signaling pathway in yeast.** *PLoS Biol* 2004, **2**(5):128.
20. Gygi SP, Rochon Y, Franza BR, Aebersold R: **Correlation between protein and mRNA abundance in yeast.** *Mol Cell Biol* 1999, **19**(3):1720–1730.
21. Rodriguez A, De La Cera, T, Herrero P, Moreno F: **The hexokinase 2 protein regulates the expression of the GLK1, HXK1 and HXK2 genes of Saccharomyces cerevisiae.** *Biochem J* 2001, **355:**625–631.
22. de Jong-Gubbels P, van den Berg, MA, Steensma HY, van Dijken JP, Pronk JT: **The Saccharomyces cerevisiae acetyl-coenzyme a synthetase encoded by the ACS1 gene, but not the ACS2-encoded enzyme, is subject to glucose catabolite inactivation.** *FEMS Microbiol Lett* 1997, **153:**75–81.
23. Lee FJ, Lin LW, Smith JA: **A glucose-repressible gene encodes acetyl-CoA hydrolase from Saccharomyces cerevisiae.** *J Biol Chem* 1990, **265:**7413–7418.
24. Navarro-Avino JP, Prasad R, Miralles VJ, Benito RM, Serrano R: **A proposal for nomenclature of aldehyde dehydrogenases in Saccharomyces cerevisiae and characterization of the stress-inducible ALD2 and ALD3 genes.** *Yeast* 1999, **15:**829–842.
25. Hohmann S: **Characterization of PDC6, a third structural gene for pyruvate decarboxylase in Saccharomyces cerevisiae.** *J Bacteriol* 1991, **173:**7963–7969.
26. Hedges D, Proft M, Entian KD: **CAT8, a new zinc cluster-encoding gene necessary for derepression of gluconeogenic enzymes in the yeast Saccharomyces cerevisiae.** *Mol Cell Biol* 1995, **15:**1915–1922.
27. Fu L, Bounelis P, Dey N, Browne BL, Marchase RB, Bedwell DM: **The posttranslational modificaon of phosphoglucomutase is regulated by galactose induction and glucose repression in Saccharomyces cerevisiae.** *J Bacteriol* 1995, **177:**3087–3094.
28. Panaretou B, Piper PW: **The plasma membrane of yeast acquires a novel heat-shock protein (Hsp30) and displays a decline in proton-pumping ATPase levels in response to both heat shock and the entry to stationary phase.** *Eur J Biochem* 1992, **206:**635–640.
29. Parrou JL, Enjalbert B, Plourde L, Bauche A, Gonzalez B, Francois JW: **Dynamic responses of reserve carbohydrate metabolism under carbon and nitrogen limitations in Saccharomyces cerevisiae.** *Yeast* 1999, **15:**191–203.
30. Hwang PK, Tugendreich S, Fletterick RJ: **Molecular analysis of GPH1, the gene encoding glycogen phosphorylase in Saccharomyces cerevisiae.** *Mol Cell Biol* 1989, **9:**1659–1666.
31. Steinkamp JA, Stewart CC: **Dual-laser, differential fluorescence correction method for reducing cellular background autofluorescence.** *Cytometry* 1986, **7:**566–574.
32. Roederer M, Murphy RF: **Cell-by-cell autofluorescence correction for low signal-to-noise systems: application to epidermal growth factor endocytosis by 3T3 fibroblasts.** *Cytometry* 1986, **7**(6):558–565.
33. Brown KS, Sethna JP: **Statistical mechanical approaches to models with many poorly known parameters.** *Phys Rev E* 2003, **68** (2 Pt 1):021904.
34. Finkenstädt B, Heron EA, Komorowski M, Edwards K, Tang S, Harper CV, Davis JRE, White MRH, Millar AJ, Rand DA: **Reconstruction of transcriptional dynamics from gene reporter data using differential equations.** *Bioinformatics* 2008, **24**(24):2901–2907.
35. Sheff MA, Thorn KS: **Optimized cassettes for fluorescent protein tagging in Saccharomyces cerevisiae.** *Yeast* 2004, **21**(8):661–670.
36. Janke C, Magiera MM, Rathfelder N, Taxis C, Reber S, Maekawa H, Moreno-Borchart A, Doenges G, Schwob E, Schiebel E, Knop M: **A versatile toolbox for PCR-based tagging of yeast genes: new fluorescent proteins, more markers and promoter substitution cassettes.** *Yeast* 2004, **21**(11):947–962.
37. Knop M, Siegers K, Pereira G, Zachariae W, Winsor B, Nasmyth K, Schiebel E: **Epitope tagging of yeast genes using a PCR-based strategy: more tags and improved practical routines.** *Yeast* 1999, **15**(10B):963–972.
38. Rasmussen CE, Williams CKI: *Gaussian Process for Machine Learning.* Cambridge, Massachusetts: MIT Press; 2006.
39. Sivia D, Skilling J: *Data Analysis a Bayesian Tutorial.* Cambridge, UK: Cambridge University Press; 2006.
40. MacKay DJC: *Information Theory, Inference, and Learning Algorithms.* Oxford, UK: Oxford University Press; 2003.

6

Natural transformation of *Thermotoga* sp. strain RQ7

Dongmei Han[†], Hui Xu[†], Rutika Puranik and Zhaohui Xu[*]

Abstract

Background: *Thermotoga* species are organisms of enormous interest from a biotechnological as well as evolutionary point of view. Genetic modifications of *Thermotoga* spp. are often desired in order to fully release their multifarious potentials. Effective transformation of recombinant DNA into these bacteria constitutes a critical step of such efforts. This study aims to establish natural competency in *Thermotoga* spp. and to provide a convenient method to transform these organisms.

Results: Foreign DNA was found to be relatively stable in the supernatant of a *Thermotoga* culture for up to 6 hours. Adding donor DNA to *T.* sp. strain RQ7 at its early exponential growth phase (OD_{600} 0.18 ~ 0.20) resulted in direct acquisition of the DNA by the cells. Both *T. neapolitana* chromosomal DNA and *Thermotoga-E. coli* shuttle vectors effectively transformed *T.* sp. strain RQ7, rendering the cells resistance to kanamycin. The *kan* gene carried by the shuttle vector pDH10 was detected by PCR from the plasmid extract of the transformants, and the amplicons were verified by restriction digestions. A procedure for natural transformation of *Thermotoga* spp. was established and optimized. With the optimized method, *T.* sp. strain RQ7 sustained a transformation frequency in the order of 10^{-7} with both genomic and plasmid DNA.

Conclusions: *T.* sp. strain RQ7 cells are naturally transformable during their early exponential phase. They acquire DNA from both closely and distantly related species. Both chromosomal DNA and plasmid DNA serve as suitable substrates for transformation. Our findings lend a convenient technical tool for the genetic engineering of *Thermotoga* spp.

Keywords: *Thermotoga*, Natural transformation, Natural competence, pDH10, pDH12

Background

Thermotoga species are a group of Gram-negative, strictly anaerobic bacteria that grow up to 90°C. Ever since their discoveries in 1980s, they have been attracting great attention due to their extreme life style and the underlining implications in both basic research and biotechnological applications. Most noticeably, they serve as excellent candidates for the production of hydrogen gas, which may be directly used as a clean fuel [1,2]. Producing biohydrogen using fermentative hyperthermophiles like *Thermotoga* has many advantages, including independence from light sources, reduced risks of contamination, lower energy requirement for agitation and cooling, and above all, higher hydrogen partial pressure. To fully release the potentials of *Thermotoga*, one often wishes to modify these bacteria at the genetic level. However, *Thermotoga* spp. are considered genetically recalcitrant. Their first transformants were isolated only recently after decades of efforts [3]. Making these species more genetically accessible remains a priority for the research community. Effective delivery of recombinant DNA stands as a critical step in any genetic modification attempt dealing with *Thermotoga*. Previous studies have demonstrated that *Thermotoga* cells are transformable by liposome-mediated transformation and electroporation [3,4]. Effective as these methods are, they are labor-intensive and time-consuming. A more convenient approach would be highly desirable if one needs to transform *Thermotoga* on a daily basis.

Many prokaryotes possess the ability to take up DNA directly from the environment without human intervention, i.e. they are naturally competent. Some bacteria efficiently import DNA from any source, whereas others prefer to acquire homologous DNA through the recognition of specific uptake signal sequences [5,6]. The process of natural

* Correspondence: zxu@bgsu.edu
[†]Equal contributors
Department of Biological Sciences, Bowling Green State University, 43403 Bowling Green, OH, USA

transformation involves the steps of DNA binding, translocation, and recombination. In most competent bacteria, the machinery responsible for DNA binding and translocation is related to Type IV pili (T4P) and Type II secretion systems (T2S) [7,8]. T4P are retractile appendages that contribute to twitching motility, biofilm formation, adhesion, and conjugation, in addition to natural transformation. A working model proposed for *Neisseria gonorrhoeae* suggests that exogenous DNA crosses the outer membrane through a pore made of the secretin protein PilQ and transverses the periplasmic space with the aid of the PilE complex and a high affinity DNA-binding protein ComE. One strand of the DNA is then degraded by a nuclease, and the other strand travels across the inner membrane through a channel comprised of the membrane protein ComA [9,10]. In Gram-positives, DNA is transported across the cell envelope in a similar manner, except that it does not need to overcome the outer membrane barrier. Once in the cytoplasm, the single-stranded DNA has to be protected by single strand DNA binding proteins to avoid degradation by host nucleases. Recent evidence suggests that DNA protection protein DprA conveys incoming ssDNA to RecA, the ubiquitous recombinase responsible for the initiation of homologous recombination [11-13]. Protein ComM influences transformation at a very late stage presumably by affecting recombination of the donor DNA into the chromosome [14].

More than 60 bacterial species have been documented as naturally transformable [15], which actually represents a minute fraction of all the bacterial species characterized thus far. Natural transformation is therefore considered a rare phenomenon. However, competence genes are widespread in bacterial species, even though many of which have not been reported to be naturally competent. Part of the reason is that natural competence usually develops only under certain physiological states that are inducible by environmental factors. Often these conditions are ill-defined and thus cannot be reproduced in laboratory settings. It is very likely that natural competence occurs much more frequently than we observe. For example, this mechanism was only recently noted in such well-studied organisms as *Vibrio cholerae* [16], *Bacillus cereus* [17], and *E. coli* [18]. Natural uptake of DNA has not been reported for any bacteria growing optimally at, or above, 80°C, although this ability has been documented for hyperthermophilic archaeon *Thermococcus kodakarensis* [19].

Analyses of the published *Thermotoga* genomes present evidence of frequent lateral gene transfer events between *Thermotoga* and other bacteria, archaea, and even eukaryotes [20-23]. The chimeric nature of *Thermotoga* genomes suggests that these bacteria have a remarkable efficiency for acquisition of foreign genes. Although natural competence has not been reported for *Thermotoga*, putative competence genes have long been noticed in their genomes, including those encoding T2S- and T4P-related proteins [20]. Moreover, mini-plasmids pRQ7, pMC24, and pRKU1 have been discovered in *T.* sp. strain RQ7 [24], *T. maritima* [25], and *T. petrophila* RKU-1 [26], respectively. Therefore, it is possible that *Thermotoga* spp. are capable of acquiring foreign DNA directly. Searching of the *T. maritima* genome sequence did not uncover any specific uptake signal sequences [6]. For this reason, if *Thermotoga* spp. were to acquire DNA naturally, they should do so indiscriminately. The aim of this study was to establish natural competency in *Thermotoga* spp. and to develop a natural transformation protocol to facilitate the genetic engineering of *Thermotoga*.

Materials and methods

Strains and cultivation conditions

The bacteria strains and vectors used in this study are summarized in Table 1. *E. coli* strains were cultivated in Luria-Bertani broth (1% tryptone, 0.5% NaCl, 0.5% yeast extract) at 37°C. Ampicillin was supplemented at 50 μg per ml when required. *Thermotoga* strains were cultivated at 77°C in SVO medium, which was named by us after the first author of the paper where the medium was first described [32]. Gelrite of 0.075% and 0.3% were included for preparation of soft and solid media, respectively. When needed, kanamycin was supplemented at 150 μg ml^{-1} for liquid and 250 μg ml^{-1} for soft and solid cultures. To make a *Thermotoga* plate, the liquid culture was mixed with hot SVO Gelrite medium and poured into petri dishes. The plates were incubated in an anaerobic jar for 48 h at 77°C, and single colonies were picked up, inoculated into selective SVO soft medium to grow for 24 h, and transferred into liquid selective media for further propagation and analyses. The techniques for effective handling of *Thermotoga* cultures under atmospheric conditions were detailed in our previous report [3]. Cell densities were measured as optical absorbance at 600 nm (OD_{600}).

Construction of the *Thermotoga-E. coli* shuttle vector pDH12

In the early stage of the study, pDH10, constructed in our previous study [3], was used as the substrate DNA to establish natural competency in *T.* sp. strain RQ7. In the late stage of the study, a new shuttle vector, pDH12 (Figure 1), was constructed. The two plasmids differ in two ways: 1) the *rep* gene and the *kan* gene are arranged in the same direction in pDH12 as opposed to a convergent orientation in pDH10; and 2) The upstream sequence of the *rep* gene is 416 bp in pDH12 and 207 bp in pDH10. The extra 209 bp fragment in pDH12, being included to enhance the expression of the *rep* gene, is a duplicate of the 3′-end of the *rep* gene.

To construct pDH12, a three-piece ligation strategy was adopted since the coding sequence of the pRQ7 Rep protein conveniently bears a HindIII site about a third

Table 1 Strains & vectors used in this study

Strain or plasmid	Description	Reference
T. neapolitana DSM 4359	Isolated from African continental solfataric springs	[27]
T. maritima MSB8	Isolated from geothermally heated sea floors in Italy and the Azores	[28]
T. sp. strain RQ7	Isolated from geothermally heated sea floors in Ribeira Quente, the Azores	[28]
Caldicellulosiruptor saccharolyticus DSM 8903	Isolated from a thermal pool of Taupo, New Zealand	[29]
E. coli DH5α	F^- *endA1 hsdR17* (r_k^-, m_k^+) *supE44 thi-1* λ^- *recA1 gyrA96 relA1 deoR* Δ(*lacZYA-argF*)- U169 ϕ80d*lacZ*ΔM15	[30]
Plasmids		
pKT1	pUC-derived plasmid, containing a *kan* cassette for thermostable kanamycin selections	[31]
pRQ7	Cryptic miniplasmid from *T.* sp. strain RQ7	[24]; GenBank: KF798180
pDH10	pRQ7 cloned between the EcoRI and XbaI sites of pKT1; Ap^r; Kan^r	[3]; GenBank: JN813374
pDH12	pRQ7 along with its duplicated fragment of 209 bp were cloned between the EcoRI and XbaI sites of pKT1; Ap^r, Kan^r	This study; GenBank: KF798179

way from its start codon. Primers 5′- GGGAATTCTG AAGTACTTATCAAAGGAG-3′ and 5′-CGTTTTTAA GCTTTTCCCAGC-3′ were used to amplify the N-terminus of the *rep* gene and its upstream sequence, and the resulting amplicon carries a EcoRI site at its 5′-end. Primers 5′-GGGAAAAGCTTAAAAAACGAATC C-3′ and 5′-GCTCTAGATAATCACACTAACCACATT C-3′ were used to amplify the C-terminus of the gene; the corresponding amplicon carries an XbaI site at its 3′-end. The PCR program for both reactions included 30 cycles of

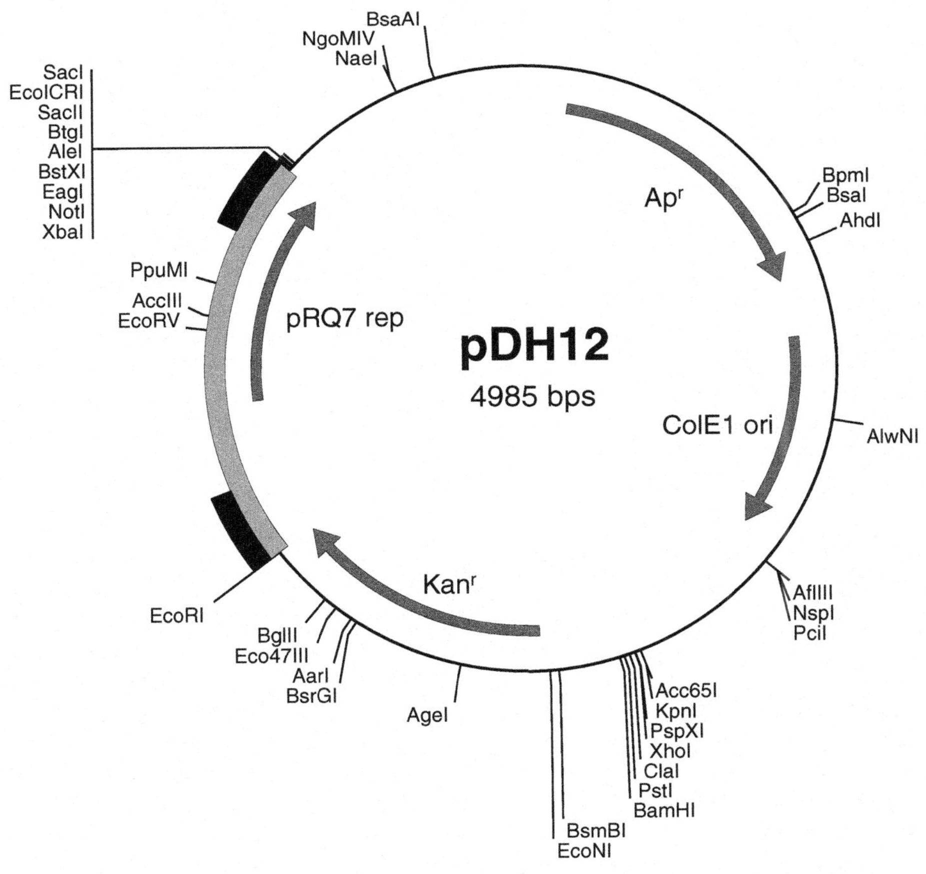

Figure 1 The genetic and physical map of pDH12. The bolded region in grey represents the sequence from pRQ7. The black bars denote the duplicated fragments. Unique enzyme sites are shown.

30 sec at 94°C, 30 sec at 54°C, and 1 min at 72°C. The PCR products were digested with EcoRI-HindIII and HindIII-XbaI, respectively, and were ligated to pKT1, pre-digested with EcoRI-XbaI, to give rise to pDH12. All DNA manipulations followed the standard methods.

Natural transformation of *Thermotoga* spp

Two slightly different approaches were used. The early stage of the study involved the utilization of a nitrogen gas bottle. Briefly, 1 ml of an overnight culture of the recipient strain was transferred into 50 ml of fresh SVO medium and incubated to various growth stages. One ml of the fresh culture was then transferred to a 25 ml sterile serum bottle filled with N_2 gas. At the same time, 50 μg of substrate DNA was added. After 6 h of incubation at 77°C with gentle agitation (100 to 125 rpm), the culture was mixed with hot SVO Gelrite medium and poured into petri dishes, with a supplement of 250 μg/ml kanamycin for selection. The plates were examined after 2 days of incubation at 77°C. Because the purpose of this study is to develop a simple method to transform *Thermotoga* spp., we intentionally did not apply DNase (to remove untransformed DNA after a designated period of time). This not only simplifies the overall procedure but also maximizes the occurrence of transformation events.

In the late stage of the study, the procedure was optimized as such: cells of 1 ml of the overnight culture of a recipient strain were collected by centrifugation, resuspended in 200 μl of fresh SVO medium, and were injected into a 100-ml serum bottle containing 10 ml of fresh SVO medium; DNA substrate was added to a final concentration ranging from 5 to 10 μg/ml; the culture-DNA mixture was incubated at 77°C for 4 to 6 h with gentle agitation and then poured into selective plates (with a culture to medium ratio of 1 to 9, v/v). Transformation efficiency was computed as transformants per microgram of DNA, and transformation frequency was expressed as transformants per microgram of DNA per viable cell. The number of viable cells was determined by plating the control culture (no DNA added) with plain SVO solid medium. Plating efficiency was calculated as the ratio of viable cells to the total number of cells counted under microscope with Hausser Bright-Line 3120 Hemacytometer (Horsham, PA, USA).

Results

Stability of DNA in the transformation environment

In natural transformation, naked DNA is directly exposed to the environment for many hours and is prone to degradation due to various physical, chemical, and biological forces, such as heat, pH, and nucleases. To maximize the chance of transformation, one would wish to incubate the DNA with the recipient cells for as long as possible. On the other hand, if the DNA has been degraded into small oligonucleotides, extended incubation will not lead to additional transformation events but instead cause the already transformed cells to age, which compromises plating efficiency and ultimately transformation frequency. Propagation of sibling transformants could be another concern. To find out how long heterologous DNA can survive under the growth conditions of *Thermotoga*, genomic DNA from *Caldicellulosiruptor saccharolyticus*, a Gram-positive bacterium with the optimal growth at 70°C, was incubated in plain SVO medium (pH 8.5) at 77°C for 6 h. Samples were withdrawn at hourly intervals and were analyzed with agarose gels (Figure 2A). Over the course of the experiment, the genomic DNA of *C. saccharolyticus* was graduately degraded into small fragments. By the end of the test the majority of the fragments were in the size range of 6.5 kb to 500 bp. Since the average length of a bacterial gene is ~ 1.1 kb, the size distribution of the *C. saccharolyticus* DNA fragments should cover most of the single genes as well as some operons, making the fragments an excellent pool of substrates for transformation.

Although DNA is fairly stable in pure SVO medium, it may not be so in a growing cell culture, where natural transformation is supposed to take place. The extracellular nucleases secreted by the host cells could expedite the degradation of foreign DNA. To take this biological factor into consideration, we next incubated the genomic DNA of *C. saccharolyticus* in the supernatant of an overnight culture of *T.* sp. strain RQ7, which contains high levels of free nucleases and is also more acidic (pH 5.5). Indeed, the foreign DNA experienced accelerated degeneration (Figure 2B). Within 3 h of incubation, the DNA fragments were around 6.5 kb or smaller, which was equivalent to 6 h of incubation in the pure growth medium. After 6 h, most DNA fragments were 4 kb or smaller. If DNA is added to a cell culture at an earlier growth stage (as was done in later experiments), its degradation profile is expected to fall between the two extreme scenarios shown here (Figure 2). Therefore, in a practical experiment, heterologous DNA may be incubated with *Thermotoga* cells for up to 6 h.

Transformation of *T.* sp. strain RQ7 with *T. neapolitana* chromosomal DNA

Among the three *Thermotoga* isolates available to us, *T. maritima*, *T. neapolitana*, and *T.* sp. strain RQ7, *T. neapolitana* is naturally resistant to kanamycin, whereas the other two are sensitive to it [3,4]. Accordingly, *T. neapolitana* served as the ideal donor strain and the other two the recipients. Transformation outcomes were measured by the acquisition of resistance to kanamycin (Kan^r). In a series of preliminary tests (data not shown), we noticed that when *T. maritima* was used as the recipient strain, a few Kan^r colony were occasionally observed, despite when or whether the sample was treated with DNA.

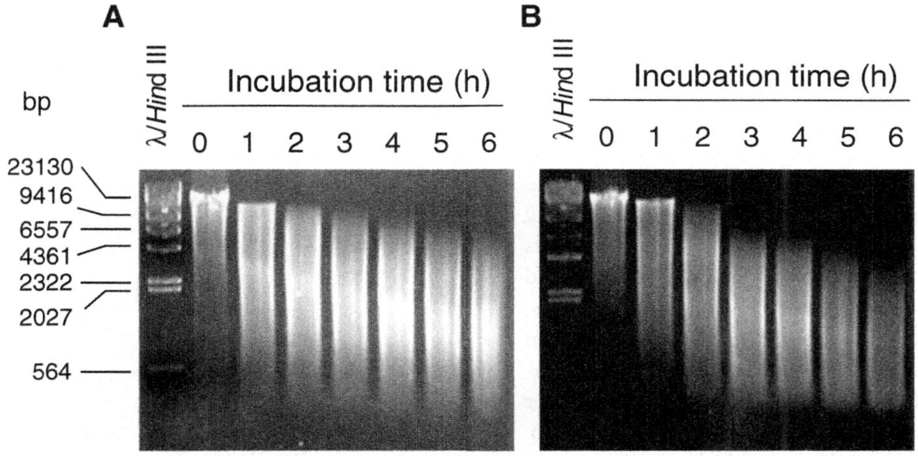

Figure 2 Degradation of *C. saccharolyticus* genomic DNA in SVO medium (A) and in the supernatant of an overnight culture of *T.* sp. strain RQ7 (B).

This indicates that these colonies were spontaneous mutants and, consequently, *T. maritima* was not naturally transformable under the tested conditions. The same was also true for the *T.* sp. strain RQ7 cells in their late log phase or stationary phase. Detailed experiments were then designed to focus on the early log phase of *T.* sp. strain RQ7 (OD_{600} = 0.1 ~ 0.25; Figure 3).

T. sp. strain RQ7 cells were tested at five time points: OD_{600} of 0.10, 0.13, 0.16, 0.19, and 0.24. The first time point did not yield any Kan^r colonies with either the treated sample or the control, probably because cells were not robust enough to be transferred to a N_2 bottle, a process inevitably introduces trace amount of oxygen to the system. For the next four time points, the samples treated with the DNA generated significantly more Kan^r colonies than the controls did (Figure 3). The colony number of the treated samples peaked to 1088 if DNA was added at a cell density of ~ 0.19. If cells were allowed to grow for another 45 min and reached OD_{600} of 0.24, the number of Kan^r colonies dropped to 700, marking the decline of natural competence. The colony numbers from the controls varied from plate to plate, indicating that the mutation events happened randomly and independently in each N_2 bottle rather than in the overnight seed culture or the fresh master culture from where samples were withdrawn.

The above results indicate that *T.* sp. strain RQ7 is most transformable when the cell densities are in the

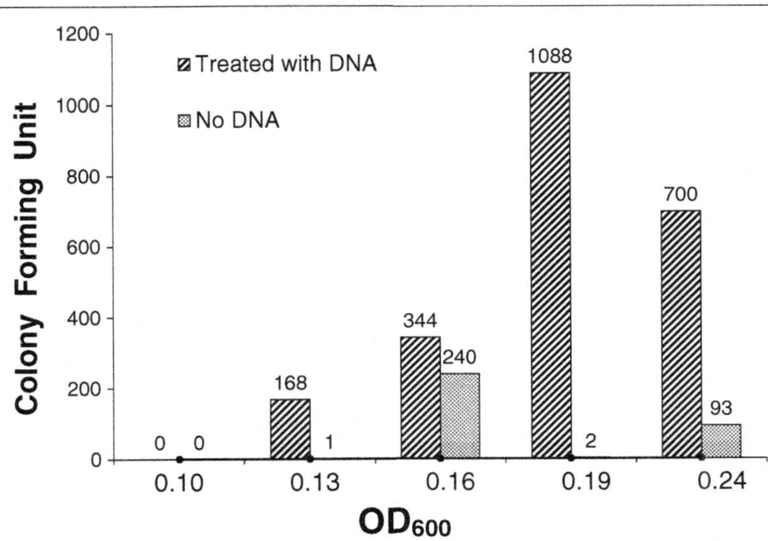

Figure 3 Transformability of *T.* sp. strain RQ7 at various time points of the early exponential phase. The number above each column represents the number of Kan^r colonies.

range of 0.18 - 0.20. We then repeated the experiment four more times under the same conditions (Figure 4). In these tests, the colony numbers fluctuated from time to time. This reflects a persisting challenge of handling *Thermotoga* cultures. Being strict anaerobes, *Thermotoga* cells are very sensitive to oxygen, especially during their active growth, which happens to be when they are mostly competent. A slight deviation in the handling time, especially in the process of transferring host cells to a N_2 bottle, could greatly affect the viability of cells and hence the final results. Nevertheless, in these experiments, at least 5 times more kan^r colonies were obtained from the treated samples compared to the corresponding controls (Figure 4). Because the specific gene(s) responsible for kanamycin resistance in *T. neapolitana* are unknown, we were unable to validate the possible *T.* sp.RQ7 transformants at the molecular level. To address this deficiency and to demonstrate the application of natural competency in *Thermotoga*, we next transformed a shuttle vector to *T.* sp. strain RQ7.

Transformation of *T.* sp. strain RQ7 with pDH10

Attempts were made to introduce the *Thermotoga-E. coli* shuttle vector pDH10 into *T.* sp. strain RQ7 and *T. maritima* through the aforementioned natural transformation procedure. The plasmid DNA was prepared from *E. coli* recombinant strain DH5α/pDH10. A total number of 132 colonies appeared on the transformant plates after 2 days, whereas the control plates had only 20 colonies arising from spontaneous mutations. No colony was observed with the *T. maritima* samples (including the controls), again indicating that this strain is not naturally transformable.

Colonies of 20 potential RQ7/pDH10 transformants were picked for further analyses. Both plasmid and genomic DNA were extracted from four best growing cultures (#2, 5, 6, and 20). An example of the plasmid extract (#20) is shown in Figure 5. Same as before [3],

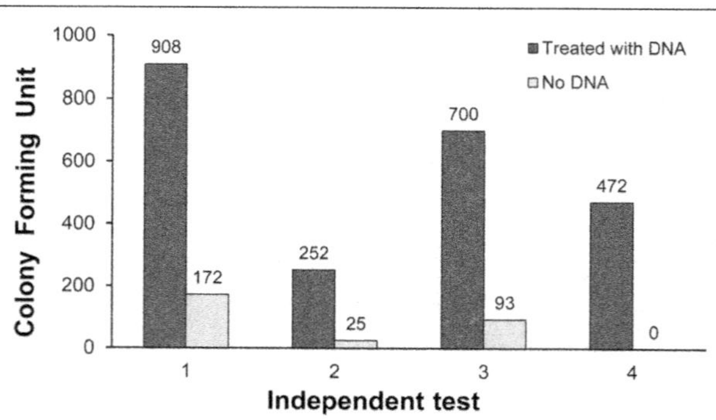

Figure 4 Four more independent experiments of transforming *T.* sp. strain RQ7 at OD_{600} of 0.18 ~ 0.20. The number above each column represents the number of Kan^r colonies.

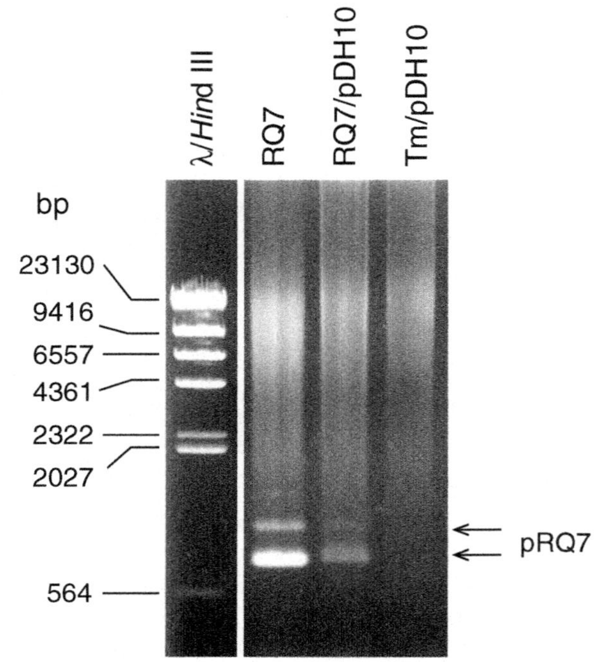

Figure 5 Plasmid extract of a *T.* sp. strain RQ7/pDH10 transformant (denoted as RQ7/pDH10). The host *T.* sp. strain RQ7 was used as the positive control (denoted as RQ7), and the recombinant strain *T. maritima*/pDH10, constructed in the previous study [3], served as the negative control (denoted as Tm/pDH10). The arrows point to the characteristic bands of pRQ7.

bands corresponding to the size of pDH10 (4762 bp) were not observed; however, the characteristic bands of pRQ7 were clearly visible, indicating the plasmid extraction was successful. The plasmid extracts as well as the genomic DNA preparations were then subjected to PCR detection. Amplification using primers unique to the *kan* gene of pDH10 generated the expected band of 778 bp from all plasmid samples. By contrast, among the four genomic samples, only one had the band clearly visible (Figure 6). This implies that the transformed pDH10 survived as an autonomous replicon rather than an integrated part of the chromosome. To further confirm the identity of the amplicons, the PCR product was purified by ethanol precipitation and was digested by AgeI or BglII, which cut the *kan* gene into fragments of 208 + 570 bp and 181 + 597 bp, respectively. The amplicon indeed generated the expected sizes (Figure 7), confirming the authenticity of the PCR product. These data are in agreement with our previous study where the same vector was transformed into *T. maritima* and *T.* sp. strain RQ7 by electroporation and liposome-mediated transformation [3]. The previous data also demonstrated that pDH10 is stably maintained by the hosts [3].

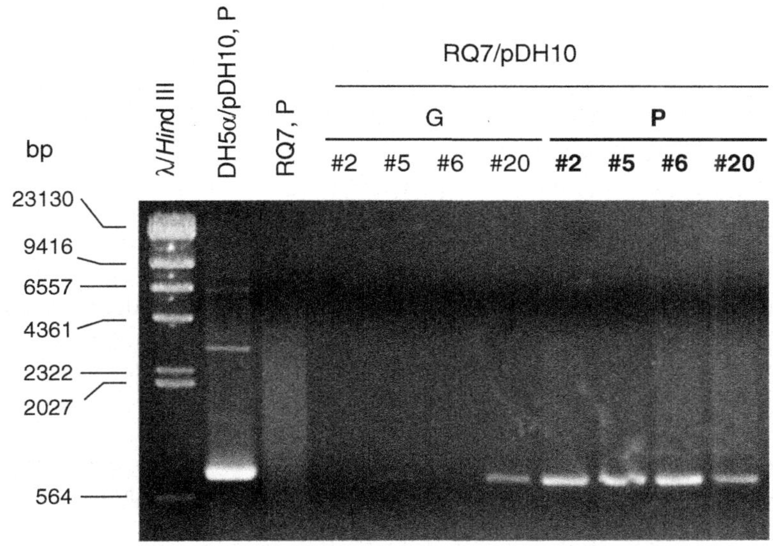

Figure 6 Amplification of the *kan* gene from the genomic DNA (G) and the plasmid DNA (P) prepared from four RQ7/pDH10 transformants. The plasmid extracts from DH5α/pDH10 (first lane after the marker) and *T.* sp. strain RQ7 (second lane after the marker) were used as positive and negative controls.

Optimization of natural transformation

We next devoted our efforts to simplifying the natural transformation procedure. The step of transferring host cultures into a N_2 bottle was eliminated, because it was the major source of oxygen invasion. Instead, DNA was added at the time of setting up the fresh culture from the overnight seed. As a precaution of carrying over excessive nucleases from the overnight supernatant, cells were pelleted by centrifugation and were resuspended in fresh SVO medium prior to inoculation. To reduce the amount of DNA used, the total volume of the recipient culture was cut back from 50 ml to 10 ml and the final concentration of substrate DNA was decreased from 50 μg/ml to 5 μg/ml. Considering that cells may have a difficult time to bind DNA if the cell-DNA mixture was constantly shaken, we attempted to transform static *T.* sp. strain RQ7 cultures but did not get any transformants. Accordingly, the practice of gentle agitation was preserved.

The plating efficiency associated with cells under these physiological conditions were determined. Briefly, cells

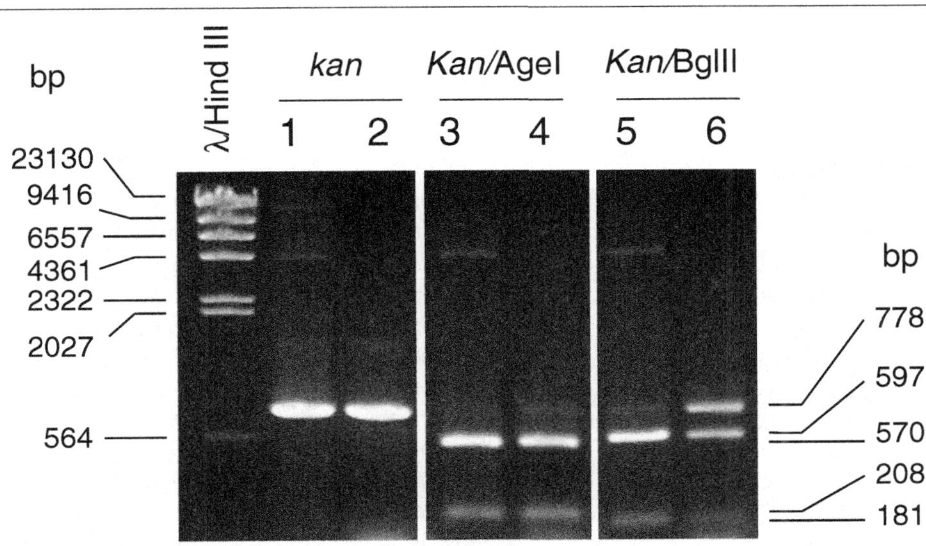

Figure 7 Restriction digestions of the *kan* gene amplicon of a RQ7/pDH10 transformant. Lanes 1, 3, and 5, positive controls prepared from DH5α/pDH10. Lanes 2, 4, and 6, RQ7/pDH10 transformant. Partial digestion is noticed in Lane 6.

from overnight cultures were collected by centrifugation and were used to inoculate plain SVO, after 4 h of growth at 77°C, the culture was then subjected to microscopic cell counting (to determine the total number of cells) and was plated with plain SVO (to find out the number of viable cells). The results of three independent tests (Table 2) indicated that by the end of the transformation there were ~1×10^8 viable cells per ml (CFU/ml), representing more than 70% of all cells found under the microscopy. The observed plating efficiency (73.4 ± 3.4%) was higher than what previously estimated (~50%) [3].

Following the optimized procedure, we measured the transformation efficiency and frequency of *T.* sp. strain RQ7. The donor DNA was prepared from two sources: the chromosome of *T. neapolitana* and the plasmid pDH12, which is another *Thermotoga-E. coli* shuttle vector derived from pKT1 (Figure 1). *T.* sp. strain RQ7 presented a transformation frequency of $(6.03 \pm 1.52) \times 10^{-7}$ (n = 3) with the genomic DNA and a frequency of $(1.45 \pm 0.02) \times 10^{-7}$ (n = 2) with the plasmid DNA; for transformation efficiency, the figures were 16.6 ± 3.9 (n = 3) and 19.3 ± 3.8 (n = 2), respectively.

It was hoped that pDH12 would have a higher copy number than pDH10 because the extra sequence upstream of the *rep* gene may enhance the expression of this gene. However, pDH12 behaved in the same way as pDH10 did. The *kan* gene of pDH12 can be readily amplified by PCR and validated by restriction digestions, but attempts to identify the transformed plasmid by retransformation or inverse PCR have proven unsuccessful. A better understanding of the replication mechanism of pRQ7 will help to formulate new strategies on improving the copy numbers of pDH10 and pDH12.

Discussion

Transformability of *T.* sp. strain RQ7 and *T. maritima*

It is unclear why *T. maritima* is not naturally competent while *T.* sp. strain RQ7 is, even though the two strains are closely related. Because the genome sequence of *T.* sp. strain RQ7 is not publically available yet, we are unable to compare the putative competence genes across the two genomes. We speculate that the putative competency genes may be differentially regulated between the two strains, or some of the genes may have accumulated mutations leading to altered functionalities in one strain. It is also possible that there are novel natural competence genes that do not exhibit significant homology to known ones and therefore are beyond recognition by sequence comparison. More information should emerge once the genome sequence of *T.* sp. strain RQ7 is available. Studies on the expression of the putative competence genes should also yield important clues.

Table 2 Plating efficiency of *T.* sp. strain RQ7 cells

Test	Viable cells (x10^8)	Total cells (x10^8)	Plating efficiency (%)
1	1.20	1.70	70.7
2	0.97	1.25	77.2
3	0.97	1.34	72.3
Average	1.05	1.43	73.4
Standard deviation	0.13	0.24	3.40

Transformation frequency of *T.* sp. strain RQ7

With our optimized procedure, *T.* sp. strain RQ7 presented a transformation frequency in the order of 10^{-7}. The chromosomal DNA and the plasmid DNA transformed this strain with comparable frequency, even though the plasmid DNA contained a higher mass ratio of the relevant gene(s), *i.e.* kanamycin-resistance gene(s). This is probably due to host restriction-modification systems, which selectively degrade DNA of distantly related species. A Type II RM system has been characterized in *T. neapolitana* and has also been found in *T. maritima*, *T. petrophila*, *T.* sp. strain RQ2, and *T. naphthophila* [33]. In addition, REBASE has identified a few putative methylases that might be involved in restriction-modification mechanisms in *Thermotoga*, such as M.TspRQ2ORF1808P in *T.* sp. strain RQ2 as well as M.TneDDamP and M.TneDORF1590P in *T. neapolitana* (http://rebase.neb.com). Because restriction nucleases have a low level of sequence conservation in general, it is much harder to predict their presence through bioinformatics analyses than finding methylases. Plasmid incompatibility is unlikely to be the primary reason for the lower than expected transformation frequency associated with pDH12, because plasmids containing the same origin of replication (*ori*) are more compatible than we previously thought. For instance, two plasmids carrying the same colE1 *ori* yet different antibiotic markers – kanamycin versus ampicillin – were found to be 100% compatible after five consecutive transfers under just kanamycin selection [34].

The level of competency observed in *T.* sp. strain RQ7 is comparable to what reported for *Ralstonia solanacearum*, whose frequency is in the order of 10^{-7} [35], but lower than those measured in well-characterized system such as *Vibrio cholera* [36] and *Helicobacter pylori* [37], where a frequency of 10^{-5} or higher can often be obtained. It did not escape our attention that none of the putative competence genes found in *Thermotoga* are clustered. In many model systems, competence genes are often co-transcribed, such as the *com* operons in *Bacillus* and *Streptococcus* [15,38] and the *pil* operons in *Haemophilus* and *Thermus* [39]. The lack of the similar operons in *Thermotoga* suggests that their relevant molecular machinery either is not well-established or, to a lesser extent, represent a different type of machinery that does not share much similarity to those previously discovered. Nevertheless, the level of competence observed in *T.* sp. strain RQ7 sufficiently satisfies our engineering needs. As a matter of fact, we have been introducing recombinant DNA into *T.* sp. strain RQ7 exclusively by natural transformation ever since we discovered this mechanism.

Conclusions

Natural competency has been established in *T.* sp. strain RQ7. A natural transformation procedure has been developed and optimized for *Thermotoga* spp.. This study provides an alternative way to transform these genetically recalcitrant yet biotechnologically attractive organisms.

Abbreviations

Ap: Ampicillin; bp: Base pairs; CFU: Colony forming unit; DNA: Deoxyribonucleic acid; Kan: Kanamycin; *ori*: Origin of replication; PCR: Polymerase chain reaction; RQ7: *Thermotoga* sp. RQ7; Tm: *Thermotoga maritima*; Tn: *Thermotoga neapolitana*.

Competing interests

The authors declare that they have no competing interests.

Authors' contributions

ZX conceived and coordinated the study and drafted the manuscript. DH discovered natural competence in *T.* sp. strain RQ7 and developed the initial transformation protocol. HX carried out the rest experiments. RP independently validated the results of transforming *T.* sp. strain RQ7 with pDH10. All authors participated in experimental designs and data analyses. All authors read and approved the final manuscript.

Acknowledgements

We thank Dr. Robert Kelly at North Carolina State University for providing *C. saccharolyticus*. This work was supported by the BGSU Commercialization Catalyst Award and the BGSU Building Strength Award.

References

1. Schroder C, Selig M, Schonheit P: **Glucose fermentation to acetate, CO2 and H2 in the anaerobic hyperthermophilic eubacterium thermotoga maritima: involvement of the embden-meyerhof pathway.** *Arch Microbiol* 1994, **161**(6):460–470.
2. Selig M, Xavier KB, Santos H, Schonheit P: **Comparative analysis of embden-meyerhof and entner-doudoroff glycolytic pathways in hyperthermophilic archaea and the bacterium thermotoga.** *Arch Microbiol* 1997, **167**(4):217–232.
3. Han D, Norris SM, Xu Z: **Construction and transformation of a Thermotoga-E coli shuttle vector.** *BMC Biotechnol* 2012, **12**:2.
4. Yu JS, Vargas M, Mityas C, Noll KM: **Liposome-mediated DNA uptake and transient expression in Thermotoga.** *Extremophiles* 2001, **5**(1):53–60.
5. Schwarzenlander C, Averhoff B: **Characterization of DNA transport in the thermophilic bacterium Thermus thermophilus HB27.** *FEBS J* 2006, **273**(18):4210–4218.
6. Smith HO, Gwinn ML, Salzberg SL: **DNA uptake signal sequences in naturally transformable bacteria.** *Res Microbiol* 1999, **150**(9–10):603–616.
7. Dubnau D: **Binding and transport of transforming DNA by Bacillus subtilis: the role of type-IV pilin-like proteins–a review.** *Gene* 1997, **192**(1):191–198.
8. Dubnau D: **DNA uptake in bacteria.** *Annu Rev Microbiol* 1999, **53**:217–244.
9. Averhoff B, Friedrich A: **Type IV pili-related natural transformation systems: DNA transport in mesophilic and thermophilic bacteria.** *Arch Microbiol* 2003, **180**(6):385–393.
10. Claverys JP, Martin B: **Bacterial "competence" genes: signatures of active transformation, or only remnants?** *Trends Microbiol* 2003, **11**(4):161–165.
11. Berge M, Mortier-Barriere I, Martin B, Claverys JP: **Transformation of Streptococcus pneumoniae relies on DprA- and RecA-dependent protection of incoming DNA single strands.** *Mol Microbiol* 2003, **50**(2):527–536.
12. Berge M, Moscoso M, Prudhomme M, Martin B, Claverys JP: **Uptake of transforming DNA in gram-positive bacteria: a view from streptococcus pneumoniae.** *Mol Microbiol* 2002, **45**(2):411–421.
13. Mortier-Barriere I, Velten M, Dupaigne P, Mirouze N, Pietrement O, McGovern S, Fichant G, Martin B, Noirot P, Le Cam E, Polard P, Claverys JP: **A key presynaptic role in transformation for a widespread bacterial protein: DprA conveys incoming ssDNA to RecA.** *Cell* 2007, **130**(5):824–836.
14. Gwinn ML, Ramanathan R, Smith HO, Tomb JF: **A new transformation-deficient mutant of Haemophilus influenzae Rd with normal DNA uptake.** *J Bacteriol* 1998, **180**(3):746–748.
15. Johnsborg O, Havarstein LS: **Regulation of natural genetic transformation and acquisition of transforming DNA in Streptococcus pneumoniae.** *FEMS Microbiol Rev* 2009, **33**(3):627–642.
16. Meibom KL, Blokesch M, Dolganov NA, Wu CY, Schoolnik GK: **Chitin induces natural competence in Vibrio cholerae.** *Sci* 2005, **310**:1824–1827.
17. Mironczuk AM, Kovacs AT, Kuipers OP: **Induction of natural competence in Bacillus cereus ATCC14579.** *J Microbial Biotechnol* 2008, **1**:226–235.
18. Tsen SD, Fang SS, Chen MJ, Chien JY, Lee CC, Tsen DH: **Natural plasmid transformation in Escherichia coli.** *J Biomed Sci* 2002, **9**(3):246–252.
19. Cava F, Hidalgo A, Berenguer J: **Thermus thermophilus as biological model.** *Extremophiles* 2009, **13**(2):213–231.
20. Nelson KE, Clayton RA, Gill SR, Gwinn ML, Dodson RJ, Haft DH, Hickey EK, Peterson JD, Nelson WC, Ketchum KA, McDonald L, Utterback TR, Malek JA, Linher KD, Garrett MM, Stewart AM, Cotton MD, Pratt MS, Phillips CA, Richardson D, Heidelberg J, Sutton GG, Fleischmann RD, Eisen JA, White O, Salzberg SL, Smith HO, Venter JC, Fraser CM: **Evidence for lateral gene transfer between archaea and bacteria from genome sequence of thermotoga maritima.** *Nature* 1999, **399**(6734):323–329.
21. Nesbo CL, Doolittle WF, Mongodin EF, Nelson KE: **Outside forces helped shape the Thermotoga metagenome.** *Microbe* 2006, **1**(5):235–241.
22. Noll KM, Thirangoon K: **Interdomain transfers of sugar transporters overcome barriers to gene expression.** *Methods Mol Biol* 2009, **532**:309–322.
23. Zhaxybayeva O, Swithers KS, Lapierre P, Fournier GP, Bickhart DM, DeBoy RT, Nelson KE, Nesbo CL, Doolittle WF, Gogarten JP, Noll KM: **On the chimeric nature, thermophilic origin, and phylogenetic placement of the thermotogales.** *Proc Natl Acad Sci U S A* 2009, **106**(14):5865–5870.
24. Harriott OT, Huber R, Stetter KO, Betts PW, Noll KM: **A cryptic miniplasmid from the hyperthermophilic bacterium thermotoga Sp strain Rq7.** *J Bacteriol* 1994, **176**(9):2759–2762.
25. Akimkina T, Ivanov P, Kostrov S, Sokolova T, Bonch-Osmolovskaya E, Firman K, Dutta CF, McClellan JA: **A highly conserved plasmid from the extreme thermophile thermotoga maritima MC24 is a member of a family of plasmids distributed worldwide.** *Plasmid* 1999, **42**(3):236–240.
26. Nesbo CL, Dlutek M, Doolittle WF: **Recombination in thermotoga: implications for species concepts and biogeography.** *Genetics* 2006, **172**(2):759–769.
27. Belkin S, Wirsen CO, Jannasch HW: **A new sulfur-reducing, extremely thermophilic eubacterium from a submarine thermal vent.** *Appl Environ Microbiol* 1986, **51**(6):1180–1185.
28. Huber R, Langworthy TA, Konig H, Thomm M, Woese CR, Sleytr UB, Stetter KO: **Thermotoga maritima sp. nov. represents a new genus of unique extremely thermophilic eubacteria growing up to 90 degrees C.** *Arch Microbiol* 1986, **144**(4):324–333.
29. Sissons CH, Sharrock KR, Daniel RM, Morgan HW: **Isolation of cellulolytic anaerobic extreme thermophiles from new zealand thermal sites.** *Appl Environ Microbiol* 1987, **53**(4):832–838.
30. Grant SG, Jessee J, Bloom FR, Hanahan D: **Differential plasmid rescue from transgenic mouse DNAs into escherichia coli methylation-restriction mutants.** *Proc Natl Acad Sci U S A* 1990, **87**(12):4645–4649.
31. Lasa I, Caston JR, Fernandez-Herrero LA, de Pedro MA, Berenguer J: **Insertional mutagenesis in the extreme thermophilic eubacteria thermus thermophilus HB8.** *Mol Microbiol* 1992, **6**(11):1555–1564.
32. Van Ooteghem SA, Beer SK, Yue PC: **Hydrogen production by the thermophilic bacterium thermotoga neapolitana.** *Appl Biochem Biotechnol* 2002, **98**:177–189.
33. Xu Z, Han D, Cao J, Saini U: **Cloning and characterization of the TneDI restriction: modification system of thermotoga neapolitana.** *Extremophiles* 2011, **15**(6):665–672.
34. Velappan N, Sblattero D, Chasteen L, Pavlik P, Bradbury AR: **Plasmid incompatibility: more compatible than previously thought?** *Protein Eng Des Sel* 2007, **20**(7):309–313.

35. Mercier A, Bertolla F, Passelegue-Robe E, Simonet P: **Influence of DNA conformation and role of comA and recA on natural transformation in Ralstonia solanacearum.** *Can J Microbiol* 2009, **55**(6):762–770.
36. Lo Scrudato M, Blokesch M: **The regulatory network of natural competence and transformation of Vibrio cholerae.** *PLoS Genet* 2012, **8**(6):e1002778.
37. Lin EA, Zhang XS, Levine SM, Gill SR, Falush D, Blaser MJ: **Natural transformation of helicobacter pylori involves the integration of short DNA fragments interrupted by gaps of variable size.** *PLoS Pathog* 2009, **5**(3):e1000337.
38. Londono-Vallejo JA, Dubnau D: **comF, a bacillus subtilis late competence locus, encodes a protein similar to ATP-dependent RNA/DNA helicases.** *Mol Microbiol* 1993, **9**(1):119–131.
39. Friedrich A, Prust C, Hartsch T, Henne A, Averhoff B: **Molecular analyses of the natural transformation machinery and identification of pilus structures in the extremely thermophilic bacterium thermus thermophilus strain HB27.** *Appl Environ Microbiol* 2002, **68**(2):745–755.

7

Flow cytometry-based functional selection of RNA interference triggers for efficient epi-allelic analysis of therapeutic targets

David R Micklem[1,2†], Magnus Blø[1,2†], Petra Bergström[3], Erlend Hodneland[1], Crina Tiron[1], Torill Høiby[1], Christine Gjerdrum[1], Ola Hammarsten[3] and James B Lorens[1*]

Abstract

Background: The dose-response relationship is a fundamental pharmacological parameter necessary to determine therapeutic thresholds. Epi-allelic hypomorphic analysis using RNA interference (RNAi) can similarly correlate target gene dosage with cellular phenotypes. This however requires a set of RNAi triggers empirically determined to attenuate target gene expression to different levels.

Results: In order to improve our ability to incorporate epi-allelic analysis into target validation studies, we developed a novel flow cytometry-based functional screening approach (CellSelectRNAi) to achieve unbiased selection of shRNAs from high-coverage libraries that knockdown target gene expression to predetermined levels. Employing a Gaussian probability model we calculated that knockdown efficiency is inferred from shRNA sequence frequency profiles derived from sorted hypomorphic cell populations. We used this approach to generate a hypomorphic epi-allelic cell series of shRNAs to reveal a functional threshold for the tumor suppressor p53 in normal and transformed cells.

Conclusion: The unbiased CellSelectRNAi flow cytometry-based functional screening approach readily provides an epi-allelic series of shRNAs for graded reduction of target gene expression and improved phenotypic validation.

Background

Delineating gene function relies on the genetic principle established by decades of genetic mapping that mutations establish causality in phenotypes. *Trans*-dominant genetic effectors such as small molecule inhibitors and small interfering RNA (siRNA) that alter the expression and function of specific genes and proteins *in trans*, have dramatically enhanced our ability to correlate the extent of cell behavioral changes with the perturbation of gene expression and function. Pharmacological dose response studies correlate the degree of compound-induced inhibition of protein function with cellular responses. Similarly, hypomorphic alleles, mimicked by RNAi, have served as valuable tools for deciphering the roles of various genes in cell function [1,2]. Such epi-allelic hypomorphic analysis using different RNAi triggering sequences empirically determined to knockdown gene expression to distinct levels, provides a unique opportunity to correlate gene expression levels with phenotypic outcome [3]. Hence, epi-allelic analysis can provide a dose response profile for putative therapeutic targets in lieu of small molecule inhibitors.

In order to improve our ability to conduct epi-allelic analysis, we developed a robust flow cytometry-based functional screening approach to enable unbiased selection of shRNAs that generate graded knockdown of target gene expression to predetermined levels. This genetic dose-response can be used to validate candidate drug targets and predict *in vivo* responses to inhibitor compounds.

* Correspondence: jim.lorens@biomed.uib.no
†Equal contributors
[1]Department of Biomedicine, University of Bergen, N-5009 Bergen, Norway
Full list of author information is available at the end of the article

Methods

Cell culture

All cells were cultured in a humidified atmosphere at 37°C with 5% CO_2. Phoenix A retroviral packaging cells (Dr. Garry Nolan, Stanford University, USA) and MDA-MB-231 cells (ATCC) were maintained in DMEM (Gibco) and F-12 K nutrient mixture (*Kaighn's modification*) respectively, both supplemented with 10% fetal bovine serum (FBS) (EuroClone, Italy), 2 mM L-glutamine, 100 U/ml penicillin and 100 μg/ml streptomycin (all from Sigma-Aldrich) [4]. Leukemic cell line HL60 (gift from Dr. Bjørn T. Gjertsen; University of Bergen, Norway) were cultured in Iscoves medium containing 10% FBS, antibiotics and glutamine (Sigma-Aldrich) [5]. Primary human umbilical vein endothelial cells (HUVEC) (C2517A, Lonza, Switzerland) were maintained in the supplier's recommended complete medium (EGM-2).

Retroviral transductions

Phoenix A cells were transfected using the calcium phosphate method [4]. Approximately 30 h after transfection, the medium was changed to the appropriate growth medium for the cells to be infected. Infectious supernatant was harvested 48 h after transfection. Target cells were spin-infected in virus-containing supernatant supplemented with 5 μg/ml protamine sulfate for 90 min at 1200 × g, and then incubated overnight before being returned to regular growth medium. Infected cells were selected with 1 μg/ml puromycin for 48 h.

Immunostaining and flow cytometry

The MDA-MB-231 cells were trypsinized using standard procedures and washed with 0.2% BSA/PBS. Cells were stained with anti-Axl antibody (MAB154; R&D Systems, USA) at a final concentration of 2 μg/ml in 0.2% BSA/PBS for 40 min. at room temperature. The cells were then washed twice in 0.2% BSA/PBS and incubated with secondary antibody (goat anti-mouse allophyocyanin-crosslinked; Molecular Probes, USA) at a final concentration of 0.2 μg/ml for 30 min. at room temperature in the dark. The cells were washed twice with 0.2% BSA/PBS and resuspended in 300 μl of 0.2% BSA/PBS before analysis on a FACSAria cell sorter (BD Biosciences, USA). Data analyses were carried out with FlowJo software (Tree Star, USA).

Constructs

Functional shRNAs that were identified from the library screen, were recloned into a new vector. The shRNAs were expressed from a modified human U6 promoter in the LTR of the retroviral vectors RRI-Red/L087 (GenBank: EU424173). For sequence information on shAxl2, shAxl278, shAxl279, shAxl280 and shLuc see [2].

The retroviral expression vector pCSI expresses a puromycin-N-acetyl-transferase-EGFP and firefly luciferase cassette (Puro2A- GFP2ALuc2), with each open-reading frame separated by a linker encoding the 2A region (XXSGLRSGQLLNFDLLKLAGDVES- NPGP) from foot-and-mouth disease virus for stoichiometric co-translational polyprotein cleavage [6].

Construction of the reporter gene vector

Reporter constructs were made by cloning the target cDNA (Axl: 2.9 kb EcoRI/BamHI fragment from IMAGE cDNA clone 5205825; p53: 1.5 kb PshA1/NotI fragment from pSVp53 (gift from B.T. Gjertsen) into the 3′-untranslated region of retroviral vector L101 (Complete sequence in Additional file 1: Figure S1).

shRNA library construction

The construction of the Axl library has been described previously [2]. The p53 library (p53 cDNA kindly provided by BT Gjertsen, University of Bergen) was produced by a new method (see Additional file 2: Figure S2), generating a larger number of longer hairpins. p53 cDNA was fragmented with each of the following frequently-cutting enzymes: HpaII/Hin6I (mixed), AluI, DpnI, BsuRI, RsaI, HpyCH4V and blunt-ended with Klenow. The fragments in each digestion mix were then capped at each end by ligation with excess Hairpin oligo DK540 (PhoCTGCTG GATCCAGAGACTCCAGAATTGTGAGCGCTCACAA TTCTGGAGATGAGCCTTCGGTCTCATCTCTGGAT CCAGCAG). PvuII and an appropriate frequent-cutter enzyme were added to each mixture to inhibit unwanted side reactions. After overnight ligation, reactions were pooled, phenol extracted, chloroform extracted and size-selectively precipitated with 15% PEG 8000/0.1 M $MgCl_2$/ 20 μg glycogen to get rid of excess DK540 [7,8] Hairpin structures comprising 27 nt of p53-derived sequence joined to hairpin oligo DK540 were released from each end of the capped fragments by digestion with EcoP15I, dephosphorylated with Antarctic Phosphatase (NEB) and cleaned up by phenol/chloroform extraction and ethanol precipitation. Following blunt-ending with Klenow the hairpin loops (apparent MW approx 80 bp) were run on a 2% agarose/5 mM sodium borate gel [9] and purified by electroelution onto GF/C paper (Whatman).

The hairpins were cloned and converted into hairpin expression cassettes in a single step: Dephosphorylated hairpins were ligated to NaeI/NarI-cut L267 pEntr-U6NaeIXcmI, a kanamycin resistant derivative of pDONR221 carrying a PacI-U6 promoter-Termination signal-HindIII cassette, and then cut with XcmI. The relatively small size of this vector compared to retroviral vector L087 simplifies the cloning and gives improved transformation efficiencies. The product of ligation contains a nick at the 5′ end of the hairpin due to the lack of 5′ phosphate. Extension from the free 3′OH at this nick with a strand-displacing polymerase (Klenow) opens out the hairpin structure into dsDNA and at the same time removes the single nucleotide 3′ XcmI overhang. The construct was then recircularised and electroporated into competent cells and plated. This primary library (many thousand colonies) was scraped up and plasmid DNA isolated. The size of the hairpin loop was reduced by digestion with GsuI, blunt-ending with Klenow, gel-purification, recircularisation and re-cloning. Finally, the complete U6:hairpin:termination signal cassette was transferred into retroviral vector L297, a slight modification of vector L087. Restriction enzymes and polymerases were from New England Biolabs or Fermentas.

Isolation of p53 targeting shRNAs

To select shRNAs that target human p53 mRNA from the approximately 100 unique shRNA sequences, we prepared virus supernatants from the p53 shRNA library and control shLuc vectors as described above. Three independent populations of 2×10^5 HL60 RNAi p53 reporter cells were infected with virus at a transduction efficiency below 10% to ensure favor single retroviral integrations per cell, with approximately 200 independent integrations per shRNA contstruct. After 48 h of Puromycin selection and another 2 days of culture, mCherry positive (as a marker for library infected cells) cells displaying reduced GFP levels were sorted using a FACSAria cell sorter. One hundred cells per well were sorted directly into a 96-well PCR microplate (FS-C, Axygen Bioscienes) containing 10 μl 1× Thermopol buffer/proteinase K (100 μg/ml) (Sigma-Aldrich) per well. After sorting, the PCR plate was incubated for 30 min at 37°C, and then the enzyme was inactivated at 95°C for 10 min. Each well was added 10 μl 2× PCR mix (see below) to a final volume of 20 μl.

Amplification of shRNA pools from sorted cells

PCR mixture

The PCR mixture for amplifying vector genomic DNA in a total volume of 20 μl contained 1.3 U and 0.1 U of Taq/Pfu DNA Polymerase (Fermentas), 1× Thermopol buffer (containing 20 mM of Tris–HCl (pH 8.8), 10 mM of KCl, 10 mM of (NH4) 2SO4, 2 mM of MgSO4, 0.1% of Triton X-100) (NEB), 0.25 μM of each primer, 0.375 mM dNTP mix (Fermentas). The reaction mixture for the second amplification round was the same as for the first one, except that 0.5 μM of the "inner" primers with tags were used instead of the "outer" primers. In the second amplification 2 μl of the first round PCR product was used and the reaction was also supplemented with 2.5 mM $MgCl_2$. See Table 1 for primer information.

Nested PCR

PCR thermal cycles were carried out using a thermal cycler (PTC-100, MJ Research) to amplify vector genomic DNA using the following sequences: preheat at 95°C for 5 min; 35 cycles of 95°C for 30 s, 58°C for 20 s and 72°C for 40 s. After 35 cycles of first round PCR, the second round PCR thermal cycles were performed: preheat at 95°C for 2 min; 20 cycles of 95°C for 20 s, 56°C for 20 s and 72°C for 30 s; additional extension at 72°C for 5 min to end the amplification for both programs. DNA was analyzed by 2% agarose gel electrophoresis and stained by ethidium bromide. 9 μl of each nested PCR were pooled, run on agarose gel and the PCR band (ca. 280 nt) was purified with Illustria GFX PCR DNA and gel band purification kit (GE Healthcare) before 454 sequencing (Roche) at The Norwegian High-Throughput Sequencing Centre (University of Oslo).

454 sequence data analysis

Raw FASTA format 454 sequence data were analysed using a series of simple Perl scripts (D. Micklem, unpublished). Initially, the sequences were scanned to determine orientation (sequencing off A or B-primer) and to identify the multiplex tag associated with each primer. Primer and constant vector sequences were then deleted and empty/too short sequences excluded. Remaining sequences were tested to determine whether they encoded a hairpin structure. Finally, the sequences were clustered into groups of near-identical sequences and the number of times each sequence was seen in each tag-group was tabulated.

Immunoblotting

Cells were lysed in RIPA buffer (25 mM Tris-HCl pH 7.6, 150 mM NaCl, 1% NP-40, 1% sodium deoxycholate, 0.1% SDS) supplemented with protease inhibitor (13457200; Roche). After addition of lysis buffer, the extract was incubated for 5 min on ice, and then passed through a 29 gauge syringe. The suspension was centrifuged for 10 min at 10000 rpm and the supernatant protein concentration was measured using a BCA protein assay kit (#23227, Pierce). SDS/PAGE (Novex NuPAGE Bis-Tris 4-12% gels and MES buffer) and protein transfer were carried out according to standard procedures given by the manufacturer (Life Technologies). 20 μg protein was loaded on the gel. Western blots were developed with ECL substrate (#32106, Pierce) and the signal was detected and analyzed using the BioRad ChemiDoc XRS + system. p53 was detected using mouse monoclonal anti-human p53 antibody (ab1101, Abcam) and Glyceraldehyde 3-phosphate dehydrogenase (GAPDH) loading control was detected with Millipore MAB379 antibody (clone 6C5). Secondary antibody Goat anti-mouse (GAM) IgG HRP (#G21040, Invitrogen).

Intensity of individual bands was quantified using ImageJ densitometry software, andexpressed relative to GAPDH signal, as a measure of protein relative abundance in the different samples.

Irradiation

L108 Luc, shp53-3, shp53-10 and shp53-13 HUVEC cells were harvested by trypsination and replated in

Table 1 Primers used in the amplification of shRNA pools from sorted cell

Primer sequence	Id	Info
CCGAGGTGGGCAGTCAATCAATCT	DK633outerr	outer primer
ACGCCATTTTGCAAGGCAT	DK637altouterf	outer primer
454-adaptor primers		
Lowercase letters provide a multiplexing code to allow separate PCR reactions to be pooled and sequenced together		
GCCTCCCTCGCGCCATCAGagcTTTCTTGGGTAGTTTGCAGTTT	DK621ACodeAGC	'A' type, nested
GCCTCCCTCGCGCCATCAGtgaTTTCTTGGGTAGTTTGCAGTTT	DK621ACodeTGA	'A' type, nested
GCCTCCCTCGCGCCATCAGcagTTTCTTGGGTAGTTTGCAGTTT	DK621ACodeCAG	'A' type, nested
Type A primers are identical except for the 3 nt's in small letters		
GCCTTGCCAGCCCGCTCAGtactgacTTGTACAAGAAAGCTGGGTAAG	DK622BCodeTActgaC	'B' type, nested
GCCTTGCCAGCCCGCTCAGtacgatcTTGTACAAGAAAGCTGGGTAAG	DK622BCodeTAcgatC	'B' type, nested
GCCTTGCCAGCCCGCTCAGtacagtcTTGTACAAGAAAGCTGGGTAAG	DK622BCodeTAcagtC	'B' type, nested
GCCTTGCCAGCCCGCTCAGtagactcTTGTACAAGAAAGCTGGGTAAG	DK622BCodeTAgactC	'B' type, nested
GCCTTGCCAGCCCGCTCAGtagcatcTTGTACAAGAAAGCTGGGTAAG	DK622BCodeTAgcatC	'B' type, nested
GCCTTGCCAGCCCGCTCAGtacgtacTTGTACAAGAAAGCTGGGTAAG	DK622BCodeTAcgtaC	'B' type, nested
GCCTTGCCAGCCCGCTCAGtagctacTTGTACAAGAAAGCTGGGTAAG	DK622BCodeTAgctaC	'B' type, nested
GCCTTGCCAGCCCGCTCAGtagtcacTTGTACAAGAAAGCTGGGTAAG	DK622BCodeTAgtcaC	'B' type, nested
GCCTTGCCAGCCCGCTCAGcactgagTTGTACAAGAAAGCTGGGTAAG	DK622BCodeCActgaG	'B' type, nested
GCCTTGCCAGCCCGCTCAGcacgatgTTGTACAAGAAAGCTGGGTAAG	DK622BCodeCAcgatG	'B' type, nested
GCCTTGCCAGCCCGCTCAGcacagtgTTGTACAAGAAAGCTGGGTAAG	DK622BCodeCAcagtG	'B' type, nested
GCCTTGCCAGCCCGCTCAGcagactgTTGTACAAGAAAGCTGGGTAAG	DK622BCodeCAgactG	'B' type, nested
GCCTTGCCAGCCCGCTCAGcagcatgTTGTACAAGAAAGCTGGGTAAG	DK622BCodeCAgcatG	'B' type, nested
GCCTTGCCAGCCCGCTCAGcacgtagTTGTACAAGAAAGCTGGGTAAG	DK622BCodeCAcgtaG	'B' type, nested
GCCTTGCCAGCCCGCTCAGcagatcgTTGTACAAGAAAGCTGGGTAAG	DK622BCodeCAgatcG	'B' type, nested
GCCTTGCCAGCCCGCTCAGcagtcagTTGTACAAGAAAGCTGGGTAAG	DK622BCodeCAgtcaG	'B' type, nested
Type B primers are identical except the 7 nt's in small letters.		

12- or 24-well plates 24 h before irradiation. The cells were irradiated on the plate with 2 or 8 Gy using a Philips RT 100 X-ray machine with an acceleration voltage of 100 kV.

mRNA extraction

Irradiated cells were lysed by addition of 360 μl Buffer MRL (Qiagen)/well directly to the 24-well plates. mRNA was extracted and purified with a MagAttract Direct mRNA M48 Kit with oligo(dT) covered magnetic beads on a GenoM-48 Robotic Workstation (Qiagen). Standard settings for mRNA extraction were used with 100 μl elution volumes.

cDNA synthesis

10 μl of mRNA was used as template for cDNA synthesis in a 20 μl reaction. The reaction buffer consisted of First Strand Buffer ×1, 10 mM dithiothretiol (DTT) and 5 U/μl SuperScriptTM II Reverse Transcriptase (Invitrogen) and 1 U/μl Protector RNase Inhibitor, 20 pmole/μl Hexanucleotide Mix and 0.25 mM of each dNTP, Li-Salt (Roche Diagnostics). The RT reaction was performed on a PTC-200 Peltier Thermal Cycler (MJ Research) at 22°C for 10 min, 42°C for 45 min and 99°C for 3 min.

Quantitative PCR

TaqMan® Gene Expression Assays for p53, p21 and POLR2a (Pol II) were purchased from Applied Biosystems (Table 2). All primer products span exon–exon

Table 2 Gene expression assays used for quantitative PCR

Gene	Exon-exon boundary	Cat. number
Cyclin-dependent kinase inhibitor 1A CDKN1A/P21	2-3	Hs00355782_m1
Tumor protein p53 TP53/p53	9-10	Hs00153349_m1
Polymerase (RNA) II (DNA directed) polypeptide A POLR2A	1-2	Hs00172187_m1

All primers/probes were purchased from Applied Biosystems.

junctions to avoid amplification of genomic DNA. QPCR reactions were performed using 5 μl of cDNA in MicroAmp™ 96-well optical microtiter plates on a 7900HT Fast QPCR System in TaqMan® Fast Universal PCR Master Mix (Applied Biosystems) according to protocol (total volume 25 μl). cDNA was diluted 10 times prior to qPCR and all samples were run in duplicate. PCR results were analyzed with SDS 2.3 software (Applied Biosystems) and relative quantity was determined using the ΔΔCT Method [10], with DMSO-treated cells as calibrator and POLR2A as endogenous reference.

Gamma-H2AX staining

Gamma-H2AX detection was performed essentially as previously described [11]. H2AX is a form of Histone 2A that is phosphorylated upon DNA double strand break induction (gamma-H2AX) and can thus be used as a measure of DNA double strand breaks after irradiation. HUVEC cells were seeded in 12-well plates 24 h pre-irradiation. In brief, after irradiation the cells were incubated at 37°C for 30 min to allow complete H2AX phosphorylation and then kept on ice during the whole procedure to avoid phosphatase activity. After trypsination and harvest, the cells were incubated with Block-9 staining buffer containing anti-phospho-H2AX FITC-conjugated antibody (Millipore) and Vybrant Dye Cycle Violet stain (Invitrogen, Molecular Probes) for 3 h on ice. The increase in phosphorylated H2AX was then detected on a FACSAria flow cytometer (BD Biosciences). Excitation wavelengths were 488 nm/405 nm respectively and emission was detected with filter/bandpass 450/40 for FITC and 530/30 for Vybrant Dye Cycle Violet Stain.

EdU staining

HUVEC cells were seeded out in 24-well plates 24 h pre-radiation. After irradiation, the cells were harvested by trypsination, diluted 1:8 and reseeded into 6-well plates. After growing for 48 h, EdU was added to the cells at a concentration of 10 μM. EdU assay, as a measure of cellular proliferation, was performed essentially as earlier described (Ref: Flow cytometry methods for DNA-damage induced clonogenic cell death calibrated to the colony forming assay, Yue Gao, Pegah Johansson, Aida Muslimovic and Ola Hammarsten, submitted manuscript), using a Click-iT® EdU Alexa Fluor® 488 Imaging Kit (Invitrogen). In brief, after 16 h incubation with EdU, the cells were washed with PBS and harvested by trypsination. After a 0.1% BSA/PBS wash, the cells were fixed for 15 min. The cells were washed again in BSA/PBS and then permeabilized for 30 min. After another wash with BSA/PBS, the cells were incubated in staining buffer according to protocol for 30 min, kept from light. After a final wash in 1× component E, the cells were incubated with Ribonuclease A for 30 min. Before Analysis on a FACSAria flow cytometer (BD Biosciences), the cells were diluted in 300 μl BSA/PBS.

Data analysis

Statistical analysis was carried out using ANOVA followed by Tukey's post hoc tests (GraphPad 4.0 Software, San Diego, CA). Significance was assumed for p value less than 0.05. Sigmoidal best fit curve generated with sigmoidal fit equation at http://zunzun.com, using the Hill with offset equation: ax^b/(c^b + x^b) + Offset.

Results and discussion

In an effort to derive RNAi-competent sequences from a target gene we prepared an shRNA-expressing retroviral library based on random 21-26 nt sequences (~250 sequences) derived from the human Axl gene sequence, a receptor tyrosine kinase frequently expressed in human cancer and correlated with poor overall survival [12]. The shRNA library was transduced into MDA-MB-231 breast carcinoma cells; individual cells with reduced Axl cell surface expression were isolated by FACS and the cognate shRNA sequences identified by PCR and deep sequencing. We noted that each sequence displayed a unique, reproducible gene knockdown level that was not predictable by analysis in available RNAi sequence algorithms (Additional file 3: Figure S3). Collectively, these Axl-targeting shRNAs comprised an epi-allelic hypomorphic series in MDA-MB-231 cells that validated the role of Axl in metastasis *in vitro* and *in vivo* [2].

Based on these results we developed an approach to rapidly identify functional shRNA sequences from high-coverage gene specific shRNA libraries to enable epi-allelic hypomorphic analysis of putative therapeutic targets [1]. The natural fluctuation in protein levels routinely displayed by flow cytometric analysis can be modeled by normal probability distributions [13]. Hence, a cell population expressing an shRNA library against a target gene can be represented as a continuous distribution comprising multiple samples. The probability that a cell will be found below a given threshold is proportional to the difference in target gene expression means of the hypomorphic and wild type cell populations (Additional file 4: Figure S4). Importantly, this leads to the prediction that the number of cells expressing a specific shRNA found below this threshold will be related to the potency of the shRNA. This implies that the frequency of different shRNA sequences recovered from a sorted cell population should be proportional to their knockdown levels.

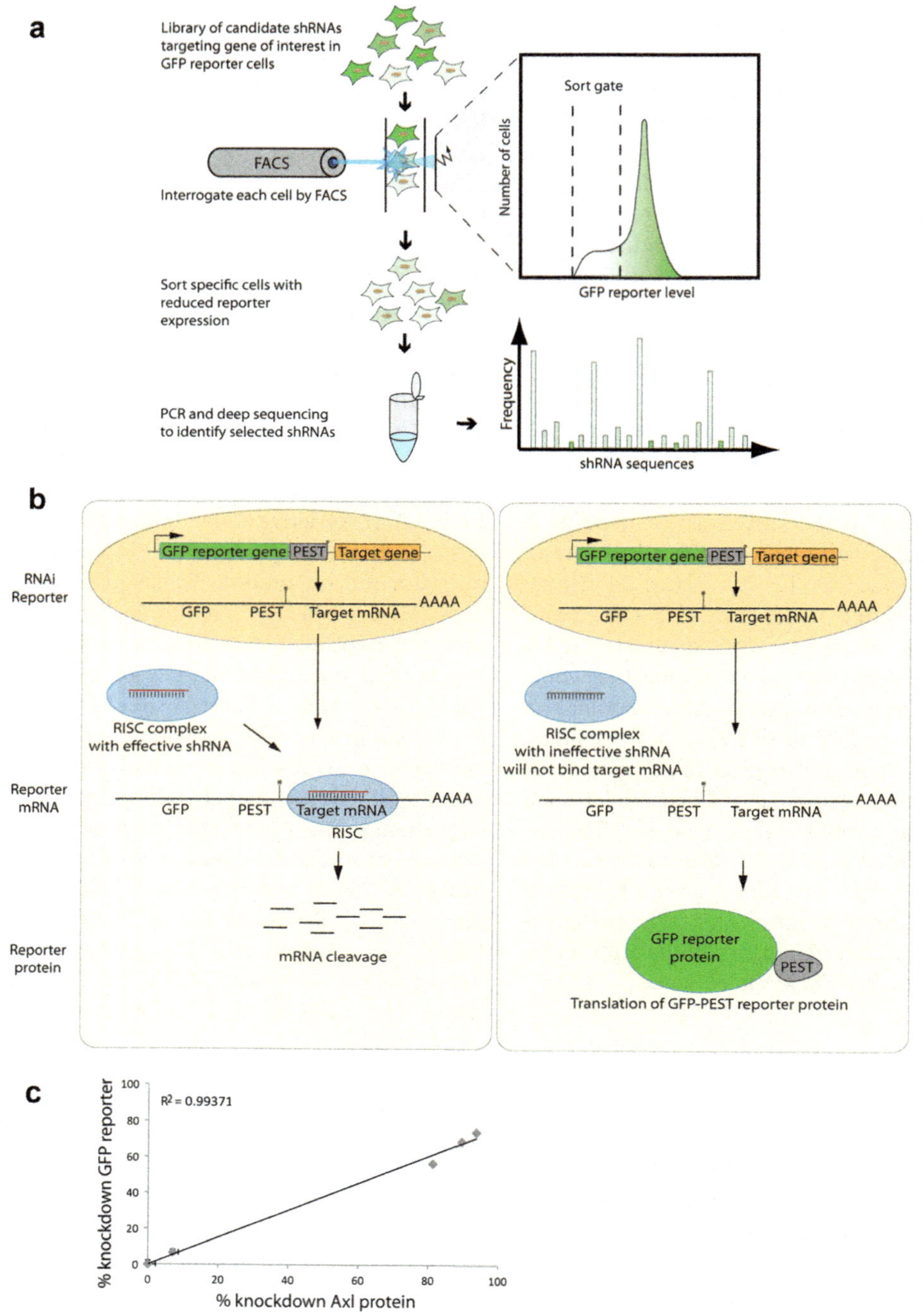

Figure 1 The CellSelectRNAi approach. (a) CellSelectRNAi principle. HL60 reporter cells are transduced with a library of candidate shRNAs targeting gene of interest. Each cell is interrogated by flow cytometry; cells with reduced reporter levels are sorted for subsequent PCR and shRNA sequence identification by 454 deep sequencing. Sequences are plotted by frequency representation in the sorted population. **(b)** Schematic of the CellSelectRNAi system. The RNAi reporter comprises a destabilized GFP (dsGFP), with the untranslated target gene cloned downstream of a translational stop codon (*). In the scenario on the left, a RISC complex is loaded with an effective shRNA that will bind to the target mRNA sequence in the reporter mRNA and initiate mRNA cleavage and prevention of dsGFP reporter translation. In the scenario on the right, the RISC complex is loaded with an ineffective shRNA that is unable to bind the target mRNA. The efficiency of each shRNA is proportional to the amount of dsGFP in the cell, and can be measured by flow cytometry. **(c)** RNAi reporter correlates with endogeneous gene expression. Flow cytometric analysis showing the level of RNAi reporter and endogeneous Axl cell surface protein in each epi-allelic MDA-MB-231 cell line demonstrates a linear relationship between RNAi reporter intensity and Axl protein stained with Axl antibody. Each point on the graph represents the geometric mean and the graph is representative of four individual experiments. The error bars represent 95% confidence-limit on the mean.

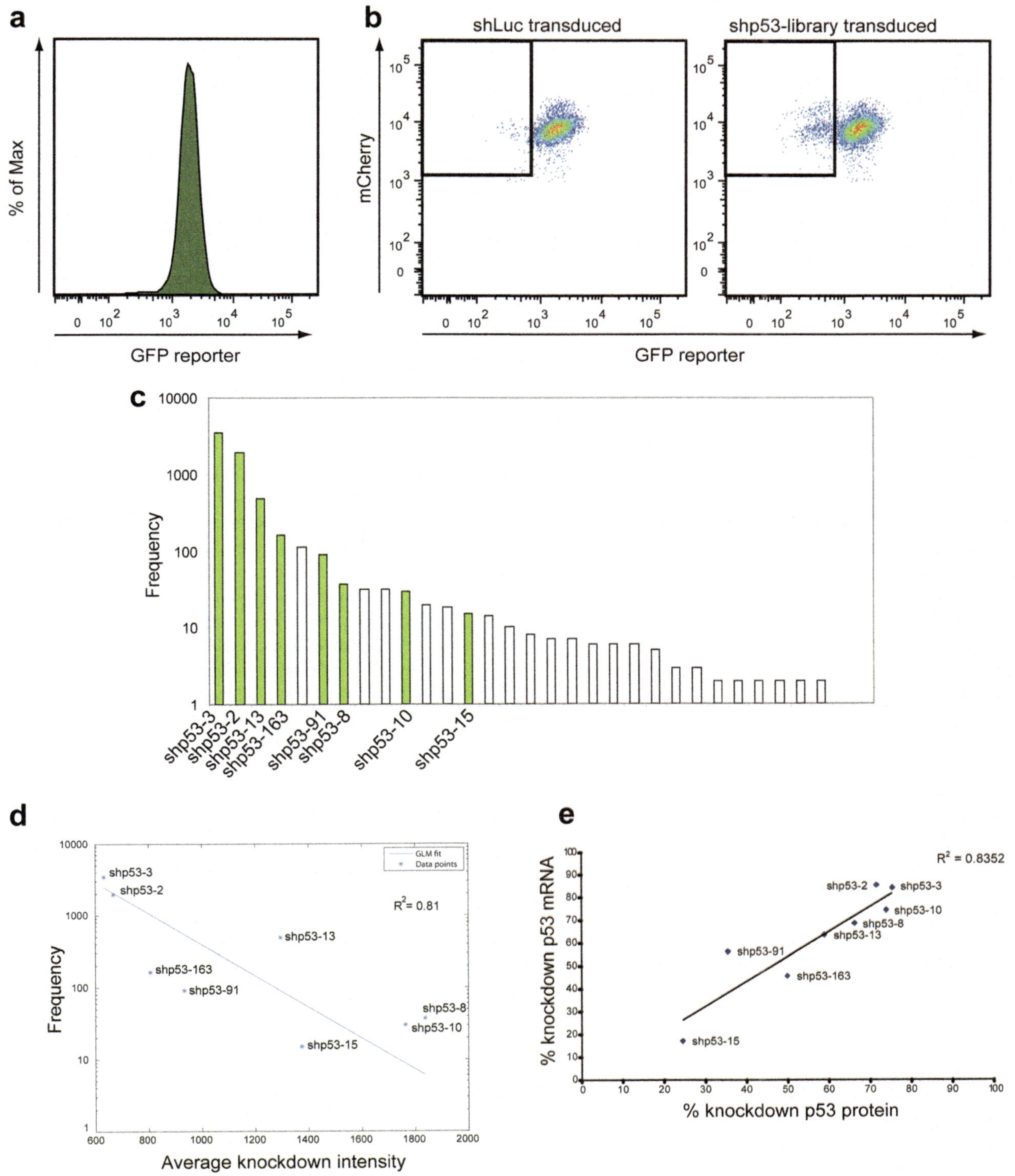

Figure 2 (See legend on next page.)

(See figure on previous page.)
Figure 2 Generation of an shp53 epi-allelic series. (a) HL60 p53-RNAi reporter cell line. A HL60 cell line was transduced with p53-RNAi reporter construct and a single cell-derived clone was isolated by FACS, expanded and reanalysed for optimal (GFP) reporter expression by flow cytometry. **(b)** HL60 p53-RNAi reporter cells transduced with shp53 library show a population with reduced RNAi reporter expression. HL60 p53-RNAi reporter cells were transduced with a library of candidate p53 shRNA or control shRNA (shLuc). All cells were monitored by FACS and cells with reporter levels within the GFP_{low}/RFP + gate (*box*) were sorted. **(c)** Functionally selected shRNA sequence frequency profile. Sequences retrieved from the 454 deep sequencing were tested to determine whether they encoded a hairpin structure and clustered into groups of near-identical sequences, and plotted by frequency. **(d)** Generalized linear model (GLM) fit of shRNA sequence frequency versus RNAi reporter level reveals a statistical relationship. Representative shp53-sequences were recloned, transduced into HL60 p53-RNAi reporter cells and analyzed by flow cytometry. A generalized linear model (GLM) with a log link function ('log(frequency) ~ 1 + knockdown') and Poisson distribution was applied to model the shRNA frequency versus RNAi reporter fluorescence intensity, generating estimated coefficients listed in Table 2, Additional file 4: Figure S4. The GLM model shows an exponential relationship between the observed shRNA frequency and measured RNAi reporter at a significance level of $\alpha = 0.05$ ($p \approx 0$). The reporter intensity data represents the mean of three independent experiments. **(e)** HUVE cells transduced with different shp53 clones generates an epi-allelic series. Transduced HUVE cells expressing different shp53 shRNA sequences show graded p53 expression levels as measured by relative quantification of Western blot versus mRNA levels.

To test this prediction, we devised a FACS-based screening approach to identify functional RNAi-triggers (CellSelectRNAi; Figure 1a). Libraries of candidate retrovirally expressed shRNAs are transduced into a population of host cells stably expressing a dedicated RNAi-GFP reporter and sorted for target gene knockdown by FACS. shRNA sequences are identified from the sorted cells by PCR and deep sequencing, and assembled into a frequency profile. The frequency of a given shRNA should correlate with knockdown efficiency. The RNAi-GFP reporter, comprising an open reading frame encoding a destabilized GFP [14], fused to hygromycin phosphotransferase (hpt) together with the untranslated target cDNA sequence serving as a 3′-UTR may be used to identify functional shRNAs for any gene (Figure 1b). To validate this RNAi-reporter system we prepared an Axl RNAi-reporter MDA-MB-231 cell line that was transduced separately with four different Axl shRNAs (shAxl2, shAxl280, shAxl279, shAxl278) in addition to a control shRNA targeting luciferase (shLuc). These five cell lines were simultaneously analyzed for reporter and endogenous Axl surface receptor protein levels by multicolor flow cytometry. Axl RNAi-reporter GFP fluorescence correlated strongly ($R^2 = 0.9931$) with endogeneous surface Axl receptor protein (Figure 1c) in the epi-allelic MDA-MB-231 cell series.

To evaluate the CellSelectRNAi approach, we generated p53 RNAi-reporter expressing HL60 cells. Flow cytometric analysis of the HL60/p53-RNAi reporter cell line displayed dsGFP fluorescence with a mean fluorescence intensity of 1721 and coefficient of variation (CV) of 36.8 (Figure 2a). A high coverage p53 shRNA library was prepared using a novel cloning approach where random sequences derived from the p53 mRNA sequence are converted into a form that encodes a hairpin structure and cloned into a U6 promoter cassette within a self-inactivating retroviral vector (Additional file 5: Figure S5). The p53 shRNA library (approximately 100 unique sequences) was transduced into the HL60/p53-RNAi reporter cell line and analyzed by FACS (Figure 2b). A low transduction rate (~10%) was used to favor single retroviral integrations per cell. A sorting gate was set at 2 standard deviations from the mean of the wild type population and HL60/p53-RNAi reporter cells carrying p53-targeting shRNAs (RFP^+) within the sorting gate were isolated (Figure 2b). 1800 sorting events were collected and proviral PCR amplification was conducted to generate a ~280 bp product that was analyzed by massively parallel pyrosequencing. The resulting 14790 shRNA sequences, of which 32 were p53 shRNA (Additional file 6: Figure S6) were rank ordered by frequency to generate a distribution plot (Figure 2c). Eight p53 shRNA sequences representing high, intermediate and low frequency levels were recloned and reintroduced into the HL60/p53-RNAi reporter cell line and monitored for reporter knockdown by flow cytometry. As predicted by the continuous distribution model, the shRNA sequence frequency derived from the sorted population showed a clear statistical relationship with the reporter intensity (Figure 2d). The most prevalent shRNA sequences identified by the CellSelectRNAi approach were the most potent also when measured at endogenous protein level and mRNA level (Figure 2e, Additional file 7: Figure S7). These results demonstrate accurate prediction of effective shRNA sequences by the CellSelectRNAi approach. Importantly, the most potent p53 shRNAs identified in the library screen are not predicted by available RNAi sequence algorithms [15] (Additional file 3: Figure S3).

We conducted epi-allelic hypomorphic analysis using selected p53-targeting shRNAs to determine the threshold of p53-dependent responses to ionizing radiation in primary human endothelial cells. p53 is a transcription factor involved in the DNA damage response. p53 activation by DNA damage, such as upon ionizing radiation,

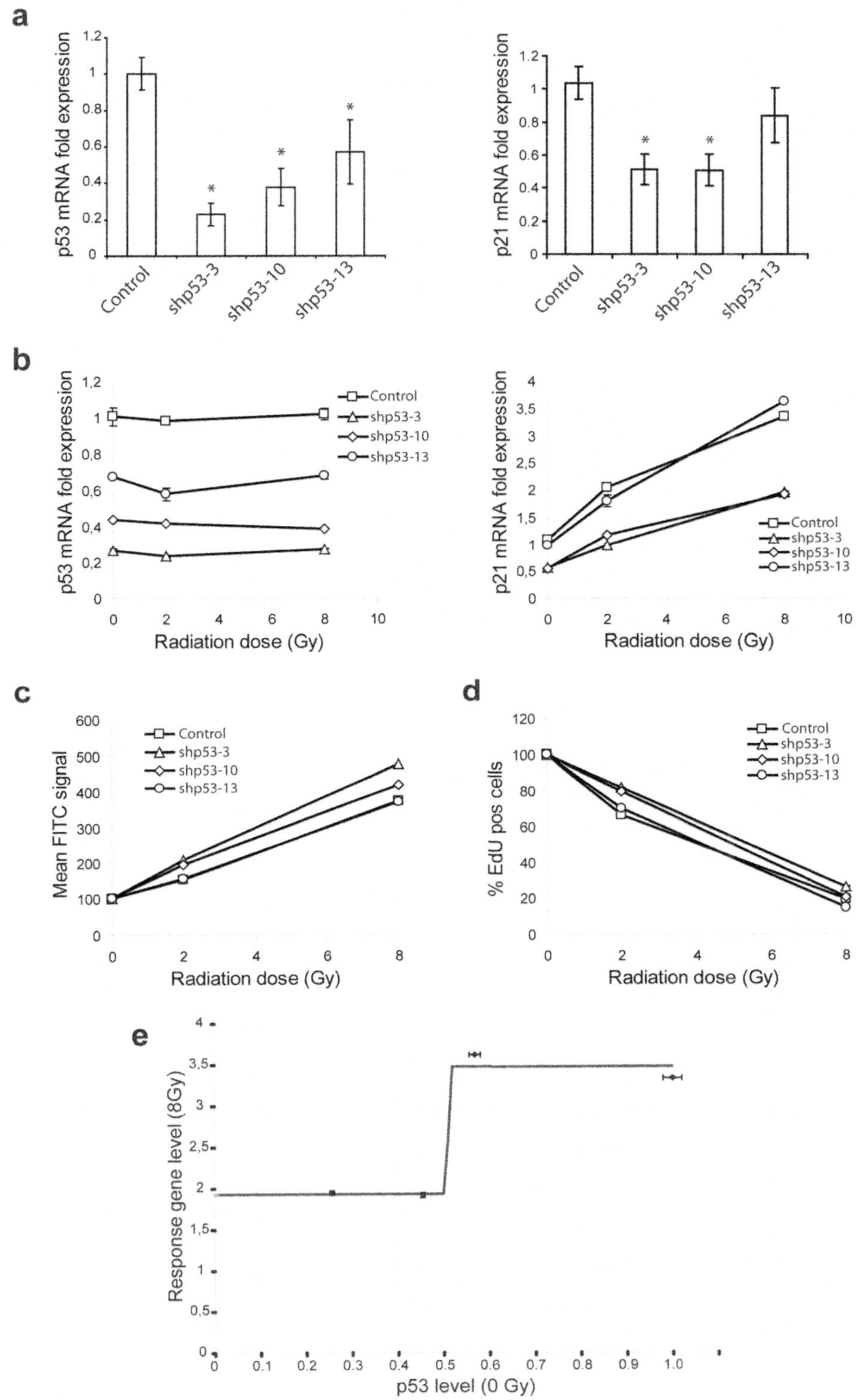

Figure 3 (See legend on next page.)

(See figure on previous page.)
Figure 3 Epi-allelic hypomorphic analysis to determine the threshold of p53-dependent responses to ionizing radiation. (a) Non-linear correlation between p53 and p21 mRNA levels in p53 epi-allelic primary human endothelial cells. Graded p53 knockdown in human umbilical cord endothelial cells (HUVEC) with shp53-3, shp53-10 shp53-13 or negative control (shLuc) (*left panel*) corresponded to a non-linear reduction in p21 gene expression (*right panel*). Data represent the mean ± SD of three (p53) and five (p21) measurements representative for two independent experiments relative to control cells (set as 100%). Statistical analysis was carried out using ANOVA followed by Tukey's post hoc tests, *P < 0,05. **(b)** Non-linear correlation between p53 and p21 mRNA levels in p53 epi-allelic primary human endothelial cells after irradiation. Irradiation of p53-epi-allelic HUVEC with 2 or 8 Gy did not affect p53 mRNA levels but increased p21 mRNA expression in a non-linear, p53-dependent manner. The data represents the mean of duplicate samples ± range. **(c)** Ionizing radiation induced H2AX phosphorylation in p53 epi-allelic primary human endothelial cells. H2AX phosphorylation, a measure of DNA double strand break induction, in HUVEC, correlates with greater p53 knockdown. The data represents the mean of duplicate samples ± range. **(d)** Cell proliferation in p53 epi-allelic primary human endothelial cells with shRNA against p53 after irradiation. The number of EdU positive p53-epi-allelic HUVEC following ionizing radiation, as a measure of cell division, increases with p53 knockdown efficiency. The data represents the mean of duplicate samples ± range. **(e)** Dose response relationship between p53-levels and radiation-induced p21 activation by epi-allelic analysis. Ionizing radiation-induced p21 activation displays a sigmoidal dose-dependency on p53 gene expression.

results in increased expression of several inducible response genes. p21 is an example of a p53-regulated gene which is induced following ionizing radiation and was used here to assess the effects of shRNA-mediated knockdown on p53 activity [16]. As shown in Figure 3a-b, graded p53 knockdown in primary human endothelial cells resulted in a non-linear attenuation of p21-activation following ionizing radiation. Radiation-induced double-strand break repair (histone H2AX phosphorylation) and DNA synthesis were similarly diminished in endothelial cells with less than 50% of wild type p53 expression levels (Figure 3c-d). Intriguingly, the observed dose-dependency of ionizing radiation-induced p21-activation on p53 gene expression levels fits a sigmoidal dose response model with an apparent IC50 around 50% p53-expression, the expected gene dosage of a $p53^{+/-}$ heterozygote (Figure 3e).

Conclusion

These results demonstrate that flow cytometric selection of RNAi knockdown cells combined with deep sequence analysis can be used to identify multiple shRNA sequences that reduce target gene expression to predetermined levels. Perhaps surprisingly, the most potent functionally-selected sequences do not follow widely used RNAi scoring algorithms and are therefore not represented in current RNAi collections. The ability to readily generate several shRNAs per gene carries significant advantages. In addition to obviating the influence of "off-target" effects [17,18], a set of shRNAs with differing silencing potentials provides an epi-allelic series of hypomorphs, a genetic dose-response, that can uniquely relate gene expression to functional response thresholds and improve target-phenotype correlations [1,2]. Epi-allelic analysis highlights the "robustness" that buffers perturbations on biological systems and delineates drug target inhibition profiles for therapeutic translation [19].

Additional files

Additional file 1: Figure S1. Retroviral vector L101 complete sequence.

Additional file 2: Figure S2. (A) L267 cloning vector with details of polylinker. Hairpins are expressed under the control of a modified U6 promoter, and cloned between the NaeI at the transcription start and the XcmI at the transcription terminator. PacI/HindIII restriction enzyme sites to facilitate cloning of the completed cassette into the final vector. **(B)** Structure of Hairpin Oligo DK540, including EcoP15I site (blue), GsuI sites (red) and a half-PvuII site (green). The loop sequence present in the completed hairpin cassette is in yellow. **(C)** Hairpin Oligo DK540 ligated to each end of a short fragment of p53. Digestion with EcoP13I releases two hairpin-tagged sequences which are then blunt-ended with Klenow fragment and dephosphorylated. **(D)** NaeI/XcmI-digested vector is ligated to haripin-tagged sequences. Because the haripin tag lacks a 5′ phosphate, the ligation product will have a 'nick' in the DNA phosphate backbone as indicated. DNA polymerisation from this nick using a strand-displacing polymerase opens out the hairpin into dsDNA. **(E)** Subsequent digestion with GsuI and recircularisation shortens the hairpin loop.

Additional file 3: Figure S3. Functionally selected shRNA sequences fail in RNAi scoring algorithms. Hairpin sequences from this study that gave good knockdown efficiency (actual efficiency) were scored with eight different RNAi scoring algorithms using the i-Score server (http://www.med.nagoya-u.ac.jp/neurogenetics/i_Score/i_score.html). *Bold italic*: score within top 100 for gene. Glt: guide strand-loop-template strand orientation. Tlg: template strand-loop-guide strand orientation.

Additional file 4: Figure S4. Predictions of knock-out effect.

Additional file 5: Figure S5. Schematic drawing of the hairpin expression cassette. shRNAs are produced from the U6/tethracyclin-inducible (TetO) promoter. The U6 promoter can transcribe short RNAs by RNA Polymerase III. This expression cassette also contains genes coding for mCherry and Puromycine separated by a linker encoding the 2A self-cleaving sequence under the control of the EF1a promoter. Abbreviations: pA, polyadenylation signal; uro, puromycin; Ψ, packaging signal; T, terminator.

Additional file 6: Figure S6. Sequences that have been scored as hairpins in screen. Lowercase letters signify hairpin loop.

Additional file 7: Figure S7. Epiallelic HUVE cells comprising p53-targeting shRNA's show graded p53 expression levels by immunoblotting.

Abbreviations

dsGFP: Destabilized GFP; FACS: Fluorescence activated cell sorter; GFP: Green fluorescent protein; HUVEC: Human umbilical vein endothelial cell; LTR: Long terminal repeat; nt: Nucleotides; RFP: Red fluorescent protein; RNAi: RNA interference; shRNA: Short hairpin RNA.

Competing interests

M.B. and D.R.M. are current employees of BerGenBio AS. J.B.L. is a founder, equity stockholder and consultant for BerGenBio AS. Aspects of the methodology presented are part of a patent (U. S. Patent No.: 8,735,064).

Authors' contributions

DRM, MB, OH and JBL designed the study; DRM and JBL developed the shRNA methodology; MB, PB, CT, TH and CG performed experiments; EH, PB and MB performed the statistical analysis; JBL, DRM and MB wrote the manuscript. All authors have read and approved the final manuscript.

Acknowledgements

This work was supported by a grant from the Norwegian Research Council to J.B.L., a University of Bergen predoctoral fellowship to M.B. and postdoctoral fellowship to DRM. The funders had no role in study design, data collection and analysis, decision to publish, or preparation of the manuscript.

Author details

[1]Department of Biomedicine, University of Bergen, N-5009 Bergen, Norway. [2]BerGenBio A/S, N-5009, Bergen, Norway. [3]Department of Clinical Chemistry and Transfusion Medicine, Sahlgrenska University Hospital, S-413 45 Göteborg, Sweden.

References

1. Hemann MT, Fridman JS, Zilfou JT, Hernando E, Paddison PJ, Cordon-Cardo C, Hannon GJ, Lowe SW: **An epi-allelic series of p53 hypomorphs created by stable RNAi produces distinct tumor phenotypes in vivo.** *Nat Genet* 2003, **33:**396–400.
2. Gjerdrum C, Tiron C, Hoiby T, Stefansson I, Haugen H, Sandal T, Collett K, Li S, McCormack E, Gjertsen B, Micklem D, Akslen L, Glackin C, Lorens J: **Axl is an essential epithelial-to-mesenchymal transition-induced regulator of breast cancer metastasis and patient survival.** *Proc Natl Acad Sci* 2010, **107:**1124–1129.
3. Wille L, Kemp ML, Sandy P, Lewis CL, Lauffenburger DA: **Epi-allelic Erk1 and Erk2 knockdown series for quantitative analysis of T cell Erk regulation and IL-2 production.** *Mol Immunol* 2007, **44:**3085–3091.
4. Swift SE, Lorens JB, Achacouso P, Nolan GP: **Rapid Production of Retroviruses for Efficient Gene Delivery to Mammalin Cells using 293T Cell-Based Systems.** In *Current Protocols in Immunology.* Edited by Coligan JE, Kruisbeek AM, Margulies DH, Shevach EM, Strober W. New York: Wiley; 1999:1–17.
5. Gallagher R, Collins S, Trujillo J, McCredie K, Ahearn M, Tsai S, Metzgar R, Aulakh G, Ting R, Ruscetti F, Gallo R: **Characterization of the continuous, differentiating myeloid cell line (HL-60) from a patient with acute promyelocytic leukemia.** *Blood* 1979, **54:**713–733.
6. Lorens JB, Pearsall DM, Swift SE, Peelle B, Armstrong R, Demo SD, Ferrick DA, Hitoshi Y, Payan DG, Anderson D: **Stable, stoichiometric delivery of diverse protein functions.** *J Biochem Biophys Methods* 2004, **58:**101–110.
7. Paithankar KR, Prasad KSN: **Precipitation of DNA by polyethylene glycol and ethanol.** *Nucleic Acids Res* 1991, **19:**1346.
8. Schmitz A, Riesner D: **Purification of nucleic acids by selective precipitation with polyethylene glycol 6000.** *Anal Biochem* 2006, **354:**311–313.
9. Brody J, Kern S: **Sodium boric acid: a Tris-free, cooler conductive medium for DNA electrophoresis.** *Biotechniques* 2004, **36:**214–216.
10. Livak KJ, Schmittgen TD: **Analysis of relative gene expression data using real-time quantitative PCR and the 2-[Delta][Delta]CT method.** *Methods* 2001, **25:**402–408.
11. Muslimovic A, Ismail IH, Gao Y, Hammarsten O: **An optimized method for measurement of gamma-H2AX in blood mononuclear and cultured cells.** *Nat Protocols* 2008, **3:**1187–1193.
12. Linger RMA, Keating AK, Earp HS, Graham DK: **Taking aim at Mer and Axl receptor tyrosine kinases as novel therapeutic targets in solid tumors.** *Expert Opin Ther Targets* 2010, **14:**1073–1090.
13. Hoffmann M, Chang HH, Huang S, Ingber DE, Loeffler M, Galle J: **Noise-driven stem cell and progenitor population dynamics.** *PLoS ONE* 2008, **3:**e2922.
14. Li X, Zhao X, Fang Y, Jiang X, Duong T, Fan C, Huang CC, Kain SR: **Generation of destabilized green fluorescent protein as a transcription reporter.** *J Biol Chem* 1998, **273:**34970–34975.
15. Ichihara M, Murakumo Y, Masuda A, Matsuura T, Asai N, Jijiwa M, Ishida M, Shinmi J, Yatsuya H, Qiao S, Takahashi M, Ohno K: **Thermodynamic instability of siRNA duplex is a prerequisite for dependable prediction of siRNA activities.** *Nucleic Acids Res* 2007, **35:**e123.
16. Prives C, Hall PA: **The p53 pathway.** *J Pathol* 1999, **187:**112–126.
17. Nature Editorial: **Whither RNAi?** *Nat Cell Biol* 2003, **5:**489–490.
18. Cullen BR: **Enhancing and confirming the specificity of RNAi experiments.** *Nat Meth* 2006, **3:**677–681.
19. Friedman A, Perrimon N: **Genetic screening for signal transduction in the era of network biology.** *Cell* 2007, **128:**225–231.

8

Differences in virulence of pneumolysin and autolysin mutants constructed by insertion duplication mutagenesis and in-frame deletion in *Streptococcus pneumoniae*

Esther Yip-Mei Liu[1], Feng-Yee Chang[2], Jen-Chang Chang[3] and Chang-Phone Fung[1,4*]

Abstract

Background: Insertion duplication mutagenesis (IDM) and in-frame deletion (IFD) are common techniques for studying gene function, and have been applied to pneumolysin (*ply*), a virulence gene in *Streptococcus pneumoniae* (D39). Discrepancies in virulence between the two techniques were observed in both the previous and present studies. This phenomenon was also observed during mutation analysis of autolysin (*lytA*).

Results: Our data showed that target gene restoration (TGR) occurred in IDM mutants, even in the presence of antibiotics, while the IFD mutants were stable. In PCR result, TGR occurred later in IDM-*ply* and -*lytA* mutants cultured in non-supplemented medium (4–5 h) compared with those grown in medium supplemented with erythromycin (erm)/chloramphenicol (cat) (3–4 h), but plateaued faster. Real-time PCR for detecting TGR had been performed. When compared with 8-h culture, TGR detection increased from Day 1 and Day 2 of IDM mutant's culture. erm-sensitive clones from IDM mutant were found. Southern blot hybridization and Western blotting also confirmed the phenomenon of TGR. The median survival of mice following intraperitoneal (IP) injection with a 3-h culture of IDM-mutants was significantly longer than that with an 8-h culture, irrespective of antibiotic usage. The median survival time of mice following IP injection of a 3-h culture versus an 8-h culture of IDM-*ply* in the absence of antibiotics was 10 days versus 2 days ($p = 0.031$), respectively, while in the presence of erm, the median survival was 5 days versus 2.5 days ($p = 0.037$), respectively. For an IDM-*lytA* mutant, the corresponding values were 8.5 days versus 2 days ($p = 0.019$), respectively, for non-supplemented medium, and 2.5 versus 2 days ($p = 0.021$), respectively, in the presence of cat. A comparable survival rate was observed between WT D39 and an 8-h IDM culture.

Conclusion: TGR in IDM mutants should be monitored to avoid inconsistent results, and misinterpretation of data due to TGR could lead to important biological meaning being overlooked. Therefore, based on these results, IFD is preferable to IDM for disruption of target genes.

Keywords: Insertion duplication mutagenesis, In-frame deletion, Pneumolysin, Autolysin, Target gene restoration

* Correspondence: cpfung@vghtpe.gov.tw
[1]Institute of Clinical Medicine, School of Medicine, National Yang-Ming University, Taipei 11217, Taiwan
[4]Division of Infectious Diseases, Department of Medicine, Taipei Veterans General Hospital and National Yang-Ming University, Taipei 11217, Taiwan
Full list of author information is available at the end of the article

Background

In the past two decades, advances in gene manipulation technologies have been applied widely in different aspects of molecular research, and have been especially useful in the functional analysis of genes. The analysis of mutant organisms generated by molecular modification is important in determining the functions of the wild-type genes. Homologous recombination in bacterial systems is the main tool for investigation of gene function, and enables the generation of targeted mutants of almost any gene. These techniques have increased our ability to investigate the temporal control of gene knockdown, analyze mutations, and express proteins through targeted gene "knock in" into another gene. Other than transposon mutagenesis and site–specific mutations, two other methods are widely used for functional analysis of bacterial genes: insertion duplication mutagenesis (IDM) and in-frame deletion (IFD) [1-5]. Comparison of a parental bacterial strain and its isogenic knockdown (KD) or knockout (KO) mutant, generated using these techniques, may reveal its specific function.

IDM has been widely used in the study of virulence genes such as pneumolysin (*ply*) and autolysin (*lytA*) in *Streptococcus pneumoniae* [6-11]. In this method, a partial target gene is amplified and ligated into a vector with a resistance marker. The constructed plasmid containing the chimeric DNA sequence then undergoes homologous recombination with the target chromosomal region following transformation. Integration and linearization, followed by insertion of the complete plasmid into the target gene, occurs in a single crossover between the two homologous gene sequences. As a result, the function of the target gene is disrupted [10,12].

The other approach is IFD, which is a replacement homologous recombination [8,13]. Unlike IDM, in which the insertion construct is integrated into the homologous site of the target gene, IFD is preceded by a double crossover event, leading to the complete replacement of the target gene. As a result, none of the sequence is duplicated in the recombinant, and the wild-type gene cannot be regenerated.

Although both methods, which have been widely used in virulence studies in *S. pneumoniae*, can achieve the goal of gene disruption, no studies have compared whether the experimental data produced by IDM are akin to that produced by IFD. In this study, the stability of IDM and IFD was evaluated. We also examined whether target gene restoration (TGR) could occur in either type of mutant.

Results

Confirmation of *S. pneumoniae* IDM-*ply* and -*lytA* mutants

Three sets of PCR primers were used to confirm the correct formation of IDM-*ply* mutants (Table 1 and Figure 1A). Positive DNA fragments were observed for IDM-*ply* using the ply-P2/pVA891-F and ply-P1/pVA891-R primer set. As expected, no amplification was observed with the ply-P1/ply-P2 primer set (Figure 2B). Bands were amplified from wild-type (WT) strain D39 using the ply-P1/ply-P2 primer set, but not with the ply-P2/pVA891-F and ply-P1/pVA891-R primer set (Figure 2A).

Incorporation of pEVP3-chloramphenicol (cat), carrying a partial *lytA* sequence, into target chromosomal *lytA* by homologous recombination, generating a IDM-*lytA* mutant, was confirmed using three PCR primer sets (Table 1 and Figure 1B). DNA fragments were obtained from IDM-*lytA* using the lytA-P2/pEVP3-F and lytA-P1/pEVP3-R primer set. No DNA amplification was observed from IDM-*lytA* DNA using the lytA-P1/lytA-P2 primer set (Figure 2D). Positive DNA fragments were only obtained from WT D39 DNA using the lytA-P1/lytA-P2 primers, while no amplification was observed using the lytA-P1/pEVP3-R and lytA-P2/pEVP3-F primer sets (Figure 2C).

Confirmation of *S. pneumoniae* IFD-*ply* and -*lytA* mutants

By transformation of linear DNA fragments comprising the 5′ and 3′ flanking regions of *ply* or *lytA* (Figure 3A and B), respectively, into IDM-*ply* and -*lytA*, IFD-*ply* and -*lytA* mutants were obtained from a second round of homologous recombination. Both IFD mutants were confirmed by PCR using primer sets ply-P1/ply-P2 for pneumolysin and lytA-P1/lytA-P2 for autolysin (Figure 3A and B). WT strain D39 was used as a control. Predicted PCR products are shown in Table 1. A 670-bp and a 616-bp fragment were obtained for the IFD-*ply* and IFD-*lytA* mutants, respectively, using the same primer pairs as above, while positive amplification of D39 produced a 1824-bp fragment for *ply* and a 1346-bp fragment for *lytA* (Figure 4).

Detection of TGR in IDM and IFD mutants by PCR

TGR was identified in both the IDM-*ply* and -*lytA* mutants, while no TGR was found in either the IFD-*ply* or -*lytA* mutants (Figure 5AI and BI). WT *ply*, formed by TGR in IDM-*ply*, was observed 4–5 h post-inoculation of mutants into plain brain-heart infusion (BHI) broth under antibiotic selection (Figure 5AI). This restoration of mutants back to the WT genotype increased to a plateau at about 8–10 h of cultivation, with or without antibiotic selection (Figure 5AI). The highest percentage increase in *ply* detection, relative to the earliest detection of TGR (4 h), in plain BHI and BHI with erythromycin (erm) was 80.57% and 23.27%, respectively, indicating a relatively slow rate of TGR under antibiotic selective pressure (Figure 5AIII). TGR was also observed in IDM-*lytA* mutants (Figure 5BI). In BHI medium, TGR was

Table 1 Primers used to assess and confirm the mutants for IDM-*ply*, IDM-*lytA*, IFD-*ply*, and IFD-*lytA*

	Sequence (5′-3′)	Accession number	PCR product size (bp)		
			D39	IDM	IFD
Pneumolysin					
ply-P1	CCTGTATATCAAGGAATTGGCTACC	CP000410.1			
ply-P2	GATAGAAGAGCCGGATCTAGCTCG	CP000410.1	1824	ND	670
ply-P1	CCTGTATATCAAGGAATTGGCTACC	CP000410.1			
pVA891-R[a]	CGCTCATCGTCATCCTCGGC	AB057644.1		1582	
ply-P2	GATAGAAGAGCCGGATCTAGCTCG	CP000410.1			
pVA891-F[a]	CCAGTAGTAGGTTGAGGCCGTTG	AB057644.1		1445	
ply-1a	CCTTTGGCTTTATCAATCGCTTTATCG	CP000410.1			
ply-1b	actcttcacttacatgcatgcCGTACGGTTTATGAAAAAACC	CP000410.1			1227
ply-2a	gcatgcatgtaagtgaagagtGCTCCGCTTCTTTCTTTCGAT	CP000410.1			
ply-2b	TTACCTGTCGCCCTTGCTC	CP000410.1			1185
ply-F91	CGTTTCATCAAAGAGGGTAA	CP000410.1			
ply-R1070	GCTGTAACCTTAGTCTCAAC	CP000410.1	980		
ply-pbF	TAGCGATTTGGCTGTGAA	CP000410.1			
ply-pbR	GCTGATTACCCTCTTTGA	CP000410.1	483		
Autolysin					
lytA-P1	CAGTACTTAAAGCTATCCGACTCGGTTTAC	CP000410.1			
lytA-P2	GTCTGGGGTGTTATTGTAGATAGAATG	CP000410.1	1346	ND	616
lytA-s1	CATGGCGCCTTCTTTAGCGTCTA	CP000410.1			
lytA-s2	GAGTTCATGACGGACTACCGCCT	CP000410.1		546	
lytA-P1	CAGTACTTAAAGCTATCCGACTCGGTTTAC	CP000410.1			
pEVP3-R[b]	GCCAGTTTGAGGGGACGACG			1234	
lytA-P2	GTCTGGGGTGTTATTGTAGATAGAATG	CP000410.1			
pEVP3-F [b]	TAACCTAACTCTCCGTCGC			1513	
lytA-1a	GTTCCTCATACTTCCCTTG	CP000410.1			
lytA-1b	cccatccactaatgcggccgcATGCCTTTATCCAGTCAGCGG	CP000410.1			1331
lytA-2a	gcggccgcattagtggatgggATTCCCAGTTGAGTGTGCGTG	CP000410.1			
lytA-2b	TGCCAGAACTCTTGCCACAG	CP000410.1			1199
Glucose kinase					
gki-up	GGCATTGGAATGGGATCACC	CP000410.1			
gki-dn	TCTCCCGCAGCTGACAC	CP000410.1	626	626	626
gki-F603	TGCCGATGAATACGAAGG	CP000410.1			
gki-R791	CGATTGTTGATGGGTTTAG	CP000410.1	189		

[a]According to pVA838 (GenBank accession number AB057644.1).
[b]pEVP3 was provided by Dr. Don Morrison, UIC Biological Sciences, 900 South Ashland Ave., Chicago, IL 60607, USA.
ND Not detectable.
Nucleotides shown in lowercase letters denote extra sequences that were added for ligation reactions.
Locations of primers are shown in Figure 1 for IDM and Figure 3 for IFD.

first detected at around 5 h post-inoculation and corresponded to a 6-fold change in detection, while in BHI supplemented with cat, TGR was detected at 3 h post-inoculation and corresponded to a 90-fold change in detection (Figure 5BIII).

Validation of TGR from the results of previous PCR assay of IDM mutants by using real-time PCR assay, southern blot hybridization and western bloting

Since only PLY antibody is commercial available for the determination of protein expression for TGR strain, valid-

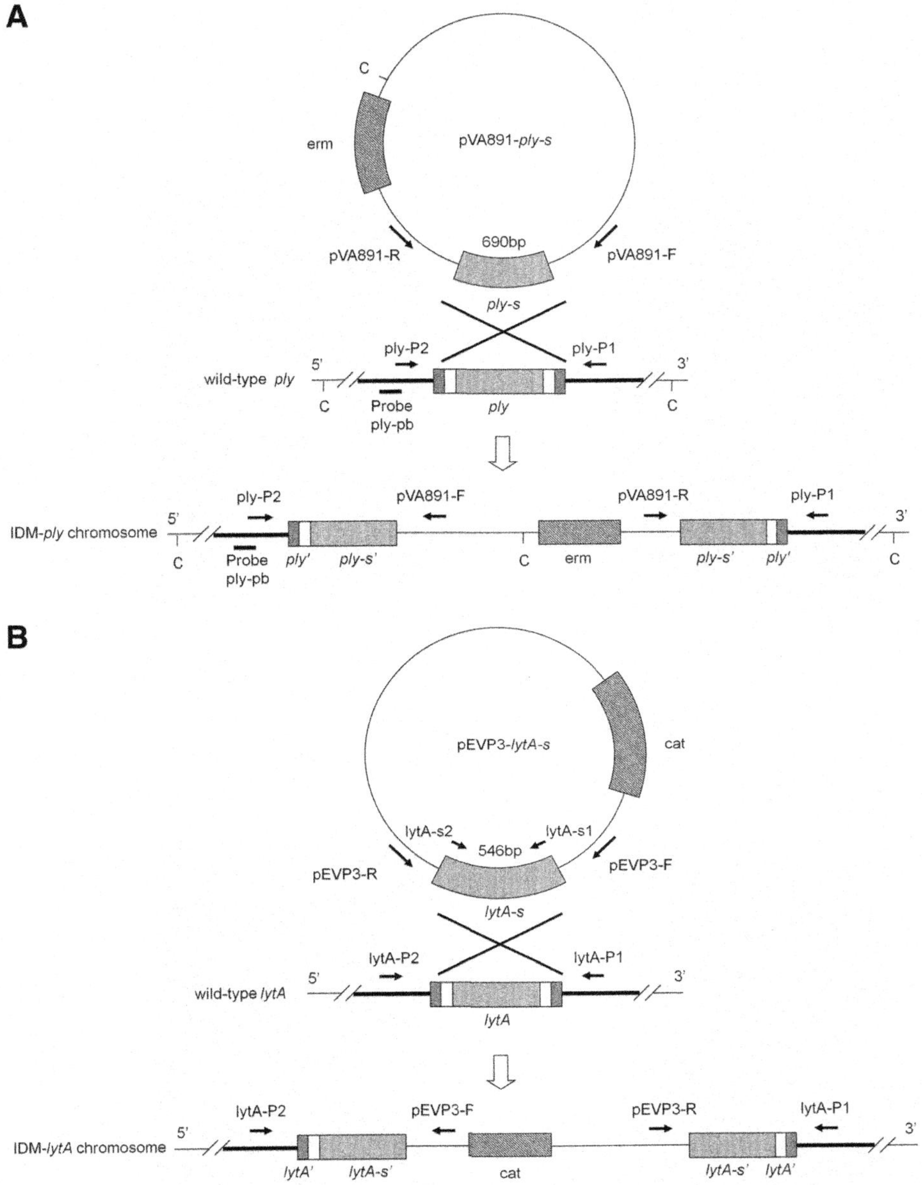

Figure 1 Construction of the *ply* and *lytA* knockdown mutants by IDM. Partial sequences of pneumolysin (*ply*-s) **(A)** and autolysin (*lytA*-s) **(B)** genes were cloned into plasmids pVA891 **(A)** and pEVP3 **(B)**, containing an erythromycin (erm)- and a chloramphenicol (cat)-resistance marker, respectively. Cloned plasmids were then used to generate homologous recombinants from their WT *ply* or *lytA* genes, respectively. Two primer sets, ply-P1/pVA891-R and ply-P2/pVA891-F or lytA-P1/pEVP3-R and lytA-P2/pEVP3-F, were used to verify the insertion of pVA891 and pEVP3 in IDM-*ply* **(A)** and -*lytA* **(B)**, respectively. C, *Cla*I.

ation of TGR from the results of the above PCR assay was used the PLY as a model. In Real time PCR assays, full length PLY expression level of Day 1 and Day 2 culture of IDM-*ply* was compared 8-h culture of IDM-*ply*. A 1.66 and 3.00 folds were respectively observed in Day 1 and Day 2 culture of IDM-*ply* indicating TGR of PLY occurred as time dependent manner (Figure 6A). In Southern blotting, *erm* susceptible isolates were selected from Day 1 culture of IDM-*ply* and hybridization with *ply* specific probe confirmed that TGR wild type strains was occurred (Figure 6B). For determination of full length protein expression of PLY, Western blotting was performed and confirmed that the PLY expression in TGR was as same as wild type D39 (Figure 6C).

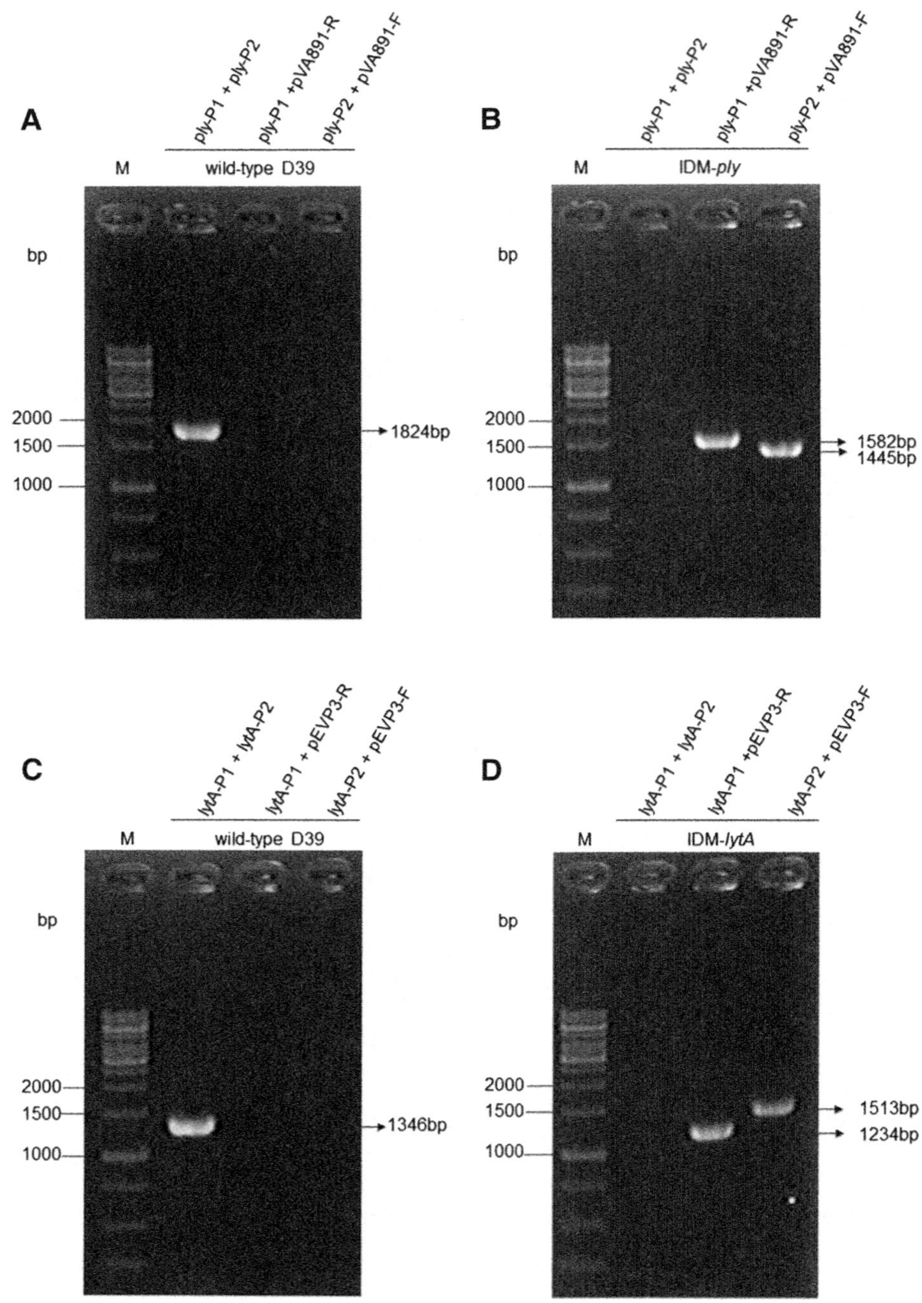

Figure 2 Verification of D39 and the IDM-*ply* and -*lytA* mutants. (A) A product could only be amplified from D39 using primer set ply-P1/ply-P2. **(B)** The IDM-*ply* mutant was confirmed by the lack of an amplification product using the ply-P1 and ply-P2 primer set, and was shown to contain an inserted pVA891 and truncated *ply* using primers pairs ply-P2/pVA891-F and ply-P1/pVA891-R, respectively. **(C)** A product was only amplified from D39 using primer pair lytA-P1/lytA-P2. **(D)** The IDM-*lytA* mutant was verified by no amplification product using primers lytA-P1 and lytA-P2 and shown to have an inserted pEVP3 and truncated *lytA* using primer sets lytA-P2/pEVP3-F and lytA-P1/pEVP3-R, respectively. The locations of primers and predicted molecular weight of the PCR products for D39, IDM-*ply*, and IDM-*lytA* are given in Figure 1 and Table 1. M, molecular DNA marker.

Effect of TGR on virulence of mutants in a mouse model of infection

For the *in-vivo* studies, mice were randomly selected and sacrified for detection of *ply* and *lytA* by PCR. No mice showed symptoms of illness at 10 min post-injection (IP), and neither the WT D39 nor the IDM mutant strain was found in the heparinized blood or liver at this time point (Figure 7A–D). However, full-length *ply* was consistently detected in the blood and liver of animals injected with IDM mutants grown with or without antibiotics at day 2

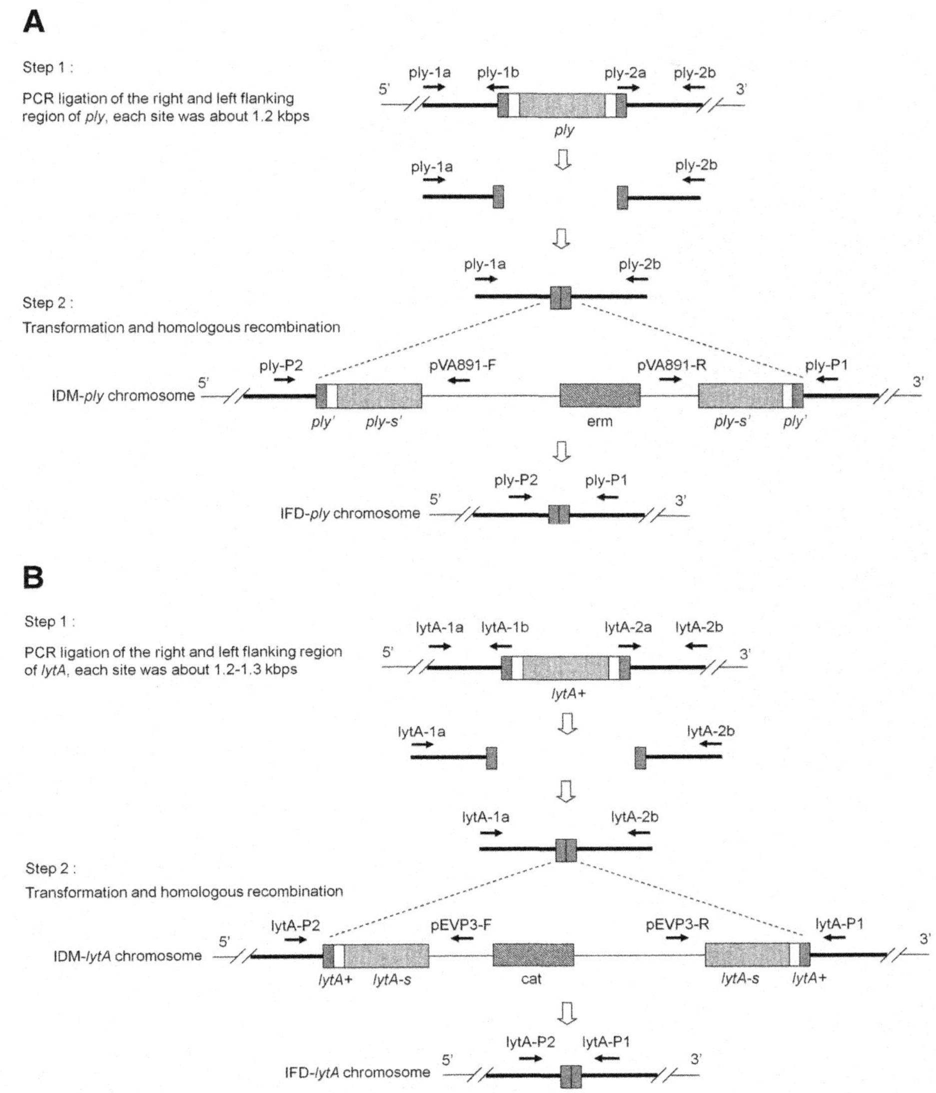

Figure 3 Construction of the *ply* and *lytA* knockout mutants by IFD. In step 1, linear DNA fragments containing flanking regions of the *ply* **(A)** and *lytA* **(B)** genes, generated by PCR, were transformed into IDM-*ply* **(A)** and -*lytA* **(B)**, respectively, for a second round of homologous recombination. The second recombination event could result in either restoration of the original gene, or mutants of IFD-*ply* **(A)** and -*lytA* **(B)** with a double-crossover. IFD-*ply* **(A)** and -*lytA* **(B)** mutants were selected by decreasing concentration of erythromycin (erm) for *ply* or chloramphenicol (cat) for *lytA*, and increasing concentrations of ampicillin. Primer sets ply-p1/ply-P2 or lytA-P1/lytA-P2 were used to verify the IFD mutants.

(Figure 7A and B), and full length *lytA* was also detected as early as day 1 (Figure 7C and D). IDM-*ply* and -*lytA* were consistently detected in all samples on day 1, indicating the coexistence of IDM mutants and TGR WT D39 during infection (Figure 7A–D).

Approximately 10^2 colony forming units (CFU) of IDM-*ply* mutant was then used in a lethality test. The median survival time of mice injected with a 3-h IDM-*ply* culture was significantly longer than that of animals injected with an 8-h IDM-*ply* culture: 10 versus 2 days, $p = 0.031$, without antibiotic selection; 5 versus 2.5 days, $p = 0.037$, with erm selection (Figure 8A). Following IP injection of a 3-h IDM-*ply* culture grown with and without antibiotics, five and four animals, respectively, out of 10 survived after 14 days of observation. All except one mouse (90%) died within 14 days when an 8-h IDM-*ply* culture was used. The survival time following injection with a 3-h IDM-*ply* culture was significantly longer ($p < 0.001$ without antibiotic and $p = 0.006$ with erm) than that with an 8-h culture of WT D39, indicating decreased virulence of the 3-h IDM-*ply* mutant. A comparable survival rate was observed between WT D39 and an 8-h IDM-*ply* culture (Figure 8A), indicating a high level of TGR in the 8-h IDM-*ply* culture.

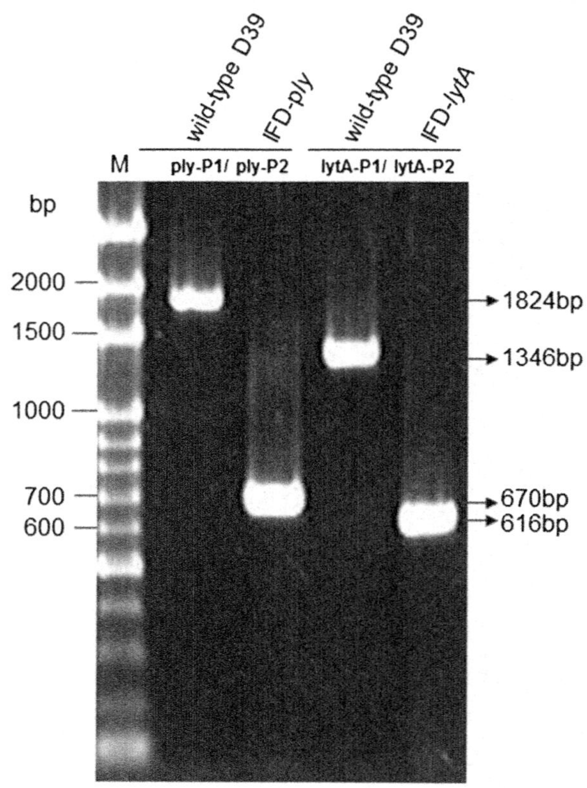

Figure 4 Verification of D39 and the IFD-*ply* and -*lytA* mutants. PCR using different primer sets specific to *ply* and *lytA* were used to confirm parental D39 and the IFD-*ply* and -*lytA* mutants. The locations of primers and predicted molecular weights of the PCR products for D39, IFD-*ply*, and IFD-*lytA* are given in Figure 3 and Table 1. M, molecular DNA marker.

Similarly, the median survival time of mice subjected to IP injection with 10^2 CFU of a 3-h culture of IDM-*lytA* was significantly longer than that following injection with the same amount of an 8-h IDM-*lytA* culture (8.5 vs. 2 days, $p = 0.019$ with no antibiotic; 2.5 vs. 2 days, $p = 0.021$ with cat) (Figure 8B). The survival rates for the two groups were 50% and 40%, respectively, at 14 days post-inoculation. None of the mice survived longer than 3 days post-inoculation with an 8-h culture of IDM-*lytA*. Like the IDM-*ply* mutant, the 3-h IDM-*lytA* culture had attenuated virulence in mice, demonstrated by a decreased mortality rate ($p = 0.045$ for non-supplemented BHI), when compared to the 8-h D39 culture. Overall, the 3-h IDM mutants were less virulent than the 8-h IDM mutants. The survival of mice injected with 8-h IDM-*ply* or -*lytA* culture was almost identical to that of mice injected with WT D39.

Discussion

A search of the literature shows that IDM is still a commonly used method for gene KD [14], and has been used to study pneumolysin and autolysin in *S. pneumoniae* [15-17]. In the present study, TGR in IDM mutants caused by intra-chromosomal recombination of a mutated target gene, was confirmed and significantly affected lethality in mice. Selective antibiotic pressure minimized the extent of the TGR but could not completely eliminate the rearrangements. Higher concentrations of antibiotics could prohibit reversion, but the growth rate of the bacteria was significantly affected. Re-circularization of inserted sequence within the IDM mutants circumvents the antibiotic selection. This type of mutant maintains its antibiotic resistance but loses the KD function of the target gene. Following injection of these mutants into mice, antibiotic selective pressure is relaxed, leading to an increased chance of TGR. This event was observed in our animal study, in which TGR-D39-type *ply* and *lytA* gene sequences were detected at days 2 and 1 post-injection, respectively (Figure 7). This effect generated the discrepancy in pathogenicity and lethality in mice, and affects the interpretation of results in our study (Figure 8).

A previous study also showed decreased survival of mice injected IP with IDM-*ply* compared with those injected with IFD-*ply* [8]. The median survival times of mice injected with IDM-*ply* ("PLN-A") and IFD-*ply* ("ΔPly") were 2.8 and 13.8 days, respectively, with a *p*-value of 0.04 (log-rank survival test), suggesting that the IDM-*ply* mutants were significantly different from the IFD-*ply* mutants, and were more virulent. Polar effects of the pVA891-mediated IDM event, along with toxicity of truncated pneumolysin polypeptides or fusion proteins resulting from insertion of the plasmid sequences into the *ply* gene cannot be excluded [8]. Our data proved that the relatively greater virulence of the IDM mutants in the mouse model may also due to TGR. The restoration of the WT gene is not specific to *ply* or plasmid pVA891, as it also occurred with the virulence gene *lytA* and plasmid pEVP3. We have previously observed this phenomenon in Gram-negative bacteria when we applied IDM to KD *ompK36*, an outer membrane porin gene of *Klebsiella pneumoniae* [18]. Nonetheless, this gene restoration effect is not common to all genes. No target gene restoration was observed in IDM-wzy_{KPK1}, a capsule synthesis polymerase gene in *K. pneumoniae* [19]. Our PCR analysis of TGR of *lytA* IDM mutants showed that TGR could be detected at as early as 3-h post-inoculation, supporting the discrepancy in results when an 8-h culture was selected for injection. Our study of lethality showed that the median survival time (2 days) of mice when using the 8-h IDM mutants was comparable to direct IP injection of the original D39 strain. We postulated that the IDM mutants could revert to their parental characteristics, resulting in WT levels of virulence (Figure 8).

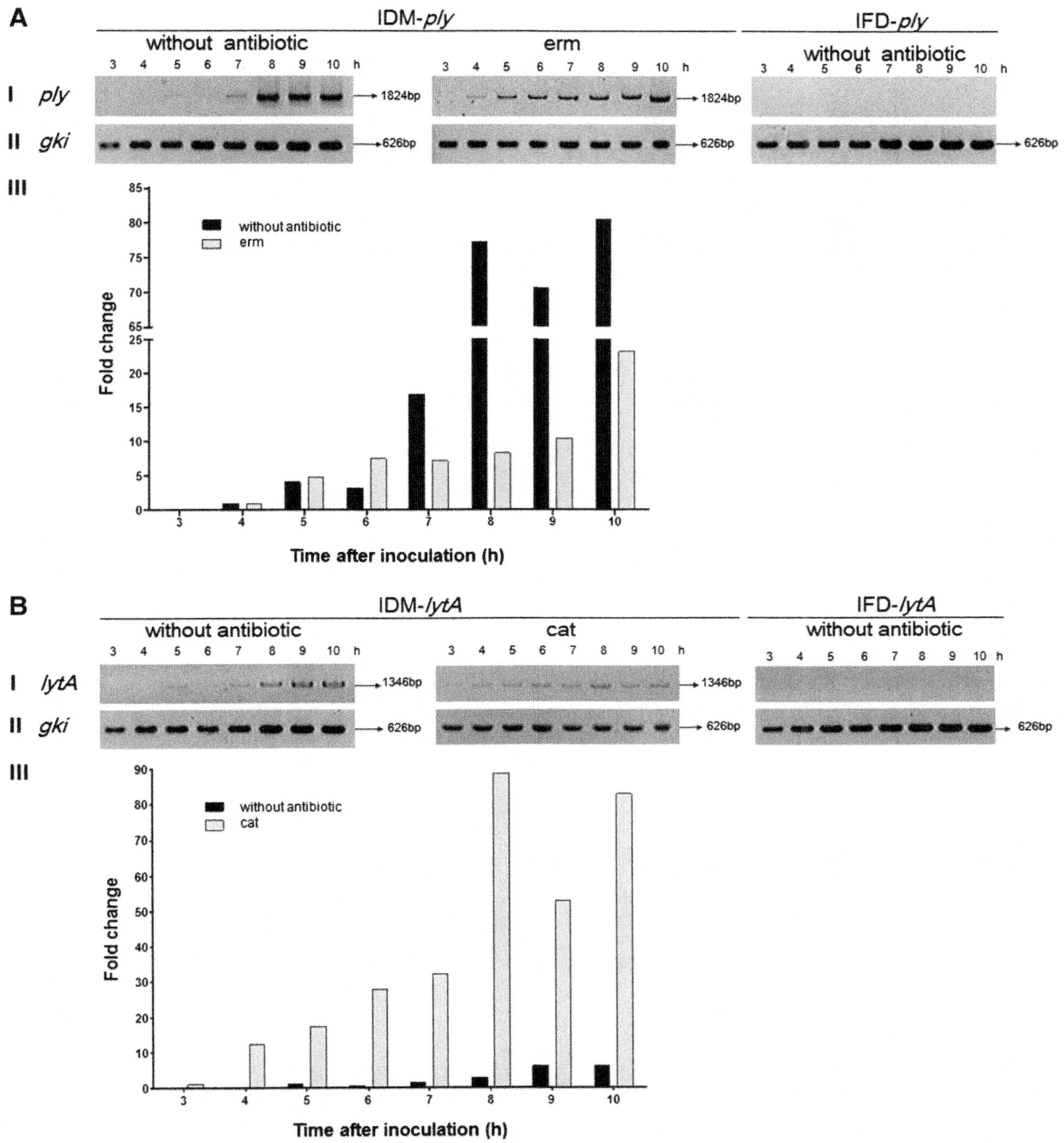

Figure 5 Target gene restoration (TGR) of wild-type *ply* (A) and *lytA* (B) was only observed in IDM mutants. (I) Full-length target genes were amplified by PCR of samples taken at 3–10 h post-inoculation. Restoration of *ply* **(A, I)** and *lytA* **(B, I)** was confirmed by PCR using the primer sets ply-P1/ply-P2 and *lytA*-P1/*lytA*-P2, respectively, in 3–10-h cultures of IDM-*ply*/-*lytA* with and without antibiotic selection. No target gene restoration was observed in IFD-*ply*/-*lytA* mutants. (II) Housekeeping gene *gki*, used as a loading control to evaluate the level of gene expression, was amplified by primer pair gki-up/gki-dn. (III) Fold change refers to ratio of the target gene restoration (*ply* or *lytA*) to *gki* in 3–10-h cultures over the aforementioned ratio found at the earliest time of target gene detection.

Based on our previous experience with IDM-*ompK36*, TGR most likely occurred at a certain time during culture in broth medium, and the effect would not be detectable at culture times of less than 2 h [18]. However, the consequence of not using serial passages of such mutants has given rise to huge inconsistencies in the data. Because the

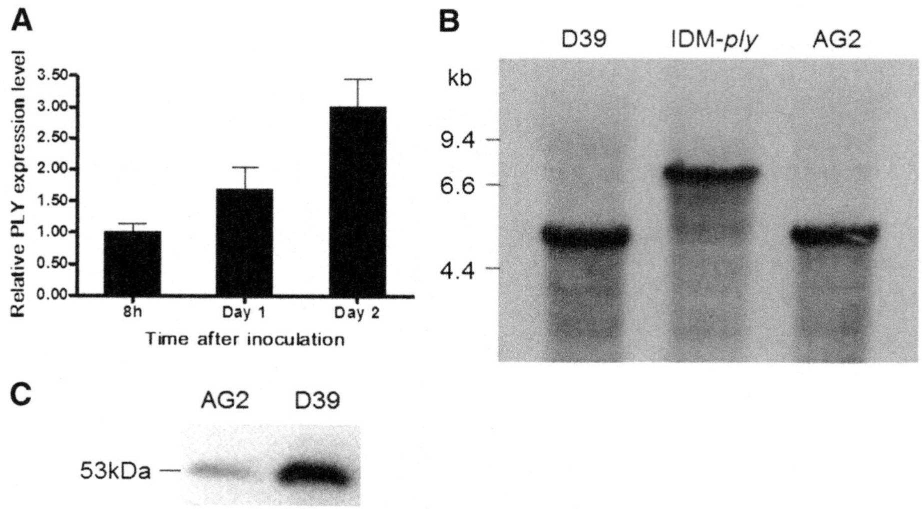

Figure 6 Validation of TGR of WT-*ply* from IDM-*ply* mutant. (A) Expression of WT-PLY in the 8 h, Day 1 and Day 2 IDM-*ply* culture by real-time PCR **(B)** *ClaI*- digested D39, IDM-*ply* and AG2, which was sensitive to erm and was obtained from Day 1 culture of IDM-*ply* mutant, were probed with a digoxigenin-labeled ply-pb in southern blot hybridization. **(C)** AG2 would produce PLY whose size was the same as PLY of D39 when incubating with a rabbit polyclonal PLY antibody.

growth rate of *S. pneumoniae* is significantly slower than many other Gram-positive bacteria in culture media, the problems associated with the unstable mutation become substantial when attempting to generate a considerable concentration of the mutant. Although this difficulty can be overcome by PCR confirmation prior to the use of cultured mutants in each experiment, it becomes labor intensive in *in vivo* studies.

In addition to the other known defects of IDM methods, TGR should be confirmed to avoid inconsistent results prior to interpretation and reporting. Misinterpretation of data due to TGR may have contributed to important biological meaning being overlooked. If inconsistencies are observed in IDM-related experimental data, previous data should be retrieved and examined for the possibility of TGR. However, the polar effect, truncated target

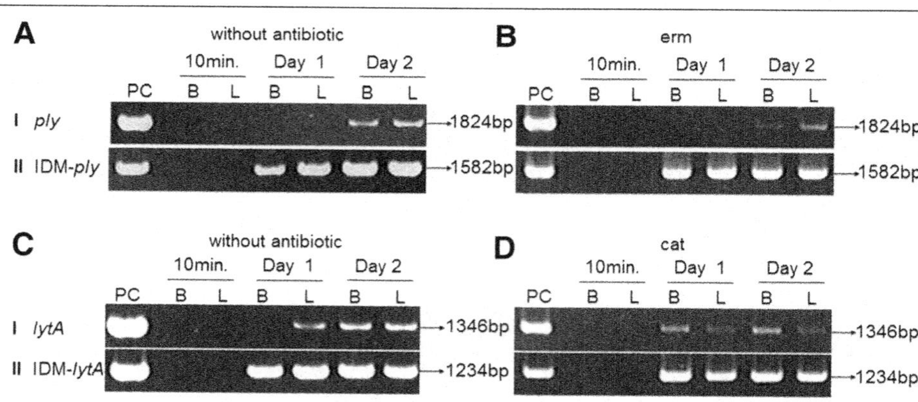

Figure 7 Detection of TGR of WT *ply* and *lytA* gene from blood and liver of BALB/c mice with 3h-IDM culture of IDM mutants. (A-D) Cultural media used for IDM mutants growth. IDM-*ply* mutant was cultivated in plain BHI broth **(A)** and BHI supplemented with erm **(B)** whilst IDM-*lytA* mutant was grown in plain BHI broth **(C)** and BHI supplemented with cat **(D)**. (I) Detection of the restored WT *ply/lytA* gene from blood and liver samples following IP injection of 3-h culture of IDM-*ply*/-*lytA* mutants at 10min, 1 day and 2 days. WT *ply* gene was amplified from samples by using the ply-P1/ply-P2 primer set for IDM-*ply* mutant grown from plain BHI (A,I) and BHI with erm (B,I). Full length of *lytA* gene was detected by the lytA-P1/lytA-P2 primer set for IDM-*lytA* mutant grown from BHI (C,I) and BHI with cat (D,I). (II) PCR analysis of IDM-*ply/lytA* gene from the same samples used in (I). Detection the genes of IDM-*ply* from IDM-*ply* mutant grown from BHI (A,II) and BHI with erm (B,II), and IDM–*lytA* gene from IDM-*lytA* mutant grown from BHI (C,II) and BHI with cat (D,II) was performed by ply-P1/pVA891-R and lytA-P1/pEVP3-R respectively. The lowest inocula of the 3-h IDM-*ply* cultures in BHI alone and BHI supplemented with erm for positive TGR of *ply* were approximately 10^4 CFU **(A)** and 10^3CFU **(B)** respectively whilst that of the IDM-lytA mutant grown in BHI and BHI supplemented with cat for positive TGR of *lytA* were 10^2 CFU **(C)** and 10^3 CFU **(D)** respectively. B, blood; L, liver; erm, erythromycin; cat, chloramphenicol.

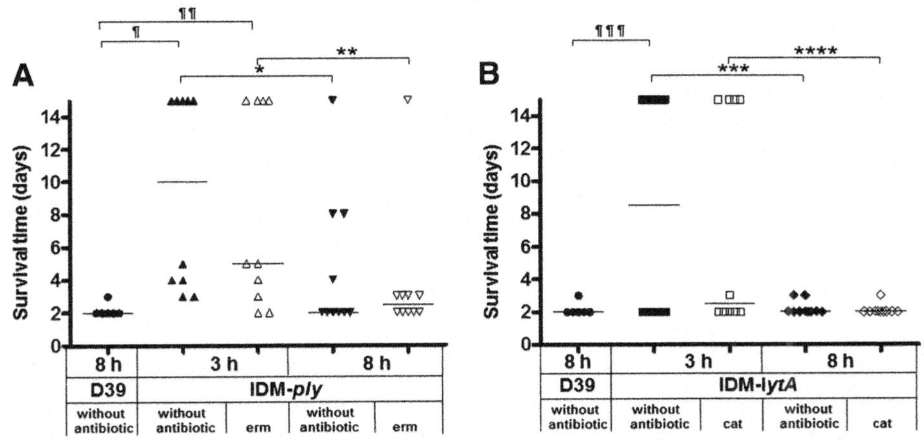

Figure 8 Survival of BALB/c mice injected intraperitoneally with WT D39 and a 3-h and an 8-h culture of IDM-*ply* (A) and -*lytA* (B). All of the inocula were approximately 10^2 CFU. The median survival time is indicated by the dashed line. Significant differences in the median survival time of mice injected with a 3-h IDM-*ply* **(A)** or -*lytA* **(B)** culture versus those for mice injected with an 8-h culture of the IDM- mutants are indicated by an asterisk (*) (*$p = 0.031$ (without antibiotic) versus **$p = 0.037$ (erm usage) in IDM-*ply*; ***$p = 0.019$ (without antibiotic) and ****$p = 0.021$(cat usage) in IDM-*lytA*). Significant differences between mice injected with 8-h WT D39 and those injected with 3-h IDM clones are indicated by a pilcrow (¶). (¶$p < 0.001$(without antibiotic) and ¶¶$p = 0.006$ (erm usage) when compared to the 3-h IDM-*ply*; ¶¶¶$p = 0.045$ (without antibiotic) when compared to the 3 h IDM-*lytA*).

polypeptides or fusion proteins resulting from insertion of the plasmid sequences of IDM are always the consideration of this method even TGR will not be occurred. The use of IFD to generate mutants in both our previous work [20] and the present study has given consistent data for both *in vitro* and *in vivo* experiments.

Conclusion

IFD is a superior KD method compared with IDM, as it produces more stable clones, even though the success rate is lower.

Methods

Bacterial isolates

Capsular type 2 *S. pneumoniae* strain D39 and a derivative of the D39 pneumolysin knock down mutant (IDM-*ply*), originally described by Berry et al. [7] and constructed using IDM with a defined truncation in the pneumolysin gene (Figure 1A), were kindly provided by Dr. David Briles, University of Alabama, Birmingham, AL, USA [21].

Construction of the autolysin KD mutant (IDM-*lytA*) by IDM

The *lytA* mutant was constructed by IDM (Figure 1B). A 546-bp DNA fragment of the autolysin gene was PCR-amplified from D39 using the primer set lytA-s1/lytA-s2 (Table 1). PCR amplification was performed in a DNA thermal cycler (Perkin-Elmer Biosystems, Foster City, CA, USA) in a 50-μl mixture containing Phusion HotStar High-Fidelity DNA polymerase, 200 μM deoxynucleotide triphosphates (GeneTeks BioScience, Taipei, Taiwan), and 0.6 μM of each oligonucleotide primer in 1× Phusion HF buffer. Template DNA (20 ng) was added to 48 μl of the master mix. The amplification profile included an initial denaturation step at 98°C for 30 s, followed by 35 cycles of 98°C for 10 s, 58°C for 10 s, and 72°C for 1 min, and a final extension of 72°C for 10 min. The suicide vector pEVP3 [22] was separately digested by *Sma*I (New England Biolabs, Ipswich, MA, USA). The pEVP3/*Sma*I digest was treated with calf intestinal alkaline phosphatase (New England Biolabs), and then ligated with the blunt-ended amplicons using T4 DNA ligase (New England Biolabs) according to the manufacturer's instructions. Transformation and homologous recombination were performed as previously described [23]. WT D39 competent cells were used as the recipient cells. Mutants with plasmid integrated into *lytA* were selected on agar plates containing 3 μg ml^{-1} cat (Figure 1B) [24].

Verification of the IDM-*ply* and -*lytA* *S. pneumoniae* mutants

To confirm the successful generation of *ply* and *lytA* knockdown mutants by the IDM technique, PCR amplification of the target sequences was performed. The primers used and predicted sizes of the PCR products of the two IDM mutants are listed in Table 1, and are shown in Figure 1A and B. For IDM-*ply*, three sets of primers were used to verify the parental and the mutant strains: (1) ply-P1/ply-P2, (2) ply-P2/pVA891-F, and (3) ply-P1/pVA891-R. Similarly, three primer sets were used to verify the IDM-*lytA* strain: (1) lytA-P1/lytA-P2, (2) lytA-P2/pEVP3-F, and (3) lytA-P1/pEVP3-R.

Construction and verification of IFD-*ply* and -*lytA* mutants

To construct the IFD mutants IFD-*ply* and -*lytA* of D39, overlap extension PCR was used to generate a linear DNA fragment containing the 5′ and 3′ flanking regions of *ply* or *lytA* (Figure 3A and B) [25,26]. Primers used to generate the DNA fragments are shown in Table 1, and the IFD mutants are shown in Figure 3A and B. Following PCR confirmation of the desired linear DNA fragments consisting of the flanking regions of *ply* or *lytA*, homologous recombination was performed using the IDM-*ply* and -*lytA* mutants as recipients. Transformation and homologous recombination for IFD were performed as previously described [23] (Figure 3A and B). Selection of IFD mutants was performed using blood agar plates with decreasing concentrations of erm for *ply* or cat for *lytA*, and increasing concentrations of ampicillin [8].

To confirm the IFD-*ply* and -*lytA* mutants were successfully obtained, PCR amplification of the target sequence was performed. Two pairs of primers, ply-P1/ply-P2 and lyt-P1/lytA-P2, were used to verify IFD-*ply* and -*lytA*, respectively. The predicted sizes of the amplicons are listed in Table 1.

***In vitro* evaluation of TGR from the IDM or IFD mutant strains**

To evaluate TGR, a single colony of each of the IDM mutants, picked from Muller-Hinton agar (MHA) plates supplemented with 5% defibrinated sheep blood and 2.5 μg ml^{-1} erm for the IDM-*ply* mutant or 3 μg ml^{-1} cat for IDM-*lytA*, was simultaneously incubated in 20 ml non-selective BHI medium, to mimic *in vivo* conditions in mice, and in 20 ml BHI with the corresponding antibiotic to compare differences in the rates of TGR. All cultures were incubated at 37°C in 5% CO_2. Samples were collected hourly from each culture over an 8-h period, beginning at 3 h post-inoculation. Genomic DNA was extracted from these samples using a Gentra Puregene DNA Purification Kit (Qiagen, Hilden, Germany). PCR detection of IDM and IFD mutants was performed immediately after sub-culturing of the mutants into BHI broth to confirm that there was no contamination with the WT D39 in the culture. The primers used to confirm the IDM and IFD mutants are listed in Table 1.

The rate of TGR was measured by semi-quantitative PCR, in which 20 ng of each template DNA was used for each measurement. The glucose kinase gene (*gki*) was used as an internal control and was amplified by primers gki-up and gki-dn (Table 1). Fold change represents the ratio of band intensity of *ply* or *lytA* having been reverted by TGR at different time points (T), divided by that of *gki* at the same time point, against the aforementioned ratio of the genes at the earliest time of TGR detection $[(ply_T/gki_T)/(ply_{\text{earliest time}}/gki_{\text{earliest time}})]$. To ensure no WT D39 contamination, PCR amplification of *ply* and *lytA* was performed concomitantly following sub-culturing of IDM and IFD mutants in BHI broth.

Validation of TGR from the results of previous PCR assay of IDM mutants by using real-time PCR assay, southern blot hybridization and western blotting

For Real time PCR assay, the 8-h culture, Day 1 and Day 2 culture of IDM-*ply*, cultured as previously mentioned, were collected. RNA was extracted by RNeasy® plus mini kit (Qiagen, Hilden, Germany) and then were converted to cDNA by SuperScipt® III First-strand Synthesis (Life Technologies). qPCR were performed by Step One™ Step One Plus™ 7500 Fast AB Biosystems and Fast SYPR Green Master Mix (Applied Biosystems, USA) and each samples were run in triplicate. Primers ply-F91 and ply-R1070 were used to detect the TGR *ply* whilst endogenous *gki* were detected by gki-F603 and gki-R791 (Table 1).

In Southern blotting, IDM-*ply* growth from different time points were collected. Serial dilution of the cultures were spread to BA to eneumate the viable colonies. Replica-plating with agar with and without erm were performed to select the colony with TGR of WT *ply*.

Southern blotting were performed similar as Ma L et al. [27]. Briefly, 10 μg of genomic DNA from D39 and TGR strain were extracted by Gentra Puregene Yeast/Bact. Kit (Qiagen, Hilden, Germany) and were then digested by *Cla*I (New England Biolabs, Ipswich, MA, USA) according to the manufactures' instruction. Samples were electrophoresed on an 0.8% agarose, which was then denatured and neutralized. Their DNA fragments were then transferred to a nylon membrane (PerkinElmer). Digoxigenin (DIG)-labeled probe ply-pb, synthesized by PCR amplification of ply-pbF and ply-pbR, were hybridized to the membrane and detected by DIG-luminescent detection kit (Roche, Mannheim Germany) (Table 1). If positive signal will be observed at 5 kb, TGR of full length of *ply* existed. In contrast, band showed at 7 kb indicating an IDM-*ply* mutant.

For the determination of protein expression of *ply*, the expression of WT PLY from IDM-*ply* mutants was checked by blotting of rabbit polyclonal pneumolysin antibody (abcam, ab71811, Cambridge, UK). Western blotting was done according to the method used by Waltman WD et al. [28]. Briefly, the 8-h of IDM-*ply* culture were harvested after overnight pneumococci grown from 5% blood agar with erm was adjusted to OD_{600} 0.05 in 20 ml BHI broth at 37°C and 5% CO_2 for eight hours until its OD_{600} equals to 0.35. Day 1 culture was obtained after adjusting the same OD_{600} 0.05 through inoculation the 8-h culture into another 20 ml BHI broth and was incubated at the same condition overnight. Cultures, centrifuged at 4000 rpm for 15 min, was washed twice with phosphate buffered saline (PBS), pH7.2. The pellet was then incubated with 200 μl lysis buffer with 100X Halt protease and phosphatase inhibitor Cocktail (Thermo Scientific, USA) at 37°C for 30 min. and sonicated to obtain the lysate. 40 µg protein in each lysates, measured by BCA method (Thermo scientific, USA), was run in a 10% SDS-PAGE and subsequently overlaid with a 0.2 µm PDVF membrane. Immobilon-PDVF membrane which was run by 100 mA for 1 hour and 20 min was blocked with 5% skim milk for 1 h at room temperature. Rabbit polyclonal pneumolysin antibody (1:2000) (Abcam, Cambridge, UK) was shaken overnight at 4°C. Goat polyclonal to rabbit IgG (HRP) antibody was used as secondary antibody (1:10000) was agitated for 1 hr at RT. WesternBright ECL reagent (Adansta, CA, USA) was used to detect WT PLY.

Evaluation of *in vivo* TGR in mice

In the *in vivo* analyses, 3-h cultures of 10^2–10^4 CFU of IDM-*ply* and IDM-*lytA* cultivated in BHI with and without antibiotics were injected intraperitoneally into five male 6-week-old BALB/c mice (National Laboratory Animal Center, Taiwan). Mice showing signs of illness at 10 min, 1 day, and 2 days were selected and sacrificed. Blood and liver samples were collected aseptically and DNA was then extracted using a DNeasy Blood and Tissue Kit (Qiagen, Hilden, Germany). Twenty nanograms of DNA from each sample were used as template in PCR analyses to detect intact *ply* or *lytA*, indicating TGR. The respective IDM-*ply* and –*lytA* genes were also amplified as controls and used for relative quantification. All animal experiments were approved by the Institutional animal care and use committee (IACUC) of National Health Research Institutes (NHRI-IACUC-101078-A) and were carried out according to their guidelines.

Effect of TGR on virulence in mice

To assess the effect of TGR on lethality in mice, 3-h and 8-h cultures of IDM mutants, incubated in the same liquid media as that used in the *in vivo* study, were investigated. D39 was used as the control. Ten-milliliter and 1-ml aliquots from 3-h and 8-h IDM cultures, respectively, were centrifuged and then washed with ice-cold phosphate buffered saline (PBS). Pellets were suspended in 300 µl of PBS. Ten 6–8-week-old male BALB/c mice were injected IP with approximately 10^2 CFU of each culture and then monitored for 14 days. Differences in survival among IDM mutants were analyzed at two time points using a log-rank test in SigmaStat 3.5 software (SPSS, Chicago, IL, USA).

Abbreviations

KD: Knockdown; KO: Knockout; *ply*: Pneumolysin gene; *lytA*: Autolysin gene; IDM: Insertion duplication mutagenesis; IFD: In-frame deletion; TGR: Target gene restoration; WT: Wild-type; qPCR: Quantitative polymerase chain reaction; DIG: Digoxigenin; cat: Chloramphenicol; erm: Erythromycin; IP: Intraperitoneal; *gki*: Glucose kinase gene; MHA: Muller-Hinton agar; BHI: Brain heart infusion; CFU: Colony-forming units; ml: Milliliter; T: Time point; IACUC: Institutional animal care and use committee; PBS: Phosphate buffered saline.

Competing interests

The authors declare that they have no competing interests.

Authors' contributions

EYL designed the study. EYL and JCC performed the laboratory work. EYL analyzed the data. CPF supervised the study. EYL, FYC, and CPF prepared the manuscript. All authors read and approved the final version of the manuscript.

Acknowledgements

We sincerely thank Dr. David Briles (Department of Microbiology, University of Alabama at Birmingham, AL, USA) for providing us with *S. pneumoniae* strain D39 and the IDM-*ply* mutant, as well as Dr. Don Morrison (UIC Biological Sciences, Chicago, IL, USA) for providing us with the suicide vector pEVP3.

Author details

[1]Institute of Clinical Medicine, School of Medicine, National Yang-Ming University, Taipei 11217, Taiwan. [2]Division of Infectious Diseases, Department of Internal Medicine, Tri-Service General Hospital, National Defense Medical Center, Taipei 11490, Taiwan. [3]National Institute of Infectious Diseases and Vaccinology, National Health Research Institutes, Zhunan, Miaoli County 35053, Taiwan. [4]Division of Infectious Diseases, Department of Medicine, Taipei Veterans General Hospital and National Yang-Ming University, Taipei 11217, Taiwan.

References

1. Hamilton HL, Schwartz KJ, Dillard JP: **Insertion-duplication mutagenesis of neisseria: use in characterization of DNA transfer genes in the gonococcal genetic island.** *J Bacteriol* 2001, **183**(16):4718–4726.
2. Ware D, Jiang Y, Lin W, Swiatlo E: **Involvement of potD in *Streptococcus pneumoniae* polyamine transport and pathogenesis.** *Infect Immun* 2006, **74**(1):352–361.
3. Koster W, Gudmundsdottir A, Lundrigan MD, Seiffert A, Kadner RJ: **Deletions or duplications in the BtuB protein affect its level in the outer membrane of Escherichia coli.** *J Bacteriol* 1991, **173**(18):5639–5647.
4. Mattos-Graner RO, Porter KA, Smith DJ, Hosogi Y, Duncan MJ: **Functional analysis of glucan binding protein B from *Streptococcus mutans*.** *J Bacteriol* 2006, **188**(11):3813–3825.
5. Joseph B, Przybilla K, Stuhler C, Schauer K, Slaghuis J, Fuchs TM, Goebel W: **Identification of *Listeria monocytogenes* genes contributing to intracellular replication by expression profiling and mutant screening.** *J Bacteriol* 2006, **188**(2):556–568.
6. Berry AM, Lock RA, Hansman D, Paton JC: **Contribution of autolysin to virulence of *Streptococcus pneumoniae*.** *Infect Immun* 1989, **57**(8):2324–2330.
7. Berry AM, Paton JC, Hansman D: **Effect of insertional inactivation of the genes encoding pneumolysin and autolysin on the virulence of *Streptococcus pneumoniae* type 3.** *Microb Pathog* 1992, **12**(2):87–93.
8. Berry AM, Ogunniyi AD, Miller DC, Paton JC: **Comparative virulence of *Streptococcus pneumoniae* strains with insertion-duplication, point, and deletion mutations in the pneumolysin gene.** *Infect Immun* 1999, **67**(2):981–985.
9. Lee MS, Seok C, Morrison DA: **Insertion-duplication mutagenesis in *Streptococcus pneumoniae*: targeting fragment length is a critical parameter in use as a random insertion tool.** *Appl Environ Microbiol* 1998, **64**(12):4796–4802.
10. Lee MS, Dougherty BA, Madeo AC, Morrison DA: **Construction and analysis of a library for random insertional mutagenesis in *Streptococcus pneumoniae*: use for recovery of mutants defective in genetic transformation and for identification of essential genes.** *Appl Environ Microbiol* 1999, **65**(5):1883–1890.
11. Berry AM, Paton JC: **Additive attenuation of virulence of *Streptococcus pneumoniae* by mutation of the genes encoding pneumolysin and other putative pneumococcal virulence proteins.** *Infect Immun* 2000, **68**(1):133–140.
12. Berry AM, Yother J, Briles DE, Hansman D, Paton JC: **Reduced virulence of a defined pneumolysin-negative mutant of *Streptococcus pneumoniae*.** *Infect Immun* 1989, **57**(7):2037–2042.
13. Graham RM, Paton JC: **Differential role of CbpA and PspA in modulation of in vitro CXC chemokine responses of respiratory epithelial cells**

to infection with *Streptococcus pneumoniae*. *Infect Immun* 2006, **74**(12):6739–6749.
14. Garcia-Sureda L, Domenech-Sanchez A, Barbier M, Juan C, Gasco J, Alberti S: **OmpK26, a novel porin associated with carbapenem resistance in *Klebsiella pneumoniae*.** *Antimicrob Agents Chemother* 2011, **55**(10):4742–4747.
15. Cruse G, Fernandes VE, de Salort J, Pankhania D, Marinas MS, Brewin H, Andrew PW, Bradding P, Kadioglu A: **Human lung mast cells mediate pneumococcal cell death in response to activation by pneumolysin.** *J Immunol* 2010, **184**(12):7108–7115.
16. Hirst RA, Gosai B, Rutman A, Guerin CJ, Nicotera P, Andrew PW, O'Callaghan C: ***Streptococcus pneumoniae* deficient in pneumolysin or autolysin has reduced virulence in meningitis.** *J Infect Dis* 2008, **197**(5):744–751.
17. Tomasz A, Moreillon P, Pozzi G: **Insertional inactivation of the major autolysin gene of *Streptococcus pneumoniae*.** *J Bacteriol* 1988, **170** (12):5931–5934.
18. Chen JH, Siu LK, Fung CP, Lin JC, Yeh KM, Chen TL, Tsai YK, Chang FY: **Contribution of outer membrane protein K36 to antimicrobial resistance and virulence in *Klebsiella pneumoniae*.** *J Antimicrob Chemother* 2010, **65**(5):986–990.
19. Yeh KM, Lin JC, Yin FY, Fung CP, Hung HC, Siu LK, Chang FY: **Revisiting the importance of virulence determinant magA and its surrounding genes in *Klebsiella pneumoniae* causing pyogenic liver abscesses: exact role in serotype K1 capsule formation.** *J Infect Dis* 2010, **201**(8):1259–1267.
20. Tsai YK, Fung CP, Lin JC, Chen JH, Chang FY, Chen TL, Siu LK: ***Klebsiella pneumoniae* outer membrane porins OmpK35 and OmpK36 play roles in both antimicrobial resistance and virulence †.** *Antimicrob Agents Chemother* 2011, **55**(4):1485–1493.
21. Briles DE, Hollingshead SK, Paton JC, Ades EW, Novak L, van Ginkel FW, Benjamin WH Jr: **Immunizations with pneumococcal surface protein A and pneumolysin are protective against pneumonia in a murine model of pulmonary infection with *Streptococcus pneumoniae*.** *J Infect Dis* 2003, **188**(3):339–348.
22. Claverys JP, Dintilhac A, Pestova EV, Martin B, Morrison DA: **Construction and evaluation of new drug-resistance cassettes for gene disruption mutagenesis in *Streptococcus pneumoniae*, using an ami test platform.** *Gene* 1995, **164**(1):123–128.
23. Chen JY, Fung CP, Chang FY, Huang LY, Chang JC, Siu LK: **Mutations of the rpoB gene in rifampicin-resistant *Streptococcus pneumoniae* in Taiwan.** *J Antimicrob Chemother* 2004, **53**(2):375–378.
24. Pestova EV, Morrison DA: **Isolation and characterization of three *Streptococcus pneumoniae* transformation-specific loci by use of a lacZ reporter insertion vector.** *J Bacteriol* 1998, **180**(10):2701–2710.
25. Gao W, Liu Y, Giometti CS, Tollaksen SL, Khare T, Wu L, Klingeman DM, Fields MW, Zhou J: **Knock-out of SO1377 gene, which encodes the member of a conserved hypothetical bacterial protein family COG2268, results in alteration of iron metabolism, increased spontaneous mutation and hydrogen peroxide sensitivity in *Shewanella oneidensis* MR-1.** *BMC Genomics* 2006, **7**:76.
26. Warrens AN, Jones MD, Lechler RI: **Splicing by overlap extension by PCR using asymmetric amplification: an improved technique for the generation of hybrid proteins of immunological interest.** *Gene* 1997, **186**(1):29–35.
27. Ma L, Lin CJ, Chen JH, Fung CP, Chang FY, Lai YK, Lin JC, Siu LK: **Widespread dissemination of aminoglycoside resistance genes armA and rmtB in *Klebsiella pneumoniae* isolates in Taiwan producing CTX-M-type extended-spectrum beta-lactamases.** *Antimicrob Agents Chemother* 2009, **53**(1):104–111.
28. Waltman WD, McDaniel LS, Gray BM, Briles DE: **Variation in the molecular weight of PspA (pneumococcal surface protein A) among *Streptococcus pneumoniae*.** *Microb Pathog* 1990, **8**(1):61–69.

9

Optimized DNA extraction from neonatal dried blood spots: application in methylome profiling

Akram Ghantous[1], Richard Saffery[2], Marie-Pierre Cros[1], Anne-Louise Ponsonby[3,4], Steven Hirschfeld[5], Carol Kasten[5], Terence Dwyer[3,4,6], Zdenko Herceg[1] and Hector Hernandez-Vargas[1*]

Abstract

Background: Neonatal dried blood spots (DBS) represent an inexpensive method for long-term biobanking worldwide and are considered gold mines for research for several human diseases, including those of metabolic, infectious, genetic and epigenetic origin. However, the utility of DBS is restricted by the limited amount and quality of extractable biomolecules (including DNA), especially for genome wide profiling. Degradation of DNA in DBS often occurs during storage and extraction. Moreover, amplifying small quantities of DNA often leads to a bias in subsequent data, particularly in methylome profiles. Thus it is important to develop methodologies that maximize both the yield and quality of DNA from DBS for downstream analyses.

Results: Using combinations of in-house-derived and modified commercial extraction kits, we developed a robust and efficient protocol, compatible with methylome studies, many of which require stringent bisulfite conversion steps. Several parameters were tested in a step-wise manner, including blood extraction, cell lysis, protein digestion, and DNA precipitation, purification and elution. DNA quality was assessed based on spectrophotometric measurements, DNA detectability by PCR, and DNA integrity by gel electrophoresis and bioanalyzer analyses. Genome scale Infinium HumanMethylation450 and locus-specific pyrosequencing data generated using the refined DBS extraction protocol were of high quality, reproducible and consistent.

Conclusions: This study may prove useful to meet the increased demand for research on prenatal, particularly epigenetic, origins of human diseases and for newborn screening programs, all of which are often based on DNA extracted from DBS.

Keywords: Blood spot, DNA extraction, Epigenetics, Methylome, HM450, Pyrosequencing, Whole bisulfitome amplification, QIAamp, GenSolve, NucleoSpin

Background

Epigenetic mechanisms, such as DNA methylation, have been suggested as possible causal pathways linking environmental exposure to disease. Many of these studies depend on the epigenome-wide analysis of prospectively collected samples, in the context of large human cohorts. As epigenome-wide technologies are becoming available, the use of such cohort studies will provide large amounts of information in the coming years. Due to the general lack of biospecimen collection in observational human studies, many of these cohorts rely on the use of dried blood spots (DBS) obtained soon after birth as the main source of biological information [1].

The use of filter paper for blood collection and analysis was implemented as early as the 1960s by Guthrie et al. using dried-blood samples for newborn phenylketonuria detection [2]. "Guthrie cards" are widely used in many types of tests, including chemical, serological, and genetic applications [3]. More recently, Flinders Technology Associates chemically treated filter papers (FTA cards) were specifically developed for DNA/RNA analyses [4]. These chemically treated cards allow long-term storage of DNA at room temperature and are impregnated with denaturants that guard against oxidation, nuclease and ultraviolet damage, and both bacterial and fungal degradation.

Neonatal DBS are routinely collected in many countries and represent a cost-effective tool to store precious

* Correspondence: vargash@iarc.fr
[1]Epigenetics Group, International Agency for Research on Cancer (IARC), 150 rue Albert-Thomas, 69008 Lyon, France
Full list of author information is available at the end of the article

biological specimens for subsequent studies. However, reliable profiling the DNA methylome in DBS has proven to be technically challenging, particularly because such techniques require stringent bisulfite preprocessing that can degrade DNA [5]. Other limitations of their use include the variable degradation of DNA due to storage and extraction, the usually small amounts of DNA that can be obtained (typical blood spots are between 6 and 10 mm in diameter), and the identification of technical artifacts potentially associated with long term storage [6].

Recently, there has been increasing interest in the use of DBS in DNA methylome analyses, using Methylated DNA Immunoprecipitaion combined with sequencing (MeDIP-seq) [7], Methyl-CpG Binding Domain (MBD) protein-enrichment combined with sequencing (MBD-seq) [8], and Infinium (Illumina) bead arrays [7,9-11]. The last version of Illumina's bead array, Infinium HumanMethylation450 (HM450) Beadchip, is cost-effective, requires DNA amounts as low as 300 ng, enables the detection and quantitation of DNA methylation levels at 486,685 CpG sites across the genome and represents one of the most comprehensive microarray methods to date for investigating the methylome [12]. Three reports have addressed the utility of HM450 on DBS [7,10,11]. The first report validated the use and the high correlation between two different methylomic platforms on DBS DNA: HM450 and MeDIP-seq [7]. The second one generated good quality methylome-wide data from DBS, as compared to their matched frozen buffy coat [10]. The third one used DBS-based HM450 analyses to study the epigenetic effects of gestational age, as recently demonstrated by our group [11]. However, none of these studies analyzed methods for optimized DNA extraction and quality verification from DBS, both of which represent major upstream steps in the pipeline for DBS-based research, including epigenetics.

In this report, we tested and developed a range of DNA extraction methods from neonatal FTA cards, individually or in combinations. We incorporated or modified protocol steps that were crucial to increase the DNA yield and quality from DBS, and additionally tested their efficiency on Guthrie cards. Moreover, we suggest an optimal protocol for both, pyrosequencing- and HM450-based, methylation studies. This work could prove useful in meeting the increased demand for research on prenatal origins of human diseases and for newborn screening programs.

Results

Optimization of Phases I and II in the DNA extraction protocols

Limited quantity and quality are important drawbacks in the use of DNA obtained from DBS, particularly for epigenome-wide studies. Initially, we ruled out the possibility of using whole bisulfitome amplification (WGA) after confirming the introduction of biases, mostly in the middle range of DNA methylation levels (Additional file 1), further confirming the recently reported finding by Bundo et al. [13]. Then, to systematically optimize DNA extraction from DBS, we divided the different steps of this process into two phases (Figure 1). Critical steps in Phase I included blood extraction off the filter papers, cell lysis and protease digestion (Figure 1A, left panel). Phase II included DNA precipitation, purification and elution (Figure 1A, right panel).

We have previously tested several genomic DNA extraction methods on DBS, including resin-based, lysis-based and magnetic bead-based [5]. Lysis- and bead-based methods were the best, but the latter is not suitable for beadchip methylation profiling, so it was not considered in this study [9]. Among lysis-based methods, several commercially available kits, including QIAamp DNA Micro Kit, GenSolve and NucleoSpin, have been shown to be efficient for DBS DNA extraction [10,14,15]. Therefore, we selected these three kits to optimize the two phases of DNA extraction. This optimization involved the combination of the different kits and modifications in several steps of the two protocol phases (Figure 1B) (described in detail in Additional file 2). A combination of GenSolve reagents in Phase I and QIAamp reagents in Phase II (referred to as GQ method) was set as a reference protocol to which other tested methods were compared (Tables 1, 2, 3, 4, 5 and 6). In all pairwise comparisons, two DBS punches from the same DBS were used, and assessment of quantity and quality were initially done using Nanodrop (with 260/280 and 260/230 spectrophotometric ratios as a measure of quality) (Tables 1, 2, 3, 4 and 5). These DBS were obtained from the National Children Study (NCS), USA, and were FTA-type, which preserves well DNA relative to other types of neonatal cards (Methods), hence, allowing comparisons across a wide range of DNA extraction protocols.

DNA yield and quality were consistently better for the reference GQ protocol when compared to Qiagen protocol (QQ) (Table 1, $p < 0.05$). Although DNA yield was drastically increased when ethanol was used in the Qiagen precipitation step (Qq, $p < 0.001$), DNA quality was still suboptimal compared to GQ, as assessed by nanodrop (Table 1). In contrast, the combination of Gensolve and NucleoSpin (GN and Gn) increased the DNA yield while preserving DNA quality, regardless of the use of ethanol in the precipitation step (Table 2, $p < 0.05$). A similar improvement was observed when using NucleoSpin kit in both phases of DNA extraction (NN and Nn) (Table 4, $p < 0.001$). The extra-small (XS) versions of NucleoSpin, with column designs specific for low elution volumes, did not consistently improve DNA quantity or quality, whether combined or not with other kits (GN-XS, Gn-XS and NN-XS; Tables 3, 5 and 6), the DNA precipitation buffer changed to

A

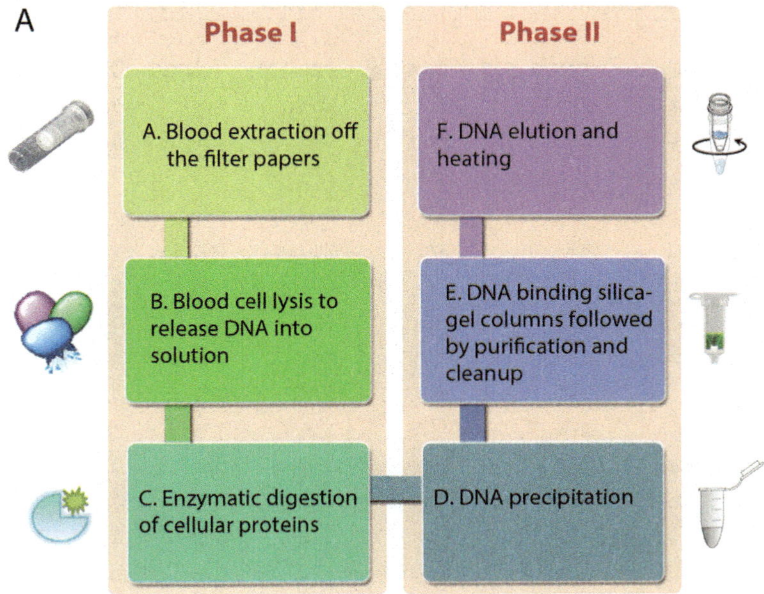

B

Protocol	Phase I Method	Phase II Method
QQ	QIAamp (DNA Micro Kit)	QIAamp (DNA Micro Kit)
Qq	QIAamp (DNA Micro Kit)	QIAamp (DNA Micro Kit) but precipitation buffer replaced by ethanol in Phase II.D
GQ	GenSolve	QIAamp (DNA Micro Kit)
GN	GenSolve	NucleoSpin (gDNA clean-up)
Gn	GenSolve	NucleoSpin (gDNA clean-up) but precipitation buffer replaced by ethanol in Phase II.D
GN-XS	GenSolve	NucleoSpin (gDNA clean-up XS)
Gn-XS	GenSolve	NucleoSpin (gDNA clean-up XS) but precipitation buffer replaced by ethanol in Phase II.D
NN	NucleoSpin (Tissue)	NucleoSpin (Tissue)
Nn	NucleoSpin (Tissue)	NucleoSpin (Tissue) but precipitation buffer replaced by ethanol in Phase II.D
NN-XS	NucleoSpin (Tissue-XS)	NucleoSpin (Tissue-XS)

Figure 1 Phases and classification of protocols used to extract DNA from DBS. Two sequential phases, each encompassing three steps, are outlined **(A)** and were optimized in the different protocols or method combinations **(B)** used to extract DNA from DBS. A spin basket is shown next to Phase I.A and consists of a tube with an embedded perforated basket used to separate blood solutions from the filter papers from which they were extracted. A silica-gel column with a funnel-shape design is shown next to Phase II.E and often used to elute small volumes (5–30 μl), as manufactured by Macherey-Nagel and supplied with the extra-small (XS) versions of NucleoSpin kits **(B)**.

ethanol, or the washing volume and frequency increased (Table 3, $p > 0.05$).

Cross-comparisons across the different tested DNA extraction protocols

DNA quality parameters assessed earlier were based on DNA 260/280 and 260/230 spectrophotometric ratios. Two other important quality parameters are DNA detectability by PCR and DNA integrity and size range, which can be assessed by gel electrophoresis and bioanalyzer analyses. DNA from all tested protocols exhibited detectable PCR bands of a housekeeping gene, *GAPDH* (Table 6 and data not shown), hence indicating that the DNA is amplifiable for specific short regions. DNA isolated by the GQ method exhibited a smear-like profile by gel electrophoresis, with peak intensity often greater than 1 kilo base pair (Kbp) (Figure 2 and Table 6). Bioanalyzer smear analyses confirmed the DNA average size peak to be greater than 1 Kbp, with an average size ranging across samples between 4.9-9.7 Kbp (Table 6 and data not shown). Compared to GQ, all tested protocols often showed more DNA degradation, except for protocol Qq which usually exhibited similar DNA smear profiles (Figure 2 and Table 6).

Overall comparison of the tested DNA quantity and quality parameters across the protocols shows that the best two protocols that at least match GQ in most of the

Table 1 Combinations of Gensolve and Qiagen protocols for DNA extraction from DBS: GQ, QQ and Qq methods

						65°C, 65 min		
	Sample code	DNA (ng/µl)	DNA (ng)	260/280	260/230	260/280	260/230	Protocol
	NCS 11a	21.9	944	2.03	1.04	1.89	1.02	GQ
	NCS 11b	3.4	147	1.55	0.19	1.35	0.29	QQ
	NCS 12a	19.1	802	1.91	1.19	1.90	1.18	GQ
	NCS 12b	6.1	257	1.58	0.36	1.73	0.37	QQ
	NCS 13a	8.8	376	1.82	0.68	2.01	0.67	GQ
	NCS 13b	2.9	120	2.48	0.25	2.00	0.21	QQ
Precipitation buffer in QQ changed to ethanol, leading to protocol Qq	NCS 8a	13.9	596	1.88	0.73	1.74	0.79	GQ
	NCS 8b	32.1	1379	1.25	0.93	1.28	0.98	Qq
	NCS 9a	11.2	493	2.06	0.72	1.93	0.75	GQ
	NCS 9b	27.4	1178	1.22	0.75	1.19	0.76	Qq
	NCS 10a	12.0	502	2.07	0.88	2.01	0.92	GQ
	NCS 10b	31.9	1371	1.47	0.99	1.44	1.01	Qq
	NCS 4-1a	15.3	644	2.10	0.83	1.96	0.89	GQ
	NCS 4-1b	26.1	1124	0.74	0.58	0.73	0.60	Qq
	NCS 4-2a	24.7	1062	1.97	1.05	1.96	1.06	GQ
	NCS 4-2b	49.9	2048	0.85	0.73	0.87	0.76	Qq
	NCS 4-3a	28.5	1225	2.08	1.22	1.87	1.15	GQ
	NCS 4-3b	46.9	1969	1.50	1.23	1.50	1.21	Qq

Two punches, each having 9 mm diameter, were analyzed per DBS. Punches labeled "a" were tested with GQ while their matched pairs, labeled "b", were tested with QQ or Qq. When the DNA precipitation buffer in QQ was changed to ethanol, the resultant protocol was termed Qq. Average eluate volume by GQ, QQ or Qq was 42 µl. Data represent averages of 2–4 readings per sample.

Table 2 Combinations of GenSolve, Qiagen and NucleoSpin protocols for DNA extraction from DBS: GQ *versus* GN and Gn methods

						65°C, 65 min		
	Sample code	DNA (ng/µl)	DNA (ng)	260/280	260/230	260/280	260/230	Protocol
	NCS 31a	5.77	242	1.70	0.42	1.55	0.41	GQ
	NCS 31b	8.87	426	1.81	0.95	1.54	0.71	GN
	NCS 30a	12.8	538	1.88	0.79	1.76	0.76	GQ
	NCS 30b	24.4	1171	1.84	1.22	1.90	1.51	GN
	NCS 29a	20.7	828	1.85	1.13	1.76	1.00	GQ
	NCS 29b	23.1	1063	1.93	1.38	1.85	1.37	GN
Precipitation buffer in GN changed to ethanol, leading to protocol Gn	NCS 22a	9.85	394	1.81	0.50	1.79	0.48	GQ
	NCS 22b	8.00	368	1.90	0.83	1.79	0.75	Gn
	NCS 21a	24.3	972	1.83	0.84	1.83	0.90	GQ
	NCS 21b	27.7	1274	1.64	0.54	1.83	0.61	Gn
	NCS 20a	15.0	600	1.78	0.56	1.77	0.60	GQ
	NCS 20b	14.4	662	1.70	0.76	1.74	0.89	Gn

Two punches, each having 9 mm diameter, were analyzed per DBS. Punches labeled "a" were tested with GQ while their matched pairs, labeled "b", were tested with GN. When the DNA precipitation buffer in GN was changed to ethanol, the resultant protocol was termed Gn. Average eluate volumes by GQ and GN/Gn were 42 µl and 47 µl, respectively. Data represent averages of 2–4 readings per sample.

Table 3 Combinations of GenSolve, Qiagen and NucleoSpin protocols for DNA extraction from DBS: GQ *versus* GN-XS and Gn-XS methods

						65°C, 65 min		
	Sample code	DNA (ng/µl)	DNA (ng)	260/280	260/230	260/280	260/230	Protocol
	NCS 28a	20.3	873	1.84	0.95	1.82	0.97	GQ
	NCS 28b	37	1813	1.87	0.11	1.78	0.13	GN-XS
	NCS 27a	19.4	834	1.85	0.85	1.80	0.90	GQ
	NCS 27b	27.9	1339	2.19	0.61	2.26	0.61	GN-XS
	NCS 26a	19.7	847	1.83	0.97	1.74	0.93	GQ
	NCS 26b	23.7	1161	1.66	0.79	1.77	0.80	GN-XS
	NCS 25a	16.3	701	1.9	1.07	1.79	1.07	GQ
	NCS 25b	21.7	1042	1.83	0.34	1.96	0.28	GN-XS
	NCS 24a	11.3	486	1.82	0.78	1.79	0.82	GQ
	NCS 24b	21.6	1037	1.53	0.08	1.48	0.07	GN-XS
	NCS 23a	11.9	512	1.18	0.49	1.27	0.54	GQ
	NCS 23b	11.4	547	1.44	0.50	1.52	0.42	GN-XS
Precipitation buffer in GN-XS changed to ethanol, leading to protocol Gn-XS	NCS 19a	30.1	1174	1.53	0.71	1.57	0.81	GQ
	NCS 19b	15.1	725	1.49	0.65	1.50	0.70	Gn-XS
	NCS 18a	12.2	488	1.81	0.62	1.78	0.64	GQ
	NCS 18b	20.4	979	1.30	0.64	1.44	0.72	Gn-XS
	NCS 17a	59	2360	1.80	1.37	1.91	1.56	GQ
	NCS 17b	42.6	2130	1.65	0.53	1.75	0.57	Gn-XS
Washing volume and frequency in GN-XS increased	NCS 16a	16.8	823	1.71	0.76	1.80	0.78	GQ
	NCS 16b	13.1	707	1.43	0.21	1.55	0.22	GN-XS
	NCS 15a	27.1	1382	1.76	1.04	1.79	0.97	GQ
	NCS 15b	12.6	668	7.83	0.39	Error *	0.45	GN-XS
	NCS 14a	30.3	1545	1.77	1.01	1.83	1.07	GQ
	NCS 14b	22.8	1208	2.37	0.60	2.60	0.61	GN-XS

Two punches, each having 9 mm diameter, were analyzed per DBS. Punches labeled "a" were tested with GQ while their matched pairs, labeled "b", were tested with GN-XS. When the DNA precipitation buffer in GN-XS was changed to ethanol, the resultant protocol was termed Gn-XS. For NCS 16b, 15a and 14b, the washing volume was increased from 100 to 500 µl, and washing was performed twice instead of once. Average eluate volumes by GQ and GN-XS/Gn-XS were 42 µl and 48 µl, respectively. Data represent averages of 2–4 readings per sample. * The error represents values out of range.

tested parameters are Qq and NN (or Nn) (Table 6). Relative to GQ, the only drawback of Qq is its 260/280 DNA ratios (Table 6). Qq 260/280 ratios were always out of range, indicating that the high Qq DNA quantity measurements recorded spectrophotometrically by nanodrop may not be accurate. However, quantification with Qubit, a fluorescent-based method, confirmed the yield in Qq to be 2.7× higher than in GQ (data not shown). The only drawback of NN (or Nn) relative to GQ was DNA integrity, with the DNA size ranges in NN or Nn being lower than in GQ (Table 6).

In conclusion, protocol GQ seems to be the most robust among the tested methods across all tested DNA quantity and quality parameters. Qq can be rather suitable for applications requiring larger DNA quantities from DBS, while maintaining large fragment sizes, but in which 260/280 ratios are not a necessity. On the other hand, protocols NN or Nn may be better suitable for applications requiring larger DNA quantities from DBS, relative to GQ, while maintaining optimal 260/280 and better 260/230 ratios, but in which large DNA fragment sizes are not a requirement.

Of note, when GQ was tested on DBS (Guthrie cards, Whatman 903) from the Tasmanian Infant Health Survey (TIHS), Australia, dating more than 20 years old, an average of 66 ± 15 ng (n = 3) of DNA could be extracted per two punches, each being 1 mm in diameter, with a mean 260/280 ratio of 1.66 ± 0.02; these DNA quantities are equivalent to 42.0 ng/mm^2 for TIHS, compared to 12.7 ng/mm^2 for NCS samples.

Performance of DNA extracted from DBS using methylome-wide analysis

Methylation probe call index

In order to assess the performance of DNA extracted from DBS in HM450 methylome-wide analyses, we used

Table 4 Combinations of GenSolve, Qiagen and NucleoSpin protocols for DNA extraction from DBS: GQ *versus* NN and Nn methods

						65°C, 65 min		
	Sample code	DNA (ng/μl)	DNA (ng)	260/280	260/230	260/280	260/230	Protocol
	NCS 1a	25.3	1062	1.84	1.15	1.88	1.18	GQ
	NCS 1b	40.0	1802	1.82	1.84	1.90	1.90	NN
	NCS 2a	28.8	1209	1.87	1.27	1.89	1.33	GQ
	NCS 2b	47.3	2127	1.89	2.00	1.90	1.91	NN
	NCS 3a	15.2	638	2.02	1.09	1.93	1.15	GQ
	NCS 3b	27.6	1271	1.82	1.59	1.91	1.61	NN
	NCS 4-4a	23.0	987	1.93	0.93	1.96	1.03	GQ
	NCS 4-4b	30.0	1348	1.85	1.72	1.90	1.81	NN
	NCS 4-5a	33.4	1435	1.82	1.01	1.83	1.02	GQ
	NCS 4-5b	35.4	1593	1.98	1.94	1.87	1.76	NN
	NCS 4-6a	19.9	856	2.01	1.14	1.97	1.10	GQ
	NCS 4-6b	33.9	1525	1.93	1.85	1.89	1.82	NN
Precipitation buffer in NN changed to ethanol, leading to protocol Nn	NCS 4-7a	18.0	774	1.76	0.87	1.86	0.84	GQ
	NCS 4-7b	25.7	1155	1.93	1.37	1.81	1.35	Nn
	NCS 4-8a	14.7	617	1.99	0.90	1.94	0.91	GQ
	NCS 4-8b	24.6	1105	1.92	1.33	1.85	1.42	Nn
	NCS 4-9a	26.1	1121	1.90	1.20	1.94	1.26	GQ
	NCS 4-9b	39.0	1754	1.95	1.89	1.88	1.77	Nn

Two punches, each having 9 mm diameter, were analyzed per DBS. Punches labeled "a" were tested with GQ while their matched pairs, labeled "b", were tested with NN. When the DNA precipitation buffer in NN was changed to ethanol, the resultant protocol was termed Nn. Average eluate volumes by GQ and NN/Nn were 42 μl and 45 μl, respectively. Data represent averages of 2–4 readings per sample.

Table 5 Combinations of GenSolve, Qiagen and NucleoSpin protocols for DNA extraction from DBS: GQ *versus* NN-XS methods

					65°C, 65 min		
Sample code	DNA (ng/μl)	DNA (ng)	260/280	260/230	260/280	260/230	Protocol
NCS 4a	17.5	734	1.98	0.96	1.84	0.98	GQ
NCS 4b	17.3	833	1.72	0.84	1.65	0.78	NN-XS
NCS 5a	17.6	741	1.82	1.05	1.81	1.07	GQ
NCS 5b	12.8	614	1.61	1.02	1.73	1.07	NN-XS
NCS 6a	8.2	328	2.02	0.81	1.79	0.76	GQ
NCS 6b	15.7	756	1.70	0.89	1.70	0.86	NN-XS
NCS 4-10a	22.2	956	1.95	1.21	2.02	1.19	GQ
NCS 4-10b	13.5	660	1.84	1.10	1.85	1.08	NN-XS
NCS 4-11a	19.1	822	1.60	0.85	1.71	0.88	GQ
NCS 4-11b	12.6	618	1.60	1.00	1.58	1.11	NN-XS
NCS 4-12a	10.4	447	1.54	0.74	1.76	0.74	GQ
NCS 4-12b	13.1	657	1.55	0.76	1.70	0.74	NN-XS

Two punches, each having 9 mm diameter, were analyzed per DBS. Punches labeled "a" were tested with GQ while their matched pairs, labeled "b", were tested with NN-XS. Average eluate volumes by GQ and NN-XS were 42 μl and 49 μl, respectively. Data represent averages of 2–4 readings per sample.

Table 6 Cross-comparisons of DNA quantity and quality parameters among the different tested DNA extraction protocols

Protocol	DNA quantity (ng)[a]	260/280 ratio (after heating)[b]	260/230 ratio (after heating)[c]	DNA integrity[d]	Detectability by PCR
GQ	808 ± 376 ng	1.83 ± 0.14	0.92 ± 0.23	Peak intensity > 1 Kbp (25/32)	Detectable (42/42)
QQ	Lower (3/3) 175 ± 73 ng	Similar (2/3); Worse (1/3)	Lower (3/3)	NA[e]	**Similar (3/3)**
Qq	**Higher (6/6) by 2.0 ×**	Worse (6/6) 0.89 ± 0.22	**Similar (6/6)**	**Similar (4/6); Better (1/6); Worse (1/6)**	**Similar (6/6)**
GN or *Gn*	**Higher (4/6) by 1.5 ×; Similar (2/6)**	**Similar (6/6)**	**Similar (4/6); Higher (2/6)**	Worse (4/4)	**Similar (6/6)**
GN-XS or *Gn-XS*	**Higher (6/9) by 1.7 ×; Similar (2/9); Lower (1/9) by 0.4 ×**	Similar (6/9); Worse (3/9)	Similar (5/9); Lower (4/9)	Worse (6/6)	Similar (9/9)
NN or *Nn*	**Higher (8/9) by 1.7 ×; Similar (1/9)**	**Similar (9/9)**	**Higher (9/9) 1.71 ± 0.20**	Worse (6/6)	**Similar (9/9)**
NN-XS	Similar (2/6); Higher (2/6) by 1.8 ×; Lower (2/6) by 0.3 ×	**Similar (5/6); Worse (1/6)**	**Similar (6/6)**	Worse (3/5); Similar (1/5); Better (1/5)	**Similar (6/6)**

The GQ protocol is set as the reference protocol above all the other protocols to which it is compared. The DNA quantity or quality parameters of the other protocols are described always in comparison to GQ. DNA yields of QQ were significantly lower than GQ ($p < 0.05$), those of Qq, GN, Gn, NN and Nn were significantly higher than GQ ($p < 0.05$), and those of GN-XS, Gn-XS and NN-XS were not different from GQ ($p > 0.05$), as compared by paired t-test. For each DNA parameter described, the counts of hits over the total number of DBS analyzed for that parameter are indicated between parentheses. Table cells highlighted in bold represent protocol performance that is at least as good as that of GQ.

[a]Quantities showing less than 20% change from GQ were considered 'similar' to GQ. This threshold exceeds the 15.7% average increase in DNA quantities observed between GQ technical replicates, with an average coefficient of variation of 12.3% (n = 6 pairs of replicates).

[b]In protocol pairwise comparisons, 260/280 ratios that were considered 'similar' were either both within the optimal 1.7-2.0 range or both outside this range. Otherwise, the ratio outside the 1.7-2.0 range was considered 'worse' relative to that within.

[c]'Lower' and 'higher' indicate 260/230 ratio differences of at least 0.30 absorbance units below or above GQ ratios, respectively; otherwise, ratios were considered 'similar'. The following guidelines were adopted for the 260/230 ratio: optimal and indicating pure DNA if higher than 2.0, acceptable if between 1.5-2.0, and tolerated if between 1.0-1.49 (Macherey-Nagel GmbH & co. KG, reference 740230).

[d]DNA integrity refers to DNA size range and level of degradation, as assessed by gel electrophoresis. Ten samples were also reassessed by bioanalyzer, showing similar relative comparisons to GQ. Every DBS, in which both punches of the tested protocol pairs exhibited high DNA degradation (size range below 1 Kbp), was excluded from the pairwise comparisons of protocols.

[e]NA: Not Applicable due to limited quantities of extractable DNA.

the DNA extracted by GQ, being the most robust protocol. DNA from two sample pairs, each representing two punches (serving as technical replicates) from the same DBS, were analyzed by HM450. In addition, DBS pairs were compared to reference DNA from neonatal blood or cell lines (Table 7). In all tested samples, whether originating from neonatal blood, DBS or cell lines, more than 99% of the 485,577 HM450 individual probes were detected, using the commonly accepted quality control detection p-value of 0.01, hence, indicating high quality data. The average beta-values were similar between the technical replicates NCS 37a and 37b (approximately 0.47 for either) and between the technical replicates NCS 38a and 38b (approximately 0.42 for either) (Table 7).

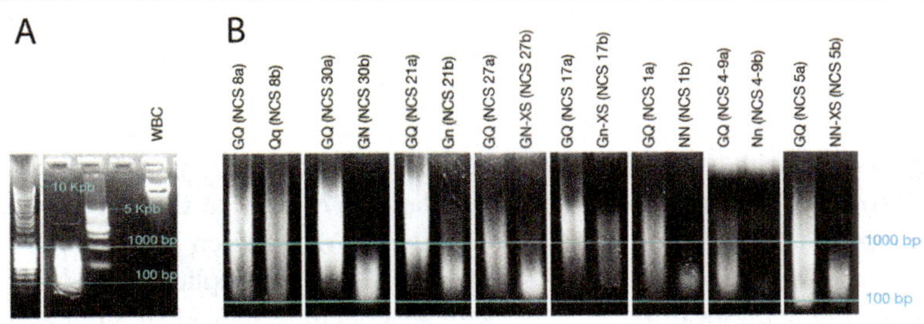

Figure 2 DNA integrity and size range as assessed by agarose gel electrophoresis. (A) DNA size markers used to estimate size ranges are shown in addition to genomic DNA that was isolated from white blood cells (WBC) and used as a positive control. **(B)** Representative DBS from each of the tested protocols are shown, except for protocol QQ in which DNA amounts were insufficient to be analyzed by gel electrophoresis. Eight different gel sections are shown and are derived from either the same gel or different gels. In each section, two punches from the same NCS spot were run on the same gel, with the first punch, labeled 'a', representing protocol GQ and the second punch, labeled 'b', representing another unique protocol from the tested set. The two blue lines, representing the 100 and 1000 base pair (bp) size ranges, were set according to the molecular size marker used in each section. The 1000 base pair limit is a minimum size range with useful applications in many genetic and epigenetic studies, including Illumina's HM450 Beadchip array. The results of other DBS analyzed by gel electrophoresis or bioanalyzer are summarized in Table 6.

Table 7 Methylation quality control probe evaluation

	Sample	Probe		Beta-value		
		Number of CpGs detected with $p < 0.01$	Percentage of CpGs detected with $p < 0.01$	Average	Minimum	Maximum
Neonatal Blood DNA	NB 1672	485405	99.96	0.4886	0.0012	0.9929
	NB 1597	485392	99.96	0.4729	0.0006	0.9947
	NB 1842	485358	99.95	0.4911	0.0009	0.9940
	NB 1645	485119	99.91	0.4704	0.0011	0.9914
DBS DNA	NCS 37a	484990	99.88	0.4712	0.0038	0.9953
	NCS 37b	484946	99.87	0.4719	0.0045	0.9916
	NCS 38a	483897	99.65	0.4226	0.0005	0.9942
	NCS 38b	482519	99.37	0.4240	0.0001	0.9935
Cell Line DNA	Cell Line 1	485124	99.91	0.4748	0.0022	0.9926
	Cell Line 2	485175	99.92	0.4813	0.0032	0.9920
	Cell Line 3	485342	99.95	0.4738	0.0006	0.9926
	Cell Line 4	485272	99.94	0.4743	0.0021	0.9934

Three sources of DNA are used in HM450 array: neonatal blood (NB), DBS and cell lines. NB and cell line DNA is of good genomic quality and serves as technical reference. NB provides DNA from the same tissue origin as DBS, being blood. The cell line DNA is of hepatic origin. The technical pairs are represented by two punches from each of two blood spots and are labeled as 'NCS 37a and 37b' and 'NCS 38a and 38b', respectively. Probe p-value was set to 0.01. The percentage of detected probes ($p < 0.01$) represents the proportion out of the total of 485,577 probes on the HM450 array. The average, minimum and maximum beta-values with detection $p < 0.01$ are shown (background was not subtracted, so minimum beta-values are not exactly zeros).

Sample–dependent and –independent HM450 internal quality control probes

For sample and array quality, HM450 array includes 850 quality control (QC) probes. Fifteen QC probes are sample-independent and 835 QC probes are sample-dependent [12]. DNA from DBS, neonatal blood or cell lines passed the described HM450 QC (Additional file 3 and Figure 3). Background probes, wherever included, produced minimal signal (maximum limit is 1000 units, as recommended by Illumina Inc.), and the intended positive signals from the experimental QC probes were above background, for all of the three tested DNA sources (Additional file 3 and exemplified in Figure 3 using non-polymorphic probes, which are indicative of overall performance). In addition, performance of DBS samples was similar to that of subsets taken from reference neonatal blood and cell line samples (Additional file 4 and Figure 3). Bisulfite conversion efficiency for both, type I and II probes, was high for all tested samples (Additional file 4) and was confirmed by PCR using primers that are specific either to bisulfite converted or to non-modified *GAPDH* DNA regions (data not shown).

Differential methylation and clustering analyses using HM450 data

Differential methylation of HM450 beta-values produced two major clusters (Figure 4). Cell line DNA samples formed one cluster while DBS and neonatal blood DNA formed another; this is expected because the cell lines used are of hepatic tissue origin while both, DBS and neonatal blood samples, are of blood tissue origin. Within the cluster of blood biospecimens, all four neonatal blood samples formed one sub-cluster, which was segregated away from the DBS sub-cluster. The two punches NCS 37a and 37b, representing the technical replicates from the same spot NCS 37, clustered together and away from the other two technical replicates, NCS 38a and 38b, which also clustered together (Figure 4). This further supports the observed higher correlations ($p < 0.001$, Steiger Z test) between technical duplicates punched from the same DBS (r^2 range = 0.963-0.990) *versus* different DBS (r^2 range = 0.949-0.951; Additional file 4A). When the analysis was limited to the top 1% of probes that showed the highest variance in M-values (transformed betas) across any of the four tested DBS, NCS_37a, NCS_37b, NCS_38a and NCS_38b, the correlation between replicates (r^2 range = 0.850-0.972; $p < 0.001$, Pearson) become significantly higher and more predictive of replication ($p < 0.001$, Steiger Z test) than the correlation between non-replicates (r^2 range = 0.365-0.371; $p < 0.001$, Pearson) (Additional file 4B). Similar observations were reported using Spearman correlations. Moreover, the frequency distributions of delta M-values (δM) between samples were centred at zero only between technical replicates (Additional file 4B). Hence, we can conclude that HM450 analyses using DBS DNA, extracted using the GQ protocol, is reproducible.

Performance of DNA extracted from DBS using sequence-specific methylation analysis

The performance of DNA extracted from DBS by the GQ method was then tested using sequence-specific methylation analyses. For this purpose, we analyzed the methylation levels of several CpG sites in Line1 and AluYb8 sequences,

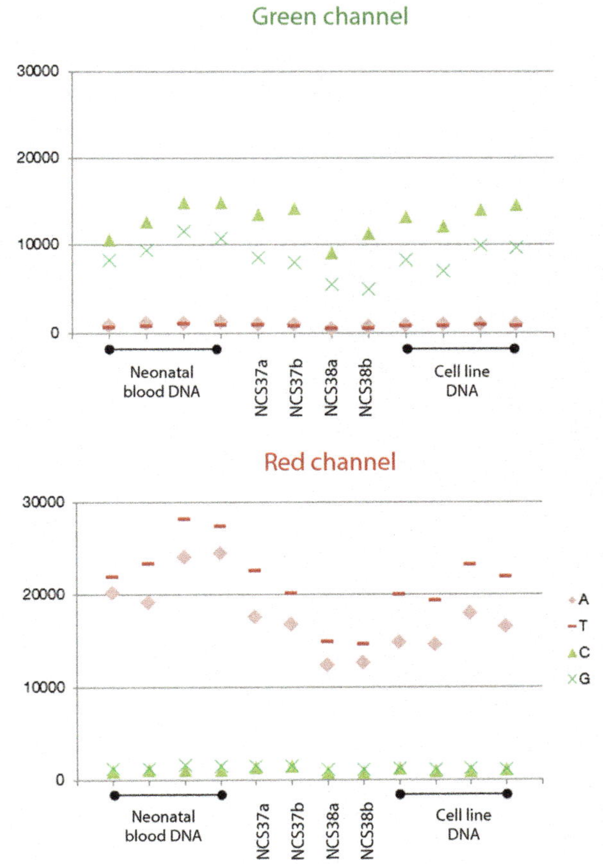

Figure 3 HM450 QC plot using Non-polymorphic probes which assess overall performance. In the green channel, background signals are shown in red and pink while positive signals in opaque and fluorescent green. In the red channel, background signals are shown in opaque and fluorescent green while positive signals in red and pink. One non-polymorphic control has been designed for each of the four nucleotides A, T, C, and G. Four DBS DNA samples are shown between four neonatal blood and four cell line DNA samples, in each of the two plots. The DBS samples represent two NCS spots, 37 and 38, each consisting of two tested punches labeled 'a' or 'b'.

both of which are proxy markers of global methylation, being transposable elements interspersed across the genome [16]. The technical replicates, NCS 37a and 37b, showed similar Line1 and AluYb8 methylation levels at each tested CpG site, and similar data were observed with the pair, NCS 38a and 38b (Figure 5A and B). These results show that inter-replicate variation in methylation levels is minimal using several CpG sites in two different loci, Line1 and AluYb8. Moreover, the observed difference in methylation levels between the pair NCS 37a and 37b *versus* NCS 38a and 38b was consistent at every single CpG tested and across both, Line1 and AluYb8 loci ($p < 0.1$ for CpG6 in Line1 and CpG3 in AluYb8 and $p < 0.05$ for all other CpGs; Mann–Whitney test) (Figure 5A and B). These results confirm that DNA extracted from DBS using GQ is suitable to detect small methylation differences in a consistent manner and with low inter-replicate variation.

Discussion

DBS have become an increasingly important tool for diagnostic purposes and for epigenetic, genetic and epidemiological research. We have previously tested a range of commercially available DNA extraction kits for purifying genomic DNA from fresh and dried blood for downstream PCR and DNA methylation applications [5]. We found that genomic DNA extraction, using the ChargeSwitch Forensic DNA Purification kit (Invitrogen), with subsequent bisulfite modification, using the MethylEasy kit (Human Genetic Signatures), was best in yielding bisulfite-converted DNA of sufficient quantity and quality for downstream candidate-gene DNA methylation analyses, such as SEQUENOM MassArray EpiTYPER analysis [5]. However, DNA extraction with ChargeSwitch was recently shown to be not suitable for beadchip methylation profiling, leading to up to 16% loss of detectable probes in Infinum HumanMethylation27 (Illumina Inc.) arrays analysis [9]. In relation to the limited amounts of DNA extracted from DBS, a recent report also pointed to the biases introduced by whole bisulfitome amplification and the need for careful data interpretation [13], as we have also observed in this study.

In this work, we have systematically compared different DNA extraction methods from DBS by dissecting different phases of extraction and optimizing several steps within each phase, using commercial and in-house extraction protocols. For these purposes, we used a homogenous set of DBS samples, spotted on the same day and stored in a similar manner, to provide a common platform for cross-protocol comparisons. Moreover, this study emphasizes DNA extraction protocols that have particular utility in a recent technology for studying methylome-wide methylation, Infinium HM450, and sequence-specific methylation, by pyrosequencing. The use of DBS for diagnostic and research purposes is not new, but there is a lack of quality standards for optimizing DNA extraction. This study suggests different DNA extraction protocols, each having specific advantages tailored for specific applications. Protocol GQ does not extract the highest DNA yield, but provides DNA in quantities and qualities sufficient for HM450 methylome-wide and sequence-specific methylation analyses. With GQ, the 260/280 ratios are consistently optimal and the extracted DNA is less fragmented relative to other protocols. Protocol Qq, on the other hand, produces twice as much DNA as GQ and with similar DNA integrity. However, 260/280 ratios in Qq are unreliable and cannot be used for sample selection, particularly for expensive downstream applications. As for protocols NN or Nn, they extract 1.7 folds more DNA than GQ and show optimal ranges for both, 260/280 and 260/230 ratios. However,

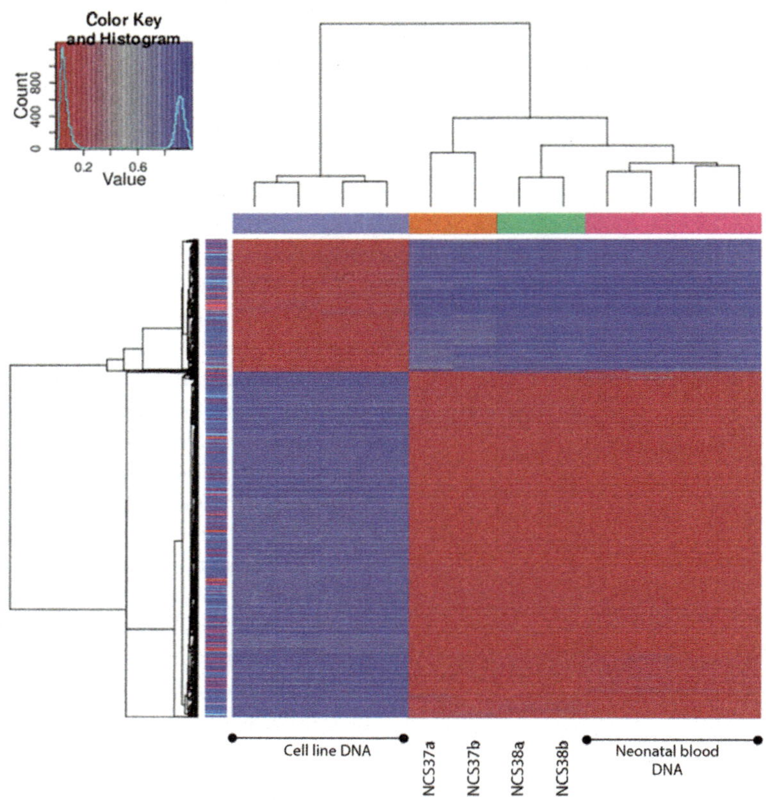

Figure 4 Differential methylation and unsupervised clustering analysis of HM450 data from neonatal blood, DBS and cell line DNA. Neonatal blood and cell line DNA samples are used as positive controls of good DNA quality for reference comparisons with DNA extracted from DBS. Neonatal blood and DBS are from different individuals. Four DBS DNA samples are shown between four different neonatal blood and four different cell line DNA samples. The DBS samples represent two NCS spots, 37 and 38, each consisting of two tested punches labeled 'a' or 'b'. HM450 beta-values were clustered using Euclidean distance as the dissimilarity index. As shown in the color key, the red and blue signals represent relatively hypomethylated and hypermethylated regions, respectively.

these protocols lead to more DNA fragmentation relative to GQ. This may represent a limitation for bead array-based assays and other DNA methylation assays where DNA integrity is a requirement. It should be noted though that 260/280 and 260/230 ratios should be treated with caution; for example, different contaminants can compensate for each other's' deviations, resulting in misleading optimal 260/280 ratios.

Because the type of purification column was identical in GQ and Qq, but different from those used in all the other methods, column nature (Phase II) could be the reason for the better DNA integrity in the two protocols. This is further supported by our results showing that in GQ, changing Phase II (includes column type) while maintaining Phase I, as in GN/n or GN/n-XS, compromises DNA integrity (Table 6). Changing the DNA precipitation buffer to ethanol reduces the need in some protocols to vigorously vortex to dissolve resultant precipitates but does not seem to enhance DNA integrity (Figure 2; compare GN *versus* Gn, GN-XS *versus* Gn-XS, and NN *versus* Nn). However, ethanol, as a precipitating buffer, was essential in some protocols to increase the DNA yield, as in Qq versus QQ (Tables 1 and 6).

Other studies have compared DNA extraction protocols from DBS, but irrespective of epigenetic applications [17-19]. It is difficult to compare DNA extraction protocols across different studies due to many reasons, such as differences in the structures of filter papers on which blood was soaked, the storage conditions and year-durations, and DNA quantifications methods used. However, these studies used QIAamp DNA Mini Kit (Qiagen) as a reference method, which is similar to the QQ protocol in this work, hence, allowing comparisons between our and their optimized methods. Sjoholm et al. reported that QIAamp DNA Mini Kit performed the best relative to EZNA (Omega Bio-Tek), Chelex 100 (AmershamBiosciences) and alkaline lysis (GenomiPhi DNA Amplification Kit, AmershamBiosciences) [19]. On the other hand, the in-house developed DNA extraction methods reported by Hue et al. and Hollegaard et al. yielded, by Nanodrop quantification, 3.3 and 2.5 fold more DNA, respectively than QIAamp DNA Mini Kit [18]. In comparison, the GQ

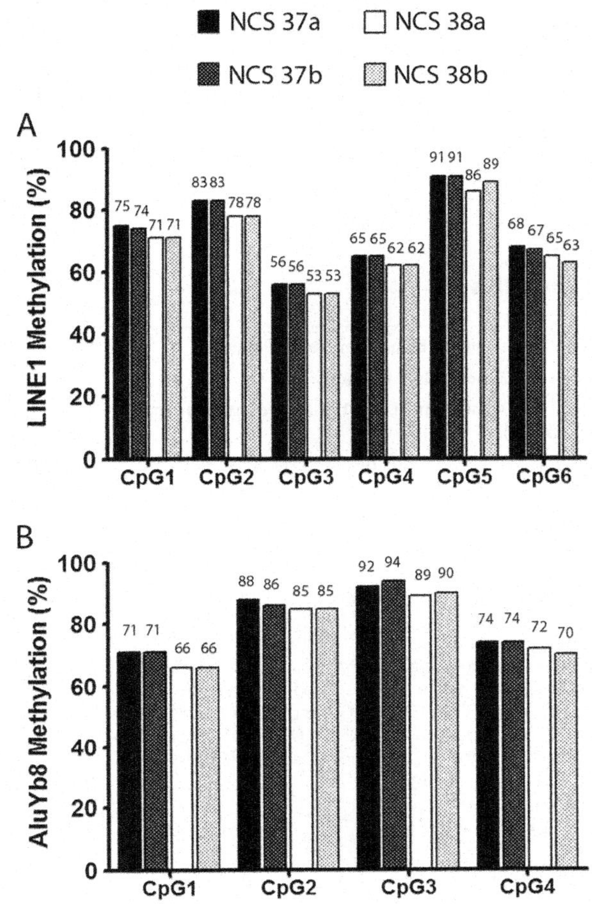

Figure 5 Methylation analyses of *Line1* and *AluYb8* loci using bisulfite pyrosequencing. The methylation levels of six and four CpG sites were analyzed for *Line1* **(A)** and *AluYb8* **(B)**, respectively, and are expressed as percent of the total number of CpGs analyzed for each individual CpG site. The DBS samples represent two NCS spots, 37 and 38, each consisting of two tested punches labeled 'a' or 'b'.

protocol in our study yielded, also by Nanodrop quantification, on average 4.6 fold more DNA than matched QQ samples; in addition, protocols Qq, NN and Nn yielded at least 1.7 folds more DNA than matched GQ samples. Moreover, the in-house protocol by Hue et al. produced a low purity 260/280 average ratio (1.50) [18], while GQ, NN and Nn ratios were optimal in every tested sample. These findings support the good performance of our optimized methods relative to many other in-house and commercial DNA extraction protocols from DBS. Interestingly, one study reported a recent method suitable for performing scalable DNA extractions simultaneously from many DBS, but with less emphasis on DNA quality and yield comparisons across different methods [20]. The scale of our tested methods can be increased by implementing the QIAcube technology (Qiagen), and, with scalable designs, laser cutting of DBS punches would eliminate cross-contamination, as has been recently reported [21].

Conclusion

This study arises from an international effort across several cohorts and working groups aiming to fulfill the need to systematize quality standards in DNA extraction and to increase the DNA yield using DBS, particularly with the advent of high-throughput epigenomic technology that require high quality and quantity of DNA. Given the emerging appreciation of DBS collected at birth as a valuable resource for epigenetic analyses prior to phenotypic onset, our optimized methods for DNA extraction with application in methylation analyses have great potential for diagnostic and research purposes.

Methods

Sample overview

DBS that were used to perform the comparisons across the DNA extraction protocols were obtained from NCS, USA, and have been spotted on Flinders Technology Associates (FTA) mini cards on the same day and dried in air-sealed containers for approximately two years at room temperature. Guthrie cards (Whatman 903) from TIHS, Australia, dating more than 20 years old, were also used to test the efficiency of the robust protocol GQ. Both NCS and TIHS samples were heel-prick without anticoagulants added. Permissions from the ethical committees of the International Agency for Research on Cancer (IARC), as well as both, NCS and TIHS, were obtained. TIHS is one of the founder cohorts of the International Childhood Cancer Cohort Consortium (I4C) [1].

DNA extraction protocols

Combinations of different commercially available DNA extraction kits were used, including QIAamp DNA Micro Kit (Qiagen 56304), GenSolve (Gen Vault, GVR110), NucleoSpin (gDNA clean-up, Macherey-Nagel 740230), and the extra-small (XS) version of NucleoSpin (gDNA clean-up XS, Macherey-Nagel 740904). Reported quantifications were done using Nanodrop, unless indicated otherwise using Qubit™ dsDNA High Sensitivity Assay, Invitrogen Q32851. Detailed protocols are included in Additional Methods (Additional File 2).

Gel electrophoresis and bioanalyzer analysis

Samples were run on a 0.8% agarose gel (Eurobio GEPAGA07-65) in 1 × Tris Acetate-EDTA buffer and stained with GelRed. 300 ng of DBS DNA were utilized per sample in electrophoresis analyses of DNA integrity and size range. The following DNA size markers were used: 80–10,000 bp ladder (Thermo SM0403), 500–5,000 bp ladder (Takara 3411A), and 100–1,000 bp ladder (Thermo SM0243). As for bioanalyzer analysis, 500 pg of DNA per sample was loaded on the chip and analyzed on Agilent 2100 Bioanalyzer, as per manufacturer's instructions (Agilent Technology, High Sensitivity DNA Kit, 5067–4626).

Bisulfite conversion and PCR

DBS DNA (300 ng) samples were bisulfite converted using EZ DNA Methylation Kit (Zymo Research D5001) according to manufacturer's instructions. To assess the efficiency of bisulfite conversion, DNA was amplified using PCR primers that were specific either to bisulfite-converted or to non-modified DNA, and spanning the region of the housekeeping gene, *GAPDH*. The primer pairs spanning bisulfite-converted *GAPDH* regions are termed GAPDH-bc and consisted of the following forward and reverse primers, respectively: 5'-GTATTTGTTGATGGGTTAAGG-3' and 5'-ATAAAAACAAATCCCCTACCC-3'. The primer pairs spanning non-modified *GAPDH* regions are termed GAPDH-nm and consisted of the following forward and reverse primers, respectively: 5'-CTCTTGCTACTCT GCTCTGG-3' and 5'-GCTAAGTTTAGCCTGCCTGG-3'. Efficient conversion is observed when PCR bands are detected with GAPDH-bc but not with GAPDH-nm for a given sample. The PCR conditions used were: 95°C 15 min, [95°C 30 s, 57°C 30 s, 72°C 30 s] × 50 cycles, 72°C 10 min, and pause 4°C.

Pyrosequencing

DNA methylation analysis by pyrosequencing of bisulfie-converted DNA was performed as described [22]. Briefly, the region of interest was amplified with forward and reverse primers, one of which is biotinylated (btn), and then the methylation levels of the amplified region were analyzed using a sequencing primer. The forward, reverse and sequencing primers used for Line1 were the following, respectively, and adopted from Daskalos et al. [16]: 5'-btn-TAGGGAGTGTTAGATAGTGG-3', 5'-AACTCCCTAACC CCTTAC-3' and 5'-CAAATAAAACAATACCTC-3'. The forward, reverse and sequencing primers used for AluYb8 were the following, respectively: 5'-AGATTATTTTGGT TAATAAG-3', 5'-btn-AACTACRAACTACAATAAC-3' and 5'-GTTTGTAGTTTTAGTTATT- 3', as previously described [23].

Illumina Infinium HM450 array and data processing

Infinium HM450 arrays were processed according to manufacturer's instructions. GenomeStudio was used to analyze quality controls. The raw colour channels were corrected using the internal control probes and converted, without background subtraction and normalization, into absolute methylation levels (beta-values). Data were then imported into R (3.0.0), using the *minfi* package version 1.2.0 (http://www.bioconductor.org). Subset-quantile Within Array Normalisation (SWAN) normalisation was performed to correct for technical discrepancies between Type I and Type II [24]. Probes with detection p-values above 0.01 were considered as background noise and omitted from further analysis. Sex chromosome-specific probes were eliminated to minimize gender-specific variation of the X *versus* Y chromosomes [25]. Logarithmic transformation of the beta-values into M values was done, as previously described [26]. Statistical tests were performed using M-values. Clustering plots were generated using the *lumi* R package and based on coefficient of variation, which is calculated by standard deviation divided by mean across samples [27]. Differentially methylated CpGs were identified using an F-test in *minfi*.

Additional files

Additional file 1: Whole bisulfitome amplification.

Additional file 2: Additional Methods.

Additional file 3: Infinium HM450 quality control probes (minfi package).

Additional file 4: Correlations between technical replicates.

Abbreviations

bp: base pair; DBS: dried blood spots; gDNA: genomic DNA; GQ: GenSolve-QIAamp; GN: GenSolve-NucleoSpin; Gn: GN with NucleoSpin's precipitation buffer replaced by ethanol; GN-XS: GN extra small; Gn-XS: GN-XS with NucleoSpin's precipitation buffer replaced by ethanol; HM450: Infinium HumanMethylation450 Beadchip; Kbp: kilo base pair; NCS: National Children Study - USA; NN: NucleoSpin-NucleoSpin; Nn: NN with NucleoSpin's precipitation buffer replaced by ethanol; NN-XS: NN extra small; QC: quality control; QQ: QIAamp-QIAamp; Qq: QQ with QIAamp's precipitation buffer replaced by ethanol; SWAN: Subset-quantile Within Array Normalisation; TIHS: Tasmanian Infant Health Survey - Australia.

Competing interests

No conflicts of interest are disclosed.

Authors' contributions

AG carried out all experiments and drafted the manuscript. MPC participated in the optimization of the protocols. ALP participated in the collection of TIHS samples. SH and KC participated in the collection of NCS samples. ZH, TD, RS and HH conceived the study and participated in its design and coordination and helped to draft the manuscript. AG and HH performed the statistical and bioinformatics analyses. All authors read and approved the final manuscript.

Acknowledgements

We thank Dr. Florence Le Calvez-Kelm and Mr. Geoffroy Durand for helping in the HM450 experimental process and bioanalyzer measurements. This work was partially supported by the Association pour le Recherche Contre le Cancer (l'ARC). AG is supported by the IARC Postdoctoral Fellowship and Marie Curie Actions-People-COFUND. We thank the International Childhood Cancer Cohort Consortium for contributing to the coordination between participating cohorts.

The findings and conclusions in this report are those of the authors and do not necessarily represent the views of the National Children's Study, National Institutes of Health, or U.S. Department of Health and Human Services. Supported in part by NICHD Contracts HHSN275200503414C, HHSN275200503411C, HHSN275200603416C, HHSN275200503415C, HHSN275200503413C, HHSN275200503410C, HHSN275200503396C, HHSN275201000121U, and HHSN275200900010C.

This manuscript is dedicated to the participating children and their families.

Author details

[1]Epigenetics Group, International Agency for Research on Cancer (IARC), 150 rue Albert-Thomas, 69008 Lyon, France. [2]Cancer and Disease Epigenetics, Murdoch Childrens Research Institute Royal Children's Hospital, Flemington Road Parkville, Victoria 3052, Australia. [3]Environmental & Genetic Epidemiology Research, Murdoch Children's Research Institute Royal Children's Hospital, Flemington Road Parkville, Victoria 3052, Australia. [4]Menzies Research Institute, University of Tasmania, Hobart TAS 7000,

Australia. [5]Department of Health and Human Services, National Children's Study, Eunice Kennedy Shriver National Institute of Child Health and Human Development, National Institutes of Health U.S., 6100 Executive Boulevard, Room 3A01, Bethesda, MD 20892, USA. [6]Chair, Steering Committtee, International Childhood Cancer Cohort Consortium (I4C).

References

1. Brown RC, Dwyer T, Kasten C, Krotoski D, Li Z, Linet MS, Olsen J, Scheidt P, Winn DM: **Cohort profile: the International Childhood Cancer Cohort Consortium (I4C).** *Int J Epidemiol* 2007, **36**:724–730.
2. Guthrie R, Susi A: **A simple phenylalanine method for detecting phenylketonuria in large populations of newborn infants.** *Pediatrics* 1963, **32**:338–343.
3. Mei JV, Alexander JR, Adam BW, Hannon WH: **Use of filter paper for the collection and analysis of human whole blood specimens.** *J Nutr* 2001, **131**:1631S–1636S.
4. He H, Argiro L, Dessein H, Chevillard C: **Improved technique that allows the performance of large-scale SNP genotyping on DNA immobilized by FTA technology.** *Infect Genet Evol* 2007, **7**:128–132.
5. Wong N, Morley R, Saffery R, Craig J: **Archived Guthrie blood spots as a novel source for quantitative DNA methylation analysis.** *Biotechniques* 2008, **45**:423–424. 426, 428 passim.
6. Cruickshank MN, Pitt J, Craig JM: **Going back to the future with Guthrie-powered epigenome-wide association studies.** *Genome Med* 2012, **4**:83.
7. Beyan H, Down TA, Ramagopalan SV, Uvebrant K, Nilsson A, Holland ML, Gemma C, Giovannoni G, Boehm BO, Ebers GC, Lernmark Å, Cilio CM, Leslie RD, Rakyan VK: **Guthrie card methylomics identifies temporally stable epialleles that are present at birth in humans.** *Genome Res* 2012, **22**:2138–2145.
8. Aberg KA, Xie LY, Nerella S, Copeland WE, Costello EJ, van den Oord EJCG: **High quality methylome-wide investigations through next-generation sequencing of DNA from a single archived dry blood spot.** *Epigenetics* 2013, **8**:542–547.
9. Hollegaard MV, Grauholm J, Nørgaard-Pedersen B, Hougaard DM: **DNA methylome profiling using neonatal dried blood spot samples: a proof-of-principle study.** *Mol Genet Metab* 2013, **108**:225–231.
10. Joo JE, Wong EM, Baglietto L, Jung C-H, Tsimiklis H, Park DJ, Wong NC, English DR, Hopper JL, Severi G, Giles GG, Southey MC: **The use of DNA from archival dried blood spots with the Infinium HumanMethylation450 array.** *BMC Biotechnol* 2013, **13**:23.
11. Cruickshank MN, Oshlack A, Theda C, Davis PG, Martino D, Sheehan P, Dai Y, Saffery R, Doyle LW, Craig JM: **Analysis of epigenetic changes in survivors of preterm birth reveals the effect of gestational age and evidence for a long term legacy.** *Genome Med* 2013, **5**:96.
12. Bibikova M, Barnes B, Tsan C, Ho V, Klotzle B, Le JM, Delano D, Zhang L, Schroth GP, Gunderson KL, Fan J-B, Shen R: **High density DNA methylation array with single CpG site resolution.** *Genomics* 2011, **98**:288–295.
13. Bundo M, Sunaga F, Ueda J, Kasai K, Kato T, Iwamoto K: **A systematic evaluation of whole genome amplification of bisulfite-modified DNA.** *Clin Epigenetics* 2012, **4**:22.
14. Hardin J, Finnell RH, Wong D, Hogan ME, Horovitz J, Shu J, Shaw GM: **Whole genome microarray analysis, from neonatal blood cards.** *BMC Genet* 2009, **10**:38.
15. Bank S, Nexø BA, Andersen V, Vogel U, Andersen PS: **High-quality and -quantity DNA extraction from frozen archival blood clots for genotyping of single-nucleotide polymorphisms.** *Genet Test Mol Biomarkers* 2013, **17**:501–503.
16. Daskalos A, Nikolaidis G, Xinarianos G, Savvari P, Cassidy A, Zakopoulou R, Kotsinas A, Gorgoulis V, Field JK, Liloglou T: **Hypomethylation of retrotransposable elements correlates with genomic instability in non-small cell lung cancer.** *Int J cancer* 2009, **124**:81–87.
17. Hollegaard MV, Thorsen P, Norgaard-Pedersen B, Hougaard DM: **Genotyping whole-genome-amplified DNA from 3- to 25-year-old neonatal dried blood spot samples with reference to fresh genomic DNA.** *Electrophoresis* 2009, **30**:2532–2535.
18. Hue NT, Chan NDH, Phong PT, Linh NTT, Giang ND: **Extraction of human genomic DNA from dried blood spots and hair roots.** *Int J Biosci Biochem Bioinforma* 2012, **2**:21–26.
19. Sjöholm MIL, Dillner J, Carlson J: **Assessing quality and functionality of DNA from fresh and archival dried blood spots and recommendations for quality control guidelines.** *Clin Chem* 2007, **53**:1401–1407.
20. Saavedra-Matiz CA, Isabelle JT, Biski CK, Duva SJ, Sweeney ML, Parker AL, Young AJ, Diantonio LL, Krein LM, Nichols MJ, Caggana M: **Cost-effective and scalable DNA extraction method from dried blood spots.** *Clin Chem* 2013, **59**:1045–1051.
21. Murphy SC, Daza G, Chang M, Coombs R: **Laser cutting eliminates nucleic acid cross-contamination in dried-blood-spot processing.** *J Clin Microbiol* 2012, **50**:4128–4130.
22. Paliwal A, Vaissière T, Krais A, Cuenin C, Cros M-P, Zaridze D, Moukeria A, Boffetta P, Hainaut P, Brennan P, Herceg Z: **Aberrant DNA methylation links cancer susceptibility locus 15q25.1 to apoptotic regulation and lung cancer.** *Cancer Res* 2010, **70**:2779–2788.
23. Choi SH, Worswick S, Byun H-M, Shear T, Soussa JC, Wolff EM, Douer D, Garcia-Manero G, Liang G, Yang AS: **Changes in DNA methylation of tandem DNA repeats are different from interspersed repeats in cancer.** *Int J cancer J* 2009, **125**:723–729.
24. Maksimovic J, Gordon L, Oshlack A: **SWAN: Subset-quantile within array normalization for illumina infinium HumanMethylation450 BeadChips.** *Genome Biol* 2012, **13**:R44.
25. Morris T, Lowe R: **Report on the Infinium 450 k methylation array analysis workshop: April 20, 2012 UCL, London, UK.** *Epigenetics* 2012, **7**:961–962.
26. Du P, Zhang X, Huang C-C, Jafari N, Kibbe WA. Hou L, Lin SM: **Comparison of Beta-value and M-value methods for quantifying methylation levels by microarray analysis.** *BMC Bioinformatics* 2010, **11**:587.
27. Du P, Kibbe WA, Lin SM: **lumi: a pipeline for processing Illumina microarray.** *Bioinformatics* 2008, **24**:1547–1548.

A single blastocyst assay optimized for detecting CRISPR/Cas9 system-induced indel mutations in mice

Takayuki Sakurai[1*], Satoshi Watanabe[2], Akiko Kamiyoshi[1], Masahiro Sato[3] and Takayuki Shindo[1]

Abstract

Background: Microinjection of clustered regulatory interspaced short palindromic repeat (CRISPR)/CRISPR-associated protein 9 (Cas9)-related RNA and DNA into fertilized eggs is a novel approach for creating gene-modified mice. Blastocysts obtained just before implantation may be appropriate for testing the fidelity of CRIPSR/Cas9-mediated genome editing because they can be individually handled *in vitro* and obtained 3 days after microinjection, thus allowing researchers to check mutations rapidly. However, it is not known whether indel mutations caused by the CRISPR/Cas9 system can be reproducibly detected in embryos. In this study, we assessed the detection of CRISPR/Cas9-induced mutations in embryos.

Results: T7 endonuclease I was more effective than Surveyor nuclease for detecting mutations in annealed fragments derived from 2 plasmids, which contained nearly identical sequences. Mouse fertilized eggs were microinjected with CRISPR/Cas9-related RNA/DNA to examine whether non-homologous end joining-mediated knockout and homologous recombination-mediated knockin occurred in the endogenous receptor (G protein-coupled) activity modifying protein 2 (*Ramp2)* gene. Individual blastocysts were lysed to obtain crude DNA solutions, which were used for polymerase chain reaction (PCR) assays. T7 endonuclease I-based PCR and sequencing analysis demonstrated that 25–100% of the embryos were knockout embryos and 7–57% of the embryos were knockin embryos. Our results also established that crude DNA from a single blastocyst was an appropriate template for Whole genome amplification and subsequent assessment by PCR and the T7 endonuclease I-based assay.

Conclusions: The single blastocyst-based assay was useful for determining whether CRISPR/Cas9-mediated genome editing worked in murine embryos.

Keywords: CRISPR/Cas9, Gene targeting, Microinjection, Blastocyst, T7 endonuclease I, Surveyor assay, Crude DNA solution, Whole genome amplification

Background

Gene-modified (GM) mice are essential resources for studying gene function, developmental processes, and human diseases [1,2]. The production of GM mice, such as conditional knockout (cKO) mice, requires several time-consuming steps. For example, a cKO vector carrying the gene of interest and a drug-resistant gene must be constructed. The constructed cKO vector is then introduced into the genome of embryonic stem (ES) cells via homologous recombination. The resulting cKO cells are used to produce a chimeric mouse, with which germ-line transmission is achieved via ES-derived germ cells. Finally, offspring with the KO phenotype are obtained after genotyping and breeding.

In early 2013, a novel genome editing method, called the "clustered regulatory interspaced short palindromic repeat (CRISPR)/CRISPR-associated protein 9 (Cas9) system", was reported. The CRISPR/Cas9 system enables the efficient production of biallelic (homozygous) KO cells/embryos [3]. The system comprises CRISPR-coded RNAs (crRNAs), trans-activating crRNAs (tracrRNAs), and Cas9 endonuclease. It was originally identified as an adaptive immune apparatus in bacteria and archaea for defense against invading foreign plasmids and phages [4-6]. A

* Correspondence: tsakurai@shinshu-u.ac.jp
[1]Department of Cardiovascular Research, Graduate school of Medicine, Shinshu University, 3-1-1 Asahi, Matsumoto, Nagano 390-8621, Japan
Full list of author information is available at the end of the article

working unit of the CRISPR/Cas9 complex (comprising crRNA-tracrRNA fusion transcripts as guide RNA (gRNA) and humanized Cas9 protein) was established to enable the system to work effectively in mammalian cells [4-6]. The system efficiently introduces double-strand breaks approximately 3 bp in front of a protospacer adjacent motif (PAM) sequence (NGG), which is downstream of the gRNA target sequence (23 nucleotides; GN(19) + PAM) [6]. When the target sequence is repaired by non-homologous end joining (NHEJ), an insertion and deletion (indel) mutation results. When a construct with homology to the sequences upstream and downstream of the gRNA target sequence, such as a cKO vector, is introduced with the Cas9 protein, homologous recombination (HR) occurs [6,7].

The most attractive aspect of the CRISPR/Cas9 system is the efficient production of biallelic KO offspring through the direct injection of mRNA encoding Cas9 and gRNA(s). Wang et al. [3] demonstrated that 57–100% of injected zygotes were biallelic KO, when fetuses derived from the microinjected zygotes were examined. Several researchers have confirmed these results and shown that a high rate of indel mutations can be achieved using the CRISP/Cas9 system in mice [7-11]. However, a major problem associated with this system is the need to confirm the presence of the indel mutation in the target sequence in order to know whether the samples possess the (c)KO genotype. To this end, DNA from fetuses or newborn mice is frequently assessed by nuclease assays and sequencing analysis [3,7-10,12]. For optimization of the CRSIPR/Cas9-based GM mouse system, examining blastocysts at the pre-implantation stage is desirable because the evaluation can be performed 3 days after microinjection and does not require egg transfer to recipients to produce offspring. Unfortunately, few studies have assessed indel mutations at this stage ([8]; 3–7 blastocysts were used in addition to pup samples), probably because extremely low amounts of DNA are recovered at this stage and the collected DNA is occasionally lost.

In this study, we used a single mouse blastocyst as a genomic DNA source for several mutational assays. To establish a sensitive and reproducible assay for the detection of mutations introduced by CRISPR/Cas9-mediated genome editing, we first checked whether T7 endonuclease I could be used as an alternative to the widely utilized Surveyor nuclease. We next microinjected CRISPR/Cas9-related RNA/DNA into fertilized eggs and prepared an optimized crude DNA solution from developed blastocysts to examine NHEJ-mediated knockout and HR-mediated knockin of the endogenous receptor (G protein-coupled) activity modifying protein 2 (*Ramp2*) gene. In addition, we used whole genome amplification (WGA) technology [13-15], capable of producing microgram quantities of genomic DNA from a very small amount of DNA, to see whether DNA amplified from the crude DNA extract of a single blastocyst provided an appropriate template for polymerase chain reaction (PCR) analysis. Our results will be useful for researchers and technicians who want to analyze indel mutations at the blastocyst stage.

Results and discussion

T7 endonuclease I is more effective than surveyor nuclease for detecting indel mutations

To evaluate which nuclease, Surveyor nuclease or T7 endonuclease I, provided clearer and more reproducible results, P1 and P2 PCR products (approximately 800 bp) were used as templates (Figure 1A). The P1 and P2 products were derived from PCR of pC2EpA and pCEpA, respectively, which are nearly identical (Figure 1A). A mixture containing P1 (100 ng) and P1 (100 ng) (lane 1), P2 (100 ng) and P2 (100 ng) (lane 2) or P1 (100 ng) and P2 (100 ng) (lane 3) was re-annealed and treated with each nuclease (Figure 1B). A mixture containing P1 (100 ng) and P2 (100 ng) that had been directly treated with nuclease without re-annealing was used as a control (Figure 1B, lane 4). When the PCR products were subjected to electrophoresis, 2 expected cleavage bands of 525 and 275 bp, generated by nuclease treatment, were observed (asterisks in Figure 1B). Notably, the bands produced by T7 endonuclease I were more distinct than those generated by Surveyor nuclease. The weak cleavage activity of Surveyor nuclease improved slightly when the amount of Surveyor nuclease added to the reaction mixture was increased or the reaction time was extended (data not shown). Furthermore, T7 endonuclease I produced clearer bands than did Surveyor nuclease when the nuclease assay was performed using PCR-amplified products derived from a single blastocyst (Figure 2B, middle and bottom panels). These results suggest that T7 endonuclease I is superior to Surveyor nuclease for detecting indel mutations in target DNA. In the following experiments, we used T7 endonuclease I to detect indel mutations.

Optimization of crude DNA template preparation from a single blastocyst for highly reproducible direct PCR amplification

To achieve highly reproducible PCR amplification using a single blastocyst as the source of the crude DNA template, we first checked several carriers, including yeast tRNA (Ambion), yeast total RNA (Ambion), herring sperm DNA (Promega), and glycogen (Life Technologies), all of which have been used to precipitate very low amounts of DNA. In preliminary tests, yeast tRNA was better than the other reagents, when evaluated in view of successful PCR amplification (data not shown). Therefore, we used yeast tRNA to prepare a crude DNA solution (10 μL) from a single

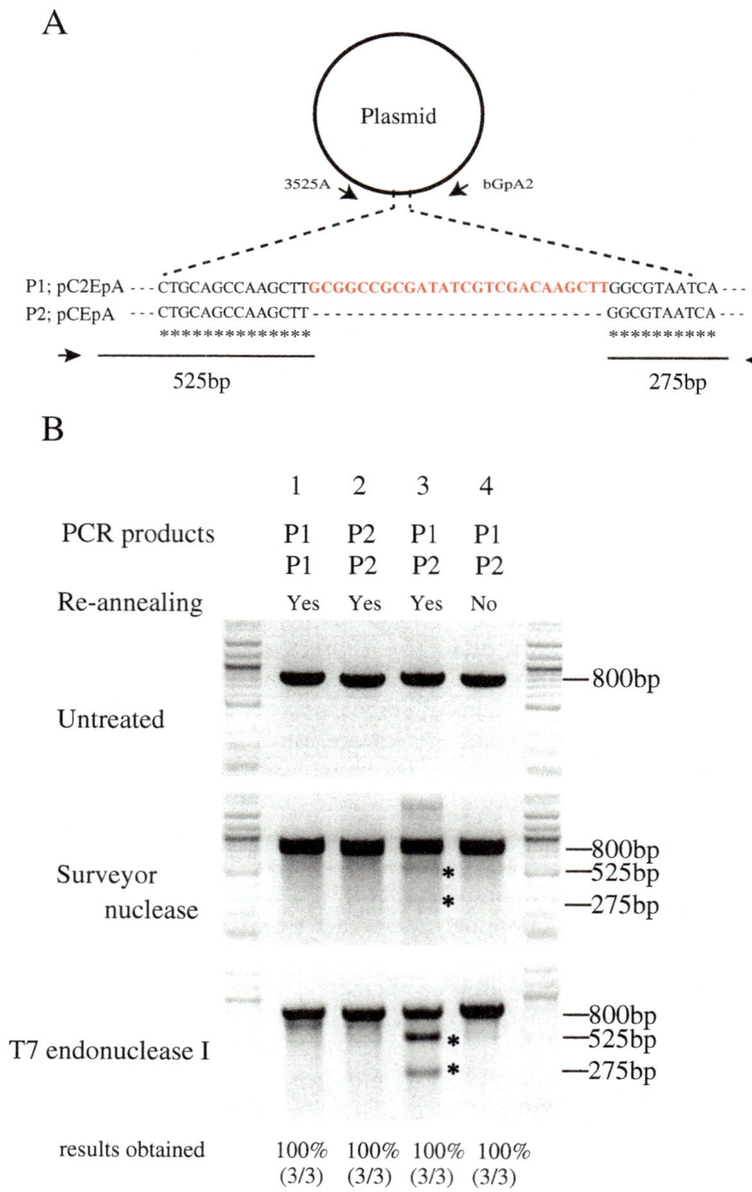

Figure 1 Comparison of assays using Surveyor nuclease and T7 endonuclease I. A: P1 and P2 polymerase chain reaction (PCR) products (approximately 800 bp) were used as template to evaluate which nuclease provide clearer and more reproducible results. The P1 and P2 PCR products were generated by PCR of pC2EpA or pCEpA, respectively, using the 3525A and bGpA2 primer set. Both 800-bp products re-annealed were expected to cleavage into two fragments (525 and 275 bp in size) by Surveyor nuclease or T7 endonuclease I. **B**: Agarose gel electrophoresis of nuclease-treated PCR products. A mixture containing P1 (100 ng) and P1 (100 ng) (lane 1), P2 (100 ng) and P2 (100 ng) (lane 2), or P1 (100 ng) and P2 (100 ng) (lane 3) was re-annealed and treated with each nuclease. In addition, a mixture containing P1 (100 ng) and P2 (100 ng) was directly treated with nuclease without re-annealing (lane 4). The 2 asterisks indicate the expected 525- and 275-bp cleavage bands. The experiments were carried out 3 times. The representative results were shown here.

blastocyst for use in the nuclease assay, as described in the Methods section. We successfully amplified the murine *Ramp2* gene from total 65 individual blastocysts (subtotal 35 in Experiment 1, 2, 3 of Table 1 and subtotal 30 in Experiment 1, 2, 3 of Table 2); see [16] for information on the function of *Ramp2*).

Use of the single blastocyst-based assay to detect indel mutations introduced by CRISPR/Cas9-mediated genome editing in endogenous murine *Ramp2*

As shown in Figure 3, we developed a single blastocyst-based mutational assay using T7 endonuclease I and a crude DNA solution. Next, we tested whether the mutational assay

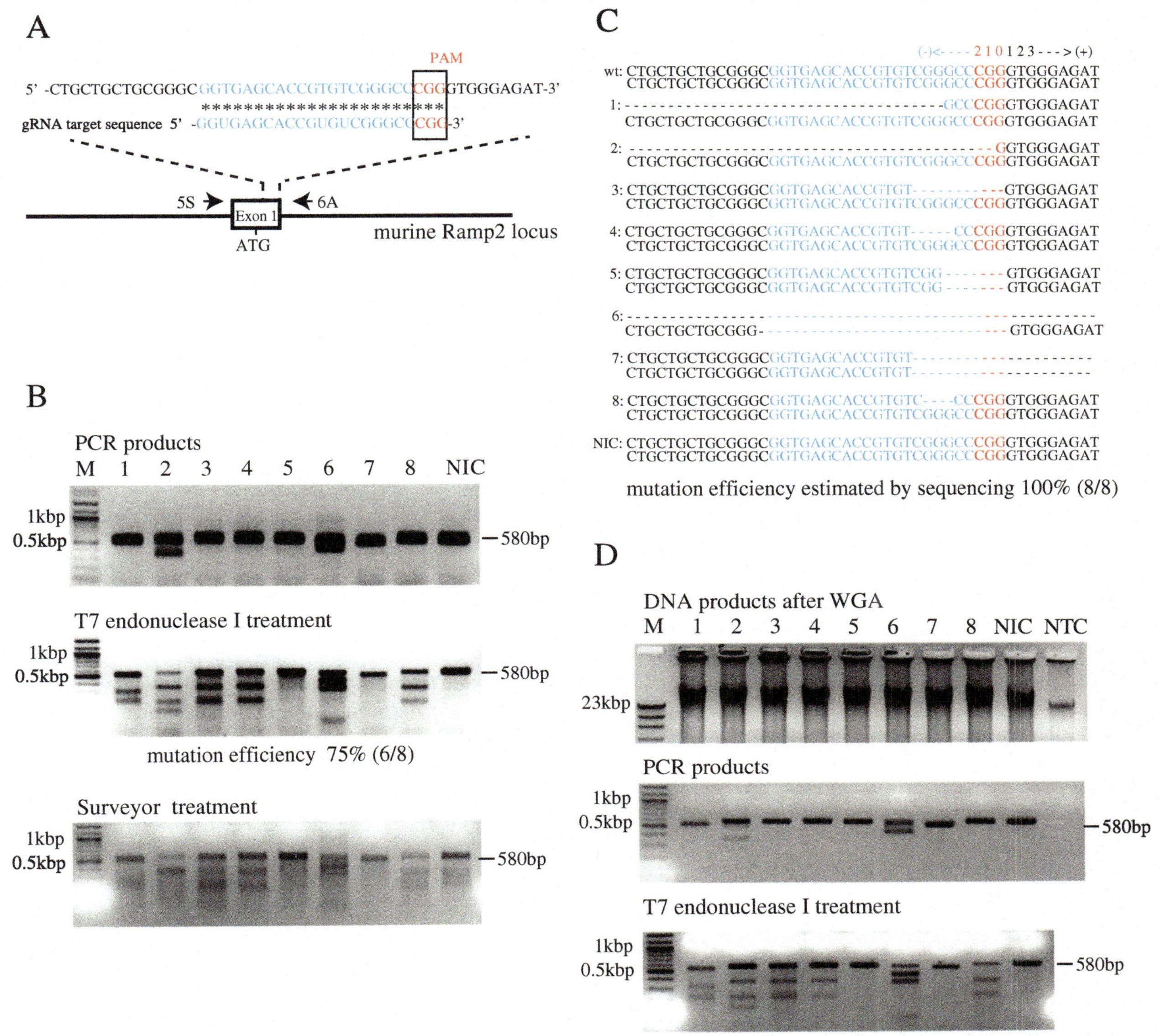

Figure 2 Detection of CRISPR/Cas9-induced indel mutations in the murine *Ramp2* gene at the single blastocyst level. A: Schema of the *Ramp2* guide RNA (R2gRNA) targeting exon 1 in the murine *Ramp2* gene. The R2gRNA-coding sequence is shown in blue. The protospacer-adjacent motif (PAM) sequence is shown in red. The arrows indicate the locations of the PCR primers (see Table 3). **B**: Agarose gel electrophoresis of T7 endonuclease I-treated PCR products and surveyor-treated PCR products derived from 8 individual blastocysts (see Table 1, experiment 3). Lanes 1–8: PCR products amplified from the crude DNA solution of each blastocyst. Lane NIC (no injection control): PCR product amplified from the crude DNA of a single uninjected control blastocyst. M: lambda *Hind*III + 100-bp ladder markers. **C**: Sequencing results for the PCR products shown in lanes 1–8 and NIC of (B). The result of each sample was summarized as the two Ramp2 alleles. "wt" indicates the wild-type sequence. "-" indicates base deletion. **D**: Gel electrophoretic pattern of DNA products after whole genome amplification (WGA) of the crude DNA shown in lanes 1–8 and NIC of (B) (top panel) and after subsequent PCR amplification (middle panel) and T7 endonuclease I digestion (bottom panel). Note that the band pattern is similar to that shown in (B). NTC (no template control) indicates the WGA product obtained using water in place of template as a negative control (top panel); no PCR products were amplified by PCR (middle panel).

could detect mutations in exon 1 of endogenous murine *Ramp2* after microinjection of Cas9 mRNA and *Ramp2* gRNA (R2gRNA) (Figure 2A) at the zygote stage. Representative results are shown in Figure 2B and C and in Table 1 (experiment 3). First, approximately 2 pL of an RNA mixture (containing 200 ng/μL of Cas9 mRNA and 50 ng/μL of R2gRNA) was injected into the pronuclei and cytoplasm of 24 fertilized mouse eggs (Figure 3, steps 1 and 2). The 17 eggs that survived after injection were cultivated to the blastocyst stage. Crude DNA was then prepared by lysing each of the 8 blastocysts derived from the RNA-injected zygotes (Figure 3, step 3) and 1 blastocyst

Table 1 Detection and efficiency of CRISPR/Cas9-mediated indel mutation at Ramp2 locus at blastocyst stage

Experiment no.	Site injected[1]	Cas9 mRNA: R2 gRNA (ng/μL)	No. zygotes injected	No. blastocysts/ no. zygotes cultures (%)	No. mutant blastocysts/ no. blastocysts tested (%)[2]
1	cyto	100:50	25	19/21 (90)	6/21 (32)
2	pn/cyto	100:50	15	8/10 (80)	2/8 (25)
3	pn/cyto	200:50	24	8/17 (47)	8/8 (100)[3]

[1]Injection was performed toward cyto: cytoplasm; pn/cyto pronuclei and cytoplasm.
[2]Mutated blastocysts were identified by both T7EndonucleaseI digestion and sequencing.
[3]Representative results from experiment 3 are shown in Figure 2B and C.

derived from a non-injected fertilized egg as a negative control (NIC: no injection control). Finally, PCR was performed using each crude DNA sample as template to specifically amplify a 580-bp region corresponding to exon 1 of murine *Ramp2* (Figure 3, step 4). PCR products were successfully amplified from all of the samples tested (Figure 2B, top panel, lanes 1–8 and NIC). Two of the samples produced 2 major bands: unexpected and expected around 580 bp (Figure 2B, top panel, lanes 2 and 6), and 1 of the samples yielded a PCR product slightly shorter than that of the others (Figure 2B, top panel, lane 7). Since Ramp2 has two alleles, these results indicated that the samples in lanes 2 and 6 had relatively long deletions in a heterozygous state or a mosaic state, while the sample in lane 7 had a short deletion in a homozygous state.

To obtain more information on the indel mutations in the 9 samples (Figure 2B, lanes 1–8 and NIC), each PCR product was digested with T7 endonuclease I as described in the Methods section, and the resulting fragments were sequenced (Figure 2C). As expected, the PCR products (580 bp) derived from NIC sample were undigestable. Sequencing of these products revealed no mutations (wild-type (wt) sequence in Figure 2C). PCR products derived from 6 blastocysts (shown in Figure 2B, lanes 1–4, 6, and 8) produced an additional 2 to 4 bands. Sequencing of the PCR product shown in lane 1 revealed a 55-bp heterozygous deletion, located at −58 to −6, where the third nucleotide (G) of the PAM motif (CGG) is considered position 0 (Figure 2C). Similarly, the PCR products in lanes 2, 3, 4, and 8 contained a 206-bp heterozygous deletion (−206 to −1), 10-bp heterozygous deletion (−8 to +1), 4-bp heterozygous deletion (−8 to −5), and 3-bp heterozygous deletion (−7 to −5), respectively. For PCR products in lane 6, both 23-bp deletion (−23 to 0) of two alleles and 372-bp deletion (−244 to +128) of one allele were revealed. The PCR products in lanes 5 and 7 were not digested by T7 endonuclease I; only 1 major band was apparent after nuclease digestion. However, sequencing demonstrated that the samples in lanes 5 and 7 had a 6-bp homozygous deletion (−5 to 0) and 43-bp homozygous deletion (−5 to + 34), respectively. Thus, all of the 8 blastocysts tested had indel mutations in the murine *Ramp2* gene (Table 1, experiment 3). Heterozygous deletions occurred in 6 blastocysts (75%; lanes 1–4, 6, and 8). Homozygous deletions occurred in 3 blastocysts (37.5%; lanes 5–7). These results showed that the T7 endonuclease I-based assay using crude DNA prepared from a single blastocyst is a sensitive and reproducible method for the detection of indel mutations in a target gene.

In addition, we tested whether the single blastocyst-based assay could detect an HR event in exon 1 of murine *Ramp2* (Figure 4A) induced by the CRISPR/Cas9 system (Figure 3, step 1). First, approximately 2 pL of an RNA mixture (containing 200 ng/μL of Cas9mRNA, 50 ng/μL of R2gRNA, and 10 ng/μL of R2HR) was injected into the pronuclei and cytoplasm of 25 fertilized mouse eggs. Of these, 20 live eggs were cultivated to the blastocyst stage. Seven embryos developed successfully to blastocysts, of which 4 were EGFP-positive. In Figure 4B, 3 of the 4 fluorescent blastocysts are shown. Next, we prepared crude DNA solutions by lysing each of the 8 blastocysts (4 EFGP-positive and 3 EGFP-negative blastocysts derived from injected fertilized eggs and 1 blastocyst derived from an uninjected fertilized egg). The presence or absence of HR in the target locus was assessed by PCR amplification of the junctional region between the

Table 2 Dectection and efficiency of CRISPR/Cas9-mediated homologous recombination at Ramp2 locus at blastocyst stage

Experiment no.	Site injected[1]	Cas9 mRNA: R2 gRNA: R2HR donor DNA (ng/μL)	No. zygotes injected	No. Blastocysts/no. zygotes cultures (%)	No. EGFP-expressing blastocysts/ no. blastocysts tested (%)
1	pn	100:50:10	22	15/22 (68)	1/15 (7)
2	cyto	100:50:10	20	8/16 (50)	2/8 (25)
3	pn/cyto	200:50:10	25	7/20 (35)	4/7 (57)[2]

[1]Injection was performed toward pn: pronuclei; cyto: cytoplasm; pn/cyto: pronuclei and cytopalsm.
[2]Representative results from experiment 3 are shown in Figure 4B and C.

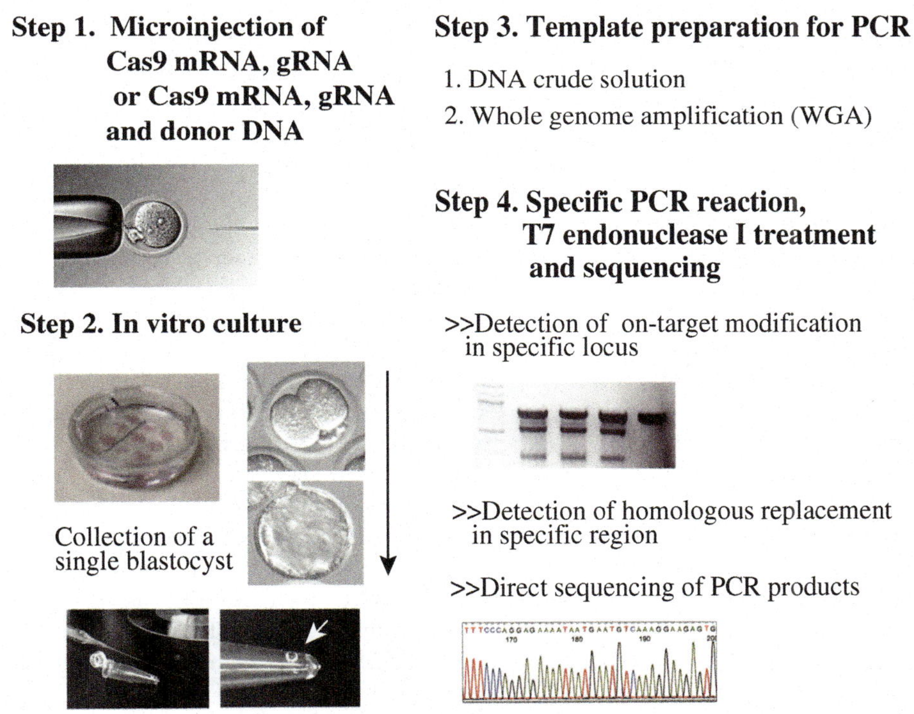

Figure 3 Overview of the nuclease assay for the detection of indel mutations in a single blastocyst.

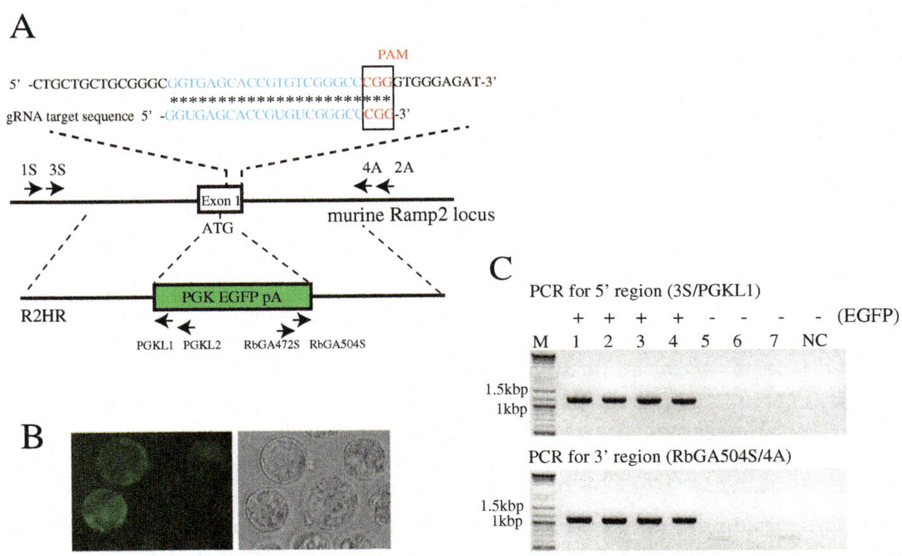

Figure 4 Detection of CRISPR/Cas9-induced homologous recombination in murine *Ramp2* at the single blastocyst level. A: Schema of the R2gRNA and R2HR DNA donor targeting sites in exon 1 of the murine *Ramp2* gene. The R2gRNA-coding sequence is shown in blue. The PAM sequence is shown in red. The R2HR DNA donor comprises 1-kb 5′ and 3′ arms corresponding to exon 1 in the *Ramp2* gene and an EGFP expression cassette. Arrows indicate the locations of the PCR primers (see Table 3). **B**: Fluorescence and bright-field photography of blastocysts developed from fertilized eggs after microinjection (Table 2, experiment 3). **C**: Gel electrophoretic pattern of PCR products corresponding to the junctional region between the 5′ (or 3′) arm and the host genome. Lanes 1–7: PCR products derived from the crude DNA of individual blastocysts. Lane NC: PCR products derived from the crude DNA of a single uninjected control blastocyst. + and –: EGFP-positive and EGFP-negative blastocysts, respectively. Note that PCR products (approximately 1.1 kb) were successfully amplified only in samples from the 4 EGFP-positive blastocysts.

5′ arm (or 3′ arm) and the host genome (Figure 4A) using crude DNA as template. As shown in Figure 4C, each of the 4 EGFP-positive blastocysts had PCR products with an expected size of approximately 1.1 kb, but the other samples, including the 3 EGFP-negative blastocysts and the 1 uninjected blastocyst, did not. Thus, HR occurred successfully in the 4 EGFP-positive samples, and the HR rate was 57.1% (4/7). These results indicate that single blastocysts are useful for the detection HR events induced by CRISPR/Cas9-mediated genome editing.

Crude DNA from a single blastocyst is an appropriate template for WGA and subsequent PCR analysis

The main problem when using the T7 endonuclease I-based assay with a single blastocyst is that the DNA content is very small, which often hampers extensive analysis of the target DNA region. For example, we typically prepare a 10-μL crude DNA solution from 1 blastocyst, which is enough for 5 PCR reactions (2 μL/reaction). Although this amount of crude DNA is presently sufficient for analyzing on-target mutations, it is not adequate for examining additional areas to identify possible off-target mutation sites. To resolve this problem, we used WGA, a method for the robust amplification of an entire genome starting with nanogram quantities of DNA [13-15]. When crude DNA solutions prepared from each of the 9 blastocysts (Figure 2B, lanes 1–8 and NIC) were subjected to WGA, 3–4 μg of genomic DNA was successfully amplified (Figure 2D, top panel). When PCR was performed using the DNA generated from WGA as template, a 580-bp product corresponding to exon 1 of murine *Ramp2* was successfully amplified (Figure 2D, middle panel). Notably, the electrophoretic pattern of the PCR products was similar to that of the PCR products obtained from direct amplification of the crude DNA solutions (see Figure 2B, top panel). When the PCR products derived from WGA of the crude DNA were re-annealed and subsequently digested with T7 endonuclease I, the digestion pattern of the PCR products was similar to that of PCR products derived from direct amplification of the crude DNA solution (Figure 2D, bottom panel vs. Figure 2B, middle panel). In the negative control, in which water was subjected to WGA (NTC: no template control), a small amount of DNA was observed, but PCR using the NTC DNA as template did not generate detectable PCR products (Figure 2D, middle panel, lane NTC), as previously suggested by Akasaka et al. [13]. These results indicate that a crude DNA solution prepared from a single blastocyst can serve as template for WGA. Amplification of genomic DNA by WGA may be an effective approach when extensive analyses of on- or off-target mutations are required. However, since WGA has a potential risk of contaminations and the introduction of unwanted mutations [13-15], WGA may require careful evaluation of the results.

Direct microinjection of CRISPR/Cas9-related nucleic acids into zygotes is a powerful tool for the efficient production of GM animals, including mice [3,7-10,12], rats [9,17,18], and monkeys [19]. In most cases, newborn mice or fetuses have been used to confirm the presence of indel mutations in the genome [3,7-10,12]. However, this requires egg transfer to recipients, which is a time-consuming process. The single blastocyst-based assay described here will allow rapid determination of whether the CRISPR/Cas9 system designed by the practitioners is able to generate GM mice, as indicated in Tables 1 and 2. Furthermore, the assay does not require preliminary tests to check the effectiveness of CRISPR/Cas9-related nucleic acids via the transfection of cultured cells. Moreover, if practitioners transfer microinjected zygotes to recipients and do not obtain GM mice, our assay provides a way to determine the reason, for example, fetal lethality caused by a CRISPR/Cas9-mediated mutation or low efficiency of the CRISPR/Cas9-related RNA/DNA.

The present study demonstrated that the crude DNA extracted from a single blastocyst is sufficient for the detection of mutations in nuclease assays that used T7 endonuclease I (see Figure 2) as an alternative to the widely utilized Surveyor nuclease [3,5,7,17]. A few researchers have reported the usefulness of T7 endonuclease I for detecting mutations in a target gene [8-10,18]. Using the single blastocyst assay with T7 endonuclease I and direct sequencing analysis, we demonstrated that microinjection of CRISPR/Cas9-related nucleic acids into zygotes leads to a relatively high degree of mutation (ranging from 25 to 100%; see Table 1), as previously described by Wang et al. [3]. We also demonstrated that the crude DNA sample is sufficient for the detection of HR in the *Ramp2* locus (ranging from 7 to 57%; see Table 2), as previously demonstrated by Yang et al. [7].

Notably, using the present assay and even one experiment with a small number (15–25) of zygotes, researchers and technicians can assess the mutation rate and post-injection embryo survival. This assay also will be helpful to support the animal welfare since the number of mice used in research can be reduced. Furthermore, we demonstrated the usefulness of WGA before PCR and the subsequent nuclease assay (see Figure 2D). WGA amplifies nanogram levels of genomic DNA from a single blastocyst to microgram levels, which enables multiple analyses of genomic DNA, as previously noted by Akasaka et al. [13].

The single blastocyst-based assay described in this study is applicable to the evaluation of mutations generated by the other gnome editing systems, such as zinc-finger nuclease (ZFN) [20,21] and transcription activator-like nuclease (TALEN) [22], because DNA modification through

NHEJ and HR with these systems is principally the same as with the CRISPR/Cas9 system.

Conclusions

We have demonstrated the usefulness of a single blastocyst-based assay for detecting mutations introduced by the CRISPR/Cas9 system. This assay, which is reproducible and convenient, allows researchers and technicians to check the effectiveness of CRSPR/Cas9-based genome editing rapidly.

Methods

Vectors and RNA synthesis

The pCEpA and pC2EpA plasmids, which are nearly identical (Figure 1A), were used as PCR templates to evaluate the effectiveness of T7 endonuclease I (New England BioLabs, Japan Inc., Tokyo, Japan) and Surveyor nuclease (Transgenomic Inc., NE, USA) in nuclease assays. To construct the pBS-NFCas9a Cas9 expression vector, cDNA corresponding to the open reading frame (ORF) of hCas9 was first PCR-amplified using phCas9 (#41815; Addgene, Cambridge, MA, USA) as template. Next, the SV40-derived nuclear location signal (NLS) and FLAG sequence were inserted in-frame immediately after the ATG site in the hCas9 cDNA by PCR. The resulting ATG-NLS-FLAG-hCas9 ORF construct was finally subcloned into pBluescript II (Agilent Technologies Japan, Ltd., Tokyo, Japan). For construction of pBS-T7-R2gRNA, the sequence from the gRNA scaffold to the TTTTTT site (83 bp) was PCR-amplified using pgRNA_GFP-T1 (#41819; Addgene) as template and then subcloned into pBluescript II. The resulting plasmid was used as template for inverse PCR. Furthermore, a candidate gRNA target sequence (5′-GGTGAGCACCGTGTCGGGCCC GG-3′) corresponding to exon 1 of the murine *Ramp2* gene [GenBank: NC_000077.6] (Figure 2A) was placed in front of the gRNA scaffold sequence in pBS-T7-R2gRNA. For construction of pR2HR, a genomic DNA sequence of approximately 1 kb (termed "5′ arm") in front of the ATG site in murine *Ramp2* (Figure 4A) was PCR-amplified using C57BL6/JJcl genomic DNA as template. Similarly, a genomic DNA sequence of approximately 1 kb (termed "3′ arm") after the ATG site was PCR-amplified. The 5′ and 3′ arms were placed before and after the PGK EGFP pA cassette (containing the mouse PGK promoter, EGFP cDNA, and poly(A) sites) in pBluescript II, respectively. Upon microinjection, the insert containing the R2HR fragment was removed from the vector backbone by digestion with *Not*I and *Sca*I. All of the constructed plasmids (pBS-NFCas9, pBS-T7-R2gRNA, and pR2HR) were sequenced to confirm their fidelity.

Cas9 mRNA was obtained using the mMESSAGE mMACHINE T3 kit (Ambion, Life Technologies Japan, Ltd., Tokyo, Japan) and *Sap* I-linearized pBS-NFCas9 as the RNA synthesis template. R2gRNA (*Ramp2* gRNA) (Figure 2A) was prepared from *Eco*RI-linearized pBS-T7-R2gRNA using the mMESSAGE mMACHINE T3 kit (Ambion).

Microinjection into fertilized mouse eggs and *in vitro* culture

BDF1 and C57BL6/JJcl (hereafter termed B6) mice were purchased from CLEA Japan Inc. (Tokyo, Japan) and bred with a light period of 7:00 to 19:00. All animal experiments were carried out according to the ethical guidelines of Shinshu University.

Fertilized eggs were collected using a standard *in vitro* fertilization (IVF) method [23]. Briefly, ovulated oocytes were collected from 4- to 10-week-old BDF1 females superovulated by injection with 5 IU of equine chorionic gonadotropin (peamex; Novartis Animal Health Inc., Tokyo, Japan) and 5 IU of human chorionic gonadotropin (Gonatropin; ASKA Pharmaceutical Co., Ltd., Tokyo, Japan) at a 48-h interval and inseminated with B6 spermatozoa obtained from the cauda epididymides of 8- to 10-week-old males. Successfully fertilized eggs were identified by the presence of both pronuclei and the second polar body at 6–7 h after insemination.

Cas9mRNA, R2gRNA, and the R2HR DNA donor fragment were microinjected using a method previously described by Sakurai et al. [23]. Briefly, approximately 2 pL of a 2-template mixture (containing 100 or 200 μg/mL of Cas9mRNA and 50 μg/mL of R2gRNA) or 3-template mixture (containing 10 μg/mL of R2HR, 100 or 200 μg/mL of Cas9mRNA, and 50 μg/mL of R2gRNA) was injected into the pronucleus and/or cytoplasm using a standard microinjection system (Narishige-Olympus MMO-202ND, MM-89, UT-2, IM-9B, and IX-70; Narishige Group, Ltd., Tokyo, Japan). Injected eggs were then cultured in 30 μL of KSOM medium [24] covered with paraffin oil at 37°C in an atmosphere of 5% CO_2 up to the blastocyst stage. EGFP fluorescence was monitored at the blastocyst stage using an Olympus IX70 inverted fluorescence microscope with the U-MWIBA2 filter set (Olympus, Tokyo, Japan).

Preparation of crude DNA from a single blastocyst and whole genome amplification (WGA)

Crude DNA derived from a single blastocyst was prepared for use as a PCR template according to the method described by Sakurai et al. [25] with some modifications. Briefly, under a stereomicroscope (SZX12; Olympus), 0.1–0.5 μL of KSOM medium containing 1 blastocyst was transferred to the wall near the bottom of a 0.2-mL PCR tube (thin wall PCR tube with cap; QSP #430-Q; Thermo Scientific, CA, USA) (Figure 3, step 2) using a glass micropipette. Thereafter, 10 μL of blastocyst lysis buffer (125 μg/mL proteinase K, 100 mM Tris–HCl (pH 8.3), 100 mM KCl, 0.02% gelatin, 0.45% Tween 20, and 60 μg/mL yeast tRNA (Ambion)) was gently added to each tube. Each PCR tube was incubated at 56°C for 10 min and then at 95°C for 10 min in a PCR machine (ABI GeneAmp PCR System 9700; Applied Biosystems, Life Technologies Japan, Ltd., Tokyo, Japan). The resulting crude DNA solution was stored at −20°C until use.

For WGA with single blastocyst-derived crude DNA, we used a REPLI-g Mini Kit (Qiagen K.K.,Tokyo, Japan) according to the manufacturer's protocol. Briefly, 4 μL of crude DNA solution was mixed with 3 μL of Buffer D. The mixture was incubated at 65°C for 10 min, and the reaction was stopped by the addition of 3 μL of stop solution. Finally, 40 μL of REPLI-g solution was added to the mixture, which was incubated at 30°C for 16 h and at 60°C for 3 min.

PCR

The sequences of the PCR primers are shown in Table 3. A DNA region spanning 800 bp that is different in pCEpA and pC2EpA was PCR-amplified using rTaq DNA polymerase (TaKaRa Bio, Inc., Shiga, Japan) with primers 3525A and bGpA2 (Figure 1A). PCR was performed with 30 cycles of 30 s at 94°C, 30 s at 58°C, and 30 s at 72°C.

Two rounds of PCR were performed to amplify a 580-bp region of the murine *Ramp2* gene (Figure 2A) using crude DNA as template. For the first round of PCR, a 23-μL reaction mixture containing 1 × Tks Gflex buffer, 0.4 μM each of 2 primers (5S and 6A; Figure 2A), and 1.25 U of a high-fidelity, DNA polymerase derived from *Thermococcus* sp.; Tks Gflex DNA polymerase (TaKaRa Bio, Inc.) was mixed with 2 μL of crude DNA extracted from a single blastocyst. PCR was performed with denaturation at 94°C for 1 min, followed by 30 cycles of 98°C for 5 s, 68°C for 20 s, and 68°C for 5 min. For the second round of PCR, 2 μL of the PCR product from the first round was added to the 23-μL reaction mixture described above. The PCR conditions were the same.

For PCR using DNA derived from WGA, a 25-μL reaction mixture containing 10 mM Tris-Cl, pH 8.3, 1.5 mM $MgCl_2$, 50 mM KCl, 0.01% (w/v) gelatin, 0.4 mM of each dNTP, 0.4 μM each of 2 primers (5S and 6A; Figure 2A), 1.25 U of rTaq DNA polymerase, and 50 ng of genomic DNA obtained from WGA was prepared. PCR was performed with denaturation at 94°C for 1 min; 30 cycles of 94°C for 20 s, 58°C for 30 s, and 72°C for 1 min; and a final extension at 72°C for 5 min.

Nested PCR was performed to test whether HR-mediated insertion of the R2HR donor fragment had occurred (Figure 4A). For the first round of PCR, to amplify the 1.1-kb 5′ region of the murine *Ramp2* gene (Figure 4A), a 23-μL reaction mixture containing 1 × Tks Gflex buffer, 0.4 μM each of 2 primers (1S and PGKL2; Figure 4A), and 1.25 U of Tks Gflex DNA polymerase was mixed with 2 μL of crude DNA extracted from a single blastocyst. PCR was performed with denaturation at 94°C for 1 min; 30 cycles of 98°C for 5 s and 68°C for 20 s; and a final extension at 68°C for 5 min. For the second round of PCR, the PCR conditions were the same as described above, except that 3S and PGKL1 were used as primers (Figure 4A), and 2 μL of the PCR solution from the first round was used as template.

The 1.1-kb 3′ region of the murine *Ramp2* gene (Figure 4A) was amplified by 2 rounds of PCR with the same conditions used to amplify the 5′ region of the murine *Ramp2* gene, except that primers RbGA472S and 4A were used for the first round of PCR, and primers RbGA504S and 2A were used for second round (Figure 4A).

Mutational assays with T7 endonuclease I and surveyor nuclease and direct sequencing

CRISPR/Cas9 -induced indel mutations were detected through mutational assays with T7 endonuclease I and

Table 3 Primer sequences used in this study

Primers	Sequences (5′-3′)	Purpose
3525A	CCGAAGGTAACTGGCTTCAGCAGAGCGCAG	Optimized nuclease treatment (Figure 1)
bGpA2	AACATATGCCATATGCTGGCTGCCATGAAC	Optimized nuclease treatment (Figure 1)
5S	GCCCGAGCTGGAAGCGAGAG	CRISPR/Cas9-mediated mutation (Figure 2A-D)
6A	AAGTAGGAAGCAGCATTAAGT	CRLSPR/Cas9-mediated mutation (Figure 2A-D)
1S	CTGAGTTGTGATAGCTGGCA	CRISPR/Cas9-mediated HR (Figure 4A-C)
3S	AACGTGACCTTAGCCAAGTC	CRISPR/Cas9-niediated HR (Figure 4A-C)
2A	AGTCCAGTTGCACCAGTCCTTG	CRISPR/Cas9-mediated HR (Figure 4A-C)
4A	GTACTCATACCAGCAAGGTAG	CRISPR/Cas9-mediated HR (Figure 4A-C)
PGKL1	GTTGGCGCCTACCGGTGGATGTGGAATGTG	CRISPR/Cas9-mediated HR (Figure 4A-C)
PGKL2	TGTGQGAGGCCAGAGGCCACTIGTGTAG	CRISPR/Cas9-mediated HR (Figure 4A-C)
RbGA472S	TTCCTCCTCTCCTGACTACTCCCAGTCATA	CRISPR/Cas9-mediated FIR (Figure 4A-C)
RbGA504S	CTGTCCCTCTTCTCTTATGAAGATCCCTC	CRISPR/Cas9-mediated HR (Figure 4A-C)

Surveyor nuclease and through direct sequencing of the PCR products. For the Surveyor assay, a crude DNA solution derived from a single blastocyst was purified by ethanol precipitation, dissolved in deionized distilled water, and digested with Surveyor nuclease, according to the manufacture's user guide (SURVEYOR Mutation Detection Kit for Standard Gel Electrophoresis; Transgenomic). The reaction volume was 11 μL. For the T7 endonuclease I assay, 10 μL of 1× NEB2 reaction buffer (New England BioLabs, Japan Inc.) containing 200–400 ng of PCR product was placed in a 0.2-mL PCR tube (QSP #430-Q; Thermo Scientific). The PCR tube was incubated at 95°C for 5 min for denaturation using an ABI GeneAmp PCR System 9700 (Applied Biosystems, Life Technologies Japan, Ltd.) and then incubated for 0.5–1 h at room temperature (24°C) for re-annealing. Next, 1 μL of T7 endonuclease I (2.5 U/μL) was added to the denatured/re-annealed sample, and the PCR tube was incubated at 37°C for 1–2 h.

The nuclease-treated solutions were electrophoresed using a 2% agarose gel/TAE buffer system, stained with ethidium bromide, and photographed under ultraviolet illumination.

For direct sequencing of the PCR products, 10 ng of PCR product was amplified using BigDye Terminator v.3.1 (Applied Biosystems, Life Technologies Japan, Ltd.) with the 5S or 6A primer and then sequenced with an ABI Genetic Analyzer 3130 (Applied Biosystems, Life Technologies Japan, Ltd.). Sequence analysis was performed using Genetyx-Mac ver.13.0.3 (Software Development Co., Ltd., Tokyo, Japan).

Abbreviations

CRISPR: Clustered regulatory interspaced short palindromic repeat; Cas9: CRISPR-associated protein 9; PAM: protospacer-adjacent motif; WGA: Whole genome amplification; cKO: Conditional knockout; HR: Homologous recombination; NHEJ: Non-homologous end joining; crRNA: CRISPR-coded RNA; gRNA: guide RNA; tracrRNA: Trans-activating crRNA; ES: Embryonic stem; GM: Gene-modified; *Ramp2*: receptor (G protein-coupled) activity modifying protein 2; NIC: No injection control; NTC: No template control; ORF: Open reading frame; NLS: Nuclear location signal; PCR: Polymerase chain reaction.

Competing interests

The authors declare they have no competing interests.

Authors' contributions

TSa designed the experiments, constructed the plasmids used in the study, performed experiments, drafted the manuscript, and edited the manuscript. AK and TSh assisted in conducting the experiments. SW constructed the plasmids used in the study and edited the manuscript. MS edited the manuscript. All authors read and approved the final manuscript.

Acknowledgments

This study was supported by a Grant-in-Aid for Scientific Research from the Ministry of Education, Culture, Sports, Science and Technology, Japan.

Author details

[1]Department of Cardiovascular Research, Graduate school of Medicine, Shinshu University, 3-1-1 Asahi, Matsumoto, Nagano 390-8621, Japan. [2]Animal Genome Research Unit, Division of Animal Science, National Institute of Agrobiological Sciences, 2-1-2 Kannondai, Tsukuba, Ibaraki 305-8602, Japan. [3]Section of Gene Expression Regulation, Frontier Science Research Center, Kagoshima University, 8-35-1 Sakuragaoka, Kagoshima, Kagoshima 890-8544, Japan.

References

1. Capecchi MR: **Gene targeting in mice: functional analysis of the mammalian genome for the twenty-first century.** *Nat Rev* 2005, **6**(6):507–512.
2. Kohan DE: **Progress in gene targeting: using mutant mice to study renal function and disease.** *Kidney Int* 2008, **74**(4):427–437.
3. Wang H, Yang H, Shivalila CS, Dawlaty MM, Cheng AW, Zhang F, Jaenisch R: **One-step generation of mice carrying mutations in multiple genes by CRISPR/Cas-mediated genome engineering.** *Cell* 2013, **153**(4):910–918.
4. Jinek M, Chylinski K, Fonfara I, Hauer M, Doudna JA, Charpentier E: **A programmable dual-RNA-guided DNA endonuclease in adaptive bacterial immunity.** *Science (New York, NY)* 2012, **337**(6096):816–821.
5. Cong L, Ran FA, Cox D, Lin S, Barretto R, Habib N, Hsu PD, Wu X, Jiang W, Marraffini LA, Zhang F: **Multiplex genome engineering using CRISPR/Cas systems.** *Science (New York, NY)* 2013, **339**(6121):819–823.
6. Mali P, Yang L, Esvelt KM, Aach J, Guell M, DiCarlo JE, Norville JE, Church GM: **RNA-guided human genome engineering via Cas9.** *Science (New York, NY)* 2013, **339**(6121):823–826.
7. Yang H, Wang H, Shivalila CS, Cheng AW, Shi L, Jaenisch R: **One-step generation of mice carrying reporter and conditional alleles by CRISPR/Cas-mediated genome engineering.** *Cell* 2013, **154**(6):1370–1379.
8. Fujii W, Kawasaki K, Sugiura K, Naito K: **Efficient generation of large-scale genome-modified mice using gRNA and CAS9 endonuclease.** *Nucleic Acids Res* 2013, **41**(20):e187.
9. Li D, Qiu Z, Shao Y, Chen Y, Guan Y, Liu M, Li Y, Gao N, Wang L, Lu X, Zhao Y, Liu M: **Heritable gene targeting in the mouse and rat using a CRISPR-Cas system.** *Nat Biotechnol* 2013, **31**(8):681–683.
10. Shen B, Zhang J, Wu H, Wang J, Ma K, Li Z, Zhang X, Zhang P, Huang X: **Generation of gene-modified mice via Cas9/RNA-mediated gene targeting.** *Cell Res* 2013, **23**(5):720–723.
11. Mashiko D, Young SA, Muto M, Kato H, Nozawa K, Ogawa M, Noda T, Kim YJ, Satouh Y, Fujihara Y, Ikawa M: **Feasibility for a large scale mouse mutagenesis by injecting CRISPR/Cas plasmid into zygotes.** *Develop Growth Differ* 2014, **56**(1):122–129.
12. Mashiko D, Fujihara Y, Satouh Y, Miyata H, Isotani A, Ikawa M: **Generation of mutant mice by pronuclear injection of circular plasmid expressing Cas9 and single guided RNA.** *Scientific Reports* 2013, **3**:3355.
13. Akasaka E, Ozawa A, Mori H, Mizobe Y, Yoshida M, Miyoshi K, Sato M: **Whole-genome amplification-based GenomiPhi for multiple genomic analysis of individual early porcine embryos.** *Theriogenology* 2011, **75**(8):1543–1549.
14. Snabes MC, Chong SS, Subramanian SB, Kristjansson K, DiSepio D, Hughes MR: **Preimplantation single-cell analysis of multiple genetic loci by whole-genome amplification.** *Proc Natl Acad Sci U S A* 1994, **91**(13):6181–6185.
15. Zhang L, Cui X, Schmitt K, Hubert R, Navidi W, Arnheim N: **Whole genome amplification from a single cell: implications for genetic analysis.** *Proc Natl Acad Sci U S A* 1992, **89**(13):5847–5851.
16. Ichikawa-Shindo Y, Sakurai T, Kamiyoshi A, Kawate H, Iinuma N, Yoshizawa T, Koyama T, Fukuchi J, Iimuro S, Moriyama N, Kawakami H, Murata T, Kangawa K, Nagai R, Shindo T: **The GPCR modulator protein RAMP2 is essential for angiogenesis and vascular integrity.** *J Clin Invest* 2008, **118**(1):29–39.
17. Li W, Teng F, Li T, Zhou Q: **Simultaneous generation and germline transmission of multiple gene mutations in rat using CRISPR-Cas systems.** *Nat Biotechnol* 2013, **31**(8):684–686.
18. Ma Y, Zhang X, Shen B, Lu Y, Chen W, Ma J, Bai L, Huang X, Zhang L: **Generating rats with conditional alleles using CRISPR/Cas9.** *Cell Res* 2014, **24**(1):122–125.
19. Niu Y, Shen B, Cui Y, Chen Y, Wang J, Wang L, Kang Y, Zhao X, Si W, Li W, Xiang AP, Zhou J, Guo X, Bi Y, Si C, Hu B, Dong G, Wang H, Zhou Z, Li T, Tan T, Pu X, Wang F, Ji S, Zhou Q, Huang X, Ji W, Sha J: **Generation of**

gene-modified cynomolgus monkey via Cas9/RNA-mediated gene targeting in one-cell embryos. *Cell* 2014, **156**(4):836–843.
20. Carbery ID, Ji D, Harrington A, Brown V, Weinstein EJ, Liaw L, Cui X: **Targeted genome modification in mice using zinc-finger nucleases.** *Genetics* 2010, **186**(2):451–459.
21. Rémy S, Tesson L, Menoret S, Usal C, Scharenberg AM, Anegon I: **Zinc-finger nucleases: a powerful tool for genetic engineering of animals.** *Transgenic Res* 2010, **19**(3):363–371.
22. Joung JK, Sander JD: **TALENs: a widely applicable technology for targeted genome editing.** *Nat Rev Mol Cell Biol* 2013, **14**(1):49–55.
23. Sakurai T, Kimura M, Sato M: **Temporary developmental arrest after storage of fertilized mouse oocytes at 4 degrees C: effects on embryonic development, maternal mRNA processing and cell cycle.** *Mol Hum Reprod* 2005, **11**(5):325–333.
24. Lawitts JA, Biggers JD: **Culture of preimplantation embryos.** *Methods Enzymol* 1993, **225**:153–164.
25. Sakurai T, Kamiyoshi A, Watanabe S, Sato M, Shindo T: **Rapid zygosity determination in mice by SYBR green real-time genomic PCR of a crude DNA solution.** *Transgenic Res* 2008, **17**(1):149–155.

11

Construction of sized eukaryotic cDNA libraries using low input of total environmental metatranscriptomic RNA

Rajiv Kumar Yadav[1*], Florian Barbi[1], Antoine Ziller[1], Patricia Luis[1], Roland Marmeisse[1], M Sudhakara Reddy[2] and Laurence Fraissinet-Tachet[1]

Abstract

Background: Construction of high quality cDNA libraries from the usually low amounts of eukaryotic mRNA extracted from environmental samples is essential in functional metatranscriptomics for the selection of functional, full-length genes encoding proteins of interest. Many of the inserts in libraries constructed by standard methods are represented by truncated cDNAs due to premature stoppage of reverse transcriptase activity and preferential cloning of short cDNAs.

Results: We report here a simple and cost effective technique for preparation of sized eukaryotic cDNA libraries from as low as three microgram of total soil RNA dominated by ribosomal and bacterial RNA. cDNAs synthesized by a template switching approach were size-fractionated by two dimensional agarose gel electrophoresis prior to PCR amplification and cloning. Effective size selection was demonstrated by PCR amplification of conserved gene families specific of each size class. Libraries of more than one million independent inserts whose sizes ranged between one and four kb were thus produced. Up to 80% of the insert sequences were homologous to eukaryotic gene sequences present in public databases.

Conclusions: A simple and cost effective technique has been developed to construct sized eukaryotic cDNA libraries from environmental samples. This technique will facilitate expression cloning of environmental eukaryotic genes and contribute to a better understanding of basic biological and/or ecological processes carried out by eukaryotic microbial communities.

Keywords: Metatranscriptomics, cDNA, cDNA library, mRNA, Gel electrophoresis

Background

The numerous eukaryotic microorganisms present in the environment potentially represent a rich source of genes encoding for novel enzymes or other proteins of interest in biotechnology. In this respect, functional metatranscriptomics has been demonstrated as a powerful tool in discovery of these genes [1-6]. Functional metatranscriptomics first requires the extraction of total RNA from environmental samples. Eukaryotic 3′ polyadenylated (poly-A) messenger RNAs can then be purified from total RNA to remove the ribosomal RNA, other non-coding RNAs as well as the bacterial mRNAs that largely dominate environmental metatranscriptomes [1,4,5]. Poly-A mRNAs are then converted into cDNAs which are cloned in an appropriate expression vector. Such eukaryotic-specific environmental cDNA libraries, first described by Grant *et al.* [7], thus encompass protein coding genes expressed by the different eukaryotic microorganisms present in the original environmental sample [1,7]. Genes of interest can then be screened by expressing them in an appropriate eukaryotic system such as the yeast *Saccharomyces cerevisiae* [1,3-5].

Expression cloning of eukaryotic genes using reverse transcribed poly-A mRNA is a fundamental technology in molecular biology. However, obtaining libraries enriched in long cDNAs remains challenging for the production of functional proteins. The first step is the reverse transcription of mRNAs into cDNAs. This step is adversely affected by

* Correspondence: rajiv-kumar.yadav@univ-lyon1.fr
[1]Ecologie Microbienne, UMR CNRS 5557, USC INRA 1364, Université Lyon 1, Université de Lyon, Villeurbanne, France
Full list of author information is available at the end of the article

many factors and as a consequence, a large proportion of long mRNAs (e.g. larger than 1 kb) is represented by 5′ truncated cDNAs [8]. Being small in size, these truncated cDNAs are preferentially amplified and cloned. Furthermore, functional metatranscriptomics most often involves the use of very low quantities of environmental eukaryotic mRNAs. This necessitates a highly efficient cDNA cloning approach which can make long eukaryotic transcripts available for functional studies.

Various library construction methods that enrich long cDNAs have been proposed. For example, several approaches that use the 5′ end-specific cap structure of eukaryotic poly-A mRNAs have been devised [9-14]. All these approaches however require many enzymatic and purification steps and as a consequence, relatively large quantities of starting poly-A mRNAs are needed which, once again, are difficult to obtain from environmental samples where eukaryotic mRNAs are diluted among predominant bacterial RNAs. cDNA size fractionation by agarose gel electrophoresis is an alternative strategy which requires few enzymatic steps and allows the preparation of different sized cDNA libraries from a single RNA sample. Libraries enriched in long cDNAs were for example constructed by agarose gel mediated size fractionation of cDNA synthesized from mouse embryo and human brain [15,16]. However, despite the application of all these strategies, all of them also require microgram amounts of poly-A mRNAs and numerous clones in these cDNA libraries still represent truncated transcripts.

Reverse transcriptase (RT) template switching is another approach for generation of cDNAs resulting from the reverse transcription of entire RNAs [17]. This technique, implemented in the commercial SMART (Clontech) or Mint (Evrogen, Moscow, Russia) kits, makes use of two activities of MMLV RT. The first one is to add a few deoxycytidines (dC) at the end of single strand (ss) cDNAs. The second one is to switch template and to reverse transcribe an oligonucleotide whose 3′ deoxyguanosine-rich sequence anneals to the dC stretch artificially added at the end of the ss cDNA. As a consequence all cDNAs are bordered at their 3′ end by the same oligonucleotide sequence which can be used for their amplification by PCR in combination with a primer sequence added to the poly-dT primer used to initiate reverse transcription the mRNA.

In the present investigation, we demonstrate an efficient method of construction of sized eukaryotic cDNA libraries from environmental samples such as soil. This method is adapted from Wellenreuther *et al.* [18] who combined template switching in combination with agarose gel mediated size fractionation prior to full cDNA PCR amplification to isolate long full-length genes using microgram amounts of purified human poly-A mRNA. In this study, we modified this method to accommodate small limiting amounts of total environmental RNA.

Results and discussion

Eukaryotic cDNA synthesis from total soil RNA

We extracted total RNA from three soil samples coming from contrasted geographic localities (Additional file 1). The presence of sharp bands in the Bioanalyser electrophoregram, corresponding to small and large subunits rRNA and of a wide mRNA smear (approx. from 0.2 kb to 5 kb) suggested that these RNA samples were not degraded (data not shown). Extraction yields ranged from 330 to 980 ng.g^{-1} of soil and at least 3 μg of total soil RNA were obtained from each sample.

According to Urich *et al.* [19], soil RNA can be constituted of up to 90% of non-coding sequences (essentially rRNA) and approximately only 7% of the coding sequences can originate from eukaryotes. Although these figures certainly differ from one soil to another, bacterial biomass seems to always dominate in soil [20]. As a consequence, purification of μg amounts of poly-A mRNAs, as recommended in most cDNA library construction protocols, can hardly be met. We therefore developed a protocol which makes use of (i) few μg of total RNA as starting material and (ii) includes long range PCR amplification for the synthesis of long cDNAs. This protocol, implemented in the Mint-2 kit (Evrogen) allowed us to obtain ds cDNAs of eukaryotic origin from as low as 3 μg of total soil RNA. Such a quantity may contain only a few ng of poly-A mRNA. Success of cDNA synthesis was demonstrated by PCR amplification of an EF1α gene fragment (data not shown).

Size fractionation and PCR amplification of eukaryotic cDNAs

Getting a high number of clones having long inserts (e.g. above 1 kb) is one of the main challenges in cDNA library construction. As suggested in [18], we performed size fractionation of cDNA prior to their amplification to facilitate both the synthesis of long cDNA and minimize the preferential amplification of short ones from complex mixtures of long and short cDNAs. In the current protocol, ds cDNAs were first amplified through three PCR cycles, to slightly increase cDNA quantities without affecting significantly their average length. Pre-amplified cDNAs were then separated by bi-dimensional agarose gel electrophoresis (Figure 1) to maximize complete separation of cDNAs and minimize cross-contamination of size fractions. The cDNA containing gel was left unstained to prevent the adverse effect of UV irradiation but also because of the low quantity of loaded cDNAs which prevented their visualization. Three different size fractions (A, 0.1–0.5 kb; B, 0.5–1 kb and C, 1–4 kb) were recovered and amplified by PCR using between 22 and 30 cycles.

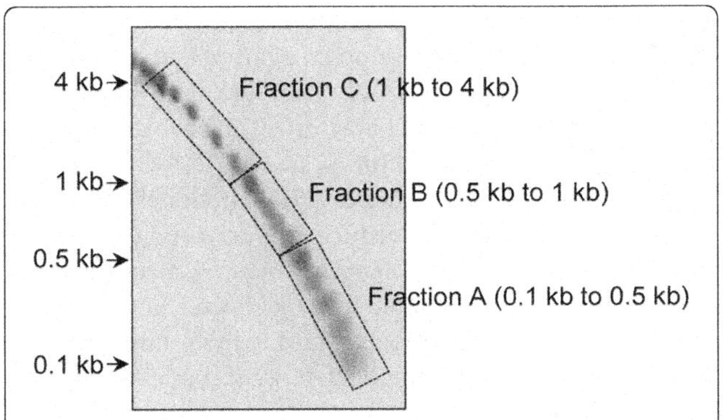

Figure 1 Bi-dimentional cDNA size fractionation. A DNA size standard was run along with cDNA on two identical but separate agarose gels to visualize the fractionation. The cDNAs unstained gel was superimposed over the ethidium bromide-stained DNA marker gel and three gel slices containing different cDNA size fractions were cut out.

While amplified cDNAs without size fractionation gave a broad smear throughout the gel lane (Figure 2A), three distinct and discrete DNA smears (broad bands) corresponding to the different cDNA fractions were clearly visible (Figure 2B and Additional file 2A & B). Fractions were tested for the absence of cross contaminations by PCR amplification of four eukaryotic genes of different length classes. For the 1–4 kb class, the conserved genes encode β-tubulin (β-Tub) and EF1-α whose coding sequences are of about 1300 and 1400 bp respectively, while for the 0.5–1 kb class, they encode 40S ribosomal protein S3 (RibS3) and peptide methionine sulphoxide reductase (MsrA) whose coding sequences are of about 800 and 600 bp respectively. As expected, positive amplifications of β-Tub and EF1-α fragments were only obtained for fractions C (1–4 kb), while RibS3 and MsrA fragments were only detected for the intermediate B fractions (0.5–1 kb) and no amplification was visible for the lowest A fractions (0.1–0.5 kb) (Figure 3). These results demonstrate that the method can not only be used for obtaining long cDNAs but also in getting high quality eukaryotic cDNA size fractions from very low amounts of starting total RNAs.

Construction of sized cDNA libraries

The three cDNA fractions from soil sample PL and the largest fraction C from sample BB were used for construction of cDNA libraries by directional cloning in the pFL61 yeast expression vector. All four libraries contained at least 10^6 independent clones (range from 1.1 10^6 for library PL-B to 2.6 10^6 for library BB-C), but many more could have been obtained. After *Sfi*I digestion of the library plasmid pools and their separation by agarose gel electrophoresis, the released cDNA insert pools of each library were detected as a smear whose size range corresponded to the original cDNA fraction size (Figure 2C and Additional file 2C). The cDNA inserts of 42 random colonies from each of the three libraries from PL soil and 30 from library BB-C were PCR amplified. Absence of inserts was found for less than 1% of the plasmids and all of the PCR products fell within their expected size range (Additional file 3). For the 3 PL libraries, ten inserts per library were sequenced from their 5′-ends. Globally, 60% of the sequences returned a positive result upon BlastX searches against the Eukaryotic GenBank protein database (80%, 70% and 30% for inserts from libraries PL-C, PL-B and PL-A respectively; Additional file 4).

Conclusions

We established a robust protocol to generate high quality eukaryotic sized cDNA libraries from total RNA extracted

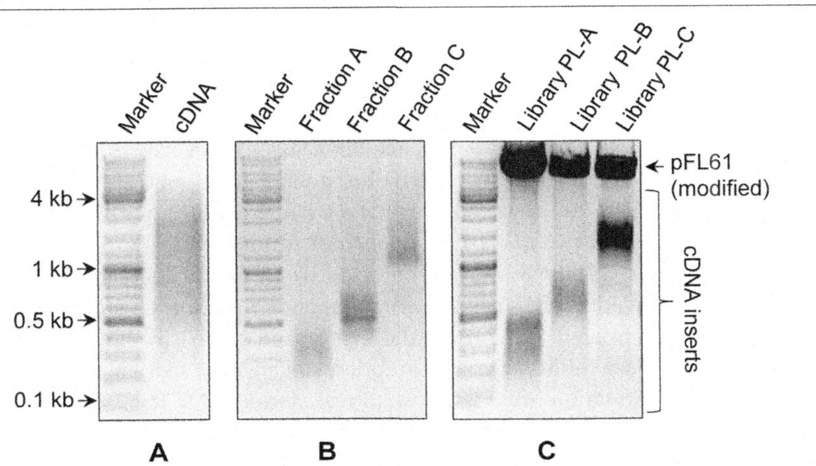

Figure 2 cDNA size fractions and libraries preparation from soil PL. Agarose gel separation of **(A)** amplified cDNA without size fractionation, **(B)** the three different cDNA fractions A (0.1–0.5 kb); B (0.5–1 kb) and C (1–4 kb) and **(C)** *Sfi*I digested libraries PL-A, PL-B and PL-C. For each sized library, the size range of the cDNA inserts is similar before **(B)** and after **(C)** cloning.

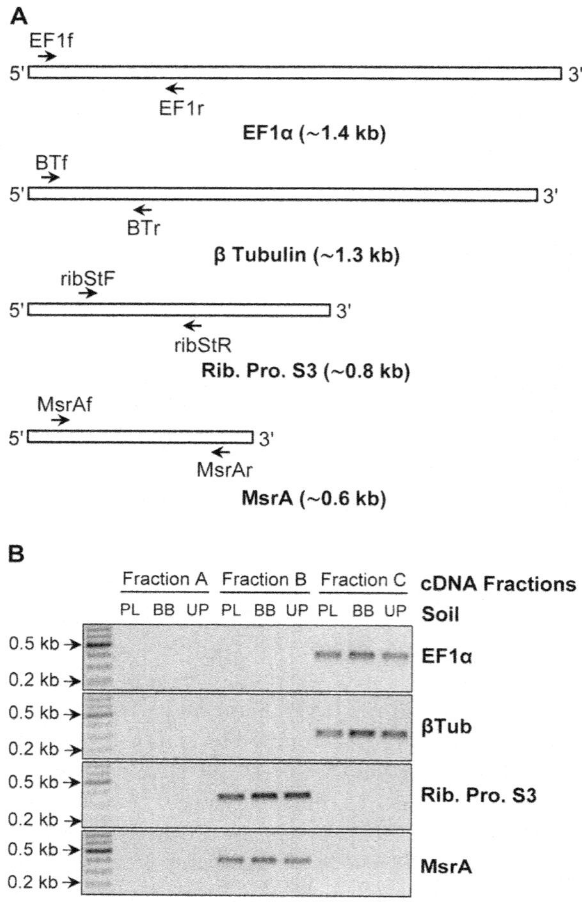

Figure 3 PCR validations of each cDNA size fraction obtained from soil samples PL, BB and UP. All three cDNA size fractions from each soil were tested for cross-contaminations by PCR amplification of conserved genes of different length. **(A)** Approximate size of coding sequences of each of the four tested genes and positions of the PCR primers. **(B)** Gel electrophoresis of the PCR products showing perfect correlation between sized cDNA fraction and detection of a PCR product from genes whose expected size is included in the corresponding fraction size range. In fraction C, only, elongation factor 1-alpha and beta-tubulin genes were amplified while in fraction B, only ribosomal protein S3 and peptide methionine sulphoxide reductase genes were detected whereas none of theses genes was amplified from fraction A.

from environmental samples. We demonstrated that the original protocol developed by Wellenreuther *et al.* [18] using micrograms amounts of purified poly-A mRNA extracted from human tissues can be considerably scaled down and implemented on total RNA from environments dominated by bacterial RNA. This protocol fulfils several requirements. Firstly yields of RNA extraction directly from environments are usually low (sometimes less than 100 $ng.g^{-1}$ of soil) and therefore isolation of μg amounts of poly-A mRNA appears, in most cases, almost unfeasible. Secondly, two dimensional electrophoretic separation of cDNAs leads to isolation of sized cDNA pools almost free from contaminations by either longer or shorter cDNAs. Most importantly these pools of amplified cDNAs allow production of large cDNA libraries necessary to capture the gene diversity which characterizes microbial communities. Thirdly, the production of sized cDNA libraries is of direct relevance in the context of functional metatranscriptomics as it should reduce the screening effort for gene categories of defined length which, as a result of size selection, will not be diluted among shorter or longer transcripts. As an example, most glycoside hydrolases (e.g. cellulases, hemicellulases), implicated in organic matter degradation, are encoded by transcripts 1–3 kb in length and screening for these genes can be limited to the corresponding sized cDNA libraries. In conclusion, we believe that the protocol presented in this paper should facilitate and promote studies of environmental eukaryotic communities, which in a context of environmental biotechnology represent a promising and almost untouched source of genes of interest.

Methods

RNA extraction from soil

Three soils, two from France (PL and BB) and one from India (UP) were used (Additional file 1). Total RNAs were isolated from soil samples according to Damon *et al.* [21] for the BB sample or by using the RNA PowerSoil® Total RNA Isolation Kit (Mo Bio laboratories, Carlsbad, CA) for the PL and UP samples. All soil RNA samples were treated with RNase-free DNase I. After a final precipitation step, all the RNAs were dissolved in nuclease-free water. RNA integrity was checked by Bioanalyzer 2100 (Agilent Technologies, USA) electrophoresis and RNA quantity and purity were determined by spectrophotometry (SAFAS UVmc2, SAFAS Monaco).

Synthesis, size fractionation and amplification of cDNAs

cDNAs were synthesized by using the Mint-2 cDNA synthesis kit (Evrogen, Moscow, Russia) according to the manufacturer's instructions. Briefly, three μg of total soil RNAs were mixed with 10 μM of two oligonucleotide adapters. The 3′end CDS-4M adapter contains (i) an oligo (dT) sequence that anneals to the poly (A) stretch of eukaryotic mRNAs, (ii) a *Sfi*IB restriction site and (iii) sequence of primer M1. The 5′-end PlugOligo-3M adapter contains (i) an oligo (dG) sequence which anneals to the complementary oligo (dC) stretch added to the 3′-end of the first-strand cDNA by Mint MMLV RT (ii) a *Sfi*IA restriction site and (iii) sequence of primer M1. The mixture was incubated at 70°C for 2 minutes. First-strand cDNAs were synthesized at 42°C by Mint RT in presence of dNTPs, DTT, first-strand buffer and IP-solution in 15 μL of total reaction volume. Second-strand cDNA synthesis was carried out by the thermostable Encyclo DNA polymerase (Evrogen) using the M1 primer which recognizes both the PlugOligo-3M and CDS-4M adapter sequences. Four μL from first-strand cDNA reaction, i.e. the equivalent of 800 ng of total soil RNA, were used in second-strand cDNA synthesis. Second strand synthesis was followed by a PCR amplification limited to 3 cycles at 95°C for 15 sec, 66°C for 20 sec and 72°C for 3 min. Resulting double-stranded cDNAs (ds cDNAs) were purified by phenol-chloroform extraction and precipitated.

Size fractionation of cDNAs was performed as described by Wellenreuther *et al.* [18]. ds cDNAs and a DNA size standard were size fractionated in two separate but identical 0.7% agarose gels at identical running conditions. After electrophoresis, the gel lanes containing ds cDNAs and DNA size standard were cut out, rotated at 90° and placed into two separate but identical gel trays. Identical volumes of 1.4% low melting point agarose (Bioprobe Systems, Montreuil, France) were cast in both the trays and the gels were subjected to electrophoresis at 2.6 $V.cm^{-1}$ for 10 h. The unstained gel containing cDNAs was then superimposed over the ethidium bromide stained gel containing the size standard which was visualized using a Dark Reader transilluminator (Clare Chemical Research, Inc., USA). Three gel slices corresponding to the cDNA size fractions A, 0.1–0.5 kb; B, 0.5–1 kb and C, 1–4 kb, were cut out from the unstained gel. cDNAs were extracted from each gel slice by using QIAEX II Gel Extraction Kit (Qiagen, Netherlands), precipitated and amplified by PCR using primer M1 as described above but using higher number of cycles. Depending on cDNA size fraction and RNA sample, between 22 to 30 cycles were performed for optimal amplification.

Validation of cDNA size fractions

Validation of the different sized cDNA fractions was performed by running each of them in standard 1% agarose gels and by PCR amplification, on each of them, of different gene families representative of the different size groups. For fraction C, the selected families encoded β-Tubulin and Elongation factor 1-alpha (EF1α). They were amplified using respectively primer pairs BTf (GGTAACCAAAT CGGTGCTGCTTTC)/BTr (ACCCTCAGTGTAGTGACC CTTGGC) [22] and EF1f (GTCGTYGTYATYGGHCAY GT)/EF1r (TGYTCNCGRGTYTGNCCRTCYTT) [23]. For fraction B, the selected families encoded 40S Ribosomal protein S3 amplified using primers ribStF (CHSKHACYG ABRTCATCATCCG) and ribStR, (AADCCRTCRGTGAA CTTCATG) (this study) and Peptide methionine sulphoxide reductase amplified using MsrAf (CGCCG CCGGCTGYTTYTGGGG) and MsrAr (ATRGTRGT YNWCATGGACCTGTTCTTGGGGC) [24]. For the smallest fraction A, no conserved gene family in this size range could be identified to design PCR primers that would work on environmental cDNA samples. Ten ng of each cDNA fraction were used as template in 35 cycles PCR reactions.

cDNA cloning

cDNA fractions (500 ng) were digested by *Sfi*I which recognizes *Sfi*IA and *Sfi*IB sites located in the sequences of PlugOligo-3M and CDS-4M, respectively. Following phenol-chloroform extraction and precipitation, cDNAs were ligated downstream of the *S. cerevisiae* PGK1 promoter in a modified pFL61 yeast expression vector containing *Sfi*IA and *Sfi*IB sites [5,25]. Recombinant plasmids were introduced into electro-competent *E. coli* cells (MegaX DH10B™ T1R Electrocomp™ cells, Invitrogen) and at least 10^6 ampicillin resistant bacterial colonies growing on agar medium were collected and pooled to constitute each of the libraries.

Sequencing

Plasmids were isolated from ten randomly selected bacterial colonies from each of the three libraries constructed from PL sample. cDNA inserts were sequenced from their 5′ end and deduced amino acid sequences were used in similarity search (BLASTX) against the GenBank nr eukaryotic protein database (as in December 2013). Resulting cDNA sequences appear in the EMBL database under accession Nos. HG964498 to HG964527.

Additional files

Additional file 1: Main characteristics of the soil sampling sites.

Additional file 2: cDNA size fractionation from different soils. *Description of data*: Agarose gel showing amplified eukaryotic cDNA fractions **A**, **B** and **C** after size fractionation from **(A)** Indian soil UP and **(B)** French forest soil BB sample. Fraction C of soil BB was cloned in modified pFL61 vector. **(C)** The corresponding library (BB-C) was digested with *Sfi*I enzyme realeasing cDNA inserts of sizes ranging, as expected, between 1 and 4 kb.

Additional file 3: Validation of insert size by colony PCR amplification. *Description of data*: cDNA inserts from 42 random colonies from each of the three libraries PL-A, PL-B and PL-C and 30 random colonies from library BB-C were amplified by colony PCR. As expected, the sizes of amplified DNA inserts of each library were confined between their expected size cut offs. Panels **A**, **B** and **C** are the gel images of separated PCR products from libraries PL-A, PL-B and PL-C respectively. Panel **D** is the gel image after migration of PCR products from library BB-C.

Additional file 4: Protein similarity search of sequenced cDNA inserts using blastx against GenBank eukaryotic protein sequences. An e-value threshold of 10^{-5} was retained for annotation. aa, amino acids.

Competing interests

The authors declare that they have no competing interests.

Authors' contributions

RKY, RM and LF designed and coordinated the study. RKY, FB, AZ and MSR carried out the experiments. RKY, RM and LF wrote the manuscript, and PL and MSR reviewed the manuscript. All authors read and approved the final manuscript.

Acknowledgements
This study is part of project 4709-1 funded by Indo-French Centre for the Promotion of Advanced Research which granted a post-doctoral fellowship to RKY. We would like to thank Jacques Ranger for access to the Breuil site, and Damien Blaudez and Michel Chalot for access to the Pierrelaye one.

Author details
[1]Ecologie Microbienne, UMR CNRS 5557, USC INRA 1364, Université Lyon 1, Université de Lyon, Villeurbanne, France. [2]Department of Biotechnology, Thapar University, Patiala 147004, Punjab, India.

References

1. Bailly J, Fraissinet-Tachet L, Verner MC, Debaud JC, Lemaire M, Wesolowski-Louvel M, Marmeisse R: **Soil eukaryotic functional diversity, a metatranscriptomic approach.** *ISME J* 2007, **1**:632–642.
2. Todaka N, Moriya S, Saita K, Hondo T, Kiuchi I, Takasu H, Ohkuma M, Piero C, Hayashizaki Y, Kudo T: **Environmental cDNA analysis of the genes involved in lignocellulose digestion in the symbiotic protist community of Reticulitermes speratus.** *FEMS Microbiol Ecol* 2007, **59**:592–599.
3. Kellner H, Luis P, Portetelle D, Vandenbol M: **Screening of a soil metatranscriptomic library by functional complementation of Saccharomyces cerevisiae mutants.** *Microbiol Res* 2011, **166**:360–368.
4. Damon C, Vallon L, Zimmermann S, Haider MZ, Galeote V, Dequin S, Luis P, Fraissinet-Tachet L, Marmeisse R: **A novel fungal family of oligopeptide transporters identified by functional metatranscriptomics of soil eukaryotes.** *ISME J* 2011, **5**:1871–1880.
5. Lehembre F, Doillon D, David E, Perrotto S, Baude J, Foulon J, Harfouche L, Vallon L, Poulain J, Da Silva C, Wincker P, Oger-Desfeux C, Richaud P, Colpaert V, Chalot M, Fraissinet-Tachet L, Blaudez D, Marmeisse R: **Soil metatranscriptomics for mining eukaryotic heavy metal resistance genes.** *Env Microbiol* 2013, **15**:2829–2840.
6. Takasaki K, Miura T, Kanno M, Tamaki H, Hanada S, Kamagata Y, Kimura N: **Discovery of glycoside hydrolase enzymes in an avicel-adapted forest soil fungal community by a metatranscriptomic approach.** *PLoS One* 2013, **8**:e55485.
7. Grant S, Grant WD, Cowan DA, Jones BE, Ma Y, Ventosa A, Heaphy S: **Identification of eukaryotic open reading frames in metagenomic cDNA libraries made from environmental samples.** *Appl Env Microbiol* 2006, **72**:135–143.
8. Malboeuf CM, Isaacs SJ, Tran NH, Kim B: **Thermal effects on reverse transcription: improvement of accuracy and processivity in cDNA synthesis.** *Biotechniques* 2001, **30**:1074–1078. 1080, 1082, passim.
9. Theissen H, Etzerodt M, Reuter R, Schneider C, Lottspeich F, Argos P, Luhrmann R, Philipson L: **Cloning of the human cDNA for the U1 RNA-associated 70K protein.** *EMBO J* 1986, **5**:3209–3217.
10. Edery I, Chu LL, Sonenberg N, Pelletier J: **An efficient strategy to isolate full-length cDNAs based on an mRNA cap retention procedure (CAPture).** *Mol Cell Biol* 1995, **15**:3363–3371.
11. Carninci P, Kvam C, Kitamura A, Ohsumi T, Okazaki Y, Itoh M, Kamiya M, Shibata K, Sasaki N, Izawa M, Muramatsu M, Hayashizaki Y, Schneider C: **High-efficiency full-length cDNA cloning by biotinylated CAP trapper.** *Genomics* 1996, **37**:327–336.
12. Suzuki Y, Yoshitomo-Nakagawa K, Maruyama K, Suyama A, Sugano S: **Construction and characterization of a full length-enriched and a 5'-end-enriched cDNA library.** *Gene* 1997, **200**:149–156.
13. Seki M, Narusaka M, Kamiya A, Ishida J, Satou M, Sakurai T, Nakajima M, Enju A, Akiyama K, Oono Y, Muramatsu M, Hayashizaki Y, Kawai J, Carninci P, Itoh M, Ishii Y, Arakawa T, Shibata K, Shinagawa A, Shinozaki K: **Functional annotation of a full-length Arabidopsis cDNA collection.** *Science* 2002, **296**:141–145.
14. Fernandez C, Gregory WF, Loke P, Maizels RM: **Full-length-enriched cDNA libraries from Echinococcus granulosus contain separate populations of oligo-capped and trans-spliced transcripts and a high level of predicted signal peptide sequences.** *Mol Biochem Parasitol* 2002, **122**:171–180.
15. Seki N, Ohira M, Nagase T, Ishikawa K, Miyajima N, Nakajima D, Nomura N, Ohara O: **Characterization of cDNA clones in size-fractionated cDNA libraries from human brain.** *DNA Res* 1997, **4**:345–349.
16. Draper MP, August PR, Connolly T, Packard B, Call KM: **Efficient cloning of full-length cDNAs based on cDNA size fractionation.** *Genomics* 2002, **79**:603–607.
17. Zhu YY, Machleder EM, Chenchik A, Li R, Siebert PD: **Reverse transcriptase template switching: a SMART approach for full-length cDNA library construction.** *Biotechniques* 2001, **30**:892–897.
18. Wellenreuther R, Schupp I, Poustka A, Wiemann S: **SMART amplification combined with cDNA size fractionation in order to obtain large full-length clones.** *BMC Genomics* 2004, **5**:36.
19. Urich T, Lanzen A, Qi J, Huson DH, Schleper C, Schuster SC: **Simultaneous assessment of soil microbial community structure and function through analysis of the meta-transcriptome.** *PLoS One* 2008, **3**:e2527.
20. Daniel R: **The metagenomics of soil.** *Nature Rev Microbiol* 2005, **3**:470–478.
21. Damon C, Barroso G, Ferandon C, Ranger J, Fraissinet-Tachet L, Marmeisse R: **Performance of the COX1 gene as a marker for the study of metabolically active Pezizomycotina and Agaricomycetes fungal communities from the analysis of soil RNA.** *FEMS Microbiol Ecol* 2010, **74**:693–705.
22. Glass NL, Donaldson GC: **Development of primer sets designed for use with the PCR to amplify conserved genes from filamentous ascomycetes.** *Appl Env Microbiol* 1995, **61**:1323–1330.
23. Rehner SA, Buckley E: **A Beauveria phylogeny inferred from nuclear ITS and EF1-alpha sequences: evidence for cryptic diversification and links to Cordyceps teleomorphs.** *Mycologia* 2005, **97**:84–98.
24. Lewis CT, Bilkhu S, Vincent R, Eberhardt U, Szoke S, Seifert KA, Lévesque CA: **Identification of fungal DNA barcode targets and PCR primers based on Pfam protein families and taxonomic hierarchy.** *Open Appl Inform J* 2011, **5**:30–44.
25. Minet M, Dufour ME, Lacroute F: **Complementation of Saccharomyces cerevisiae auxotrophic mutants by Arabidopsis thaliana cDNAs.** *Plant J* 1992, **2**:S-422.

12

Targeted gene disruption by use of transcription activator-like effector nuclease (TALEN) in the water flea *Daphnia pulex*

Chizue Hiruta[1], Yukiko Ogino[1,2], Tetsushi Sakuma[3], Kenji Toyota[1,2], Shinichi Miyagawa[1,2], Takashi Yamamoto[3] and Taisen Iguchi[1,2*]

Abstract

Background: The cosmopolitan microcrustacean *Daphnia pulex* provides a model system for both human health research and monitoring ecosystem integrity. It is the first crustacean to have its complete genome sequenced, an unprecedented ca. 36% of which has no known homologs with any other species. Moreover, *D. pulex* is ideally suited for experimental manipulation because of its short reproductive cycle, large numbers of offspring, synchronization of oocyte maturation, and other life history characteristics. However, existing gene manipulation techniques are insufficient to accurately define gene functions. Although our previous investigations developed an RNA interference (RNAi) system in *D. pulex*, the possible time period of functional analysis was limited because the effectiveness of RNAi is transient. Thus, in this study, we developed a genome editing system for *D. pulex* by first microinjecting transcription activator-like effector nuclease (TALEN) mRNAs into early embryos and then evaluating TALEN activity and mutation phenotypes.

Results: We assembled a TALEN construct specific to the *Distal-less* gene (*Dll*), which is a homeobox transcription factor essential for distal limb development in invertebrates and vertebrates, and evaluated its activity *in vitro* by single-strand annealing assay. Then, we injected TALEN mRNAs into eggs within 1 hour post-ovulation. Injected embryos presented with defects in the second antenna and altered appendage development, and indel mutations were detected in *Dll* loci, indicating that this technique successfully knocked out the target gene.

Conclusions: We succeeded, for the first time in *D. pulex*, in targeted mutagenesis by use of Platinum TALENs. This genome editing technique makes it possible to conduct reverse genetic analysis in *D. pulex*, making this species an even more appropriate model organism for environmental, evolutionary, and developmental genomics.

Keywords: *Daphnia pulex*, *Distal-less*, Platinum TALEN, Gene disruption, Knock-out, Targeted mutagenesis, Gene manipulation, Genome editing

Background

Environmental cues constantly influence gene expression; hence interactions among them are indispensable for animal adaptations against changing environments. The water flea *Daphnia pulex* shows a striking ability to contend with environmental changes, resulting in various adaptive phenotypes involving traits such as body size, longevity, behavior, and morphology [1]. It is well known that daphnids occupy a key position as the intermediate link between primary productivity and top predators in the aquatic food chain, and they also serve as an environmental indicator organism because of their high sensitivity to water quality [2]. These characteristics, in addition to their short life cycle, large brood sizes, and synchronization of oocyte maturation, make them valuable for environmental, evolutionary, and developmental genomics studies. Furthermore, the *D. pulex* genome has been sequenced [3], which facilitates the identification of candidate genes involved in

* Correspondence: taisen@nibb.ac.jp
[1]Okazaki Institute for Integrative Bioscience, National Institute for Basic Biology, National Institutes of Natural Sciences, Higashiyama 5-1, Myodaiji, Okazaki, Aichi 444-8787, Japan
[2]Faculty of Life Science, Graduate University for Advanced Studies (SOKENDAI), 5-1 Higashiyama, Myodaiji, Okazaki, Aichi 444-8787, Japan
Full list of author information is available at the end of the article

their unique biological attributes. The analysis of transcriptomics, proteomics, and metabolomics also showed profiles of a large number of putative factors involved in their unique life history characteristics (cf. [4,5]). The *D. pulex* genome possesses as many as about 31,000 genes; this large suite of genes may provide the arsenal responsible for the organism's responsiveness to environmental challenges [3]. So far, however, there are no effective methods for manipulating genes, and only transient analysis of gene function by the RNA interference (RNAi) method [6] is available. The RNAi system in *D. pulex* suffers from several weaknesses, including incomplete silencing, transient effects, and limited analyzable stages. Because of these situations, even in the post-genomic era, the establishment of a gene manipulation technique has been eagerly anticipated to address the characterization of gene function in *D. pulex*.

There are three main types of artificial sequence-specific nucleases based on the source mechanism of DNA binding that guides nuclease activity to a genomic target: zinc-finger nucleases (ZFNs), transcription activator-like effector nucleases (TALENs), and clustered regularly interspaced short palindromic repeats (CRISPR) [7]. In this study, we chose TALEN to edit the *D. pulex* genome, although each method has both advantages and disadvantages in terms of cost, sequence-specificity, off-target effects, and so on [8,9]. The transcription activator-like effectors (TALEs) were first discovered in the plant pathogen *Xanthomonas* sp., and are composed of a conserved central domain for site-specific DNA binding [10,11]. The DNA-binding domain consists of varying numbers of tandem repeats of a 34-amino acid monomer, which specifies the DNA-binding sequence by its 12th and 13th amino acids, called repeat-variable di-residues (RVDs). RVDs specifically recognize a single nucleotide with the following codes: NG = T, HD = C, NI = A, NN = G or A [10,12,13]. Sakuma et al. [14] demonstrated that TALENs with periodically-patterned repeat variants harboring non-RVD variations, called Platinum TALENs, showed higher activities than TALENs without non-RVD variations (so-called Golden TALENs). TALENs are artificially generated by fusing TAL effector DNA-binding domains to a Fok I nuclease domain, and successfully harnessed to custom-designed sequence-specific nucleases [15,16]. TALENs induce DNA double-strand breaks (DSBs) that can be repaired by the error-prone non-homologous end joining (NHEJ) system to cause insertion and/or deletion mutations at targeted genomic loci.

TALEN-mediated gene targeting has been applied in a great number of vertebrates and invertebrates [7,9,17]. In arthropods, it has been reported in insect species including flies [18], mosquitos [19,20], crickets [21], and silkworms [22]. Very recently, mutagenesis of an eye development gene was achieved in one congener *Daphnia magna* using the CRISPR/Cas system [23]. However, the conditions developed in *D. magna* are not directly applicable to *D. pulex*, possibly because of the difference in egg size and contents, and developmental duration, as described in our previous study [6].

Distal-less (*Dll*) and its homologs, *Dlx* genes, which function as homeodomain transcription factors, play one of the major roles in distal limb development throughout the animal kingdom [24]. Reduction of Dll activity caused defects of distal appendage segments in arthropods, resulting in the production of an easily recognizable phenotype (cf. [25,26]). In addition, the results of *Dll* RNAi in *D. pulex* [6] provide possible comparative data regarding phenotype, which is why this endogenous developmental gene was selected as a target for proof-of-principle TALEN in *D. pulex*.

The goal of this study was to develop a targeted gene disruption system by TALEN-mediated artificial DSBs in *D. pulex*. A TALEN target site was designed and assembled, and then TALEN mRNAs were microinjected into embryos to successfully induce insertion and/or deletion mutations.

Results and discussion

Construction and evaluation of *Dll* TALENs

We first cloned and sequenced the partial sequence of the *Dll* gene in *D. pulex*, and then designed two pairs of TALENs, *Dll*_A and *Dll*_B, targeted to the first exon to induce a nonsense mutant protein (Figure 1). After constructing the TALENs by the Platinum Gate TALEN Kit, the activities of TALENs were evaluated by human cell-based single-strand annealing (SSA) assay [27] (Figure 2). Both *Dll*_A and *Dll*_B TALENs showed higher scores than the positive control ZFN [28], suggesting that both TALENs have DSB-inducing activities. We have evaluated the TALEN activities using human cells for various animal applications including sea urchin [29] and fruit fly [27,30], suggesting that the activity score measured in human cells is, in principle, a good indication of the level of inducing DSBs in a wide range of animals. SSA assay can directly be performed with the constructed TALEN plasmids in human cells, so that we can easily know the quality of the constructed TALENs before making mRNAs and injecting them into the animals. We thus think the validation process with human cell-based SSA assay is beneficial to apply TALENs in animals including *D. pulex*.

Detection of mutations and their sequences

Two pairs of *Dll* TALEN mRNAs were microinjected into eggs just after ovulation, and the genome modification efficiencies *in vivo* were evaluated by a T7 endonuclease I (T7EI) assay. We found that a signal intensity of undigested band in the TALEN-injected embryos was weaker than that in uninjected embryos (Figure 3A). Furthermore, multiple digested bands were detected in *Dll*_B-injected embryos, whereas only a slightly separated band was detected in

```
                  Dll_A_Left                          Dll_A_Right
Dll_dp     1:ATGTCGTCGACGACGCCCAATTCCGTCGGCGCCGGAGATTGCGACCAGCAACAACAACAA 60
                                               Dll_B_Left
Dll_dp    61:CAACAACAACAAGATCCATCCAATTCATCCGGACAGCAGCAGCAGCAGGCGGCAGCGGCT 120
             Dll_B_Right
Dll_dp   121:GCAGCGGCCGCCGCATCCAAGAACGCCTTCCTGGAACTGCAACAGGCCCACCAGCACCTG 180

Dll_dp   181:GCCAGTCAGGCCGGCATCGGTCCGCCACCACCGCACATGGGCGGAGGTGGCGGCGGCGGC 240

Dll_dp   241:GGAGGGGGGCCGTACGGCATGAGCCGCAACCCTTACGGCCAGATGCAGCAGCAGCATCCG 300

Dll_dp   301:TCGGCCAACAGTTTGAGCGTCAACCACCAGCTGGGCGGCTACCCCTTCTCCCACATGACG 360

Dll_dp   361:TCACAAAACTCGTACGCCGCAGCGGCGGCCGCCGCCGCCGCCGGATACCACCACCTCAGT 420

Dll_dp   421:CCTTATCCGTCGCAATGTCCATCACCTCCCAGAGACG 457
```

Figure 1 The first exon nucleotide sequence of the *D. pulex Dll* gene. Two pairs of TALENs, *Dll*_A and *Dll*_B, were designed in the first exon (highlighted in gray). The spacer regions of *Dll*_A and *Dll*_B each contain 15 bp nucleotides. Underline indicates ATG start codon.

*Dll*_A-injected embryos. We thus concluded that *Dll*_B exhibits higher genome modification efficiencies rather than *Dll*_A.

To investigate the successful genome modifications of the F_0 founder line, the genomic region surrounding the TALEN target was amplified and subcloned. Sequencing analysis revealed that TALENs predominantly induce small indel mutations ranging in size from 6-13 bp (Figure 3B), which is similar to results obtained in other arthropods [18-22]. In *Dll*_A-injected F_0 founders, only two types of mutations, 6- and 9-nucleotide deletions, were detected. These mutations did not cause a frame shift and growth defect (see next section), suggesting that the gene function was not compromised. On the other hand, seven types of mutations were detected in *Dll*_B-injected F_0 founders, five of which induced a frame shift leading to the appearance of a stop codon in the first exon. These mutations are predicted to result in a very short protein (83 amino acids) without a homeodomain, which is a functional DNA-binding domain and highly conserved among other arthropods. The variations of mutation patterns in *Dll*_B-injected F_0 founders were higher than those in *Dll*_A-injected F_0 founders, as expected from the T7EI assay. Based on these results, we conclude that the T7EI assay is a quick and easy method for estimating TALEN *in vivo* activity in F_0 embryos of *D. pulex*.

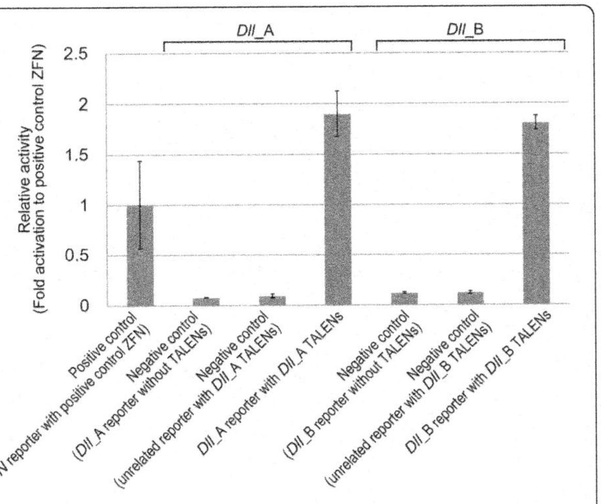

Figure 2 Activity evaluation of *Dll*_A and *Dll*_B TALENs by SSA assay. TALEN plasmids and the corresponding reporter plasmid, TALEN plasmids and unrelated reporter plasmid, or the reporter plasmid without TALEN plasmids were transfected into HEK293T cells. After the luciferase assay, fold activation scores against the positive control ZFN (pSTL-ZFA36 with the corresponding reporter) [28] were calculated. Data are expressed as means ± SEM (n = 3).

To investigate whether these mutations were heritable, we then cloned and sequenced the target genomic region from F_2 embryos. DNA sequencing revealed that inherited modifications existed in the two and one F_2 lines injected with *Dll*_A and B, respectively (Figure 3C). Progeny from the same F_0 founder had the same mutation genotype, indicating that the F_0 founder carried a single germline mutation, not a mosaic germline mutation. Because cleavage proceeds without cytokinesis during injection period [31] and TALEN mRNAs can easily diffuse throughout the egg, it is considered that this method is able to induce mutation at very early developmental stages and easily establish mutant lines having a single germline

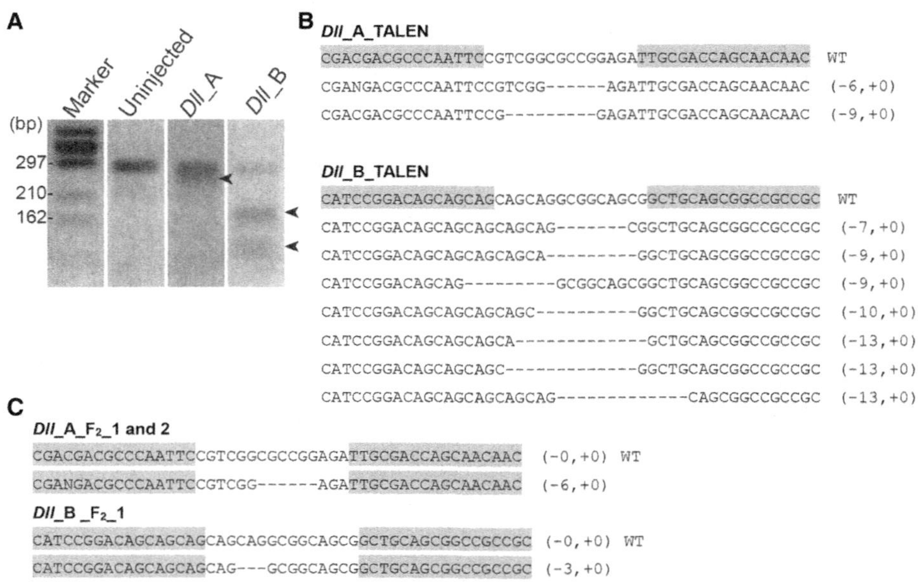

Figure 3 Detection of TALEN-induced mutations and their sequences. (A) Mutations in *Dll* were detected by T7EI assay. Arrowheads indicate the cleavage products generated in T7EI assay of F_0 founders. **(B)** Sequence analysis of *Dll* mutant alleles in F_0 founders induced by TALENs. The wild type (WT) sequence is shown at the top with the TALEN-binding sites (highlighted in gray). The number of nucleotides deleted (-) and inserted (+) is indicated in parentheses. **(C)** Sequence analysis of *Dll* mutant alleles in F_2 embryos. Binding sites and insertions/deletions indicated as in **(B)**.

mutation, without a need for complicated screening. In other words, mutation analysis can be carried out in a mere generation.

Phenotype of *Dll* TALEN in *D. pulex*

Results from TALEN experiments are summarized in Table 1. Compared to uninjected embryos, the low viability of the embryos injected with *Dll* TALEN mRNAs may be partly attributed to damage due to microinjection and/or toxicity of TALEN mRNAs, as well as the embryonic lethality of *Dll* knock-out mutants. The difference of viability and/or morphological phenotype among embryos injected with *Dll*_A or *Dll*_B might depend on their genetic phenotype, such as non-mosaic/mosaic, monoallelic/biallelic, and non-frame shift/frame shift mutations. When a half concentration of *Dll*_B (250 ng/μl) was injected, viability of injected embryos increased by 7.5% and two individuals did not have any prominent altered phenotypes, suggesting dose dependency of the TALENs on viability and phenotypes.

Morphologically, the *Dll*_A injected embryos did not show any prominent altered phenotypes, and both homozygous and heterozygous non-frame shift mutations were detected in *Dll* loci. Hence, we assessed the phenotype of *Dll*_B-injected F_0 founders hereafter. Similar to the RNAi knockdown phenotype of *Dll* [6], *Dll*_B-injected F_0 founders displayed various degrees of defects in second antennae, appendages, ocellus, abdominal claw, and abdominal setae, all tissues in which *Dll* is normally expressed (Figure 4 and Additional file 1: Figure S1). In the second antennae, the degree of the segment truncation was variably detected, ranging from severe (peduncle alone, lacking dorsal and ventral rami) to mild (peduncle and deficient rami). The first to fifth thoracic appendages, including each exopodite, were shortened. Moreover, a loss of the ocellus and abdominal setae, and minimized abdominal claw, were observed in *Dll*_B-injected embryos. We found that the mutants with severe defects in the second antenna and appendage development have biallelic mutations, failed to molt, and died before becoming adults. The lethality of *Dll* knock-out mutants was in accordance with previous studies on insects [25,32]. This easy and efficient induction of biallelic mutations by TALENs provides a valuable tool for parthenogenetic

Table 1 Summary of TALEN results

Platinum TALENs	mRNAs concentration (ng/μl)	Injected embryos	Juveniles	Viability (%)	Shortened 2nd antennae
*Dll*_A	500	88	36	40.9	0
*Dll*_B	500	206	63	30.6	63
	250	134	51	38.1	49
Uninjected	-	130	107	82.3	0

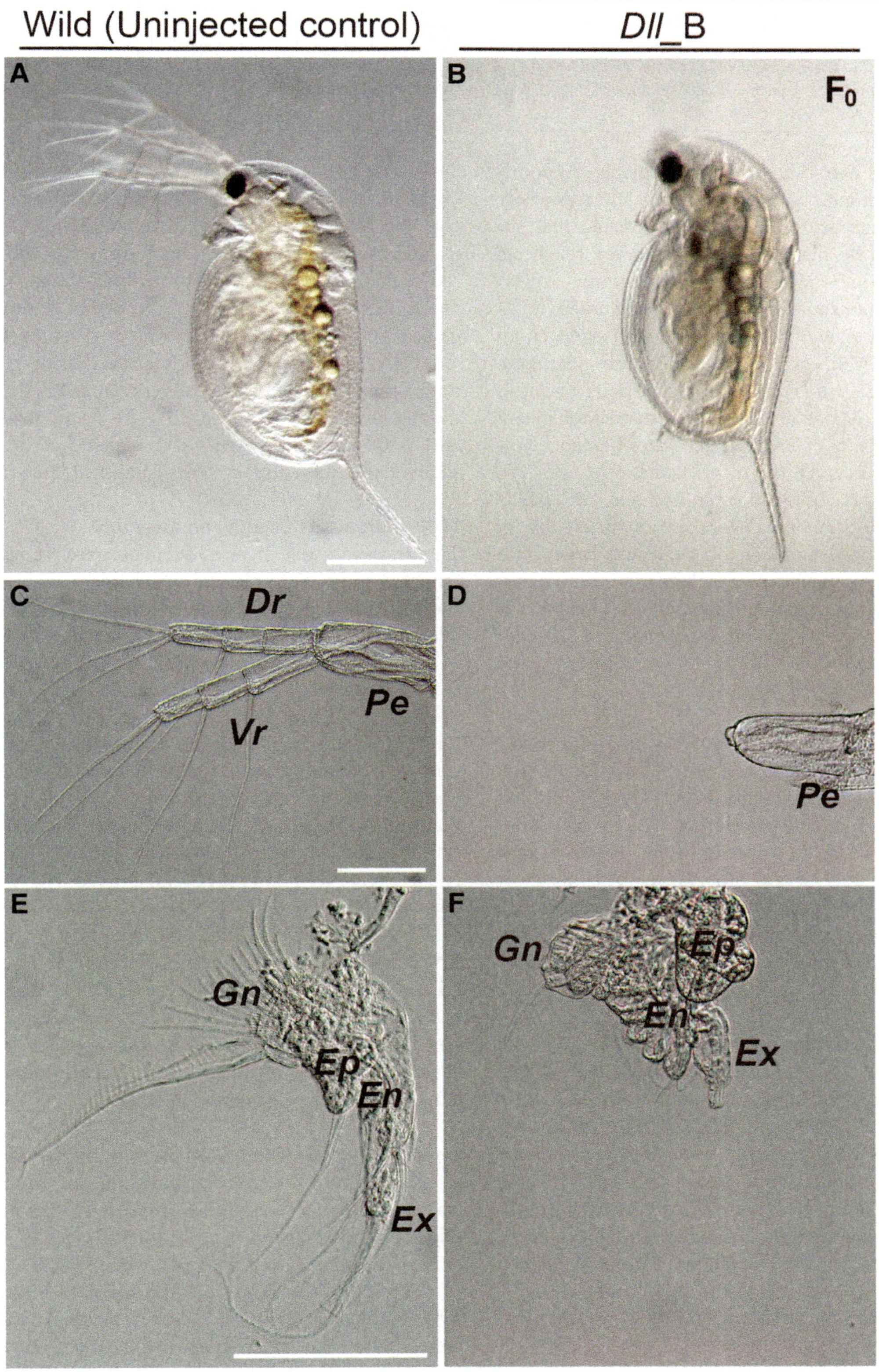

Figure 4 (See legend on next page.)

(See figure on previous page.)
Figure 4 Major *Dll*_B TALEN phenotypes. The left and right columns show representative phenotypes of uninjected controls and individuals injected with *Dll*_B mRNAs, respectively. **(A, B)** Lateral view of first instar juvenile. **(C, D)** Lateral view of the second antennae. **(E, F)** Second thoracic limb (T2). *Dr*, dorsal ramus; *Ep*, epipodite; *En*, endopodite; *Ex*, exopodite; *Gn*, gnathobase; *Pe*, peduncle; *Vr*, ventral ramus. Scale bars =200 μm in **A**, **B**; 100 μm in **C**, **D** and **E**, **F**.

animals, including *Daphnia*, which are unable or poorly suited for conducting crossing experiments. In cases where monoallelic mutants are obtained, it is possible that the mutant line could be maintained and embryos reinjected with TALEN mRNAs to induce biallelic mutations.

Our results demonstrated that TALENs worked in *D. pulex* to induce heritable mutations into the endogenous genes. Moreover, TALENs can be used to induce more complicated mutagenesis other than micro-deletions or insertions. For example, Ma et al. [22] showed that simultaneous expression of two pairs of TALENs generates heritable large chromosomal deletions. Furthermore, Bedell et al. [33] and Zu et al. [34] demonstrated that co-injection of TALEN mRNAs and single-stranded oligodeoxyribonucleotides or donor vectors induced knock-in of small and large DNA fragments, respectively, into target loci via homologous recombination. All these technique based on TALENs will enable us to accurately define gene functions in *D. pulex* in future investigations.

Conclusions

We developed a TALEN-mediated gene targeting model in *D. pulex*. Targeted mutagenesis can be attained in early embryos by injecting the right and left TALEN mRNAs into eggs within 1 hour post-ovulation. The genome editing by Platinum TALENs presented here, together with the full-genome sequence and other developed molecular tools, may significantly promote the genetic tractability of *D. pulex* as an important model for environmental, evolutionary, and developmental genomics.

Methods

Daphnia strain and culture conditions

The *D. pulex* strain [West Trenton (WTN6), collected in May 2006 by Sarah Schaak] was obtained from the Center for Genomics and Bioinformatics (Indiana University, USA). The strains were maintained in dechlorinated tap water, which was aerated and filtered through activated carbon for 2 weeks, at 18°C under artificial light conditions of 14 h light and 10 h dark to maintain reproduction. A 0.01-ml suspension of 10^9 cells/ml *Chlorella vulgaris* was added every day to each culture (20-25 individuals/L). For rearing embryos, M4 culture medium (M4) was prepared using MilliQ water [35].

Cloning of *Distal-less*

Total RNA was extracted using an ISOGEN kit (NIPPON GENE, Tokyo, Japan), and converted to cDNA using Superscript III and random primers (Life Technologies, Carlsbad, CA, USA) according to the manufacturer's protocol. The *Dll* fragment was PCR amplified from the cDNA using a set of primers designed from the *Dll* sequence retrieved from wFleaBase http://wfleabase.org/ (Additional file 2: Table S1). Subsequently, the cDNA fragments were cloned into the pGEM-T Easy vector (Promega, Madison, WI, USA) according to the manufacturer's instructions. Plasmids were sequenced using Sanger techniques that included the Big Dye terminator Ver. 3.1 (Life Technologies) on an ABI 3100 Avant or ABI 3130 Genetic Analyser DNA sequencer (Applied Biosystems Japan Ltd, Tokyo, Japan).

TALEN target site design and assembly

The first exon sequences of *Dll* gene were scanned for potential TALEN target sites, which were identified using the TALEN Targeter program at https://tale-nt.cac.cornell.edu/node/add/talen. The following parameters were used: 1) spacer length: 12-16; 2) repeat array length of 15-20; 3) G substitute: NN; 4) Filter Options: Show all TALEN pairs (include redundant TALENs); 5) Streubel et al. guidelines: On. The Platinum Gate TALEN construction system described in Sakuma et al. [14] was used to assemble two pairs of TALENs, *Dll*_ A and *Dll*_ B (Figure 1), using the Platinum Gate TALEN Kit (Addgene, cat#1000000043). Assembled RVD repeats were as follows: HD NN NI HD NN NI HD NN HD HD HD NI NI NG NG HD for the left *Dll*_A TALEN, NN NG NG NN NG NG NN HD NG NN NN NG HD NN HD NI NI for the right *Dll*_A TALEN, HD NI NG HD HD NN NN NI HD NI NN HD NI NN HD NI NN for the left *Dll*_B TALEN, and NN HD NN NN HD NN NN HD HD NN HD NG NN HD NI NN HD for the right *Dll*_B TALEN. ptCMV-153/47-VR vectors (Addgene, included in the Platinum Gate TALEN Kit) were used as destination vectors.

Single-strand annealing (SSA) assay in human cells

Construction of the reporter plasmids and the SSA assay was performed as previously reported [27]. Briefly, the oligonucleotides listed in Additional file 3: Table S2 were annealed and inserted into pGL4-SSA vector (Addgene, Plasmid 42962) to construct the reporter plasmids. The TALEN plasmids and/or reporter plasmids were transfected into HEK293T cells using Lipofectamine LTX (Life Technologies) according to the manufacturer's instruction. At 24 hours post-transfection, a luciferase assay was performed using Dual-Glo Luciferase Assay System (Promega) and intensity of luminescence was measured with TriStar LB 941 plate reader (Berthold Technologies, Bad Wildbad, Germany).

Injection of TALEN mRNAs into embryos

TALEN plasmid templates were linearized with Sma I (TaKaRa Bio, Shiga, Japan) and then transcribed *in vitro* using the mMESSAGE mMACHINE T7 Ultra kit (Ambion, Austin, TX, USA) according to the manufacturer's protocol. The resultant mRNAs were purified using a NucAway Spin Columns (Ambion) and extracted with water-saturated phenol/chloroform, and finally resuspended in RNase-free water. Equal amounts of the right and left TALEN mRNAs were mixed to a final concentration of 500 and 250 ng/μl for microinjection and stored at −80°C until use. TALEN mRNAs were injected into the embryos as described previously [6].

T7 endonuclease I (T7EI) assay

T7EI is a mismatch-resolving enzyme that can recognize heteroduplex DNAs and cleave DNA at single base pair mismatches. Thus TALEN-induced indel mutations can be detected. A pool of thirteen embryos collected at 48 h following injection with *Dll*_ A or *Dll*_B was used. Genomic DNA was isolated from TALEN-injected and uninjected embryos using DNeasy Blood and Tissue kit (Qiagen, Hilden, Germany). Additional file 2: Table S1 shows primers used to amplify the region containing the *Dll* target site from genomic DNA for T7EI assay and mutation analysis. PCR was performed using TaKaRa Ex Taq (TaKaRa Bio). Amplification conditions were: 98°C for 10 s, 56°C for 30 s, 72°C for 20 s, 45 cycles. T7 endonuclease I (New England Biolabs, Beverly, MA, USA) was added to PCR fragments and incubated at 37°C for 60 min. The samples were electrophoresed using 2% agarose gels.

Mutation analysis

To confirm the presence of TALEN-mediated mutations, genomic DNA of each individual was isolated separately from twelve F_0 embryos and nine F_2 embryos injected with *Dll* _A or *Dll* _B as described above. The target genomic region was amplified with TaKaRa Ex Taq (TaKaRa Bio). Amplification conditions were: 98°C for 1 min; 35 cycles of 98°C for 10 s, 56°C for 30 s, 72°C for 20 s; 72°C for 1 min. The resulting fragments were subcloned into the pGEM-T Easy vector (Promega) and then sequenced as described above. For observation of morphological changes, first instar animals were examined under a stereomicroscope (M165 FC; Leica Microsystems GmbH, Wetzlar, Germany), and then fixed in ethanol and dissected with a pair of needles. The specimens were mounted in glycerin, observed by Nomarski differential interference contrast microscopy (Axioplan 2; Zeiss, Oberkochen, Germany), and recorded by digital camera (DP72; Olympus). The terminology used herein for morphology follows the usage of Stachowitsch [36].

Additional files

Additional file 1: Figure S1. Phenotypes of *Dll*_B mRNAs injected juveniles. The left and right columns show representative phenotypes of uninjected controls and individuals injected with *Dll*_B mRNAs, respectively. **(A, B)** First thoracic limb (T1). The exopodite and endopodite were shortened by *Dll*_B TALEN mRNAs. **(C, D)** Third and fourth thoracic limbs (T3/4), having the same morphology. The exopodite was shrunk in *Dll*_B-injected juveniles. **(E, F)** Fifth thoracic limb (T5). The exopodite was shortened by *Dll*_B mRNAs. **(G, H)** Lateral view of the rostrum and head; an arrowhead indicates an ocellus. **(I, J)** Lateral view of abdomen; an arrow and arrowhead show an abdominal claw and abdominal seta, respectively. *Ep*, epipodite; *En*, endopodite; *Ex*, exopodite; *Fc*, filter comb. Scale bars =100 μm.

Additional file 2: Table S1. The list of primer sequences.

Additional file 3: Table S2. Oligonucleotides for construction of reporter plasmids.

Competing interests

The authors declare that they have no competing interests.

Authors' contributions

CH and YO designed the work, performed and analyzed the experiments; CH, TS, TY and TI wrote the paper. TS and TY performed TALEN construction and SSA assay. KT and SM discussed and commented on results. All authors have read and approved the final manuscript.

Acknowledgments

Daphnia pulex genomic sequence data were produced by The Center for Genomics and Bioinformatics at Indiana University and distributed via wFleaBase in collaboration with the Daphnia Genomics Consortium https://wiki.cgb.indiana.edu/display/DGC/Home. We thank Brenna Doheny for reviewing and editing the manuscript; members of the Iguchi laboratory for helpful advice and discussions; members of Yamamoto's laboratory for assistance in TALEN construction. This work was supported by a JSPS Research Fellowship for Young Scientists to CH (No.12J04065), KT (No.12J05579), JSPS KAKENHI Grant Number 26840115 to CH, a grant from Ministry of the Environment, Japan and MEXT KAKENHI Grant Numbers 24657022, 24370029 to TI. Our work benefits from, and contributes to, the Daphnia Genomics Consortium.

Author details

[1]Okazaki Institute for Integrative Bioscience, National Institute for Basic Biology, National Institutes of Natural Sciences, Higashiyama 5-1, Myodaiji, Okazaki, Aichi 444-8787, Japan. [2]Faculty of Life Science, Graduate University for Advanced Studies (SOKENDAI), 5-1 Higashiyama, Myodaiji, Okazaki, Aichi 444-8787, Japan. [3]Department of Mathematical and Life Sciences, Graduate School of Science, Hiroshima University, Higashi-Hiroshima 739-8526, Japan.

References

1. Tollrian R, Harvell CD: *The Ecology And Evolution Of Inducible Defences.* Princeton: Princeton University Press; 1999.
2. Dodson SI, Hanazato T: **Commentary on effects of anthropogenic and natural organic chemicals on development, swimming behavior, and reproduction of *Daphnia*, a key member of aquatic ecosystems.** *Environ Health Perspect* 1995, **103**:7–11.
3. Colbourne JK, Pfrender ME, Gilbert D, Thomas WK, Tucker A, Oakley TH, Tokishita S, Aerts A, Arnold GJ, Basu MK, Bauer DJ, Cáceres CE, Carmel L, Casola C, Choi J-H, Detter JC, Dong Q, Dusheyko S, Eads BD, Fröhlich T, Geiler-Samerotte KA, Gerlach D, Hatcher P, Jogdeo S, Krijgsveld J, Kriventseva EV, Kültz D, Laforsch C, Lindquist E, Lopez J, *et al*: **The ecoresponsive genome of *Daphnia pulex*.** *Science* 2011, **331**:555–561.

4. Gard AL, Lenz PH, Shaw JR, Christie AE: **Identification of putative peptide paracrines/hormones in the water flea *Daphnia pulex* (Crustacea; Branchiopoda; Cladocera) using transcriptomics and immunohistochemistry.** *Gen Comp Endocr* 2009, **160:**271–287.
5. Dircksen H, Neupert S, Predel R, Verleyen P, Huybrechts J, Strauss J, Hauser F, Stafflinger E, Schneider M, Pauwels K, Schoofs L, Grimmelikhuijzen CJP: **Genomics, transcriptomics, and peptidomics of *Daphnia pulex* neuropeptides and protein hormones.** *J Proteome Res* 2011, **10:**4478–4504.
6. Hiruta C, Toyota K, Miyakawa H, Ogino Y, Miyagawa S, Tatarazako N, Shaw JR, Iguchi T: **Development of a microinjection system for RNA interference in the water flea *Daphnia pulex*.** *BMC Biotechnol* 2013, **13:**96.
7. Sakuma T, Woltjen K: **Nuclease-mediated genome editing: At the front-line of functional genomics technology.** *Develop Growth Differ* 2014, **56:**2–13.
8. DeFrancesco L: **Move over ZFNs.** *Nat Biotechnol* 2011, **29:**681–684.
9. Gaj T, Gersbach CA, Barbas CF: **ZFN, TALEN, and CRISPR/Cas-based methods for genome engineering.** *Trends Biotechnol* 2013, **31:**397–405.
10. Boch J, Scholze H, Schornack S, Landgraf A, Hahn S, Kay S, Lahaye T, Nickstadt A, Bonas U: **Breaking the code of DNA binding specificity of TAL-Type III effectors.** *Science* 2009, **326:**1509–1512.
11. Boch J, Bonas U: ***Xanthomonas* AvrBs3 family-type III effectors: discovery and function.** *Annu Rev Phytopathol* 2010, **48:**419–436.
12. Moscou MJ, Bogdanove AJ: **A simple cipher governs DNA recognition by TAL effectors.** *Science* 2009, **326:**1501.
13. Bogdanove AJ, Voytas DF: **TAL effectors: Customizable proteins for DNA targeting.** *Science* 2011, **333:**1843–1846.
14. Sakuma T, Ochiai H, Kaneko T, Mashimo T, Tokumasu D, Sakane Y, Suzuki K, Miyamoto T, Sakamoto N, Matsuura S, Yamamoto T: **Repeating pattern of non-RVD variations in DNA-binding modules enhances TALEN activity.** *Sci Rep* 2013, **3:**3379.
15. Miller JC, Tan S, Qiao G, Barlow KA, Wang J, Xia DF, Meng X, Paschon DE, Leung E, Hinkley SJ, Dulay GP, Hua KL, Ankoudinova I, Cost GJ, Urnov FD, Zhang HS, Holmes MC, Zhang L, Gregory PD, Rebar EJ: **A TALE nuclease architecture for efficient genome editing.** *Nat Biotechnol* 2011, **29:**143–148.
16. Cermak T, Doyle EL, Christian M, Wang L, Zhang Y, Schmidt C, Baller JA, Somia NV, Bogdanove AJ, Voytas DF: **Efficient design and assembly of custom TALEN and other TAL effector-based constructs for DNA targeting.** *Nucleic Acids Res* 2011, **39:**e82.
17. Joung JK, Sander JD: **TALENs: a widely applicable technology for targeted genome editing.** *Nat Rev Mol Cell Biol* 2013, **14:**49–55.
18. Liu J, Li C, Yu Z, Huang P, Wu H, Wei C, Zhu N, Shen Y, Chen Y, Zhang B, Dend WM, Jiao R: **Efficient and specific modifications of the *Drosophila* genome by means of an easy TALEN strategy.** *J Genet Genomics* 2012, **39:**209–215.
19. Aryan A, Anderson MA, Myles KM, Adelman ZN: **TALEN-based gene disruption in the dengue vector *Aedes aegypti*.** *PLoS One* 2013, **8:**e60082.
20. Smidler AL, Terenzi O, Soichot J, Levashina EA, Marois E: **Targeted mutagenesis in the malaria mosquito using TALE nucleases.** *PLoS One* 2013, **8:**e74511.
21. Watanabe T, Ochiai H, Sakuma T, Horch HW, Hamaguchi N, Nakamura T, Bando T, Ohuchi H, Yamamoto T, Noji S, Mito T: **Non-transgenic genome modifications in a hemimetabolous insect using zinc-finger and TAL effector nucleases.** *Nat Commun* 2012, **3:**1017.
22. Ma S, Zhang S, Wang F, Liu Y, Liu Y, Xu H, Liu C, Lin Y, Zhao P, Xia Q: **Highly efficient and specific genome editing in silkworm using custom TALENs.** *PLoS One* 2012, **7:**e45035.
23. Nakanishi T, Kato Y, Matsuura T, Watanabe H: **CRISPR/Cas-mediated targeted mutagenesis in *Daphnia magna*.** *PLoS One* 2014, **5:**e98363.
24. Panganiban G, Rubenstein JLR: **Developmental functions of the *Distal-less*/Dlx homeobox genes.** *Development* 2002, **129:**4371–4386.
25. Beermann A, Jay DG, Beeman RW, Hulskamp M, Tautz D, Jurgens G: **The *Short antennae* gene of *Tribolium* is required for limb development and encodes the orthologue of the *Drosophila* Distal-less protein.** *Development* 2001, **128:**287–297.
26. Liubicich DM, Serano JM, Pavlopoulos A, Kontarakis Z, Protas ME, Kwan E, Chatterjee S, Tran KD, Averof M, Patel NH: **Knockdown of *Parhyale Ultrabithorax* recapitulates evolutionary changes in crustacean appendage morphology.** *Proc Natl Acad Sci U S A* 2009, **106:**13892–13896.
27. Sakuma T, Hosoi S, Woltjen K, Suzuki K, Kashiwagi K, Wada H, Ochiai H, Miyamoto T, Kawai N, Sasakura Y, Matsuura S, Okada Y, Kawahara A, Hayashi S, Yamamoto T: **Efficient TALEN construction and evaluation methods for human cell and animal applications.** *Genes Cells* 2013, **18:**315–326.
28. Ochiai H, Fujita K, Suzuki K, Nishikawa M, Shibata T, Sakamoto N, Yamamoto T: **Targeted mutagenesis in the sea urchin embryo using zinc-finger nucleases.** *Genes Cells* 2010, **15:**875–885.
29. Hosoi S, Sakuma T, Sakamoto N, Yamamoto T: **Targeted mutagenesis in sea urchin embryos using TALENs.** *Dev Growth Differ* 2014, **56:**92–97.
30. Kondo T, Sakuma T, Wada H, Akimoto-Kato A, Yamamoto T, Hayashi S: **TALEN-induced gene knock out in *Drosophila*.** *Dev Growth Differ* 2014, **56:**86–91.
31. Hiruta C, Nishida C, Tochinai S: **Abortive meiosis in the oogenesis of parthenogenetic *Daphnia pulex*.** *Chromosome Res* 2010, **18:**833–840.
32. Cohen SM, Jurgens G: **Proximal-distal pattern formation in *Drosophila*: cell autonomous requirement for *Distal-less* gene activity in limb development.** *EMBO J* 1989, **8:**2045–2055.
33. Bedell V, Wang Y, Campbell JM, Poshusta TL, Starker CG, Krug RG II, Tan W, Penheiter SG, Ma AC, Leung AYH, Fahrenkrug SC, Carlson DF, Voytas DF, Clark KJ, Essner JJ, Ekker SC: ***In vivo* genome editing using a high-efficiency TALEN system.** *Nature* 2012, **491:**114–118.
34. Zu Y, Tong X, Wang Z, Liu D, Pan R, Li Z, Hu Y, Luo Z, Huang P, Wu Q, Zhu Z, Zhang B, Lin S: **TALEN-mediated precise genome modification by homologous recombination in zebrafish.** *Nat Methods* 2013, **10:**329–331.
35. Elendt BP, Bias WR: **Trace nutrient deficiency in *Daphnia magna* cultured in standard medium for toxity testing: effects of the optimization of culture conditions on life history parameters of *D. magna*.** *Water Res* 1990, **24:**1157–1167.
36. Stachowitsch M: *The Invertebrates: An Illustrated Glossary.* New York: Wiley-Liss Division; 1992.

13

Targeted genome modifications in soybean with CRISPR/Cas9

Thomas B Jacobs[1,3,5*], Peter R LaFayette[2,3], Robert J Schmitz[4] and Wayne A Parrott[1,2,3]

Abstract

Background: The ability to selectively alter genomic DNA sequences *in vivo* is a powerful tool for basic and applied research. The CRISPR/Cas9 system precisely mutates DNA sequences in a number of organisms. Here, the CRISPR/Cas9 system is shown to be effective in soybean by knocking-out a green fluorescent protein (GFP) transgene and modifying nine endogenous loci.

Results: Targeted DNA mutations were detected in 95% of 88 hairy-root transgenic events analyzed. Bi-allelic mutations were detected in events transformed with eight of the nine targeting vectors. Small deletions were the most common type of mutation produced, although SNPs and short insertions were also observed. Homoeologous genes were successfully targeted singly and together, demonstrating that CRISPR/Cas9 can both selectively, and generally, target members of gene families. Somatic embryo cultures were also modified to enable the production of plants with heritable mutations, with the frequency of DNA modifications increasing with culture time. A novel cloning strategy and vector system based on In-Fusion® cloning was developed to simplify the production of CRISPR/Cas9 targeting vectors, which should be applicable for targeting any gene in any organism.

Conclusions: The CRISPR/Cas9 is a simple, efficient, and highly specific genome editing tool in soybean. Although some vectors are more efficient than others, it is possible to edit duplicated genes relatively easily. The vectors and methods developed here will be useful for the application of CRISPR/Cas9 to soybean and other plant species.

Keywords: CRISPR/Cas9, Plant transformation, Soybean, Genomic engineering, Gene targeting, Hairy roots

Background

Methods to specifically target and modify DNA sequences are indispensable for basic and applied research. Recently, the type II bacterial clustered, regularly interspaced, short palindromic repeats (CRISPR) system emerged as a simple and efficient tool to target and modify DNA sequences of interest in a variety of organisms, including; cultured human cells [1,2], zebrafish embryos [3], yeast [4], mice [5], and plants such as rice [6-9], *Arabidopsis thaliana* [10], maize [11] and liverwort [12].

There are two components to the CRISPR system: a nuclear-localized CRISPR-associated (Cas) 9 protein and a guide RNA (gRNA). Cas9 is a large protein containing two nuclease domains, and the most commonly used one is derived from *Streptococcus pyogenes*. The gRNA is a synthetic 100 nucleotide (nt) RNA molecule, of which the first approximately 20 nt are the targeting site, and the 3′ end forms a hairpin structure that interacts with the Cas9 protein [13]. Cas9 and the gRNA interact to identify DNA sequences complementary to the gRNA and generate a DNA double-strand break (DSB).

When a DNA DSB occurs in eukaryotic cells, the imprecise repair mechanism, non-homologous end joining (NHEJ), can result in the insertion and/or deletion of sequences at the breakage site, typically resulting in frameshift mutations [14]. In plants, such targeted DSBs can be used to knock-out genes [15,16], modify gene expression by disrupting promoter sequences [17], or insert transgenes at a specific location via homologous recombination [18-22].

This work characterizes and further extends the use of CRISPRs for the genetic modification of soybean genes. CRISPR vectors targeting 11 loci were introduced into soybean via *Agrobacterium rhizogenes* to generate transgenic

* Correspondence: tom.b.jacobs@gmail.com
[1]Institute for Plant Breeding, Genetics and Genomics, University of Georgia, Athens, Georgia 30602, USA
[3]Center for Applied Genetic Technologies, University of Georgia, Athens, GA 30602, USA
Full list of author information is available at the end of the article

hairy roots. Custom-amplicon sequencing of DNA from these roots show that genetic modifications were made in 95% of the tested events. Modifications were also detected in somatic embryo cultures, and these should result in soybean lines with germinal modifications. Differences between *Agrobacterium*- and particle bombardment-mediated transformation were observed and may be important considerations for transformation experiments. To facilitate CRISPR mutagenesis efforts, a series of CRISPR vectors and a novel gRNA cloning method were produced.

Results and discussion

Knock-out of a GFP transgene

The first test of the CRISPR system in soybean was with a GFP (*Green Fluorescent Protein*)-expressing soybean line, as GFP knock-outs are easily observed by a loss of fluorescence. Two GFP-targeting gRNA vectors were designed; one gRNA was designed to target the 5′ end of *GFP* (5′-target) and a second was designed to target the 3′ end (3′-target) (Figure 1A). The vectors were introduced into the GFP line via *A. rhizogenes* to produce hairy roots.

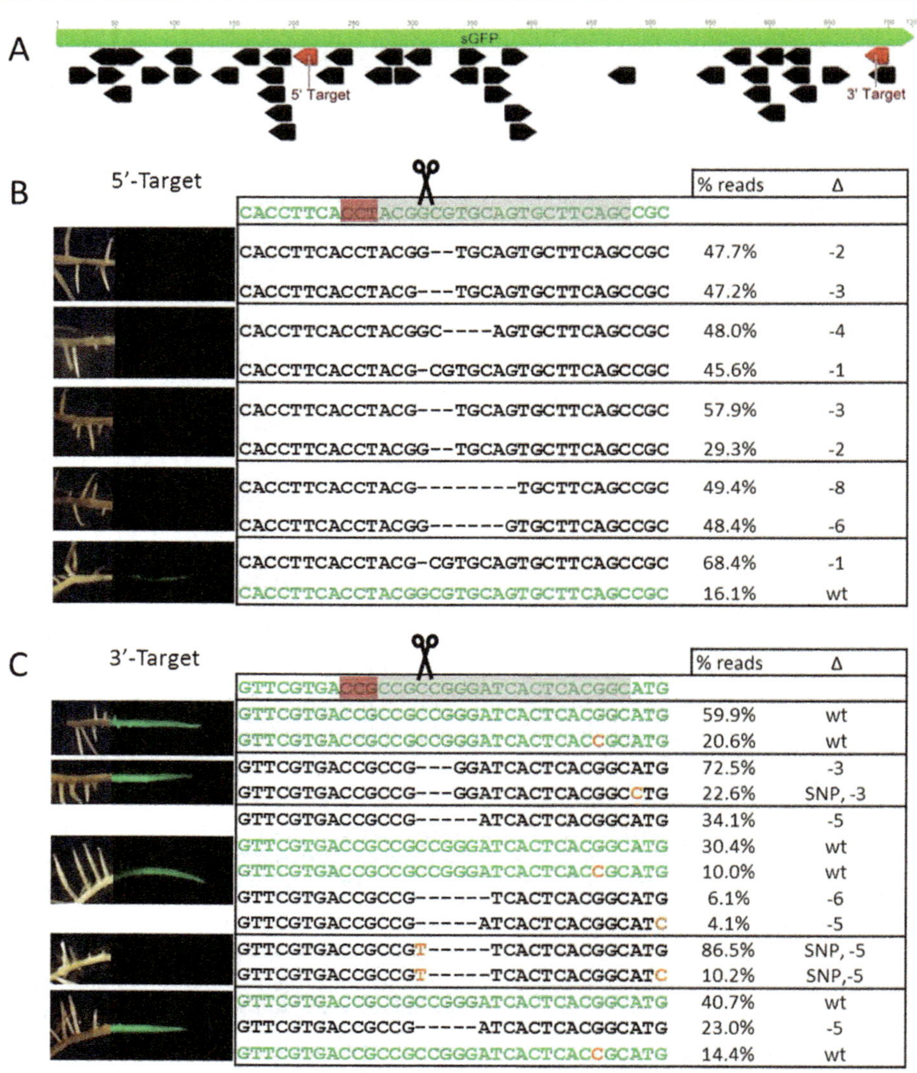

Figure 1 Cas9 targeting of a GFP gene in soybean hairy roots. (A) Schematic showing the targeted GFP sequences. The targets were designed to the negative strand of GFP. Black arrows are all possible $GN_{20}GG$ target motifs. GFP imaging and amplicon sequencing of representative **(B)** C9 + GFP 5′ target events and **(C)** C9 + GFP 3′ target events. Each panel is an independent event and blue-light images were overlaid onto white-light images of roots. The same magnification was used for all images. Wild-type sequences are in green, deletions are shown as dashes, and SNPs are shown in orange. The targeted sequences are highlighted in grey and the PAM is highlighted in red. Percentages next to sequences indicate the number of reads with sequence over the number of total reads sequenced. On average, there were 4,282 and 8,409 reads per event from the 5′-target and 3′-target events, respectively.

Fifteen out of 17 5′-target events and four of the 22 3′-target events were knock-outs as evident by a loss of fluorescence under blue-light (Additional file 1). Controls containing either Cas9 or the gRNAs alone, all fluoresced (Additional file 1). Since the GFP soybean line used is homozygous for GFP, these results show that the CRISPR system is able modify both GFP alleles, which is the only way to get loss of fluorescence.

Custom-amplicon sequencing was used to determine the genetic modifications at the *GFP* transgene. The most abundant mutations at the 5′-target were short (1-21-nt) deletions (Figure 1, Additional file 2). For event 10, a wild-type sequence was observed in 16% of the reads, which is consistent with fluorescent imaging (Figure 1 and Additional file 1). The 3′-target is less efficient; wild-type sequences were observed in seven of the events, with one event being completely unmodified (Additional file 2). Events with wild-type and modified sequences may be due to a single GFP allele being modified, or to the presence of chimeric tissues. Four of the 3′-target events contained SNPs and one event contained a T insertion, whereas the 5′-target events did not contain any SNPs or insertions. A single SNP at the 3′-target was routinely observed in the modified events and Cas9 control and may be due to errors during library preparation or sequencing.

Modifying a soybean gene

Given the successful modifications of the GFP targets, the next attempt was to modify the single-copy soybean gene, Glyma07g14530, which is a putative glucosyl-transferase. Glyma07g14530 custom amplicons from ten independent events were sequenced, and these showed a variety of mutations, including deletions, SNPs, insertions, and replacements (Additional file 2). Replacements are defined as two or more bases that were incorporated after a deletion event. Three events contained only modified sequences, six events had both wild-type and modified sequences, and one event had no modifications. These results indicate that both mono- and biallelic modifications were made and/or chimeric tissues were present.

Targeting gene pairs

Soybean is a paleopolyploid [23] and thus most genes have a homoeolog. For functional genomic studies, it would be beneficial if the CRISPR system could be used to target a homoeologous gene-pair singly and at the same time. To test this, the soybean genes Glyma01g38150 and Glyma11g07220 (orthologs of the *A. thaliana DDM1* gene) were targeted. Three gRNAs were designed; one to target Glyma01g38150 (01gDDM1), one to target Glyma11g07220 (11gDDM1), and a third to target both (01g + 11gDDM1). Both single-targeting gRNAs resulted in average indel frequencies greater than 70% (Figure 2). For 01gDDM1, eight events had indel frequencies between 87-97%. Two events only had indel frequencies of 1-2%, but these were still higher than the Cas9 control (0.14%). All but one of the 11gDDM1 events had indel frequencies greater than 95% (Figure 2). The 01gDDM1 gRNA was specific for the intended chr1 target, but the 11gDDM1 gRNA led to a small but detectable level (2-13%) of off-target modifications at the chr1 sequence (Figure 3).

Genetic modifications at both DDM1 genes were detected in events containing the 01g + 11gDDM1 gRNA, but the average indel frequency was only 21% for chr1 and 8.9% for chr11 (Figure 2). Average indel frequencies greater than 97% were observed in events targeting a different homoeologous gene pair Glyma04g36150 and Glyma06g18790 (*A. thaliana MET1* orthologs), suggesting that the lower indel frequency of the 01g + 11gDDM1 vector is due to the gRNA itself and not a result of targeting multiple genes at once.

It is noteworthy that unique insertions of the *A. rhizogenes* root-inducing (Ri) plasmid [GenBank: AJ271050] were present in two 11gDDM1 events. The Ri insertions were identified in 4.8% of the reads from event 3 and 79.2% of the reads from event 4. Both insertions are from the left-border end of the Ri plasmid, approximately 1 kb apart from each other. Cloning and sequencing of event 4 showed a 252-bp insertion from the Ri plasmid (Additional file 3). These results are particularly interesting since it should be possible to increase the chances of obtaining targeted insertions, as has been shown with other nuclease systems [24-27].

Targeting *MIR* genes

MicroRNAs (miRNAs) are small RNA molecules responsible for regulating a wide range of processes in plants [28]. MicroRNAs are encoded by *MIR* genes that are typically short (~500 bp), non-coding sequences. These features, coupled with the genetic redundancy of *MIR* families, may decrease the likelihood of isolating *MIR* mutants in mutagenesis screens [29]. Thus, the specific targeting of Cas9, and the large number of targets for any given gene, may make the Cas9 system well suited for generating *MIR* mutants. Two soybean miRNAs, miR1514 and miR1509 were targeted with Cas9. The short length of the *MIR* genes limited the number of possible Cas9 targets. Finding a *MIR1514* target near the mature miRNA was particularly difficult. Since mismatches are tolerated on the 5′ end of the gRNA [13], a C to G mismatch between the target and gRNA was made on the 5′ base (Figure 2) to get a target close to the mature miRNA. Indel frequencies greater than 95% were observed in all four miR1509-, and three out of four miR1514-targeted events. None of the short deletions (1-16 bp) were within the mature miRNA sequences, thus, none of the mutations are expected to alter the production of the miRNAs. However, these results demonstrate

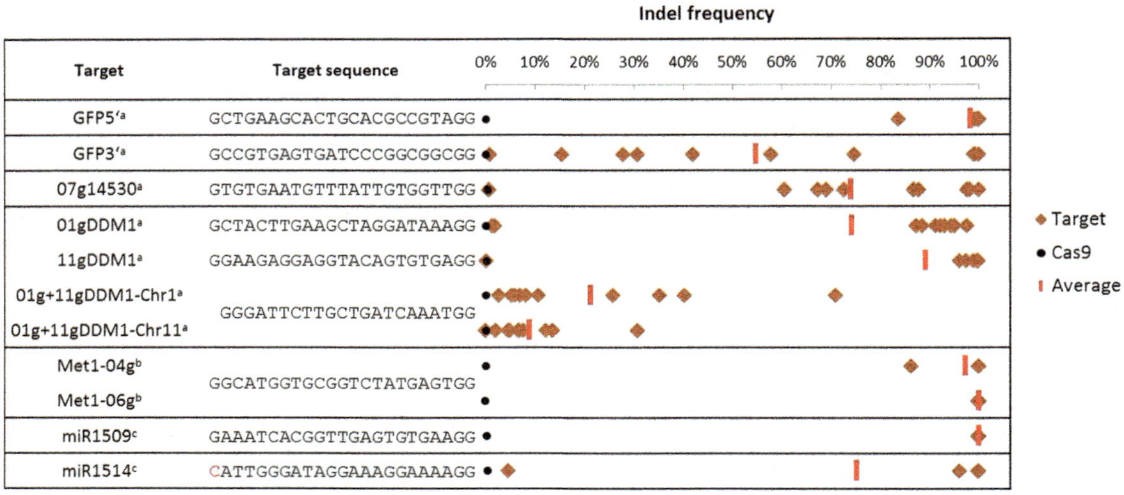

Figure 2 Modification efficiency for hairy root events. Custom-amplicon sequencing was used to measure indel frequency for each of the targeting constructs. Individual events are in orange triangles, the Cas9-tranformed control is in black circles, and average indel frequencies are vertical red bars. The miR1514 target sequence has a single mismatch to the gRNA in red. [a] n = 10, [b] n = 5, [c] n = 4.

that short, non-coding sequences, such as *MIRs*, can be readily targeted by the CRISPR/Cas system.

Genetic modification of somatic embryos

Hairy roots are an excellent transgenic model system for soybean, however, they cannot generate whole plants, and therefore heritable mutations cannot be made. To evaluate CRISPR mutagenesis in whole plants, somatic embryo cultures of soybean were biolistically transformed with Cas9 constructs. Eight Glyma07g14530 and 24 01g + 11gDDM1 hygromycin-resistant events were recovered. Although each event contained portions of the gRNA and Cas9 genes as determined by PCR (data not shown), only two Glyma07g14530 and three 01g + 11gDDM1 events contained a complete Cas9 gene as determined by long-distance PCR (Figure 4A). When hairy-root events (*Agrobacterium* transformation) were screened, a full Cas9 product was observed in all ten events (Additional file

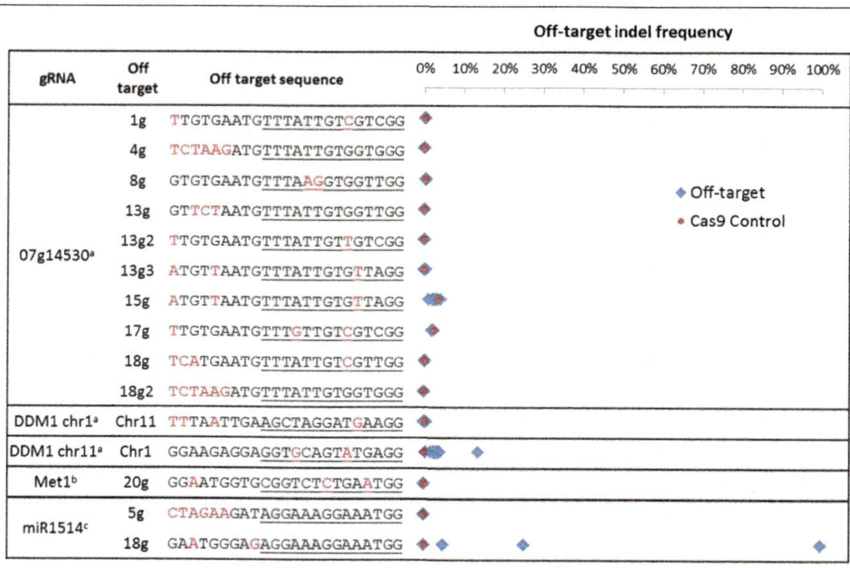

Figure 3 Off-target indel frequency for hairy-root events. The measured indel frequency is represented by a blue diamond for each event and a red dot for the Cas9 control. Mismatches between the gRNA and the off-target sequence are in red. The critical 'seed' region is underlined. [a] n = 10, [b] n = 5, [c] n = 4.

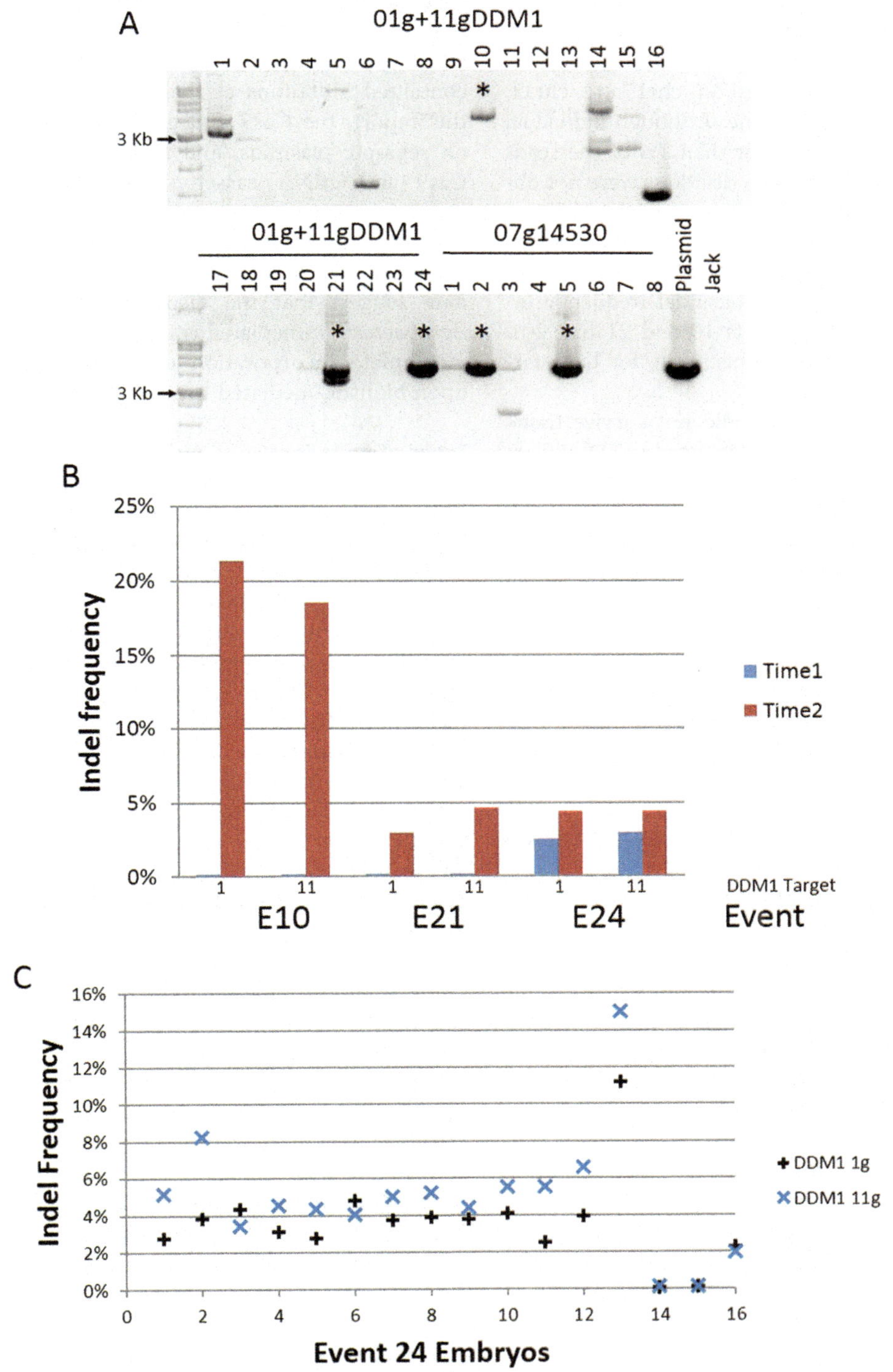

Figure 4 DNA modifications in somatic embryos. (A) Long-distance PCR for the Cas9 gene in recovered events with 01g + 11gDDM1 and 07g14530. Marker is a 1 Kb DNA ladder. Asterisks (*) indicate events with an intact Cas9. **(B)** Modifications were detected in three events transformed with the 01g + 11gDDM1 vector. At the initial time-point, modifications were only detected in event 24. When samples were taken approximately 2 weeks later, modifications were detected in all three events. **(C)** Modifications were detected in 14 out of 16 individual regenerating embryos from event 24.

4A). These results suggest that the Cas9 gene fragmented during biolistic-mediated transformation, but not upon *Agrobacterium*-mediated transformation.

As with other Cas9 systems [10], the continued activity of Cas9 in the somatic embryos resulted in additional genetic modifications. DNA samples were taken from all events once there was enough tissue, approximately 2-4 weeks after selection, and used for amplicon sequencing. At this first sequencing time-point, event 24 had approximately 2.5 % modified sequences on chr1 and chr11, whereas events 10 and 21 had none. Although individual modified sequences made up fewer than 1% of the reads in event 24 (Additional file 2), such deletions were not observed in any of the other 23 events sequenced, indicating that these deletions were not due to sequencing errors. When DNA was collected approximately two weeks after the first sequencing experiment, the indel frequency increased to 4.3% in event 24. Events 10 and 21 had 20% and 4-5% modified sequences, respectively, for both targets (Figure 4B).

The two Glyma07g14530 events did not survive tissue culture and no modifications were detected in DNA from somatic embryos (data not shown). Individual embryos from event 24 range in indel frequency from 0-14%, with most of the events at 4% (Figure 4C). Therefore continued expression of Cas9 leads to additional mutations during the development of these embryos.

Mutation efficiency

Of the nine targeting vectors used in this study, seven resulted in average indel frequencies greater than 70% (GFP 5′, 01gDDM1, 11gDDM1, Glyma04g36150, Glyma06g18790, miR1509, and miR1514). This mutation efficiency is ten-fold higher than the 3-7 % obtained with transcription-activator like effector nucleases (TALENs) in soybean hairy roots [30].

In hairy roots, the 01g + 11gDDM1 vector had the lowest average, with 21% and 8.9% for the chr1 and chr11 targets, respectively. A similar frequency was observed in the somatic embryos (Figure 4B, C). It should be noted that the 01g + 11gDDM1 gRNA is one base shorter than the rest of the gRNAs in this study ($GN_{19}GG$). However, this target length has been used in plants [31], and shorter gRNAs ($GN_{18}GG$) have been shown to be as effective as the commonly used gRNA ($GN_{20}GG$) in cultured human cells [32]. It seems unlikely that a shorter gRNA led to a decrease it indel frequency, but a thorough testing of gRNA lengths in plants has not been reported. Although each of the vectors had a range of indel frequencies, only four out of 88 (5%) hairy-roots were unmodified, demonstrating that CRISPR mutagenesis in soybean is a robust system.

The three 01g + 11gDDM1 somatic-embryo events with the complete Cas9 gene contained targeted genetic modifications. These were three out of 24 hygromycin-resistant lines. These data demonstrate that when the complete Cas9 is incorporated, genetic modifications are made, although the complete Cas9 gene is only incorporated in 12.5% biolistically-transformed events. Of the recent reports of CRISPRs being used in plants, several have shown the recovery of whole-plants. One publication reported the biolistic transformation of rice, in which 9.4% and 7.1% of the T0 rice plants recovered contained mutations at their respective targets [31]. In this report, the Cas9 and gRNA cassettes were located on separate plasmids, and it is unclear if the complete Cas9 and gRNA cassettes were incorporated in all events. In contrast, transgenic *A. thaliana* and rice plants transformed with *Agrobacterium tumefaciens* had efficiencies of 20-90% for several targets [6,7,9,33]. Our data suggest that the disparity between biolistic and *Agrobacterium*-mediated transformation could be due to incomplete incorporation of the complete Cas9 gene upon biolistic-mediated transformation.

Types of mutations

The types of mutations obtained here are similar to those observed in soybean and other plants obtained with ZFNs [15,21], TALENs [17,18,30] and CRISPRs [31,33-35]; small deletions were the most frequent mutations; SNPs were less common (Additional file 2).

The different targeting sequences tested led to a distinctive gamut of mutations. The seven most effective vectors almost exclusively generated short deletions, whereas the lower efficiency vectors contained more insertions/SNPs (Additional file 2). Of the ten 07g14530 events, seven had insertions of one or more bases. These results suggest that the differences were determined by either the target sequence or the gRNA. Therefore, multiple targeting vectors may be needed for any potential target sequence, depending on the frequencies/types of mutations desired. Obtaining a greater variety of mutations may be desirable when the intent is to produce an allelic series.

The types of mutations between the hairy-root events and somatic embryos are consistent between chromosomal targets and between transformation methods. Within the ten 01g + 11gDDM1 hairy-root events, six contained an A insertion on chr1 at the same position. From those same ten events, five contained an A insertion on the homoeologous target on chr11 (Additional file 2). Each of the somatic-embryo events has the same A insertion for both chr1 and chr11, and in many cases, it is the most abundant read (Additional file 2). Given the consistent insertion pattern, it is tempting to speculate that there may be rules governing the types of mutations that are possible for a given target.

Evaluation of off-target modifications

One limitation of the CRISPR system is the potential for off-target modifications, i.e., the modification of sequences similar to the intended target sequence [13,36]. To determine the extent to which there may be off-target modifications, putative off-target sites were identified for the Glyma07g14530, DDM1, MET1, and miR1514 vectors.

Each putative off-target site has two to six mismatches relative to the gRNA (Figure 3).

Two gRNAs created off-target mutations. The 11gDDM1 chr1 off target was modified in 2-13% of the sequenced reads, which is considerably lower than the indel frequency at the intended chr11 target (95-100%). When off-targeting occurred at miR1514 18g, there was a range of frequencies; 100%, 25%, and 5%. The 07g14530-15g and -17g off-target loci had indel frequencies of 2.8% and 2.2%, respectively. However, the increased indel frequencies were also observed in the Cas9 control, showing that they were due to sequencing errors caused by long stretches of T's in the amplicons. These results indicate that while off-targeting does occur, at least for the tested gRNAs, it is not common, and was generally at a much lower frequency than at the intended target.

gRNA vector construction

In this work, a rapid cloning method (Additional file 5) was developed to create new gRNAs. It consists of a single PCR reaction with two 41-bp primers and an In-Fusion® reaction and can be used to clone any gRNA target sequence. The pUC gRNA shuttle vector makes the construction of gRNAs simple and inexpensive. The use of the In-Fusion® cloning system has the benefit of reducing handling steps, to the point where it should be simple to automate the entire cloning process. Binary Cas9 vectors with four different selectable makers (*nptII*, GFP, hygromycin, *bar*) were also created to facilitate plant transformation experiments.

Conclusions

This work shows that the Cas9 system is functional in two stably transformed plant systems, hairy roots and somatic embryos. It was possible to efficiently mutate all 11 loci chosen for testing; only two of the targeting vectors resulted in detectable off-target mutations at predicted off-target loci. The different gRNA targets produced different types of mutations. Combined with a vector system developed to efficiently assemble the necessary gRNAs, these results confirm that the CRISPR system will be a simple and inexpensive method for genome editing in soybean, thus facilitating the use of genome editing to confirm candidate genes, develop novel alleles/phenotypes, and engineer plants with important agronomic or quality traits.

Methods

Vector construction

The human codon-optimized Cas9 gene [2] was obtained from Addgene (plasmid 41815). Two flanking primers with added *Nhe*I and *Sac*II sites were used to amplify the coding sequence, including the SV40 nuclear localization signal, with the KAPA HiFi polymerase (KAPA BioSystems). The amplicon was digested with the two restriction enzymes and ligated to the vector, pM35S, between the double-enhancer 35S promoter and nopaline synthase (nos) terminator (Additional file 6). The entire cassette is flanked with I-*Sce*I restriction sites, which were used to move the Cas9 cassette into p201N to create p201N:Cas9 (Addgene plasmid 59175). The p201N vector is a p201BK [37] vector modified to include an *nptII* selectable marker cassette and I-*Sce*I and I-*Ppo*I restriction sites (Additional file 6).

For biolistic transformation of soybean, a pSMART HC Kan (Lucigen Corporation, [GenBank: AF532107]) cloning vector was modified to contain a *hygromycin phosphotransferase* (*hph*) gene under the control of the *Solanum tuberosum* Ubi3 promoter and terminator [38] and the meganuclease I-*Ppo*I site, and is referred to as pSPH2. The vector pSPH2 was digested with I-*Ppo*I and DNA overhangs were removed with T4 DNA polymerase. To prepare the Cas9 insert, p201N:Cas9:gRNA-Glyma07g14530 was digested with *Spe*I and *Pme*I and DNA overhangs were removed with T4 DNA polymerase. The vector and insert were ligated to create the plasmid pSPH2:Cas9:gRNA-Glyma07g14530. The Glyma07g14530 gRNA was then replaced with the 01g + 11gDDM1 (Glyma01g38150 and Glyma11g07220) gRNA via I-*Ppo*I to produce pSPH2:Cas9:gRNA-01g + 11gDDM1.

Additional binary Cas9 vectors were produced by replacing *npt*II from p201NCas9, with *hph*, *bar* (phosphinothricin resistance), or GFP. The *hph* cassette was moved from pSPH2 into the p201N Cas9 vector with the *Pac*I and *Spe*I restriction sites to produce p201H:Cas9 (Addgene plasmid 59176). The bar and GFP cassettes (double-enhancer 35S promoter, nos terminator) were amplified with the *Spe*I 35SF and *Pac*I nosR primers (Additional file 7), and moved into the p201N Cas9 vector with the *Pac*I and *Spe*I restriction sites to produce p201B:Cas9 (Addgene plasmid 59177) and p201G:Cas9 (Addgene plasmid 59178).

The gRNA targets were designed as previously described [2], with the exception of the U6 promoter, which was replaced with the *Medicago truncatula* U6.6 polymerase III promoter [39] for efficient transcription in soybean. For the gRNA targets, 22-23-bp targets were chosen that had the $GN_{19-20}GG$ motif as previously described [2]. The GFP 5′- and 3′-targets were chosen because they contain restriction sites that can be used for downstream analysis; however, given the high DNA-modification frequencies, such analyses were not performed. The GN_{18-19} portion of the genomic target motif was incorporated into the gRNA target molecule. The GFP, Glyma07g14530, and DDM1 gRNA target sequences were synthesized by IDT using gBlocks. The gBlocks were amplified by PCR with flanking primers containing I-*Ppo*I restriction sites. All primer sequences can be found in Additional file 7. The products were then digested with I-*Ppo*I and inserted into the p201N vector. The MET1 (Glyma04g36150 and Glyma06g18790), miR1514, and miR1509 gRNA target sequences were produced with the pUC gRNA shuttle vector system described below. Plasmids were electroporated into *Agrobacterium rhizogenes* strain K599 and used for hairy-root transformation. Vectors containing both the Cas9 and gRNA target cassettes were combined by inserting the gRNA target cassette into the p201N Cas9 I-*Ppo*I site.

Hairy-root transformation of soybean

Soybean 'Jack-GFP [40]' and 'Jack' germinating seeds were used for transformation with slight modifications from the protocol previously described [41]. Briefly, soybean seeds were germinated for approximately one week under sterile conditions on a filter paper wetted with a ½ MSO liquid germination medium [42] supplemented with B5 vitamins [43]. *A. rhizogenes* (strain K599) containing the vectors-of-interest were streaked from glycerol stocks onto YM medium [44] supplemented with 50 mg L^{-1} kanamycin. Soybean cotyledons were prepared in a manner similar to that described for cotyledonary node transformation [45]; the root and lower hypocotyl were removed from the cotyledons, leaving approximately 5 mm of hypocotyl. The apical shoot and hypocotyl were cut longitudinally to produce two symmetrical cotyledons with a short hypocotyl piece. The apical meristem was removed and 1-mm-deep cuts were made in the cotyledons on the adaxial surface with a scalpel dipped in a solution of *A. rhizogenes* (PB Buffer (0.01 M Na2HPO4, 0.15M NaCl, pH 7.5) + 100 μM acetosyringone). Cotyledons were co-cultivated with *A. rhizogenes* for 3 days on filter paper wetted with 2 mL of liquid germination medium + 100 μM acetosyringone. Cotyledons were transferred to a hairy-root growth (HRG) medium according to Cho et al. [41] with the following modifications: ½ MS salts, 2 g L^{-1} Phytagel, and 500 mg L^{-1} timentin to inhibit *A. rhizogenes*. Each root was treated as an individual event and transferred to HRG medium with 10 mg L^{-1} of Geneticin (G418). Those roots that grew on HRG + G418 were considered events, and a 2-cm portion of a root tip was collected for CTAB DNA extraction [46]. PCR was performed to confirm the presence of the Cas9 and gRNA genes with the primers listed in Additional file 2. Long-distance PCR was performed with a Promega long-distance PCR master mix according to manufacturer's instructions.

GFP imaging

After selection on HRG + G418, root tips were imaged with an Olympus MVX10 microscope with a GFP filter cube and the imaging software DP controller version 2.2.1.227 (Olympus America Inc.). Blue-light images were taken with a 5 ms exposure.

Custom-amplicon sequencing and analysis

Genomic DNA was amplified with the KAPA HiFi polymerase (KAPA Biosystems) with tailed primers under the conditions 95°C for 3 min; 30 cycles (98°C for 15 sec, 60°C for 15 sec, 72°C for 30-45 sec); and 72°C for 5 min. PCR products were run on a 1% agarose, 1X TBE gel and visualized on a UV transilluminator to verify amplification. PCR products were pooled across amplicons, diluted 1:100, and used as a template for a second PCR with the conditions 95°C for 3 min; 10 cycles (98°C for 15 sec, 60°C for 15 sec, 72°C for 30-45 sec); and 72°C for 10 min. The second PCR was used to add the final Illumina adapters and indexes. PCR products were again visualized to ensure amplification. All products were pooled and concentrated with DNA clean and concentrator columns (Zymo Research). The pooled samples were run on a 1.5% agarose, 1X TAE + cytidine gel and the proper fragments were gel extracted with the Zymoclean Gel DNA Recovery Kit. Purified libraries were quantified with the KAPA Library Quantification Kit (KAPA Biosystems) and run on an Illumina MiSeq (Illumina Inc.). Reads were de-multiplexed with the MiSeq reporter software version 2.3.32.

Reads were imported into the software Geneious (Biomatters Ltd.) version 7. Reads were trimmed for quality and separated by amplicon using the separate-reads-by-barcode function using the forward sequencing primer + five bases downstream as the barcode. The five downstream bases were essential to remove primer-dimers from the analysis. After quality and barcode trimming, only reads within five bases of the expected length were extracted for analysis. Reads were trimmed to regions approximately 20-bp upstream and downstream of the gRNA target site (Additional file 7). Sequences that were the length of wild-type sequences were extracted. Indel frequency was then calculated by subtracting the number of wild-type sequences from the total number of extracted reads.

For each of the targeted loci, unique sequences were extracted from the trimmed total extracted reads using the find-duplicates function. The most abundant, unique reads are reported in Additional file 2.

Off-target sequence identification

Potential off-target sites were identified by comparing the 23-bp gRNA target sequences using BLAST to the soybean reference genome (Glyma v1.1), on Phytozome, setting the e-value threshold to 5 since the query sequence is only 23 nt. Only loci that had the required protospacer-adjacent motif (PAM) NGG motif at the 3′ end of the sequence were considered for analysis. Primers used for amplifying the off-target loci are in Additional file 7.

Biolistic transformation of somatic embryos

Biolistic transformation of soybean was performed as previously described [47]. DNA was isolated from somatic embryo cultures for PCR and custom-amplicon sequencing.

gRNA shuttle plasmid

To facilitate the construction of gRNA targets, a shuttle plasmid was created that makes construction quick and inexpensive. The *Medicago truncatula* U6.6 promoter was fused to the gRNA scaffold [2], and the entire gRNA is flanked by I-*Ppo*I restriction sites. To produce a novel gRNA target, forward and reverse primers were designed with tails that encode the new target sequence (Additional file 5). Fifteen bp of homology on the primer tails allowed for In-Fusion® cloning (Clonetech Laboratories Inc., Mountain View, CA). After transformation, the new gRNA target molecule was inserted between the promoter and gRNA scaffold. Sanger sequencing was performed with the commonly used M13-reverse primer to confirm the sequence of the gRNA. I-*Ppo*I was then used to move the functional gRNA target cassette into a vector of choice. The pUC gRNA Shuttle plasmid can be obtained from Addgene (plasmid 47024).

Additional files

Additional file 1: GFP imaging of modified GFP events and controls. Each panel is an independent event and blue-light images were overlaid onto white-light images of roots. Scale bar is shown as 5mm and all images are taken with the same magnification.

Additional file 2: Unique sequences from all events in this study. The most abundant reads for each event are reported. The number of reads, the respective percentages, and the type of modification (Δ) is listed for each event. Wild-type sequences are in green, dashes are deletions, SNPs are orange, insertions are red, replacements are orange, an insertion of Ri plasmid is pink, and an inversion is in purple.

Additional file 3: Cloned sequence from modified 11gDDM1 event containing a 252-bp insertion of the Ri plasmid. Red is gRNA target, underline is insertion.

Additional file 4: Long-distance PCR for the Cas9 gene in somatic embryos and hairy-root events. (A) All hairy-root events are positive for Cas9. (B) Events positive for Cas9 from Figure 4A were re-run together to get appropriate sizing. Three 01g + 011gDDM1 and two 07g14530 biolistic-events have the correct 4.3kb band.

Additional file 5: Cas9/gRNA targeting and cloning scheme to produce gRNAs. $GN_{20}GG$ motifs are identified in a genomic region of interest. Tailed forward and reverse primers are designed to amplify the entire 3 kb gRNA Shuttle Plasmid. The primer tails contain sequences for the target (blue) and share 15 bp of homology (X's) for the In-Fusion® protocol. PCR products can then undergo In-Fusion® cloning, resulting in the creation of the gRNA Target Plasmid. The gRNA cassette is in the middle of a multiple-cloning site for easy transfer to a final vector. This pUC gRNA Shuttle plasmid can be used for plant modifications, but the cloning scheme will work for any gRNA target.

Additional file 6: Vectors used in this study. The plasmid p201N Cas9 gRNA is the binary vector used for hairy-root transformations. The pSPH2 Cas9 gRNA vector was used for biolistic transformation. The pUC Shuttle vector was used to create additional gRNA targets. The targets were moved into the binary or biolistic vectors via the I-*Ppo*I restriction sites.

Additional file 7: Primers used in this study. Underline denotes restriction site. X's are 15 bp of homology between primers required for In-Fusion® Cloning. Blue nucleotides are part of the gRNA target.

Abbreviations

GFP: Green-fluorescent protein; CRISPR: Clustered, regularly interspaced, short palindromic repeats; Cas: CRISPR associated; gRNA: Guide RNA; nt: nucleotide; DSB: Double-strand break; NHEJ: Non-homologous end joining; SNP: Single nucleotide polymorphism; miRNAs: microRNAs; bp: Base pair; ZFNs: Zinc-finger nuclease; TALENs: Transcription activator-like effector nucleases; nos: nopaline synthase; hph: hygromycin phosphotransferase; PAM: Protospacer-adjacent motif.

Competing interests

The authors declare that they have no competing interests.

Authors' contributions

TJ performed experiments, analyzed data, and drafted the manuscript. PL designed and created some of the vectors used. RS designed some of the gRNAs. WP helped conceive the study and draft the manuscript. All authors read and approved the final manuscript.

Acknowledgements

This work was supported by the United Soybean Board, monies allocated to the Georgia Agricultural Experiment Stations and a research grant to RS from the National Science Foundation (IOS-1339194). We would like to thank John Finer from The Ohio State University for the GFP soybean line. Laboratory members M. Ortega, A. Umansky, N. Lawler, A. Disharoon, B. Spaid, D. Mehalakes, C. Niederhuth, and N. Rohr were instrumental for the collection and processing of materials. We would also like to thank L. Kanizay for a critical review of the manuscript.

Author details

[1]Institute for Plant Breeding, Genetics and Genomics, University of Georgia, Athens, Georgia 30602, USA. [2]Department of Crop and Soil Sciences, University of Georgia, Athens, GA 30602, USA. [3]Center for Applied Genetic Technologies, University of Georgia, Athens, GA 30602, USA. [4]Department of Genetics, University of Georgia, Athens, GA 30602, USA. [5]Present address: Boyce Thompson Institute for Plant Research, Ithaca, NY 14853, USA.

References

1. Cong L, Ran FA, Cox D, Lin SL, Barretto R, Habib N, et al. Multiplex genome engineering using CRISPR/Cas systems. Science. 2013;339(6121):819–23.
2. Mali P, Yang LH, Esvelt KM, Aach J, Guell M, DiCarlo JE, et al. RNA-Guided human genome engineering via Cas9. Science. 2013;339(6121):823–6.
3. Hwang WY, Fu YF, Reyon D, Maeder ML, Tsai SQ, Sander JD, et al. Efficient genome editing in zebrafish using a CRISPR-Cas system. Nat Biotech. 2013;31(3):227–9.
4. DiCarlo JE, Norville JE, Mali P, Rios X, Aach J, Church GM. Genome engineering in *Saccharomyces cerevisiae* using CRISPR-Cas systems. Nucleic Acids Res. 2013;41(7):4336–43.
5. Wang HY, Yang H, Shivalila CS, Dawlaty MM, Cheng AW, Zhang F, et al. One-step generation of mice carrying mutations in multiple genes by CRISPR/Cas-mediated genome engineering. Cell. 2013;153(4):910–8.
6. Xu R, Li H, Qin R, Wang L, Li L, Wei P, et al. Gene targeting using the *Agrobacterium tumefaciens*-mediated CRISPR-Cas system in rice. Rice. 2014;7(1):5.
7. Zhang H, Zhang J, Wei P, Zhang B, Gou F, Feng Z, et al. The CRISPR/Cas9 system produces specific and homozygous targeted gene editing in rice in one generation. Plant Biotechnol J. 2014;12(6):797–807.
8. Jiang WZ, Zhou HB, Bi HH, Fromm M, Yang B, Weeks DP. Demonstration of CRISPR/Cas9/sgRNA-mediated targeted gene modification in Arabidopsis, tobacco, sorghum and rice. Nucleic Acids Res. 2013;41(20):12.
9. Miao J, Guo DS, Zhang JZ, Huang QP, Qin GJ, Zhang X, et al. Targeted mutagenesis in rice using CRISPR-Cas system. Cell Res. 2013;23(10):1233–6.
10. Feng ZY, Mao YF, Xu NF, Zhang BT, Wei PL, Yang DL, et al. Multigeneration analysis reveals the inheritance, specificity, and patterns of CRISPR/Cas-

induced gene modifications in Arabidopsis. Proc Natl Acad Sci U S A. 2014;111(12):4632–7.

11. Liang Z, Zhang K, Chen KL, Gao CX. Targeted mutagenesis in *Zea mays* using TALENs and the CRISPR/Cas System. J Genetics Genomics. 2014;41(2):63–8.
12. Sugano SS, Shirakawa M, Takagi J, Matsuda Y, Shimada T, Hara-Nishimura I, et al. CRISPR/Cas9-mediated targeted mutagenesis in the liverwort *Marchantia polymorpha* L. Plant Cell Physiol. 2014;55(3):475–81.
13. Jinek M, Chylinski K, Fonfara I, Hauer M, Doudna JA, Charpentier E. A programmable dual-RNA-guided DNA endonuclease in adaptive bacterial immunity. Science. 2012;337(6096):816–21.
14. Mladenov E, Iliakis G. Induction and repair of DNA double strand breaks: The increasing spectrum of non-homologous end joining pathways. Mutat Res Fundam Mol Mech Mutagen. 2011;711(1–2):61–72.
15. Curtin SJ, Zhang F, Sander JD, Haun WJ, Starker C, Baltes NJ, et al. Targeted mutagenesis of duplicated genes in soybean with zinc-finger nucleases. Plant Physiol. 2011;156(2):466–73.
16. Osakabe K, Osakabe Y, Toki S. Site-directed mutagenesis in Arabidopsis using custom-designed zinc finger nucleases. Proc Natl Acad Sci U S A. 2010;107(26):12034–9.
17. Li T, Liu B, Spalding MH, Weeks DP, Yang B. High-efficiency TALEN-based gene editing produces disease-resistant rice. Nat Biotech. 2012;30(5):390–2.
18. Zhang Y, Zhang F, Li XH, Baller JA, Qi YP, Starker CG, et al. Transcription activator-like effector nucleases enable efficient plant genome engineering. Plant Physiol. 2013;161(1):20–7.
19. Fauser F, Roth N, Pacher M, Ilg G, Sanchez-Fernandez R, Biesgen C, et al. *In planta* gene targeting. Proc Natl Acad Sci U S A. 2012;109(19):7535–40.
20. Shukla VK, Doyon Y, Miller JC, DeKelver RC, Moehle EA, Worden SE, et al. Precise genome modification in the crop species *Zea mays* using zinc-finger nucleases. Nature. 2009;459(7245):437–U156.
21. Townsend JA, Wright DA, Winfrey RJ, Fu FL, Maeder ML, Joung JK, et al. High-frequency modification of plant genes using engineered zinc-finger nucleases. Nature. 2009;459(7245):442–U161.
22. D'Halluin K, Chantal V, Jolien VH, Joanna R, Ilse VDB, Anouk P, et al. Targeted molecular trait stacking in cotton through targeted double-strand break induction. Plant Biotechnol J. 2013;11(8):933-41.
23. Schlueter JA, Lin JY, Schlueter SD, Vasylenko-Sanders IF, Deshpande S, Yi J, et al. Gene duplication and paleopolyploidy in soybean and the implications for whole genome sequencing. BMC Genomics. 2007;8:16.
24. Qi Y, Zhang Y, Zhang F, Baller JA, Cleland SC, Ryu Y, et al. Increasing frequencies of site-specific mutagenesis and gene targeting in Arabidopsis by manipulating DNA repair pathways. Genome Res. 2013;23(3):547–54.
25. Salomon S, Puchta H. Capture of genomic and T-DNA sequences during double-strand break repair in somatic plant cells. Embo J. 1998;17(20):6086–95.
26. Tzfira T, Frankman LR, Vaidya M, Citovsky V. Site-specific integration of *Agrobacterium tumefaciens* T-DNA via double-stranded intermediates. Plant Physiol. 2003;133(3):1011–23.
27. Chilton MDM, Que QD. Targeted integration of T-DNA into the tobacco genome at double-stranded breaks: New insights on the mechanism of T-DNA integration. Plant Physiol. 2003;133(3):956–65.
28. Bartel DP. MicroRNAs: Target recognition and regulatory functions. Cell. 2009;136(2):215–33.
29. Jones-Rhoades MW, Bartel DP, Bartel B. MicroRNAs and their regulatory roles in plants. In: Annual Review of Plant Biology, vol. 57. Palo Alto: Annual Reviews; 2006. p. 19–53.
30. Haun W, Coffman A, Clasen BM, Demorest ZL, Lowy A, Ray E, et al. Improved soybean oil quality by targeted mutagenesis of the *fatty acid desaturase 2* gene family. Plant Biotechnol J. 2014;12(7):934–40.
31. Shan Q, Wang Y, Li J, Zhang Y, Chen K, Liang Z, et al. Targeted genome modification of crop plants using a CRISPR-Cas system. Nat Biotech. 2013;31(8):686–8.
32. Fu YF, Sander JD, Reyon D, Cascio VM, Joung JK. Improving CRISPR-Cas nuclease specificity using truncated guide RNAs. Nat Biotech. 2014;32(3):279–84.
33. Feng Z, Zhang B, Ding W, Liu X, Yang D-L, Wei P, et al. Efficient genome editing in plants using a CRISPR/Cas system. Cell Res. 2013;23(10):1229-32.
34. Li J-F, Norville JE, Aach J, McCormack M, Zhang D, Bush J, et al. Multiplex and homologous recombination-mediated genome editing in Arabidopsis and *Nicotiana benthamiana* using guide RNA and Cas9. Nat Biotech. 2013;31(8):688–91.
35. Nekrasov V, Staskawicz B, Weigel D, Jones JDG, Kamoun S. Targeted mutagenesis in the model plant *Nicotiana benthamiana* using Cas9 RNA-guided endonuclease. Nat Biotech. 2013;31(8):691–3.
36. Fu Y, Foden JA, Khayter C, Maeder ML, Reyon D, Joung JK, et al. High-frequency off-target mutagenesis induced by CRISPR-Cas nucleases in human cells. Nat Biotech. 2013;31(9):822–6.
37. Covert SF, Kapoor P, Lee MH, Briley A, Nairn CJ. *Agrobacterium tumefaciens*-mediated transformation of *Fusarium circinatum*. Mycol Res. 2001;105:259–64.
38. Garbarino JE, Belknap WR. Isolation of a ubiquitin-ribosomal protein gene (Ubi3) from potato and expression of its promoter in transgenic plants. Plant Mol Biol. 1994;24(1):119–27.
39. Kim GB, Nam YW. Isolation and characterization of *Medicago truncatula* U6 promoters for the construction of small hairpin RNA-mediated gene silencing vectors. Plant Mol Biol Report. 2013;31(3):581–93.
40. Hernandez-Garcia CM, Martinelli AP, Bouchard RA, Finer JJ. A soybean (*Glycine max*) polyubiquitin promoter gives strong constitutive expression in transgenic soybean. Plant Cell Rep. 2009;28(5):837–49.
41. Cho HJ, Farrand SK, Noel GR, Widholm JM. High-efficiency induction of soybean hairy roots and propagation of the soybean cyst nematode. Planta. 2000;210(2):195–204.
42. Murashige T, Skoog F. A revised medium for rapid growth and bio assays with tobacco tissue cultures. Physiol Plant. 1962;15(3):473–97.
43. Gamborg OL, Miller RA, Ojima K. Nutrient requirements of suspension cultures of soybean root cells. Exp Cell Res. 1968;50(1):151.
44. Lin JJ. Optimization of the transformation efficiency of *Agrobacterium tumefaciens* cells using electroporation. Plant Sci. 1994;101(1):11–5.
45. Olhoft PM, Flagel LE, Donovan CM, Somers DA. Efficient soybean transformation using hygromycin B selection in the cotyledonary-node method. Planta. 2003;216(5):723–35.
46. Murray MG, Thompson WF. Rapid isolation of high molecular-weight plant DNA. Nucleic Acids Res. 1980;8(19):4321–5.
47. Hancock CN, Zhang F, Floyd K, Richardson AO, LaFayette P, Tucker D, et al. The rice miniature inverted repeat transposable element *mPing* is an effective insertional mutagen in soybean. Plant Physiol. 2011;157(2):552–62.

Production of knockout mice by DNA microinjection of various CRISPR/Cas9 vectors into freeze-thawed fertilized oocytes

Yoshiko Nakagawa[1†], Tetsushi Sakuma[2*†], Takuya Sakamoto[2], Masaki Ohmuraya[1], Naomi Nakagata[1] and Takashi Yamamoto[2]

Abstract

Background: Clustered regulatory interspaced short palindromic repeats (CRISPR)/CRISPR-associated protein 9 (Cas9)-mediated genome editing permits the rapid production of genetically engineered mice. To make the most of this innovative technology, a streamlined procedure is needed for the robust construction of CRISPR/Cas9 vectors, the efficient preparation of mouse oocytes, and refined genotyping strategies. Although we previously demonstrated the applicability of oocyte cryopreservation technologies and various genotyping methods in the production of transcription activator-like effector nuclease-mediated genome editing in mice, it has not yet been clarified whether these techniques can be applied to the CRISPR/Cas9-mediated generation of knockout mice. In this study, we investigated easy, efficient, and robust methods of creating knockout mice using several CRISPR/Cas9 systems.

Results: We constructed three types of CRISPR/Cas9 vectors expressing: 1) single guide RNA (gRNA) and Cas9 nuclease, 2) two gRNAs and Cas9 nickase, and 3) two gRNAs and FokI-dCas9, targeting the same genomic locus. These vectors were directly microinjected into the pronucleus of freeze-thawed fertilized oocytes, and surviving oocytes were transferred to pseudopregnant ICR mice. Cas9 nuclease resulted in the highest mutation rates with the lowest birth rates, while Cas9 nickase resulted in the highest birth rates with the lowest mutation rates. FokI-dCas9 presented well-balanced mutation and birth rates. Furthermore, we constructed a single all-in-one FokI-dCas9 vector targeting two different genomic loci, and validated its efficacy by blastocyst analysis, resulting in highly efficient simultaneous targeted mutagenesis.

Conclusions: Our report offers several choices of researcher-friendly consolidated procedures for making CRISPR/Cas9-mediated knockout mice, with sophisticated construction systems for various types of CRISPR vectors, convenient preparation of *in vitro* fertilized or mated freeze-thawed oocytes, and an efficient method of mutant screening.

Keywords: Knockout mouse, Pronuclear microinjection, *In vitro* fertilization, Freeze-thawing, CRISPR/Cas9, Double-nicking, FokI-dCas9

Background

Genetically engineered mice (GEM) have played an essential part in elucidating the functions of specific genes, the mechanisms of disease, embryogenesis and differentiation over recent decades. Although massive numbers of GEM have been generated and analyzed throughout the world [1], there is still a demand for more efficient methods of generating *de novo* GEM.

Genome editing using programmable nucleases, such as zinc-finger nucleases (ZFNs), transcription activator-like effector nucleases (TALENs), and clustered regularly interspaced short palindromic repeats (CRISPR)/CRISPR-associated protein 9 (Cas9), is an easy and efficient strategy to generate GEM [2]. In particular, CRISPR/Cas9 provides the most convenient method by which to produce GEM because of its simple construction and high DNA double-strand break (DSB)-inducing activity [3]. A

* Correspondence: tetsushi-sakuma@hiroshima-u.ac.jp
†Equal contributors
[2]Department of Mathematical and Life Sciences, Graduate School of Science, Hiroshima University, Hiroshima 739-8526, Japan
Full list of author information is available at the end of the article

conventional CRISPR/Cas9 system consists of Cas9 nuclease and a single guide RNA (gRNA) that targets a specified genomic locus containing a 20-base sequence defined by the gRNA and a protospacer adjacent motif (PAM) sequence defined by Cas9, the complex then cleaves the double-stranded DNA. DSBs are mainly repaired by error-prone non-homologous end-joining (NHEJ), randomly inducing insertions and/or deletions, which cause targeted gene disruption.

The conventional CRISPR/Cas9 complex does not form a dimer, unlike ZFNs and TALENs, and it often leads to off-target mutations with high frequency [4-6]. Although the frequencies of off-target mutations in animal embryos, especially in mice and rats, are reportedly not so high [7,8], they do exist and therefore present a potential risk even in animals. To improve the specificity, two derivative technologies have been developed; double-nicking using Cas9 nickase [9-13] and FokI-dimerization using nuclease-deficient Cas9 fused to FokI (FokI-dCas9) [14,15]. In both strategies, paired gRNAs are used to recruit two molecules of Cas9 nickase or FokI-dCas9 on juxtaposed positions of the target genomic locus, resulting in DNA cleavage. Importantly, single gRNA-guided Cas9 nickase can only induce a nick, which is less mutagenic than a DSB. Furthermore, single gRNA-guided FokI-dCas9 does not damage the genomic DNA at all. Therefore, off-target mutations are significantly reduced in these two derivative strategies.

To generate knockout mice using the CRISPR/Cas9 system, Cas9 mRNA and gRNA are generally synthesized using *in vitro* transcription and used for microinjection [16-21]. However, Mashiko and colleagues described a simplified protocol using pronuclear microinjection of circular plasmid expressing Cas9 and a gRNA [7,22]. Direct use of plasmids avoids the need for laborious *in vitro* transcription and purification steps, and plasmids offer high stability compared with RNAs enabling the robust production of knockout mice. In addition, we recently established an all-in-one CRISPR/Cas9 vector system for the assembly of multiple gRNA cassettes and a Cas9 nuclease/nickase cassette in a single vector [23]. When combined with these procedures, efficient production of knockout mice mediated by paired gRNA-guided CRISPR systems has been thought to be achieved.

Regarding the preparation of oocytes for microinjection to generate knockout mice with genome editing technologies, fertilized oocytes were usually collected by mating a male mouse and a superovulated female mouse. However, preparing fresh fertilized oocytes for every experiment is time-consuming and laborious work. We have previously established a procedure of using freeze-thawed fertilized oocytes created by *in vitro* fertilization (IVF) for the TALEN-mediated generation of knockout mice [24], as well as previous reports of transgenic mouse production [25,26]. In addition, Li and colleagues recently reported that fresh fertilized oocytes created by IVF could be used for conventional CRISPR/Cas9-mediated mouse genome editing [21]. However, Han and colleagues reported that they failed to produce genome edited mice by pronuclear injection of CRISPR/Cas9 plasmid, while cytoplasmic injection of Cas9 mRNA and purified gRNA resulted in successful genome editing [27]. Thus, the applicability of cryopreservation and IVF technologies for various CRISPR systems is another important advance in establishing the streamlined CRISPR/Cas9 plasmid injection-mediated production of knockout mice.

Here, we expand the multiplex gRNA assembly system to include FokI-dCas9, and construct Cas9 nuclease-, Cas9 nickase- and FokI-dCas9-type vectors for the generation of knockout mice. The gRNAs for each system are designed to target the same locus of an endogenous gene, and comparisons of the birth rate and the mutation rate are conducted. The applicability of cryopreservation and IVF is also investigated. We further examined whether pronuclear injection of a single all-in-one plasmid expressing FokI-dCas9 and four gRNAs could achieve multiplex genome editing in mouse embryos.

Results and discussion

Construction of an all-in-one vector system for FokI-dCas9-mediated genome editing

We previously established a system for creating all-in-one CRISPR/Cas9 vectors expressing Cas9 nuclease or nickase and up to seven gRNAs [23], and have distributed these plasmids as a "Multiplex CRISPR/Cas9 Assembly System Kit" via Addgene (Kit #1000000055). To expand the system for FokI-dCas9-mediated genome editing, we first created pX330A_FokI-1x(2–7) vectors. Using these newly constructed plasmids and pX330s-(2–7) vectors included in the Addgene Kit, we can create an all-in-one vector for FokI-dCas9-mediated genome editing, not only for single but also for multiplex gene editing (Figure 1). The new vectors will expectedly be distributed by Addgene as a supplemental package of the current kit (Multiplex CRISPR dCas9/FokI-dCas9 Accessory Pack).

Design of CRISPR/Cas9 nuclease, Cas9 nickase, and FokI-dCas9 vectors to target the interleukin-11 (*IL11*) gene

To compare the applicability of the three CRISPR/Cas9 systems with Cas9 nuclease, Cas9 nickase, or FokI-dCas9 in mouse genome editing, we designed and constructed CRISPR/Cas9 vectors as shown in Figure 2. Although the three systems require different design parameters and it is difficult to compare the genome editing efficacy with exactly the same gRNAs, we tried to target almost the same genomic region, which is around exon 3 of the *IL11* gene. For Cas9 nuclease, two gRNAs, gRNA_A and gRNA_B, were designed within the exon and

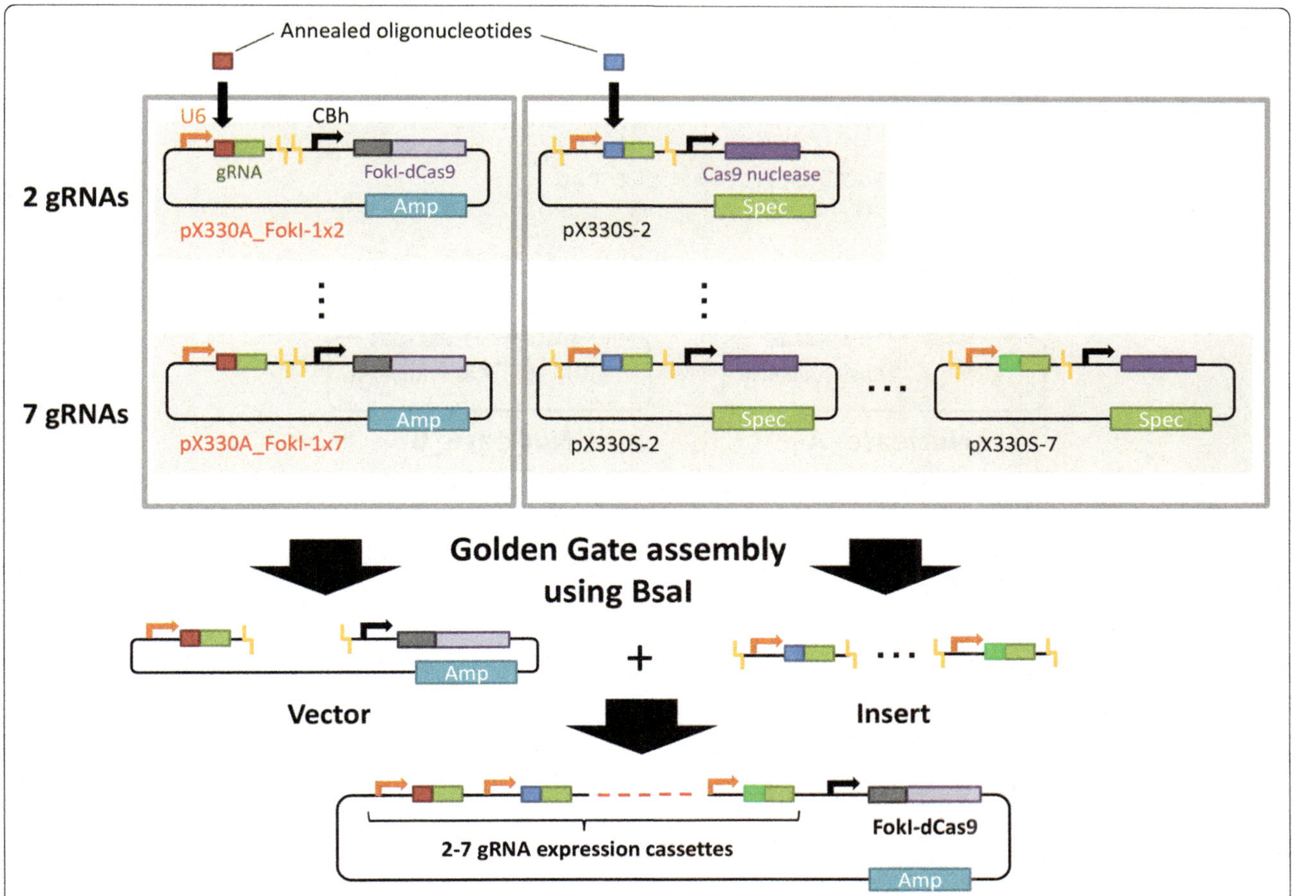

Figure 1 Schematic illustration of the all-in-one FokI-dCas9 vector construction system. Single gRNA-expressing vectors are constructed by the insertion of annealed oligonucleotides. Subsequently, the gRNA cassettes are assembled using Golden Gate cloning. Amp, ampicillin; Spec, spectinomycin; U6, human U6 promoter; CBh, chicken beta-actin short promoter.

oligonucleotides for these gRNAs were cloned separately (Figure 2A; Nuclease_A and Nuclease_B vectors). For Cas9 nickase, gRNA_B and gRNA_C were designed with a 7-bp offset (Figure 2B: Nickase_BC vector). When inducing double nick-mediated genome modification, the optimal range of the offset length is around 0–10 bp [23], whereas FokI-dCas9 requires 13- to 18-bp offsets for highly-efficient targeted mutagenesis [14]. Thus, we designed another gRNA, gRNA_D, in combination with gRNA_B for FokI-dCas9 (Figure 2C; FokI-dCas9_BD vector).

Activity validation of the all-in-one FokI-dCas9 vector

Since the genome editing potential of the all-in-one vector containing the FokI-dCas9 cassette was unknown, we evaluated the activity of the FokI-dCas9_BD vector in mouse embryos. The FokI-dCas9_BD vector was microinjected into the pronucleus of mouse fertilized oocytes. Twenty-one survival oocytes were cultured for 3.5 days. We observed seven blastocysts, and six of these embryos developed into expanded blastocysts. Each blastocyst was collected into a microtube, and then PCR was performed to amplify the targeted locus. We verified that at least one mutation had been induced in three embryos by direct sequencing (Figure 3A). In addition, bacterial cloning of the PCR products followed by DNA sequencing revealed high mutation rates (≥50%) (Figure 3B), confirming the powerful potential of the all-in-one FokI-dCas9 vector for GEM generation.

GEM generation by microinjection of all-in-one CRISPR/Cas9 plasmids into freeze-thawed fertilized oocytes

We next conducted microinjection of the three types of CRISPR/Cas9 vectors into freeze-thawed fertilized oocytes for efficient and convenient genome engineering in mice. Using reproductive engineering techniques, such as IVF and freeze-thawing, fertilized oocytes could be obtained with high efficiency (Additional file 1). Each CRISPR/Cas9 vector was directly injected into the pronucleus of freeze-thawed fertilized oocytes prepared by mating or IVF. The survival oocytes after microinjection were transferred into the oviducts of pseudopregnant

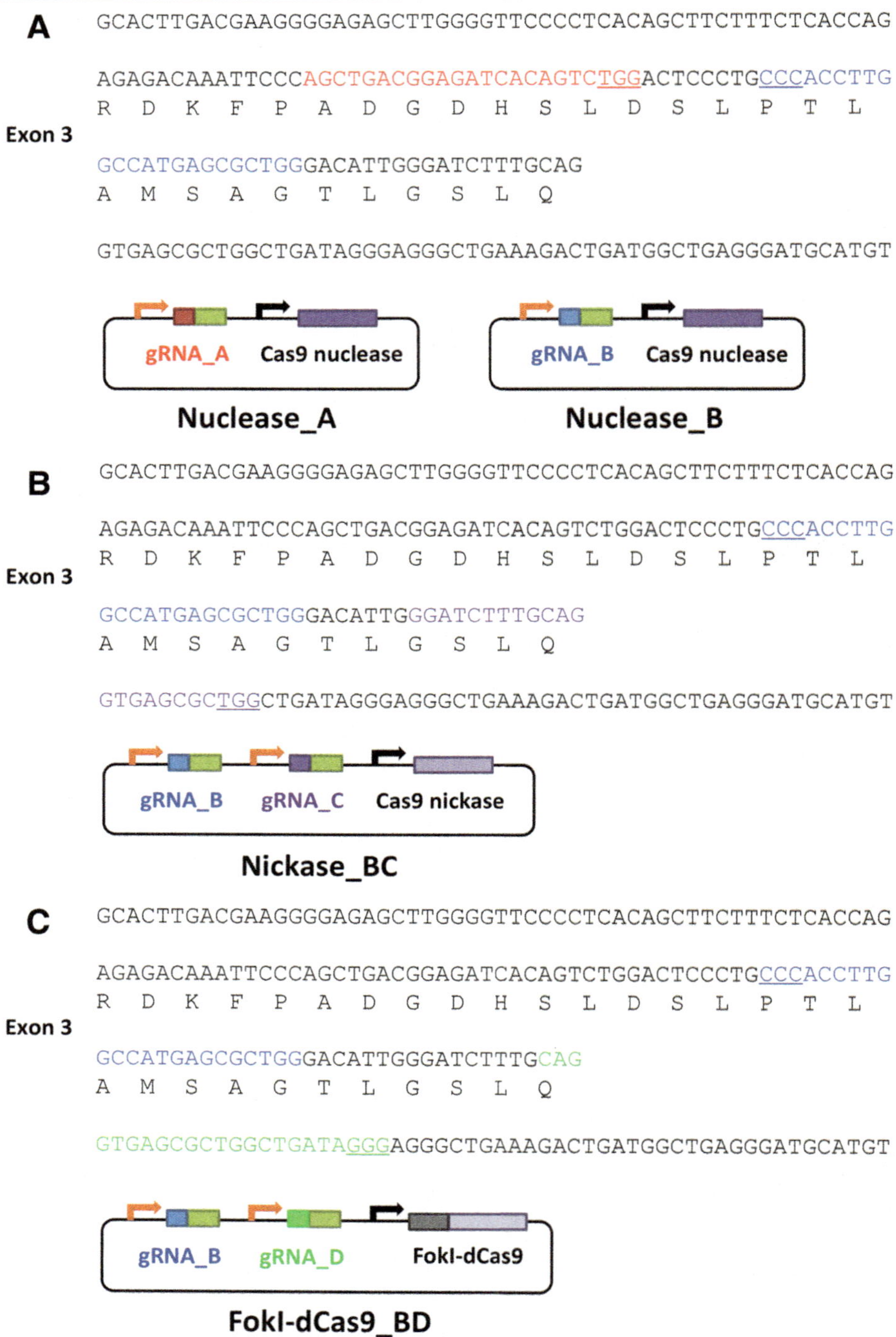

Figure 2 The target sequence and the constructed vectors. The genomic region around exon 3 of the *IL11* gene was targeted by Cas9 nuclease **(A)**, Cas9 nickase **(B)**, or FokI-dCas9 **(C)** vectors. The target sequence of each gRNA is indicated by colored bases. The PAM sequence is underlined.

ICR female mice. All the surrogate female mice gave birth to pups (Table 1), and all the pups excluding one pup died from cannibalism were genotyped by direct sequencing analysis (Table 1 and Additional file 2). Although we have shown that the combinatorial screening strategy is required for complete genotyping [28], direct sequencing is reportedly able to screen mutant mice with high mutation rates [7].

Sequencing analysis revealed that all three types of CRISPR/Cas9 vectors could produce mutant mice, but the birth rates and mutation rates of each method were not comparable. In the Cas9 nuclease group, the birth

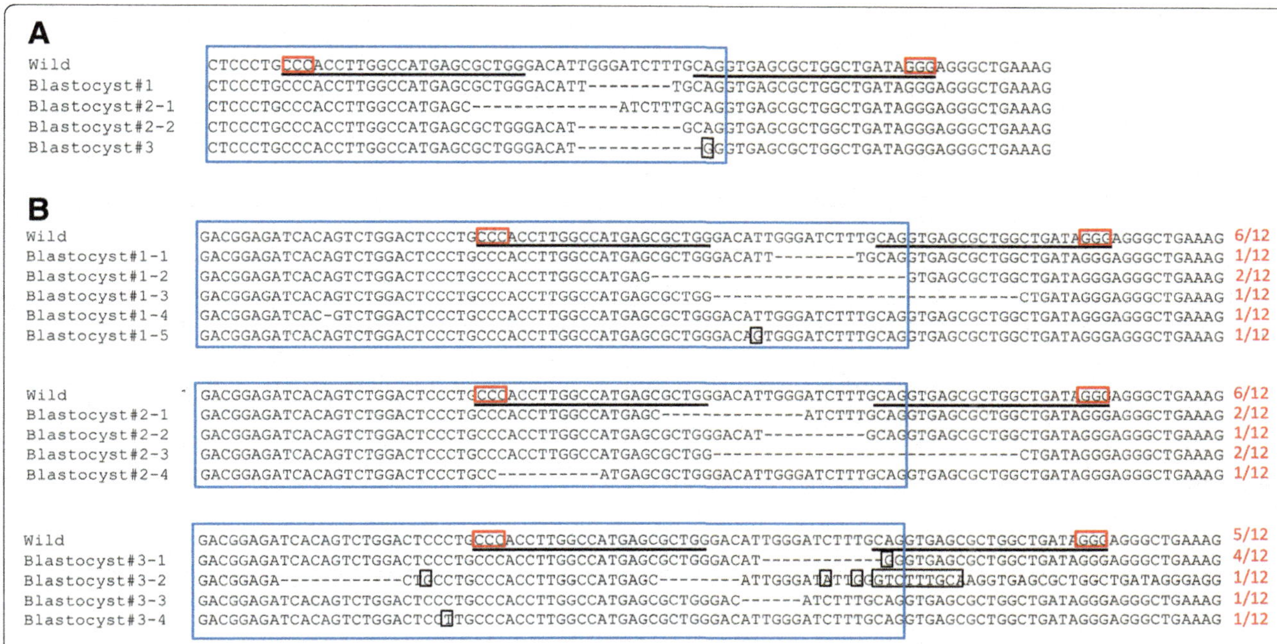

Figure 3 Sequence analysis of blastocysts injected with the FokI-dCas9_BD vector. The PCR products were analyzed by direct sequencing **(A)**, followed by sequencing of subcloned plasmids **(B)**. The wild-type *IL11* sequence is shown at the top (Wild) with the gRNA target sequences (underlined). The PAM sequence is enclosed in red boxes. Deletions are indicated by dashes and substitutions and insertions are enclosed in a black box. A blue box indicates the partial sequence of exon 3. Only one or two types of mutations were detected in direct sequencing **(A)**, whereas various mutation patterns were observed in subcloned sequencing **(B)**. Numbers in red letters on the right side of each sequence indicate the frequencies of the corresponding alleles (number of the allele/number of total clones).

rates were relatively low (7.4–12.0%), whereas the mutation rates were high (80–100%). Conversely, more pups were born in the Cas9 nickase group (19.1% and 22.2%), while the number of mutants were fewer than in the Cas9 nuclease group (5.9% and 0%). Interestingly, the FokI-dCas9 group exhibited moderate birth rates (18.2% and 17.6%) and mutation rates (66.6% and 33.3%). To confirm the superiority of FokI-dCas9-mediated method, we additionally injected the FokI-dCas9_BD vector into freeze-thawed fertilized oocytes collected from mature female mice (8 weeks of age) mated with male mice. Consistent with the results using the oocytes collected from immature female mice (5 weeks of age), the birth rate and the mutation rate were sufficiently high (38.2% and 84.6%, respectively). These results suggest not only the robustness of FokI-dCas9-mediated mouse genome editing, but also the applicability of oocytes from female mice at various weeks of age, as shown previously using TALENs [24]. Although these results do not necessarily reflect the general properties of the three CRISPR/Cas9

Table 1 Generation of *IL11* mutants using freeze-thawed fertilized oocytes

Method	Vector	Injected	Survived	Transferred	Pups	Analyzed	Mutants
Mating	Nuclease_A	76	54 (71.1)	54	4 (7.4)	4	4 (100)
	Nuclease_B	72	55 (76.4)	55	5 (9.1)	5	4 (80.0)
	Nickase_BC	108	94 (87.0)	94	18 (19.1)	17	1 (5.9)
	FokI-dCas9_BD	55	46 (83.6)	33	6 (18.2)	6	4 (66.6)
	FokI-dCas9_BD	58*	55 (94.8)	34	13 (38.2)	13	11 (84.6)
IVF	Nuclease_A	32	25 (78.1)	25	3 (12.0)	3	3 (100)
	Nuclease_B	34	22 (64.7)	22	2 (9.1)	2	2 (100)
	Nickase_BC	32	27 (84.4)	27	6 (22.2)	6	0 (0)
	FokI-dCas9_BD	39	35 (89.7)	34	6 (17.6)	6	2 (33.3)

Numbers in brackets represent percentages.
*Fertilized oocytes were collected from females at 8 weeks of age, whereas 5-week-old females were used for the other injections that oocytes were fertilized by mating.

systems, they represent an important practical example of CRISPR/Cas9-mediated GEM generation. Importantly, the low birth rates of Cas9 nuclease plasmid-injected mice are comparable with previous studies using inbred strains [29,30], although high birth rates can be achieved using B6D2F1 (BDF1) hybrid mice [7]. We thus concluded that FokI-dCas9 plasmid injection might be a powerful strategy for generating knockout mice, especially for inbred strains.

Transgene analysis and off-target analysis

Since plasmid DNA was used for microinjection, we investigated whether DNA vectors were integrated into the genome. Genomic PCR of the coding sequence of Cas9 variants and FokI was performed to detect genomic integrants. In the two founders generated by the injection of Nuclease_B vector, the Cas9 gene fragment was amplified (Additional file 3). Altogether, 10% (2/20) of the mutant mice carried the transgene. Although the observed transgenicity is slightly higher than in Mashiko's report (4.3%) [7], this may not be a major issue because the transgene can be removed by mating before expansion of the mutant strain.

Finally, potential off-target sites for each gRNA target site were analyzed by direct sequencing. The top three candidates for each site were selected using the CRISPR design tool (http://crispr.mit.edu/) (Additional file 4). We then sequenced all of the 20 founders (seven founders for Nuclease_A, six founders for Nuclease_B, one founder for Nickase_BC, and six founders for FokI-dCas9_BD), but no off-target mutations were detected. However, our analysis and all of the previously reported off-target analyses in mice provide limited genomic context, and the potential risk of off-target mutagenesis in disparate parts of the genome is undeniable. In fact, Tsai and colleagues recently revealed that CRISPR/Cas9 can induce various genomic rearrangements including megabase-scale large deletions and chromosomal translocations in cultured cells [31]. They also showed that most of these experimentally demonstrated off-target sites were not identified as off-target candidates using any computational prediction tool including the CRISPR design tool. Also in mice, these concealed off-target effects can exist, but they are difficult to detect because they cannot be predicted by *in silico* analysis. We thus believe that a highly specific gene targeting strategy, such as FokI-dCas9, is important not only for cultured cell applications but also for the creation of GEM.

Multiplex genome editing in mouse embryos using a single all-in-one FokI-dCas9 vector

We finally examined whether multiple gene targeting could be applied using microinjection of a single all-in-one FokI-dCas9 vector simultaneously expressing four gRNAs and FokI-dCas9. We designed gRNAs to target exon 3 of regenerating islet-derived 3 beta (*Reg3b*) and regenerating islet-derived 3 gamma (*Reg3g*) genes, which are located in the chromosome 6 with 96.1-kb distance (Figure 4A and B). The all-in-one FokI-dCas9 vector was microinjected into the pronucleus of fertilized oocytes. Twenty-eight survival oocytes were cultured for 3.5 days, and nine embryos developed into expanded blastocysts. Each blastocyst was collected into a microtube, and then PCR was performed to amplify around the both target loci. Direct sequencing analyses revealed that more than 50% of the embryos possessed mutated alleles in the both target loci (8/9 in *Reg3b* and 5/9 in *Reg3g*) (Figure 5A and B). More importantly, five embryos (#1, 3, 6, 7, and 8) possessed mutations in the both genes, showing high potential for creating double knockout mice mediated by microinjection of all-in-one FokI-dCas9 vector. We further performed out-out PCR to investigate whether chromosomally deleted alleles exist, but no intended amplicons were observed using blastocyst PCR genotyping. Further examinations are needed to clarify the possible existence of chromosomally deleted alleles in pups.

Conclusions

Our study compared the genome editing efficiency in one-cell mouse embryos microinjected with various CRISPR/Cas9 vectors expressing gRNA(s) and Cas9 nuclease, Cas9 nickase, or FokI-dCas9, constructed by an all-in-one vector system. In addition, we demonstrated the applicability of IVF and freeze-thawing of oocytes in GEM generation mediated by the methods described above, which makes preparation of fertilized oocytes more convenient. Our findings suggest that the FokI-dCas9-mediated approach is a highly reliable strategy for creating GEM, including double knockout mice, and we provide accessible platforms for the CRISPR/Cas9-based generation of GEM.

Methods

Plasmids

All-in-one CRISPR/Cas9 vectors for Cas9 nuclease and nickase were constructed using the Multiplex CRISPR/Cas9 Assembly System Kit (Addgene, Cambridge, MA, USA; Kit #1000000055) as previously described [23]. For FokI-dCas9 vectors, a H840A mutation was introduced in pX330A_D10A vectors using inverse PCR and the In-Fusion cloning method (Clontech, Palo Alto, CA, USA) (pX330A_dCas9 vectors). Subsequently, the coding sequence of FokI was amplified by PCR from ptCMV-136/63-VR vector contained in the Platinum Gate TALEN Kit (Addgene; Kit #1000000043) [32] using the FokI-dCas9-ins-F and FokI-dCas9-ins-R primers listed in Additional file 5. The amplified fragment was cloned using the In-Fusion cloning method into pX330A_dCas9 vectors, which were amplified by inverse PCR using the FokI-dCas9-vec-F and FokI-dCas9-vec-R primers

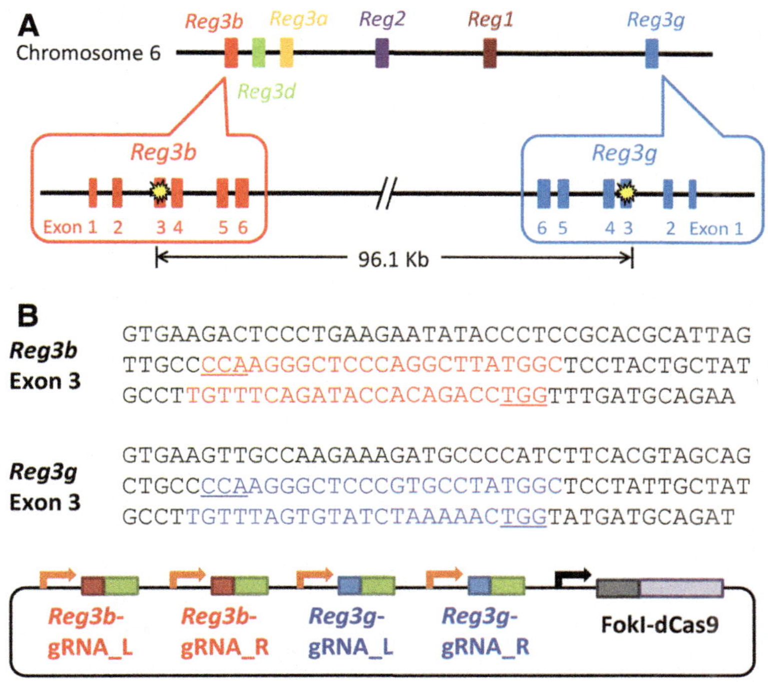

Figure 4 Schematic design of multiple gene targeting using all-in-one FokI-dCas9 vector. **(A)** The genomic context around the targeted loci. The target sites are located in exon 3 of *Reg3b* and *Reg3g* genes. **(B)** The target sequence and the constructed vectors. The target sequence of each gRNA is indicated by colored bases. The PAM sequence is underlined.

(Additional file 5) (pX330A_FokI vectors). After the construction of pX330A_FokI vectors, oligonucleotides were inserted and Golden Gate cloning was performed as previously described [23] to construct the FokI-dCas9_BD vector.

Animals

All animal experiments were approved by the Animal Care and Experimentation Committee of the Center for Animal Resources and Development, Kumamoto University, and were carried out in accordance with the approved guidelines.

IVF and freeze-thawing of fertilized oocytes

The IVF and freeze-thawing procedure was performed according to a previous report [24]. The cauda epididymides were obtained from C57BL/6 male mice over 12 weeks of age, and spermatozoa were suspended in FERTIUP® Mouse Sperm Preincubation Medium (Kyudo, Saga, Japan). Subsequently, the sample was incubated for 1 h at 37°C in 5% CO_2 and 95% humidified air until insemination. C57BL/6 female mice were superovulated using 7.5 IU of pregnant mare serum gonadotropin (Serotropin; Aska Pharmaceutical, Tokyo, Japan) and 7.5 IU of human chorionic gonadotropin (Gonatropin; Aska Pharmaceutical, Tokyo, Japan) at 8–12 weeks of age, and their oviducts were transferred to fertilization dishes containing a drop of CARD MEDIUM (Kyudo, Saga, Japan). The cumulus-oocyte complexes were introduced into the drop of CARD MEDIUM. After preincubation of the fresh sperm, the sperm suspension was added to the drop containing cumulus-oocyte complexes. The fertilization dishes were incubated at 37°C in 5% CO_2 and 95% humidified air. After 3 h, the inseminated oocytes were rinsed three times with human tubal fluid medium. After 2 h, the oocytes were cryopreserved by a simple vitrification method [33,34]. At later time points, the cryopreserved oocytes were thawed by a thawing method for vitrified oocytes using warmed PB1 containing 0.25 M sucrose [33,34]. The freeze-thawed oocytes were used for microinjection.

Mating and freeze-thawing of fertilized oocytes

C57BL/6 female mice were induced to superovulate using pregnant mare serum gonadotropin and human chorionic gonadotropin at 5 or 8 weeks of age, and then mated with C57BL/6 male mice. The fertilized oocytes were collected from females displaying vaginal plugs. The oocytes were cryopreserved as described above. The frozen oocytes were thawed and used for microinjection.

Single blastocyst assay

Fertilized oocytes were collected from C57BL/6 superovulated female mice mated with C57BL/6 male mice. After freeze-thawing of oocytes, FokI-dCas9 vector

A

```
Reg3b Wild   GCATTAGTTGCCCCAAGGGCTCCCAGGCTTATGGCTCCTACTGCTATGCCTTGTTTCAGATACCACAGACCTGGTTTGATGCAGAA
Reg3b #1-1   GCATTAGTTGCCCCAAGGGCTCCCAGG--------------------------------ATACCACAGACCTGGTTTGATGCAGAA
Reg3b #1-2   GCATTAGTTGCCCCAAGGGCTCCCAGGCTTATGGCTCCTA--------------------TACCACAGACCTGGTTTGATGCAGAA
Reg3b #1-3   GCATTAGTTGCCCCAAGGGCTCCCAGGCTTATGGCTCCTACTGCTATGCCTTGTTTCAGATACCA-------------ATGCAGAA
Reg3b #1-4   GCATTAGTTGCCCCAAGGGCTCCCAGGCTTATGGCTCCTACTGCTATG------TTCAGATACCACAGACCTGGTTTGATGCAGAA
Reg3b #2-1   GCATTAGTTGCCCCAAGGGCTCCCAGGCTTAT---------------------------ATACCACAGACCTGGTTTGATGCAGAA
Reg3b #2-2   GCATTAGTTGCCCCAAGGGCTCCCAGGCTTATG---------------CCTTGTTTCAGATACCACAGACCTGGTTTGATGCAGAA
Reg3b #2-3   GCATTAGTTGCCCCAAGGGCTCCCAGGCTTATGGCTCCTACTGCTAT------TTTCAGATACCACAGACCTGGTTTGATGCAGAA
Reg3b #3-1   ------------------------ (93 bp) --------------------------TACCACAGACCTGGTTTGATGCAGAA
Reg3b #3-2   GCATTAGTTGCCCCAAGGGCTCCCA-------------------TATGCCTTGTTTCAGATACCACAGACCTGGTTTGATGCAGAA
Reg3b #3-3   GCATTAGTTGCCCCAAGGGCTCCCA-----------------GCTATGCCTTGTTTCAGATACCACAGACCTGGTTTGATGCAGAA
Reg3b #4-1   GCATTAGTTGCCCCAAGGGCT--------------------------GCCTTGTTTCAGATACCACAGACCTGGTTTGATGCAGAA
Reg3b #4-2   GCATTAGTTGCCCCAAGGGCTCCCAGGCTTATGGCTCCTACTGC--------------------------CTGGTTTGATGCAGAA
Reg3b #5     GCATTAGTTGCCCCAAGGGCTCCCAGG--------------------------------ATACCACAGACCTGGTTTGATGCAGAA
Reg3b #6-1   GCATTAGTTGCCCCAAGGGCTCCCAG---------------------------------ATACCACAGACCTGGTTTGATGCAGAA
Reg3b #6-2   GCATTAGTTGCCCCAAGGGCTCCCAGGCTTATGGC--------------CTTGTTTCAGATACCACAGACCTGGTTTGATGCAGAA
Reg3b #7-1   GCATTAGTTGCCCC-------------------------------------------------CACAGACCTGGTTTGATGCAGAA
Reg3b #7-2   GCATTAGTTGCCCCAAGGGCTCCCAGGCTTATG---------------CCTTGTTTCAGATACCACAGACCTGGTTTGATGCAGAA
Reg3b #8-1   GCATTAGTTGCCCCAAGGGCTCCCAGGCTTATG---------------CCTTGTTTCAGATACCACAGACCTGGTTTGATGCAGAA
Reg3b #8-2   GCATTAGTTGCCCCAAGGGCTCCCAGGCTTATGGCTCCTA------------GTTTCAGATACCACAGACCTGGTTTGATGCAGAA
Reg3b #8-3   GCATTAGTTGCCCCAAGGGCTCCCAGGCTTATGGCTCCTA------TGCCTTGTTTCAGATACCACAGACCTGGTTTGATGCAGAA
```

B

```
Reg3g Wild   AGCAGCTGCCCCAAGGGCTCCCGTGCCTATGGCTCCTATTGCTATGCCTTGTTTAGTGTATCTAAAAACTGGTATGATGC
Reg3g #1     AGCAGCTGCCCCAAGGGCTCCCGTGCCT---------------------TGTTTAGTGTATCTAAAAACTGGTATGATGC
Reg3g #3     AGCAGCTGCCCCAAGGGCTCCCGTGCCTATGGCTCCTAT------GCCTTGTTTAGTGTATCTAAAAACTGGTATGATGC
Reg3g #6     AGCAGCTGCCCCAAGGGCTCCCGTGCCTATGGCTCCTAT------GCCTTGTTTAGTGTATCTAAAAACTGGTATGATGC
Reg3g #7     AGCAGCTGCCCCAAGGGCT-------------------------TGCCTTGTTTAGTGTATCTAAAAACTGGTATGATGC
Reg3g #8     AGCAGCTGCCCCAAGGGCTCCCGTGCCTATGG-----------ATGCCTTGTTTAGTGTATCTAAAAACTGGTATGATGC
```

Figure 5 Sequence analysis of blastocysts injected with all-in-one FokI-dCas9 vector simultaneously targeting *Reg3b* and *Reg3g* genes. The PCR products were analyzed by direct sequencing to identify mutations in *Reg3b* **(A)** and *Reg3g* **(B)** loci. The wild-type sequence of *Reg3b* and *Reg3g* are shown at the top (Wild) with the gRNA target sequences (underlined). The PAM sequence is enclosed in red boxes. Deletions are indicated by dashes. Blastocyst numbers on the left side of each sequence in **(A)** and **(B)** are identical. Most blastocysts had multiple types of mutations at the *Reg3b* locus. The wild-type *Reg3b* sequence was not detected in blastocyst #4 and #6. In the other blastocysts, the wild-type *Reg3b* sequence was detected along with mutant sequence. Regarding *Reg3g* locus, the wild-type sequence was detected in all blastocysts with mutant sequence.

was diluted in DNase-free PBS at 5 ng/μl, and then microinjection was performed. Survival oocytes were cultured in potassium simplex optimized medium with amino acids (KSOM-AA) at 37°C in 5% CO_2 and 95% humidified air for 3.5 days. For PCR of single blastocysts, 1 μl of KSOM-AA containing 1 blastocyst was transferred to the wall near the bottom of a 0.2-mL PCR tube. Blastocyst lysates were prepared by an alkaline lysis method. To detect the targeted *IL11* gene, PCR was performed using KOD FX (TOYOBO, Osaka, Japan) with the primers listed in Additional file 5 under the following conditions: 94°C for 2 min; followed by 40 cycles of 98°C for 10 s, 60°C for 30 s, and 68°C for 25 s. Each PCR product was subjected to automatic electrophoresis using MultiNA (Shimadzu Corporation, Kyoto, Japan). The PCR products were purified and analyzed by direct sequencing, and then the products harboring mutations were subcloned into pTA2 vector using a Target Clone -Plus- (TOYOBO). Sequencing was performed using an ABI 3130 Genetic Analyzer (Life Technologies, Carlsbad, CA, USA) with a BigDye Terminator v1.1 Cycle Sequencing Kit (Life Technologies). Also for *Reg3b* and *Reg3g*, PCR was performed using KOD FX with the primers listed in Additional file 5. Subsequently, automatic electrophoresis and direct sequencing were performed as described above.

Microinjection and transfer

Each CRISPR/Cas9-plasmid DNA was diluted in DNase-free PBS at 5 ng/μl for microinjection. The plasmid DNA was injected into the pronucleus of fertilized oocytes. The injected oocytes were cultured in KSOM-AA at 37°C in 5% CO_2 and 95% humidified air for 1 h. Surviving oocytes were transferred to the oviducts of pseudopregnant ICR female mice.

Mutation analysis of the *IL11* gene

Tail lysates were prepared by an alkaline lysis method and PCR was performed using KOD FX and the primers listed in Additional file 5 under the following conditions: 94°C for 2 min; followed by 38 cycles of 98°C for 10 s, 60°C for 30 s, and 68°C for 25 s. Each PCR product was analyzed by direct sequencing as described above.

Transgene analysis

To examine integration of vector DNA, we designed primers for Cas9, dCas9, and FokI. Each founder was examined the integration of Cas9 or dCas9, and FokI by genomic PCR using KOD FX and the primers listed in Additional file 5. The PCR products were analyzed by automatic electrophoresis using MultiNA.

Off-target analysis

Candidates for off-target sequences were selected with the CRISPR Design tool (http://crispr.mit.edu/) (Additional file 3). Genomic regions around the top three potential off-target sites were amplified by PCR from founders using KOD FX, the primers used are listed in Additional file 5.

Additional files

Additional file 1: Summary of the results of freeze-thawing in C57BL/6 mice.

Additional file 2: Mutation analysis of PCR products from founders. Sequence analysis of PCR products from founders injected with Nuclease_A (A), Nuclease_B (B), Nickase_BC (C), and FokI-dCas9_BD (D,E) vectors. Founder numbers are shown on the left side. IA, IB, and IBD were generated from oocytes fertilized by IVF. MA, MB, MBC, and MBD were generated from oocytes fertilized by mating. 8W were generated from oocytes of mature females by mating. The wild-type *IL11* sequence is shown at the top (Wild) with the gRNA target sequences (underlined). The PAM sequence is enclosed in a red box. Deletions are indicated by dashes, and insertions or substitutions are enclosed in a black box. The exon 3 sequence is enclosed in a blue box.

Additional file 3: Transgene analysis of founders. (A) Schematic drawing of the primers for each Cas9 variant. Red bars indicate primers for FokI detection. Blue bars indicate primers for Cas9 or dCas9 detection. The lengths of PCR products for FokI and Cas9/dCas9 are 395 and 473 bp, respectively. (B) Pseudo-gel images of transgene analysis. When vector DNA was integrated, a band of approximately 473-bp or 395-bp appears. Founder numbers shown on the upper side of each image are identical to Additional file 2. Red circles indicate transgene-positive founders. A P, B P, BC P, and BD P indicate positive controls for PCR amplification with Nulcease_A, Nuclease_B, Nickase_BC, and FokI-dCas9_BD plasmids, respectively, as templates. M, molecular weight maker (pGEM DNA Markers; Promega, Madison. WI, USA). W, negative control using founders without vector injection. N, negative control without template DNA.

Additional file 4: Top three candidates for potential off-target sites (5′-3′).

Additional file 5: Primer sets for PCR.

Abbreviations

CRISPR: Clustered regulatory interspaced short palindromic repeats; Cas9: CRISPR-associated protein 9; gRNA: Guide RNA; GEM: Genetically engineered mice; ZFN: Zinc-finger nuclease; TALEN: Transcription activator-like effector nuclease; DSB: DNA double-strand break; PAM: Protospacer adjacent motif; NHEJ: Non-homologous end-joining; IVF: *In vitro* fertilization; *IL11*: Interleukin-11; *Reg3b*: Regenerating islet-derived 3 beta; *Reg3g*: Regenerating islet-derived 3 gamma; KSOM-AA: Potassium simplex optimized medium with amino acids.

Competing interests

The authors declare that they have no competing interests.

Authors' contributions

YN carried out the mouse experiments and wrote the manuscript. TSakuma designed the study, created CRISPR/Cas9 vectors, and wrote the manuscript. TSakamoto constructed the all-in-one FokI-dCas9 vector system. MO and NN provided instructions. TY supervised the work. All authors read and approved the manuscript.

Acknowledgements

This study was supported by a Grant-in-Aid for the Fundamental Technologies Upgrading Program, National BioResource Project (NBRP) of the Ministry of Education, Culture, Sports, Science and Technology (MEXT), Japan. This study was also supported by KAKENHI (25890014) to TSakuma from the Japan Society for the Promotion of Science.

Author details

[1]Center for Animal Resources and Development, Kumamoto University, Kumamoto 860-0811, Japan. [2]Department of Mathematical and Life Sciences, Graduate School of Science, Hiroshima University, Hiroshima 739-8526, Japan.

References

1. International Mouse Knockout Consortium, Collins FS, Rossant J, Wurst W. A mouse for all reasons. Cell. 2007;128:9–13.
2. Aida T, Imahashi R, Tanaka K. Translating human genetics into mouse: the impact of ultra-rapid *in vivo* genome editing. Dev Growth Differ. 2014;56:34–45.
3. Singh P, Schimenti JC, Bolcun-Filas E. A Mouse Geneticist's Practical Guide to CRISPR Applications. Genetics. 2015;199:1–15.
4. Fu Y, Foden JA, Khayter C, Maeder ML, Reyon D, Joung JK, et al. High-frequency off-target mutagenesis induced by CRISPR-Cas nucleases in human cells. Nat Biotechnol. 2013;31:822–6.
5. Hsu PD, Scott DA, Weinstein JA, Ran FA, Konermann S, Agarwala V, et al. DNA targeting specificity of RNA-guided Cas9 nucleases. Nat Biotechnol. 2013;31:827–32.
6. Pattanayak V, Lin S, Guilinger JP, Ma E, Doudna JA, Liu DR. High-throughput profiling of off-target DNA cleavage reveals RNA-programmed Cas9 nuclease specificity. Nat Biotechnol. 2013;31:839–43.
7. Mashiko D, Fujihara Y, Satouh Y, Miyata H, Isotani A, Ikawa M. Generation of mutant mice by pronuclear injection of circular plasmid expressing Cas9 and single guided RNA. Sci Rep. 2013;3:3355.
8. Hu X, Chang N, Wang X, Zhou F, Zhou X, Zhu X, et al. Heritable gene-targeting with gRNA/Cas9 in rats. Cell Res. 2013;23:1322–5.
9. Mali P, Aach J, Stranges PB, Esvelt KM, Moosburner M, Kosuri S, et al. CAS9 transcriptional activators for target specificity screening and paired nickases for cooperative genome engineering. Nat Biotechnol. 2013;31:833–8.
10. Ran FA, Hsu PD, Lin CY, Gootenberg JS, Konermann S, Trevino AE, et al. Double nicking by RNA-guided CRISPR Cas9 for enhanced genome editing specificity. Cell. 2013;154:1380–9.
11. Cho SW, Kim S, Kim Y, Kweon J, Kim HS, Bae S, et al. Analysis of off-target effects of CRISPR/Cas-derived RNA-guided endonucleases and nickases. Genome Res. 2014;24:132–41.
12. Fujii W, Onuma A, Sugiura K, Naito K. Efficient generation of genome-modified mice via offset-nicking by CRISPR/Cas system. Biochem Biophys Res Commun. 2014;445:791–4.
13. Shen B, Zhang W, Zhang J, Zhou J, Wang J, Chen L, et al. Efficient genome modification by CRISPR-Cas9 nickase with minimal off-target effects. Nat Methods. 2014;11:399–402.
14. Tsai SQ, Wyvekens N, Khayter C, Foden JA, Thapar V, Reyon D, et al. Dimeric CRISPR RNA-guided FokI nucleases for highly specific genome editing. Nat Biotechnol. 2014;32:569–76.
15. Guilinger JP, Thompson DB, Liu DR. Fusion of catalytically inactive Cas9 to FokI nuclease improves the specificity of genome modification. Nat Biotechnol. 2014;32:577–82.
16. Wang H, Yang H, Shivalila CS, Dawlaty MM, Cheng AW, Zhang F, et al. One-step generation of mice carrying mutations in multiple genes by CRISPR/Cas-mediated genome engineering. Cell. 2013;153:910–8.
17. Zhou J, Wang J, Shen B, Chen L, Su Y, Yang J, et al. Dual sgRNAs facilitate CRISPR/Cas9-mediated mouse genome targeting. FEBS J. 2014;281:1717–25.
18. Fujii W, Kawasaki K, Sugiura K, Naito K. Efficient generation of large-scale genome-modified mice using gRNA and CAS9 endonuclease. Nucleic Acids Res. 2013;41, e187.
19. Fujii W, Onuma A, Sugiura K, Naito K. One-step generation of phenotype-expressing triple-knockout mice with heritable mutated alleles by the CRISPR/Cas9 system. J Reprod Dev. 2014;60:324–7.

20. Yasue A, Mitsui SN, Watanabe T, Sakuma T, Oyadomari S, Yamamoto T, et al. Highly efficient targeted mutagenesis in one-cell mouse embryos mediated by the TALEN and CRISPR/Cas systems. Sci Rep. 2014;4:5705.
21. Li F, Cowley DO, Banner D, Holle E, Zhang L, Su L. Efficient genetic manipulation of the NOD-Rag1−/−IL2RgammaC-null mouse by combining *in vitro* fertilization and CRISPR/Cas9 technology. Sci Rep. 2014;4:5290.
22. Mashiko D, Young SA, Muto M, Kato H, Nozawa K, Ogawa M, et al. Feasibility for a large scale mouse mutagenesis by injecting CRISPR/Cas plasmid into zygotes. Dev Growth Differ. 2014;56:122–9.
23. Sakuma T, Nishikawa A, Kume S, Chayama K, Yamamoto T. Multiplex genome engineering in human cells using all-in-one CRISPR/Cas9 vector system. Sci Rep. 2014;4:5400.
24. Nakagawa Y, Sakuma T, Nakagata N, Yamasaki S, Takeda N, Ohmuraya M, et al. Application of oocyte cryopreservation technology in TALEN-mediated mouse genome editing. Exp Anim. 2014;63:349–55.
25. Nakagata N. Use of cryopreservation techniques of embryos and spermatozoa for production of transgenic (Tg) mice and for maintenance of Tg mouse lines. Lab Anim Sci. 1996;46:236–8.
26. Keskintepe L, Agca Y, Pacholczyk GA, Machnicka A, Critser JK. Use of cryopreserved pronuclear embryos for the production of transgenic mice. Biol Reprod. 2001;65:407–11.
27. Han Y, Slivano OJ, Christie CK, Cheng AW, Miano JM. CRISPR-Cas9 Genome Editing of a Single Regulatory Element Nearly Abolishes Target Gene Expression in Mice—Brief Report. Arterioscler Thromb Vasc Biol. 2015;35:312–5.
28. Nakagawa Y, Yamamoto T, Suzuki K, Araki K, Takeda N, Ohmuraya M, et al. Screening methods to identify TALEN-mediated knockout mice. Exp Anim. 2014;63:79–84.
29. Li D, Qiu Z, Shao Y, Chen Y, Guan Y, Liu M, et al. Heritable gene targeting in the mouse and rat using a CRISPR-Cas system. Nat Biotechnol. 2013;31:681–3.
30. Horii T, Arai Y, Yamazaki M, Morita S, Kimura M, Itoh M, et al. Validation of microinjection methods for generating knockout mice by CRISPR/Cas-mediated genome engineering. Sci Rep. 2014;4:4513.
31. Tsai SQ, Zheng Z, Nguyen NT, Liebers M, Topkar VV, Thapar V, et al. GUIDE-seq enables genome-wide profiling of off-target cleavage by CRISPR-Cas nucleases. Nat Biotechnol. 2015;33:187–97.
32. Sakuma T, Ochiai H, Kaneko T, Mashimo T, Tokumasu D, Sakane Y, et al. Repeating pattern of non-RVD variations in DNA-binding modules enhances TALEN activity. Sci Rep. 2013;3:3379.
33. Nakagata N, Takeo T, Fukumoto K, Kondo T, Haruguchi Y, Takeshita Y, et al. Applications of cryopreserved unfertilized mouse oocytes for *in vitro* fertilization. Cryobiology. 2013;67:188–92.
34. Nakao K, Nakagata N, Katsuki M. Simple and efficient vitrification procedure for cryopreservation of mouse embryos. Exp Anim. 1997;46:231–4.

15

Reduction in C-terminal amidated species of recombinant monoclonal antibodies by genetic modification of CHO cells

Mihaela Škulj[1*], Dejan Pezdirec[1], Dominik Gaser[1], Marko Kreft[2,3,4] and Robert Zorec[2,3]

Abstract

Background: During development of recombinant monoclonal antibodies in Chinese hamster ovary (CHO) cells, C-terminal amidated species are observed. C-terminal amidation is catalysed by peptidylglycine α-amidating monooxygenase (PAM), an enzyme known to be expressed in CHO cells. The significant variations between clones during clone selection, and the relatively high content of amidated species (up to 15%) in comparison to reference material (4%), led us to develop a cell line with reduced production of C-terminal amidated monoclonal antibodies using genetic manipulation.

Results: Initial target validation was performed using the RNA interference approach against *PAM*, which resulted in a CHO cell line with C-terminal amidation decreased to 3%. Due to the transient effects of small-interfering RNAs, and possible stability problems using short-hairpin RNAs, we knocked-down the *PAM* gene using zinc finger nucleases. Plasmid DNA and mRNA for zinc finger nucleases were used to generate a *PAM* knock-out, which resulted in two CHO cell lines with C-terminal amidation decreased to 6%, in CHO *Der2* and CHO *Der3* cells.

Conclusion: Two genetically modified cell lines were generated using a zinc finger nuclease approach to decrease C-terminal amidation on recombinant monoclonal antibodies. These two cell lines now represent a pool from which the candidate clone with the highest comparability to the reference molecule can be selected, for production of high-quality and safe therapeutics.

Keywords: C-terminal amidation, Genetic modification, Recombinant monoclonal antibody

Background

The production of biopharmaceuticals for human use began in 1982 with recombinant insulin, and the development of new biopharmaceuticals has grew almost exponentially ever since. Over the past two decades, Chinese hamster ovary (CHO) cells have become the standard mammalian host cell line, with the expression and production of nearly 70% of all biopharmaceuticals [1,2]. CHO cells provide efficient post-translational modifications, which allow the production of recombinant proteins with glycoforms that are both compatible with, and bioactive in humans [1]. CHO cells can also be easily manipulated genetically, which has become of great importance more recently [3]. These two characteristics are especially important in the production of biosimilars, where achieving the correct extent of similarity to the reference molecule is a great challenge. The nucleotide sequence of the gene that encodes amino-acid sequence of the desired protein is the same as for the reference molecule. In contrast, post-translational modifications that are the consequence of metabolic pathways can differ between host cell lines, clones, cultivation conditions, medium composition, specific productivity and physiologic state of the cell [4]. Consequently, these need to be fine-tuned during development. In addition to posttranslational modifications, different charge variants can result in heterogeneity in the production of monoclonal antibodies (mAbs). These modifications potentially result in changes to the bioactivity, bioavailability or immunogenicity of the mAbs, and they therefore need to be additionally characterised to ensure the safety, quality and efficacy of the product. Among these, C-terminal

* Correspondence: mihaela-1.skulj@sandoz.com
[1]Sandoz Biopharmaceuticals, Mengeš, Lek Pharmacetucals d.d, Kolodvorska 27, 1234 Mengeš, Slovenia
Full list of author information is available at the end of the article

amidated structures on the heavy chains of mAbs have recently attracted particular attention [5-8].

C-terminal α-amidation is catalysed by peptidylglycine α-amidating monooxygenase (PAM), and this protein modification is often required to confer full biological activity to peptide hormones [6,9-11]. Amidation is catalysed starting from a glycine-extended prohormone, by two sequential actions of two enzymes, peptidylglycine α-hydroxylating monooxygenase and peptydilamidoglycolat lyase. In mammals, both of these enzymes are derived from a single gene, which gives rise to the bi-functional PAM protein [12]. PAM thus catalyses the conversion of peptidylglycine substrates into α-amidated products in a two-step reaction, and it is the only enzyme known to catalyse the formation of amidated peptides [13].

Recently, large proteins like immunoglobulins have been reported to be substrates for PAM [6,7], and the expression of PAM in CHO cells was reported previously [14]. Tsubaki et al. reported that C-terminal α-amidation was detected in 8 of 12 recombinant mAbs, with ratios from 0.3% to 25.9% [6]. During our studies, we have also observed the presence of C-terminal amidated species in recombinant mAbs produced in CHO cells. Prolinamide was detected in up to 14% of all mAb molecules, which was too high to accomplish the desired similarity to the reference molecule.

It was previously shown that the level of mAb amidation in CHO cells can be affected via bioprocesses and medium optimisation, with the addition of copper to the culture medium [7]. On the other hand, metabolic engineering is becoming a powerful tool to manipulate expression hosts for improved product quality, and the introduction of a PAM knocked-down cell line that can produce mAbs with desired comparability to a reference molecule would be a state-of-the-art solution.

In the present study, the expression of PAM, and consequently the C-terminal amidation of recombinant mAbs, was reduced by two approaches: gene manipulation using RNA interference (RNAi) and zinc finger nucleases (ZFN). RNAi has been efficiently used for down-regulation of desired genes, and it can be performed using chemically synthesised small-interfering (si)RNA molecules, or via the endogenous expression of short-hairpin (sh)RNA molecules that are encoded by plasmid vectors [15,16]. While siRNAs can provide transient knock-down of expression of a target gene, shRNA vectors can induce long-term expression of RNAi silencing in a target cell [15]. The ease and rapidity of these RNAi approaches have made them the method of choice for initial target validation.

However, using RNAi, complete elimination of expression of the target gene is rarely obtained. To achieve permanent gene knock-out, ZFNs can offer a distinct advantage. Moreover, the growing number of reports using ZFNs across different species suggests that ZFN-mediated gene disruption is a robust and general method for targeted gene knock-out. As this approach results in an alteration of the genome itself, these mutations are transmitted stably through all of the subsequent generations of the cell line, as is the case with conventional gene targeting [3,17]. Rapid gene knock-out can be achieved by using a ZFN approach to create a double-stranded break in the locus of interest, which allows it to be repaired by non-homologous end-joining; this ligates the two broken ends, with the occasional loss of genetic information. ZFNs can therefore be used to introduce small deletions at the site of such a break, an outcome that can be exploited to disrupt a target gene [18,19]. The high frequencies of gene disruption strongly support the likelihood of achieving a desired genotype, in which each copy of the target gene is functionally knocked out in 1% to 50% of all cells [17,18].

Results and discussion

siRNA and shRNA experiments

C-terminal amidation is a common posttranslational modification that is seen on therapeutic mAbs [6], and which has also been observed during recombinant mAb development (data not shown). The significant variations between clones during clone selection, and the relatively high content of amidated species in comparison to reference molecules, directed us towards the development of a cell line with reduced production of C-terminal amidated species.

RNAi has already been efficiently used for the improvement of cellular productivity and the quality of recombinant proteins produced in CHO cells. RNAi had thus been used for silencing apoptosis-associated gene expression [20-23], glycosylation-associated gene expression [24-26], and gene expression of lactate dehydrogenase [27] and dihydrofolate reductase [28,29].

Fifteen different siRNAs were designed on the basis of the *C. griseus PAM* nucleotide sequence and tested on the CHO *Der2* parental cell line. Up to an 8-fold decrease was observed in PAM mRNA expression levels using siRNAs from Invitrogen, and up to a 5-fold decrease using siRNAs from Ambion (Figure 1). On the basis of these data, and due to shRNA design limitations, siRNAs si5 and si6 (Ambion) were selected for the design of shRNAs, to obtain long-term silencing of *PAM*. siRNA knock-down of the target lasts for 3 to 5 days, and therefore to induce long-term expression of RNAi silencing in the target cells, an shRNA vector has to be used [15]. The shRNA silencing effect was tested on two different CHO parental cell lines, CHO *Der2* and CHO *Der3*, and on two mAb-expressing clones derived from the CHO *Der3* cell line, clone K62 with high (14%), and

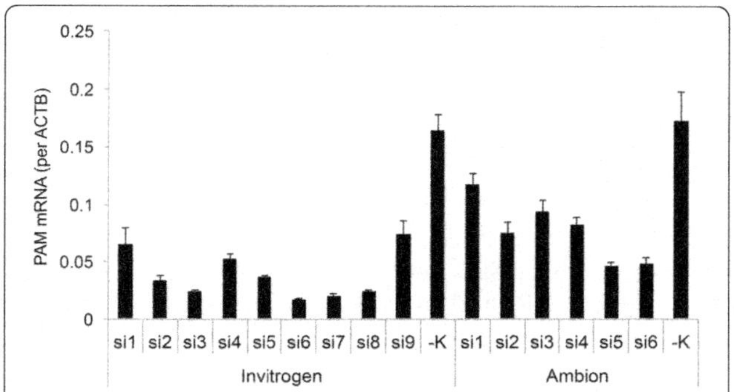

Figure 1 Silencing of *PAM* using siRNA. *PAM* mRNA expression level after transfections using the siRNAs (Invitrogen, Ambion, as indicated) in the CHO *Der2* parental cell line. In comparison to the negative control (−K), there was up to 8-fold reduction in the mRNA expression levels using the si6 Invitrogen siRNA, and up to 5-fold reductions using the si5 and si6 Ambion siRNAs. Overall, better silencing effects were obtained using the Invitrogen siRNAs. The mRNA expression levels were determined using qPCR (calculated per *ACTB* housekeeping gene), and the data are means ± standard deviations of the two biological replicates.

clone K25 with low (4%), prolinamide contents in the mAb that was produced (Figure 2). After shRNA transfection and antibiotic selection, all of the generated pools were analysed by cation-exchange chromatography (CEX), for evaluation of the prolinamide content (Figure 2). The shRNA designed on the basis of the si6 siRNA was shown to have the most potent silencing effect on all of the transfected cell lines (Figure 2).

The data presented in Figure 2 show the correlation between *PAM* mRNA expression levels and C-terminal amidation of the recombinant mAb. Up to a 4-fold decrease in mRNA and a 3-fold decrease in prolinamide content were observed. As can be seen from Figure 2, the prolinamide content for clone K62 was decreased to 4.6%, which represents a 3-fold decrease, and for clone K25, where the initial starting point was 3.5% prolinamide content, only a minimal reduction was obtained. Nevertheless, there is an interesting observation here that should be considered. No matter which clone is considered, as one with a previously high (14%) or low (4%) prolinamide content, the reduction in the prolinamide content after shRNA knock-down never decreased below 4%, which is the same as the level of the reference molecule.

ZFN experiments

The shRNA experiments gave very promising results, as they yielded PAM levels that were comparable to the reference molecule. However, possible toxicity effects on long-term expression and the additional metabolic load on the cells due to the overexpression of these factors during times of stress might also influence cell performance. In addition, shRNA-mediated knock-down relies on the constant expression of repressor molecules, which can be unstable in the knocked-down cells in the long term [17]. RNAi instability was reported by Lim et al. in the silencing of the *BAX* and *BAK* genes, where 1% of the screened clones was silenced for both genes,

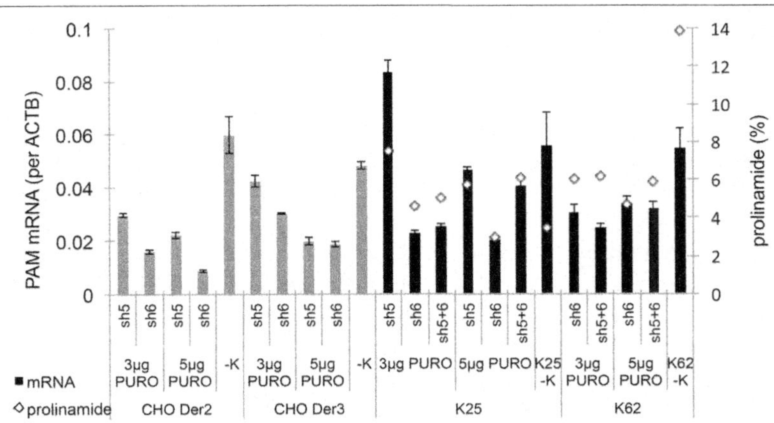

Figure 2 Silencing of *PAM* using shRNA. The mRNA expression levels are shown after the transfections using the Ambion sh5 and sh6 shRNAs on the CHO *Der2* and CHO *Der3* parental cell lines and on the K25 and K62 clones derived from the CHO *Der3* parental cell line. The data obtained for the parental cell lines are in grey. In comparison to the negative controls (CHO *Der2* -K, and CHO *Der3* -K), the highest silencing effects were achieved with shRNA sh6 and 5 µg/ml antibiotic selection (5 µg puromycin [PURO]). Up to 3.7-fold reduction in the mRNA expression levels was seen for the CHO *Der2* cell line, and up to 2.6-fold reduction for the CHO *Der3* cell line. The data for the mAb-expressing clones K25 and K62 are in black. In this case, the *PAM* expression level and the prolinamide content (%) are presented. In comparison to the negative controls (K25 -K, and K62 -K), the different antibiotic selections did not show any differences in mRNA expression levels and prolinamide content. Up to 3.5-fold reduction in mRNA expression levels was observed for K25, and up to 2.2-fold reduction in K62. Prolinamide was decreased from 3.5% to 3% for K25, and from 14% to 4.6% for K62, which represents a 3-fold decrease. The mRNA expression levels for parental cell lines and the K25 and K62 clones were determined using qPCR (calculated per *ACTB* housekeeping gene), and the data are means ± standard deviations of two biological replicates.

but there was instability of the generated knock-down clones [20]. This is an undesirable feature during the mAb production process, as it can cause heterogeneity of the product [20]. To avoid such problems, rapid and permanent gene knock-out using ZFNs can offer distinct advantages. Moreover, the growing number of reports using ZFNs across different species have suggested that ZFN-mediated gene disruption is a robust and general method for targeted gene knock-out. ZFN-mediated gene knock-out requires only transient expression of the ZFNs, and it can result in a permanent genetic mutation that is stably transmitted through all of the subsequent generations of the cell line [17].

The first reported example of the use of engineered ZFNs to disrupt an endogenous locus in a mammalian cell was a knock-out of the dihydrofolate reductase gene in CHO cells. The observed bi-allelic mutation rate was 2% to 3%. In comparison to traditional methods, this frequency of gene disruption is very high, and it increases the possibility of achieving the desired genotype in which each copy of the target gene is functionally knocked out [17].

ZFNs have also already been successfully used for the generation of many different knocked-out CHO cell lines. They have been reported as used to create triple knock-out CHO cells, with the disruption of the two selectable marker genes of glutamine synthetase and dihydrofolate reductase, as well as the gene encoding α-1,6-fucosyltransferase (FUT8) [30]. *BAK* and *BAX* deletion was also achieved, to produce apoptosis-resistant CHO cells [31]. Additionally a *FUT8* knock-out CHO cell line has been reported [32,33].

Thus, ZFNs were used to create a cell line with stably knocked-down *PAM* expression, so as to reduce C-terminal amidated species on the target mAb. These ZFNs were designed on the basis of a sequence that had been shown to generate silencing effects using RNAi. Initially, ZFN plasmid DNA was used for the transfections of the parental CHO *Der2* and CHO *Der3* cell lines. The knock-out effect was evaluated by determination of the *PAM* gene copy number using qPCR. This method was evaluated using the Cell-I Nuclease Mismatch assay (data not shown), which is usually used to evaluate knock-out clones. The qPCR method has been shown to be less laborious and time consuming, and therefore it enables efficient screening of the high numbers of clones generated.

In all, 85 clones were generated from the CHO *Der2* cell line, and 109 clones from the CHO *Der3* cell line. In contrast to the CHO *Der2* cell line, which is diploid, the CHO *Der3* cell line was shown by karyotyping to be a triploid (data not shown), and a knock-out was therefore more difficult to achieve. A partial reduction of the *PAM* copies in the CHO *Der2* and CHO *Der3* cell lines was achieved (Figure 3). Reported knock-out efficiencies for single copy genes have been higher than 1% [17,30,34], and have reached 5% [33]. Thus, to detect a single clone with a total knock-out that would be reflected in a mAb without functional *PAM*, around 100 clones would need to be generated, and in the case of the triploid CHO *Der3* cell line, this would be higher.

The *PAM* mRNA expression was evaluated for the 10 best clones by qPCR, and high correlation was observed with the copy number determination at the genomic DNA level (data not shown).

To determine the mRNA expression levels, the *PAM* mRNA was calculated per housekeeping gene. The choice of a housekeeping gene can often be quite challenging, as expression is usually not the same across different CHO cell lines and clones, and can also vary according to the physiological state of the cells. To avoid these effects, four different housekeeping genes were tested in the present study: *ACTB, GAPDH, G6PD* and *EF1* (Table 1). All four of these housekeeping genes were shown to be suitable for the calculation of the mRNA expression levels (data not shown), with the data below are based on the use of *ACTB* as the housekeeping gene.

The three best clones were transfected with a plasmid containing the mAb expression cassette. After 4 weeks of antibiotic selection and 14 days of fed-batch cultivation, the mAbs produced from these three clones were evaluated for the presence of prolinamide, using CEX. The data from the *PAM* gene copy numbers, mRNA expression levels, and prolinamide content are presented in Figure 4. The prolinamide levels for both of the CHO *Der2* and CHO *Der3* cell lines after evaluation using CEX were determined to have decreased to 6%. As observed with the shRNA (see above), there was correlation between copy number, mRNA expression, and prolinamide content of the recombinant mAb (Figure 4).

Transfections using ZFN plasmid DNA result in its random integration into the genome of CHO cells and continuous expression of the ZFN. Genome integration is not desired, due to the continuous ZFN activity, and thus potential off-target effects. The mechanism of ZFN target recognition has been reported to be highly specific, with usually an 18-bp site recognized and the dimerisation of two ZFNs needed to cause a double-stranded break [18,35]. Despite this, some off-target effects have also been observed. Off-target effects of 5.4% were reported for the disruption of the CCR5 HIV receptor by Klug et al. [19].

To avoid this, the transfections were performed using ZFN mRNA, for transient ZFN expression. The ZFN mRNA was prepared in-house using *in-vitro* transcription from the two ZFN plasmid DNAs. The cells were then transfected with the two ZFN mRNAs; i.e., ZFN1 and ZFN2. Various amounts for the combination of

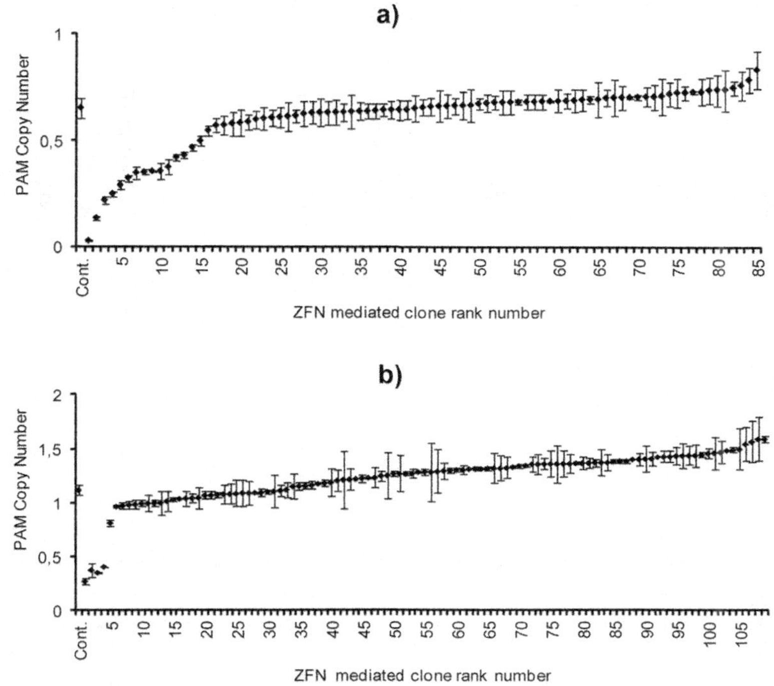

Figure 3 ***PAM*** **gene copy numbers for the clones derived from the CHO** ***Der2*** **and CHO** ***Der3*** **cell lines transfected using the ZFN plasmid DNA.** The PAM gene copy numbers for the 85 ZFN clones derived from the CHO *Der2* parental cells **(a)** and the 109 ZFN clones derived from the CHO *Der3* parental cells **(b)**, transfected using the ZFN plasmid DNA. The copy numbers were determined using qPCR, and the data are means ± standard deviations of two technical replicates.

these two mRNAs were tested: total mRNAs of 30 μg to 60 μg per transfection. The 50 μg mRNAs transfection (25 μg ZFN1 mRNA, 25 μg ZFN2 mRNA) was the most promising combination. The concentrations of the mRNAs used for the transfections correlated with the silencing effects, although the higher mRNA concentrations decreased the cell viability to <80% (data not shown). The pools were evaluated by qPCR on day 4 following transfection. A 50% knock-out was achieved for the generated pools, and the cell viability was >80%, which is needed for seeding into the semi-solid medium for clone generation using ClonePix.

These mRNA transfections were used only with the CHO *Der2* cell line. Here, 154 clones were evaluated and two were determined to be completely knocked-out at the genomic DNA level by qPCR (Figure 5), which represented a 1% knock-out efficiency. The 10 best clones were again tested for *PAM* mRNA expression (data not shown), and afterwards the best three of these 10 clones were transfected with a plasmid containing the mAb expression cassette, in duplicates, and evaluated by CEX (using the same procedure as described for the plasmid DNA transfections). These data are presented in Figure 6.

The C-terminal amidated structures in all of these tested clones decreased to a limit of 6%, which was the same as observed using ZFN plasmid DNA. mRNA expression and C-terminal amidation were observed for all of these tested clones, even though this was not expected for clones K62 and K68.

Table 1 Nucleotide sequences of the qPCR primers and probes used in the present study

	Forward primer	Reverse primer	Probe
PAM	GGCCGGATCCAATGTTTCAGAA	TCCCAAATGACGCATGTTTAATCTCT	CTGACACCAAAGAATTT
ACTB	AGCCACGCTCGGTCAG	CATCCTGCGTCTGGACCT	CCGGGACCTGACAGACT
GAPDH	CTGATTTTTCTTGCGTCGAGTTT	GAGTTGTGTTTGTGGACGAAGTAC	TCCGGTAAGACCTTTCG
EF1	CCTGGAATGGTGGTTACCTTTG	CATGGTGCATTTCAACAGACTTTACT	CCAGTCAACGTTACAACAGA
G6PD	ACAACATTGCCTGTGTGATCCT	CCCAAATTCATCAAAGTAGCCC	CCACGACCCTCAGTACCA

PAM, peptidylglycine amidating monooxygenase; ACTB, β-actin; GAPDH, glyceraldehyde-3-phosphate dehydrogenase; EF1, elongation factor 1; G6PD, Glucose-6-phosphate dehydrogenase.

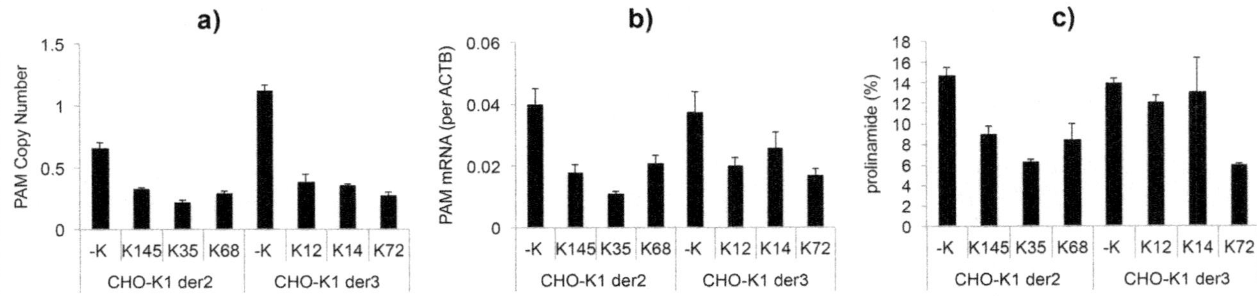

Figure 4 ***PAM*** **gene copy numbers, mRNA expression and C-terminal amidation for clones transfected using ZFN pDNA.** The *PAM* gene copy numbers **(a)**, *PAM* mRNA expression levels (**b**; calculated per *ACTB* housekeeping gene), and prolinamide content on the recombinant mAb **(c)** for the three best ZFN knock-out clones derived from the CHO *Der2* and CHO *Der3* parental cell lines, where ZFN pDNA was used for the transfection. Prolinamide was reduced to 6%. The copy numbers and mRNA expression levels were determined using qPCR, and the data are means ± standard deviations of two technical replicates. The prolinamide content was determined using CEX, and these data are means ± standard deviations of two biological replicates.

Down-regulation of *PAM* did not have any influence on cell growth, in comparison to the non-silenced controls. All of the cells showed similar cell growth, which did not differ from that expected for CHO cells. Additionally, there was no correlation between cell productivity and C-terminal amidation of the produced mAb (data not shown).

The silencing effects that were obtained using the RNAi approach showed more promising results in terms of prolinamide reduction on the recombinant mAb. In this case, prolinamide was reduced to 3%. As indicated above, siRNA effects are only transient [15] and the shRNA approach is also questionable, due to long-term stability problems and the additional metabolic burden on the cells. Using the ZFN approach, the decrease in prolinamide was not as effective as with shRNA, as it was reduced to 6%. On the other hand, this ZFN approach has the advantage of producing a stable genetic modification, which is very important for the production scale.

After transfection with the mAb-expressing plasmid, each of these two generated ZFN knocked-down cell lines (CHO *Der2* and CHO *Der3* cells) represented a mixed cell population with a mean prolinamide content of 6%. In comparison to the negative controls, which generated 15% (CHO *Der2*) and 14% (CHO *Der3*) prolinamide content, this represents a decrease in the prolinamide content of >50%. These pools with mean prolinamide content of 6% thus represent good starting points for the isolation of candidate clones that express different amounts of C-terminal amidated species, thus providing the choice for the one that is the most comparable to the reference molecule.

The higher efficiency of the knock-down achieved using shRNA in comparison to ZFN might be because the shRNA targets the degradation of the *PAM* mRNA, which will reduce its expression overall. For the ZFNs, the knock-out effects are at the level of the genomic DNA, and if total knock-out is not achieved in all of the copies, the expression of a functional mRNA will still take place, and thus the translation into the functional PAM protein will not be fully blocked. Disruption of the target gene can occur by causing a shift in the reading

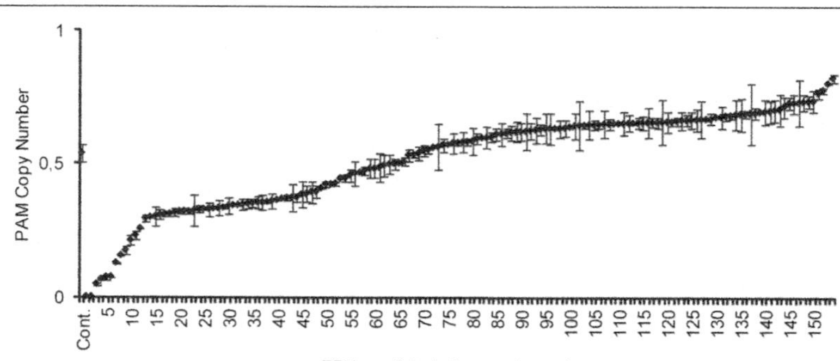

Figure 5 ***PAM*** **gene copy numbers for 154 ZFN clones from CHO** ***Der2*** **parental cell line transfected using ZFN mRNA.** The two clones (K62 and K68) identified as knocked-out for all *PAM* gene copies at the genomic DNA level. The copy number was determined by qPCR, and the data are means ± standard deviations of two technical replicates.

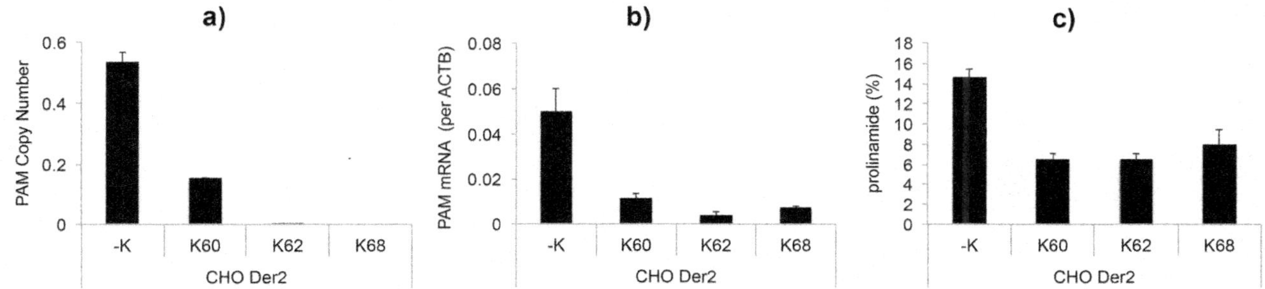

Figure 6 ***PAM*** **gene copy numbers, mRNA expression, and C-terminal amidation for the clones transfected using ZFN mRNA.** The *PAM* gene copy numbers **(a)**, *PAM* mRNA expression levels (**b**; calculated per *ACTB* housekeeping gene), and prolinamide contents of the recombinant mAb determined by CEX **(c)**, for the three best ZFN knock-out clones derived from the CHO *Der2* parental cell line, where ZFN mRNA was used for the transfection. The two clones identified (K62 and K68) were knocked-out for all *PAM* gene copies at the genomic DNA level, with mRNA expression and a prolinamide content of 6% still observed. The copy numbers and the mRNA expression levels were determined using qPCR, and the data are means ± standard deviations of two technical replicates. Prolinamide content was determined by CEX, results are presented as an average of two biological replicates.

frame, the generation of a premature termination codon, or the deletion of critical amino-acid residues. However, the result of non-homologous end joining repair might be only an amino-acid point mutation or a deletion of the protein molecule, which might result in a partially or fully functional enzyme.

However, the goal of the present study was not to generate a completely knocked-out CHO cell line, but to decrease the content of the C-terminal amidated structures to a level that is sufficiently comparable to the reference molecule, and this was successfully achieved using both approaches.

Conclusions

During development, C-terminal amidated species are detected on recombinant mAbs, an observation that was also recently reported in other studies [6]. C-terminal amidation can be reduced by development and optimisation of the grwth medium, and use of cofactors such as copper [7], although these approaches are project specific and time consuming. Due to its extensive development, metabolic engineering has provided a powerful tool to diminish the expression of the *PAM* gene through gene alterations using RNAi and ZFNs. In the present study, both of these approaches were tested. The RNAi was conducted for the initial target validation, for the later ZFN knock-out. Using the shRNA approach, decreases in C-terminal amidation were achieved for two mAb-producing clones: the K25 with low prolinamide content (4%), and the K62 clone with high (14%) prolinamide content, which were decreased to 3.5% and 5%, respectively. The prolinamide levels of the reference molecule were 4%, and therefore a comparability level was successfully achieved. In this case, the gene silencing was performed on two clones that were already expressing the desired mAb.

For recombinant mAb production, a parental cell line with stably reduced *PAM* expression is desired. To achieve this, genome alterations using ZFNs are the more appropriate tool, as breaks are created at the genomic DNA level that are stably transferred through the subsequent generations of the cell line [3,17]. Two CHO cell lines were generated, as CHO *Der2* and CHO *Der3* cells, and they were evaluated for *PAM* at the genomic DNA and mRNA expression levels. After the selection of the three most promising clones, these were transfected with a mAb-producing plasmid and evaluated for the C-terminal amidated species. The C-terminal amidation was decreased to 6% in both of these cell lines, which represents a 50% lower content in comparison to the ZFN non-treated controls. The two generated cell lines now represent two pools from which candidate clones with the highest comparability to the reference molecule can be selected.

Thus, this use of genetically manipulated cell lines now gives us the opportunity to go even further towards providing high-quality and safe therapeutics.

Methods

Design of siRNAs, shRNAs and ZFNs

The Chinese hamster *(Cricetulus griseus) PAM* nucleotide sequence was derived from a public database (Table 2). The siRNAs, shRNAs and ZFNs were constructed on the basis of this sequence.

The online design tools of Invitrogen and Ambion were used to design the siRNA sequences (nine according to Invitrogen, six according to Ambion; Table 3). After the siRNA evaluation, shRNAs were designed using the Ambion online design tool. Two complementary oligonucleotides were synthesised for each shRNA by Methabion (Table 4), and then these were annealed in-house, to generate the double-stranded oligonucleotides.

Table 2 The *Cricetulus griseus PAM* gene sequence used for the RNAi and ZFN construction

***PAM* gene nucleotide sequence**	GGGAGTGCTCCTAAGCCAGGCCAGTTCAGTGTTCCTCACAGTTTGGCCCTTGTGCCTCATTTGGACCAGTTGTGTGTGGCAGACAGGGAA AATGGCCGGATCCAATGTTTCAGAACTGACACCAAAGAATTTGTGAGAGAGATTAAACATGCGTCATTTGGGAGAAATGTATTCGCAATT TCATATATATCAGGTTTGCTCTTTGCAGTAAATGGGAAGCCTTACTTTGGAGACCATGAACCTGTGCAAGGCTTTGTGATGAACTTTTCCAG TGGGGAAATTATAGATGTCTTCAAGCCAGTACGCAAGCACTTTGACATGCCTCACGATGTGGTTGCCTCTGACGATGGGAATGTGTACATT GGAGACGCACACACGAACACGGTGTGGAAGTTCACCCTGACTGAAAAAATGGAGCATCGATCGGTTAAAAAGGCAGGCATTGAGGCTC AGGAAATCAAAGAAACCGAGGCAGTTGTTGAATCCAAAATGGAGAACAAACCCACCTCCTCAGAATTGCAGAAGATGCAAGAGAAACA GAAACTGATCAAAGAGCCAGGTTCGGGAGTGCCCGTGGTTCTCATTACAACCCTTCTGGTTATTCCTGTGGTTGTCCTGCTGGCCATTGTC ATGTTTATTCGGTGGAAAAAATCAAGGGCCTTTGGAGGAAAA

Subsequently, these annealed oligonucleotides were cloned into the *pSilencer* 2.1-U6 puro vector (Ambion). DNA sequencing was performed to verify the sequences of the oligonucleotide inserts.

For the ZFN experiments, two ZFN plasmids that targeted the *PAM* gene (pZFN PAM1, pZFN PAM2) were designed by Sigma (CompoZr^{TM} Custom Zinc Finger Nuclease, PN. SAFCZFN-1kt). These two ZFN plasmids were later used as templates for the preparation of PAM1 and PAM2 mRNA, by *in-vitro* transcription using MessageMax T7 ARCA-Capped Transcription Kit (Cellscript). The transcripts were poly-A tailed using A-Plus Poly(A) Polymerase Tailing Kit (Cellscript), and purified using RNeasy Micro Kit (Qiagen). The RNA concentrations were determined by the NanoDrop, and the transcript sizes were evaluated using a Bioanalyser.

Cell culture and transfection

siRNA and shRNA experiments

Two different CHO parental cell lines (CHO *Der2* and CHO *Der3*) and two different mAb-producing clones that were derived from the CHO *Der3* cell line (clone K25 and clone K62) were used in this study. Clones K25 and K62 were prepared by nucleofection of the parental CHO *Der3* cell line using a plasmid vector containing the mAb expression cassette. The cells were cultivated at 37°C, under 10% CO_2, and with agitation at 90–110 rpm, in animal-component-free growth medium. The C-terminal amidation of the mAb in clone K25 was determined as 4%, and in clone K62, as 14%, using CEX. It was shown that 4% of all of the reference molecules contained C-terminal amidated structures.

Table 3 Nucleotide sequences of the siRNAs used in the present study

Name	siRNA oligonucleotide	Supplier
siRNA_si1	CAGUUGUGUGUGGCAGACAGGGAAA	Invitrogen
siRNA_si2	CGGAUCCAAUGUUUCAGAACUGACA	Invitrogen
siRNA_si3	CCAAUGUUUCAGAACUGACACCAAA	Invitrogen
siRNA_si4	GAGAGAGAUUAAACAUGCGUCAUUU	Invitrogen
siRNA_si5	CAUGCGUCAUUUGGGAGAAAUGUAU	Invitrogen
siRNA_si6	UGGGAGAAAUGUAUUCGCAAUUUCA	Invitrogen
siRNA_si7	GGGAGAAAUGUAUUCGCAAUUUCAU	Invitrogen
siRNA_si8	CACACGAACACGGUGUGGAAGUUCA	Invitrogen
siRNA_si9	CAAAGAAACCGAGGCAGUUGUUGAA	Invitrogen
siRNA_si1	CAGUAAAUGGGAAGCCUUATT	Ambion
siRNA_si2	AGGCAGUUGUUGAAUCCAATT	Ambion
siRNA_si3	AACAGAAACUGAUCAAAGATT	Ambion
siRNA_si4	CAAGAGAAACAGAAACUGATT	Ambion
siRNA_si5	GAACUGACACCAAAGAAUUTT	Ambion
siRNA_si6	UUUCAGAACUGACACCAAATT	Ambion

For the siRNA experiments, CHO *Der2* cells were transfected with siRNAs, in duplicate, using nucleofection, and cultivated for 4 days. On day 4, the cell pellets were collected for analysis of gene expression by qPCR.

For the shRNA experiments, all four cell lines (CHO *Der2*, CHO *Der3*, clone K25, and clone K62) were transfected with shRNAs, in duplicate, using nucleofection. Antibiotic selection of all of the transfected pools was performed using 3 μg/ml puromycin. The puromycin was added to the cell culture 2 days after the transfection, when the cell viability exceeded 60%. The cells were split on a 2-2-3-day schedule at 2-3E5 cells per ml, in the appropriate pre-warmed medium to maintain exponential cell growth. After reaching the appropriate cell density and viability, the cell pellets were collected for gene expression analysis by qPCR, and a 10-day simple batch containing 3 μg/ml puromycin was inoculated (after 10 days, the supernatants were collected, purified on protein A columns, and evaluated for prolinamide content using CEX). The cells were further cultivated in medium containing 5 μg/ml puromycin. After reaching the appropriate cell density and viability, this procedure was repeated.

ZFN experiments

For the ZFN experiments, the CHO *Der2* and CHO *Der3* parental cell lines were transfected using 3 μg of each ZFN plasmid (PAM1, PAM2), by nucleofection. After 4 days, the cells were seeded into the semi-solid medium, and on day 10 the colonies were transferred into 96-well plates by ClonePix FL. After colony formation in the 96-well plates, the individual colonies were transferred into 24-well plates, and later into 6-well plates. Finally, the clones were transferred into 125-ml shaking flasks, and after achieving the appropriate cell density and viability, the cell pellets were collected for

Table 4 Nucleotide sequences of the shRNAs used in the present study

Name	shRNA top strand oligonucleotide	shRNA bottom strand oligonucleotide
shRNA_sh5	GATCCGAACTGACACCAAAGAATTCTCAAGAGAAAT TCTTTGGTGTCAGTTCTGTTTTTTGGAAA	AGCTTTTCCAAAAAACAGAACTGACACCAAAGAATTTCTCTT GAGAATTCTTTGGTGTCAGTTCG
shRNA_sh6	GATCCGTTTCAGAACTGACACCAAACTCAAGAGATTT GGTGTCAGTTCTGAAACATTTTTTGGAAA	AGCTTTTCCAAAAAATGTTTCAGAACTGACACCAAATCTCTT GAGTTTGGTGTCAGTTCTGAAACG

the determination of gene copy number and gene expression by qPCR. The same procedure was repeated on the CHO *Der2* cell line using 25 μg of each ZFN mRNA (PAM1, PAM2), which was prepared in-house from the PAM1 and PAM2 ZFN plasmids.

After the selection of the three clones with the highest knock-down efficiency generated using ZFN plasmid DNA and mRNA, the transfections were performed using 3 μg mAb-expressing vector, in duplicate. After 4 weeks of antibiotic selection with geneticin, a 14-day fed-batch was inoculated. After 14 days, the supernatants were collected and purified using protein A columns. The prolinamide contents were evaluated using CEX.

DNA/RNA purification and qPCR template preparation

The genomic DNA and the total RNA were isolated from pellets containing 5E6 cells on an automated QIAcube workstation. The genomic DNA was isolated using Blood & Cell Culture DNA Mini Kit (Qiagen), and the total RNA using RNeasy Mini Kit (Qiagen). After isolation, the nucleic acid concentrations were measured using the NanoDrop. After the determination of the total RNA concentrations, DNase I (Ambion) was added to 5 μg total RNA and incubated according to the manufacturer instructions. After this DNase treatment, the RNA was transcribed into cDNA using SuperScript VILO Kit (Invitrogen).

The genomic DNA and cDNA were used later as templates for the qPCR. The cDNA was used for evaluation of the siRNA and shRNA silencing effects, while the genomic DNA was used for evaluation of the knock-out effects of clones produced using the ZFNs.

qPCR experiments

siRNA and shRNA experiments

The silencing effects of the siRNAs and shRNAs were examined by qPCR, using TaqMan chemistry and absolute quantification. The standard curve was constructed from CHO *Der1* genomic DNA, which was amplified using the *PAM* and *ACTB* primers (Table 1). The *PAM* and *ACTB* mRNA copy numbers for each sample were extrapolated from the standard curve, and the *PAM* mRNA expression levels were expressed as the number of *PAM* mRNA transcripts per reference *ACTB* gene. The standard curves were constructed from five dilution points, each as three parallel determinations. The *PAM* mRNA expression levels were determined as four dilution points, each as two parallel determinations for each sample.

ZFN experiments

The efficiencies of the ZFNs were determined by qPCR, using TaqMan chemistry and absolute quantification. The TaqMan probe was designed on the ZFN cutting site, and therefore no amplification of the *PAM* gene was expected in completely knocked-out clones. The standard curve was constructed from the CHO *Der1* genomic DNA amplified using the *GLUC* and *PAM* primers (Table 5). The copy numbers for both of these genes were extrapolated from the standard curve for each sample, and the ratio between them was calculated. *GLUC* is a single-copy gene in CHO cells, and was used as the reference gene for the *PAM* gene copy number determination. Standard curves were constructed from five dilution points, each as three parallel determinations. The *PAM* copy numbers were determined as two dilution points, each as three parallel determinations for each sample.

After the selection of the 10 clones with the highest knock-out efficiency, the *PAM* expression was determined as described above, using four different housekeeping genes: *GAPDH*, *ACTB*, *G6PD* and *EF1* (Table 1).

Cation-exchange chromatography

The mAb concentrations were determined using the ForteBio system. The mAbs were purified using protein A columns and analysed using CEX, with an analytical HPLC chromatographic system. Using this method, Lys and prolinamide were eluted in the same peak. The amount of prolinamide was then determined by C-terminus treatment with carboxypeptidase. Following the further analysis using the same CEX system, the

Table 5 Nucleotide sequences of the qPCR primers and probes used in the present study

	Forward primer	Reverse primer	Probe
GLUC	ATTGCCAAACGCCACGAT	CCAAGCAATGAATTCCTTTGC	CTGAAGGGACCTTTACCA
PAM	GCCCCAGCTCGGACATG	CCTTGCAGGGCGAGCA	CCGGCTGCTGCTGCT

GLUC, glucagon; PAM, peptidylglycine amidating monooxygenase.

remaining peak defined the prolinamide levels. The prolinamide levels are given as percentages of the C-terminal amidated species of the total mAb content.

After carboxipeptidase treatment, in addition to prolinamide, the basic peak can include multiple species, like N-terminal cyclised species, succinimide species, hybrid glycan-containing species, and others. Therefore the CEX analysis was always performed in parallel with the non-silenced and non-knockout controls, represented by samples from the same cell lines that had undergone blank transfections and were grown under the same conditions. The effects of RNAi and the ZFNs on the treated samples are highly specific, and the probability of other post-translational modifications other than reductions in prolinamide is very low. The reductions in the C-terminal amidation were determined relative to the controls, and therefore the reduction in the basic peak defined the reduction in C-terminal amidation.

Competing interests

This study was performed at Sandoz Biopharmaceuticals, Mengeš, a Member of the Novartis group, which also financed the article-processing charges. The patent application entitled "Reduction of formation of amidated amino acids in cell lines for protein expression" (No. 12164264.9 - 1212) was filed in 2012. The authors confirm that there are no known conflicts of interest associated with this publication, and there has been no significant financial support for this study that might have influenced its outcome. The authors confirm that they have given due consideration to the protection of the intellectual property associated with this study, and that there are no impediments to its publication, including the timing of the publication, with respect to the intellectual property. In so doing, the authors confirm that they have followed the regulations of their institutions concerning intellectual property. The authors also confirm that the manuscript has been read and approved by all of the named authors, and that there are no other persons who satisfy the criteria for authorship who are not listed. All of the authors declare they have no competing interests of financial or non-financial natures.

Authors' contributions

MS and DP carried out all of the experiments and MS drafted the manuscript. DG participated in the planning the design of the study. DG, MK and RZ critically revised the manuscript and all authors approved the final version for publication.

Acknowledgments

The authors express their sincere thanks to Holger Laux for help with the siRNA and shRNA design, and to Uroš Jamnikar for helpful information and discussions. They would also like to thank Petja Šušteršič for handling the cells during the RNAi experiments, and Irena Filipovič for performing part of the qPCR analysis.

Author details

[1]Sandoz Biopharmaceuticals, Mengeš, Lek Pharmacetucals d.d, Kolodvorska 27, 1234 Mengeš, Slovenia. [2]Celica Biomedical Centre, Tehnološki Park 24, 1000 Ljubljana, Slovenia. [3]Laboratory of Neuroendocrinology – Molecular Cell Physiology, Institute of Pathophysiology, Faculty of Medicine, University of Ljubljana, Zaloška 4, 1000 Ljubljana, Slovenia. [4]Biotechnical Faculty, University of Ljubljana, Večna Pot 111, 1000 Ljubljana, Slovenia.

References

1. Kim JY, Kim YG, Lee GM: **CHO cells in biotechnology for production of recombinant proteins: current state and further potential.** *Appl Microbiol Biotechnol* 2012, **93:**917–930.
2. Butler M: **Animal cell cultures: recent achievements and perspectives in the production of biopharmaceuticals.** *Appl Microbiol Biotechnol* 2005, **68:**283–291.
3. Warner TG: **Enhancing therapeutic glycoprotein production in Chinese hamster ovary cells by metabolic engineering endogenous gene control with antisense DNA and gene targeting.** *Glycobiology* 1999, **9:**841–850.
4. Werner RG, Kopp K, Schlueter M: **Glycosylation of therapeutic proteins in different production systems.** *Acta Paediatr Suppl* 2007, **96:**17–22.
5. Harris RJ: **Processing of C-terminal lysine and arginine residues of proteins isolated from mammalian cell culture.** *J Chromatogr A* 1995, **705:**129–134.
6. Tsubaki M, Terashima I, Kamata K, Koga A: **C-terminal modification of monoclonal antibody drugs: amidated species as a general product-related substance.** *Int J Biol Macromol* 2013, **52:**139–147.
7. Kaschak T, Boyd D, Lu F, Derfus G, Kluck B, Nogal B, Emery C, Summers C, Zheng K, Bayer R, Amanullah A, Yan B: **Characterization of the basic charge variants of a human IgG1: effect of copper concentration in cell culture media.** *MAbs* 2011, **3:**577–583.
8. Johnson KA, Paisley-Flango K, Tangarone BS, Porter TJ, Rouse JC: **Cation exchange-HPLC and mass spectrometry reveal C-terminal amidation of an IgG1 heavy chain.** *Anal Biochem* 2007, **360:**75–83.
9. Cuttitta F: **Peptide amidation: signature of bioactivity.** *Anat Rec* 1993, 87–93. 172–173; discussion 193–175.
10. Merkler DJ: **C-terminal amidated peptides: production by the in vitro enzymatic amidation of glycine-extended peptides and the importance of the amide to bioactivity.** *Enzyme Microb Technol* 1994, **16:**450–456.
11. In Y, Fujii M, Sasada Y, Ishida T: **Structural studies on C-amidated amino acids and peptides: structures of hydrochloride salts of C-amidated Ile, Val, Thr, Ser, Met, Trp, Gln and Arg, and comparison with their C-unamidated counterparts.** *Acta Crystallogr B* 2001, **57:**72–81.
12. Bolkenius FN, Ganzhorn AJ: **Peptidylglycine alpha-amidating mono-oxygenase: Neuropeptide amidation as a target for drug design.** *Gen Pharmacol* 1998, **31:**655–659.
13. Prigge ST, Mains RE, Eipper BA, Amzel LM: **New insights into copper monooxygenases and peptide amidation: structure, mechanism and function.** *Cell Mol Life Sci* 2000, **57:**1236–1259.
14. Hayashi N, Kayo T, Sugano K, Takeuchi T: **Production of bioactive gastrin from the non-endocrine cell lines CHO and COS-7.** *FEBS Lett* 1994, **337:**27–32.
15. Wu SC: **RNA interference technology to improve recombinant protein production in Chinese hamster ovary cells.** *Biotechnol Adv* 2009, **27:**417–422.
16. Amarzguioui M, Rossi JJ, Kim D: **Approaches for chemically synthesized siRNA and vector-mediated RNAi.** *FEBS Lett* 2005, **579:**5974–5981.
17. Santiago Y, Chan E, Liu PQ, Orlando S, Zhang L, Urnov FD, Holmes MC, Guschin D, Waite A, Miller JC, Rebar EJ, Gregory PD, Klug A, Collingwood TN: **Targeted gene knockout in mammalian cells by using engineered zinc-finger nucleases.** *Proc Natl Acad Sci U S A* 2008, **105:**5809–5814.
18. Urnov FD, Rebar EJ, Holmes MC, Zhang HS, Gregory PD: **Genome editing with engineered zinc finger nucleases.** *Nat Rev Genet* 2010, **11:**636–646.
19. Klug A: **The discovery of zinc fingers and their applications in gene regulation and genome manipulation.** *Annu Rev Biochem* 2010, **79:**213–231.
20. Lim SF, Chuan KH, Liu S, Loh SO, Chung BY, Ong CC, Song Z: **RNAi suppression of Bax and Bak enhances viability in fed-batch cultures of CHO cells.** *Metab Eng* 2006, **8:**509–522.
21. Sung YH, Hwang SJ, Lee GM: **Influence of down-regulation of caspase-3 by siRNAs on sodium-butyrate-induced apoptotic cell death of Chinese hamster ovary cells producing thrombopoietin.** *Metab Eng* 2005, **7:**457–466.
22. Sung YH, Lee JS, Park SH, Koo J, Lee GM: **Influence of co-down-regulation of caspase-3 and caspase-7 by siRNAs on sodium butyrate-induced apoptotic cell death of Chinese hamster ovary cells producing thrombopoietin.** *Metab Eng* 2007, **9:**452–464.
23. Wong DC, Wong KT, Nissom PM, Heng CK, Yap MG: **Targeting early apoptotic genes in batch and fed-batch CHO cell cultures.** *Biotechnol Bioeng* 2006, **95:**350–361.
24. Ngantung FA, Miller PG, Brushett FR, Tang GL, Wang DI: **RNA interference of sialidase improves glycoprotein sialic acid content consistency.** *Biotechnol Bioeng* 2006, **95:**106–119.
25. Mori K, Kuni-Kamochi R, Yamane-Ohnuki N, Wakitani M, Yamano K, Imai H, Kanda Y, Niwa R, Iida S, Uchida K, Shitara K, Satoh M: **Engineering Chinese**

hamster ovary cells to maximize effector function of produced antibodies using FUT8 siRNA. *Biotechnol Bioeng* 2004, **88**:901–908.

26. Imai-Nishiya H, Mori K, Inoue M, Wakitani M, Iida S, Shitara K, Satoh M: **Double knockdown of alpha1,6-fucosyltransferase (FUT8) and GDP-mannose 4,6-dehydratase (GMD) in antibody-producing cells: a new strategy for generating fully non-fucosylated therapeutic antibodies with enhanced ADCC.** *BMC Biotechnol* 2007, **7**:84.
27. Kim SH, Lee GM: **Down-regulation of lactate dehydrogenase-A by siRNAs for reduced lactic acid formation of Chinese hamster ovary cells producing thrombopoietin.** *Appl Microbiol Biotechnol* 2007, **74**:152–159.
28. Hong WW, Wu SC: **A novel RNA silencing vector to improve antigen expression and stability in Chinese hamster ovary cells.** *Vaccine* 2007, **25**:4103–4111.
29. Wu SC, Hong WW, Liu JH: **Short hairpin RNA targeted to dihydrofolate reductase enhances the immunoglobulin G expression in gene-amplified stable Chinese hamster ovary cells.** *Vaccine* 2008, **26**:4969–4974.
30. Liu PQ, Chan EM, Cost GJ, Zhang L, Wang J, Miller JC, Guschin DY, Reik A, Holmes MC, Mott JE, Collingwood TN, Gregory PD: **Generation of a Triple-Gene Knockout Mammalian Cell Line Using Engineered Zinc-Finger Nucleases.** *Biotechnol Bioeng* 2010, **106**:97–105.
31. Cost GJ, Freyvert Y, Vafiadis A, Santiago Y, Miller JC, Rebar E, Collingwood TN, Snowden A, Gregory PD: **BAK and BAX Deletion Using Zinc-Finger Nucleases Yields Apoptosis-Resistant CHO Cells.** *Biotechnol Bioeng* 2010, **105**:330–340.
32. Yamane-Ohnuki N, Kinoshita S, Inoue-Urakubo M, Kusunoki M, Iida S, Nakano R, Wakitani M, Niwa R, Sakurada M, Uchida K, Shitara K, Satoh M: **Establishment of FUT8 knockout Chinese hamster ovary cells: an ideal host cell line for producing completely defucosylated antibodies with enhanced antibody-dependent cellular cytotoxicity.** *Biotechnol Bioeng* 2004, **87**:614–622.
33. Malphettes L, Freyvert Y, Chang J, Liu PQ, Chan E, Miller JC, Zhou Z, Nguyen T, Tsai C, Snowden AW, Collingwood TN, Gregory PD, Cost GJ: **Highly Efficient Deletion of FUT8 in CHO Cell Lines Using Zinc-Finger Nucleases Yields Cells That Produce Completely Nonfucosylated Antibodies.** *Biotechnol Bioeng* 2010, **106**:774–783.
34. Klug A: **The discovery of zinc fingers and their development for practical applications in gene regulation and genome manipulation.** *Q Rev Biophys* 2010, **43**:1–21.
35. Durai S, Mani M, Kandavelou K, Wu J, Porteus MH, Chandrasegaran S: **Zinc finger nucleases: custom-designed molecular scissors for genome engineering of plant and mammalian cells.** *Nucleic Acids Res* 2005, **33**:5978–5990.

16

Development of late blight resistant potatoes by cisgene stacking

Kwang-Ryong Jo[1,2,3], Chol-Jun Kim[3], Sung-Jin Kim[3], Tok-Yong Kim[3], Marjan Bergervoet[1], Maarten A Jongsma[4], Richard GF Visser[1], Evert Jacobsen[1] and Jack H Vossen[1*]

Abstract

Background: *Phytophthora infestans,* causing late blight in potato, remains one of the most devastating pathogens in potato production and late blight resistance is a top priority in potato breeding. The introduction of multiple resistance (*R*) genes with different spectra from crossable species into potato varieties is required. Cisgenesis is a promising approach that introduces native genes from the crops own gene pool using GM technology, thereby retaining favourable characteristics of established varieties.

Results: We pursued a cisgenesis approach to introduce two broad spectrum potato late blight *R* genes, *Rpi-sto1* and *Rpi-vnt1.1* from the crossable species *Solanum stoloniferum* and *Solanum venturii,* respectively, into three different potato varieties. First, single *R* gene-containing transgenic plants were produced for all varieties to be used as references for the resistance levels and spectra to be expected in the respective genetic backgrounds. Next, a construct containing both cisgenic late blight *R* genes (*Rpi-vnt1.1* and *Rpi-sto1*), but lacking the bacterial kanamycin resistance selection marker (*NPTII*) was transformed to the three selected potato varieties using *Agrobacterium*-mediated transformation. Gene transfer events were selected by PCR among regenerated shoots. Through further analyses involving morphological evaluations in the greenhouse, responsiveness to *Avr* genes and late blight resistance in detached leaf assays, the selection was narrowed down to eight independent events. These cisgenic events were selected because they showed broad spectrum late blight resistance due to the activity of both introduced *R* genes. The marker-free transformation was compared to kanamycin resistance assisted transformation in terms of T-DNA and vector backbone integration frequency. Also, differences in regeneration time and genotype dependency were evaluated.

Conclusions: We developed a marker-free transformation pipeline to select potato plants functionally expressing a stack of late blight *R* genes. Marker-free transformation is less genotype dependent and less prone to vector backbone integration as compared to marker-assisted transformation. Thereby, this study provides an important tool for the successful deployment of *R* genes in agriculture and contributes to the production of potentially durable late blight resistant potatoes.

Keywords: Potato, Late blight, Resistance gene, Cisgenesis, Marker-free transformation

Background

Genetic disease resistance is an effective tool for sustainable management of late blight, caused by *Phytophthora infestans,* which is economically the most important disease of potato. Breeding at the beginning of the twentieth century concentrated on major dominant late blight resistance (*R*) genes from the Mexican wild species *Solanum demissum* and eleven of these *R* genes were introgressed in potato [1-4]. However, rapid breakdown of resistance in potato varieties containing *S. demissum R1, R2, R3,* and *R10* [3,5] has sparked an increased focus on the introgression of multiple broad spectrum *R* genes in order to impart durability to commercial varieties. It has turned out in various crops and pathosystems that stacking of multiple *R* genes is necessary to provide satisfactory resistance in the field [6]. Although the used *R* genes provide resistance to broad spectra of late blight strains, the predominant agricultural deployment of only one *R* gene can drive the evolution of new virulent strains. In the absence of chemical controls this might even result in the destruction

* Correspondence: jack.vossen@wur.nl
[1]Wageningen UR Plant Breeding, Wageningen University & Research Centre, P.O. Box 386, 6700 AJ Wageningen, The Netherlands
Full list of author information is available at the end of the article

of an entire harvest [7]. Therefore, the use of combinations of *R* genes with different spectra must be pursued to increase durability of resistance and thereby providing food security under no or little fungicide application. *R* gene stacking might be achieved by genetic crossings but the desired variety characteristics will never be fully recovered due to the high level of heterozygosity in potato. Sarpo Mira is an example of a durably late blight resistant potato variety which contains a stack of at least four *R* genes [8,9]. Unfortunately, the variety has not acquired a large market share yet because established varieties are preferred by farmers, processors and consumers.

Addition of stacks of cloned *R* genes [10-17] to existing varieties (resistant or susceptible) through genetic modification (GM) technology is therefore an attractive alternative. Moreover, GM technology circumvents the problem of linkage drag and can speed up the introgression of the *R* gene [18,19]. GM technology has, however, met various types of opposition and a major point of criticism concerns the introduction of "foreign" genes into the food chain and environment. However, within the framework provided by cisgenesis only natural genes from the same or crossable species are used [20,21]. Cisgenes are, therefore, already present in the natural gene pool of the crop plant and cisgenesis only facilitates their introduction into crops. Indeed a majority of a broad panel of European consumers find cisgenic apples safe and not harmfull for the environment [22].

Recently, the transformation of three broad spectrum potato late blight resistance genes (*Rpi-sto1, Rpi-vnt1.1* and *Rpi-blb3*) was described in potato [23]. *Rpi-sto1, Rpi-vnt1.1* and *Rpi-blb3* are native genes from crossable species and are therefore considered as cisgenes for potato. However, the plants in the study from Zhu et al. [23] are "transgenic" as the selectable marker gene, NPTII, was of bacterial origin. Also beyond the cisgenesis framework it is not desired to introduce antibiotic resistance genes into the environment and in this study, we established a pipeline for *Agrobacterium*-mediated transformation of potato in the absence of a selectable marker gene (marker-free transformation). After the absence of vector backbone integration was confirmed, these potatoes were designated as "cisgenic" because of the absence of any foreign (non-potato) genes. This is the first scientific report on the production and functional evaluation of cisgenic *R* gene stacking in different potato varieties.

Results

Transformation and functional expression of single late blight *R* genes in potato varieties

The resistance spectra of three potato varieties (the American variety Atlantic, the Dutch variety Bintje and the Korean variety Potae9) were tested with five *P. infestans* isolates with variable virulence spectra and aggressiveness. Atlantic and Bintje were susceptible to all tested isolates while Potae9 was resistant to two isolates (EC1 and 90128; Table 1). These two isolates are a-virulent on plants carrying *R2* type of resistance genes. The presence of *R2* or a functional homolog in Potae9 was confirmed using AVR2 response experiments (data not shown). In order to make Atlantic and Bintje resistant to late blight and to broaden the resistance spectrum of Potae9, these three varieties were transformed with two constructs (pBINPLUS:*Rpi-vnt1.1* and pBINPLUS:*Rpi-sto1* harbouring the kanamycin resistance gene *NPTII*), each containing a single late blight *R* gene . The transgenic events were collected using selection for kanamycin resistance and, successively, the functional expression of the introduced *R* genes was tested using agroinfiltration of the cognate a-virulence (*Avr*) genes. Also the transgenic events were subjected to *P. infestans* inoculation using a detached leaf assay (DLA; Table 1). As an example, the interactions of a representative set of transgenic events with the selected isolates are shown in Figure 1. As expected, the majority of the transgenic events showed resistance to at least four of the five tested *P. infestans* isolates. EC1 and pic99189 were described previously to break the *Rpi-vnt1.1* and *Rpi-sto1* mediated resistances, respectively [13,24]. Indeed, transgenic Atlantic and Bintje events harbouring the *Rpi-vnt1.1* gene were susceptible to isolate EC1. The Potae9 transgenic events containing *Rpi-vnt1.1* were resistant to EC1, due to the presence of *R2* or a functional homolog in Potae9. The *Rpi-sto1*-containing events were susceptible to isolate pic99189. It is concluded that both *Rpi-vnt1.1* and *Rpi-sto1* were able to confer resistance in the selected varieties and these two genes may, therefore, be combined as a cisgenic *R* gene stack in the selected varieties.

Selection and validation of cisgenic potato plants with two late blight *R* genes

Cisgenesis excludes antibiotic resistance marker-assisted transformation since the genes encoding the selection markers are derived from non-crossable species. We, therefore, pursued marker-free transformation of the cisgenes *Rpi-vnt1.1* and *Rpi-sto1* in combination with PCR selection (Table 2). Two hundred stem explants from each of the three selected varieties were prepared and co-cultivated with an *A. tumefaciens* strain carrying only the cisgenic late blight *R* genes *Rpi-vnt1.1* and *Rpi-sto1* between the T-DNA borders of a binary plasmid (Figure 2). Between 31 and -110 days after transformation, over 1515 shoots were collected in five rounds of harvesting (Table 3). During the experiment, the shoot regeneration potential of the callus gradually dropped and at 130 days after transformation no more shoots could be harvested. These 1515 shoots were screened by PCR with *Rpi-vnt1* and *Rpi-sto1* primers and 27 PCR positive shoots were selected (Table 2). All PCR

Table 1 List of transgenic reference plants obtained by single *R* gene transformation

Variety	Introduced *R* gene	Plant ID	PCR		Agroinfiltration		DLA				
			vnt1.1	*sto1*	*Avrvnt1*	*Avrsto1*	EC1	IPO-C	DHD11	90128	pic99189
Atlantic	n	H	-	-	-	-	S	S	S	S	S
	Rpi-vnt1.1	H13-2	+	-	+	-	S	R	R	R	R
	Rpi-sto1	H9-10	-	+	-	+	R	R	R	R	S
Bintje	n	F	-	-	-	-	S	S	S	S	S
	Rpi-vnt1.1	F13-10	+	-	+	-	S	R	R	R	R
	Rpi-sto1	F9-4	-	+	-	+	R	R	R	R	S
Potae9	n	W	-	-	-	-	R	S	S	R	S
	Rpi-vnt1.1	W13-8	+	-	+	-	R	R	R	R	R
	Rpi-sto1	W9-1	-	+	-	+	R	R	R	R	S

n, no transformation; - = not detected or not responsive to agroinfiltation; + = PCR positive or responsive to agroinfiltration; R, resistant; S, susceptible; DLA: detached leaf assays with the indicated *P. infestans* isolates.

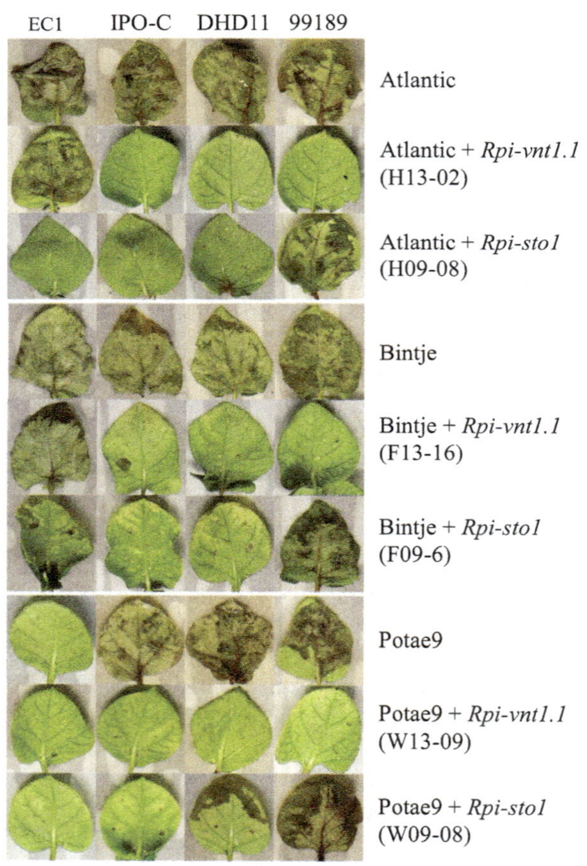

Figure 1 Detached leaf assays of transgenic potatoes obtained by marker-assisted transformation with single *R* gene constructs. Non transformed Atlantic or Bintje were susceptible to four *P. infestans* isolates. *Rpi-vnt1.1*-containing transgenic plants were susceptible to EC1 and *Rpi-sto1*-containing transgenic plants were susceptible to pic99189.

positive shoots were originating from different explants, indicating that they were independent transformation events. Two Bintje events only contained the *Rpi-vnt1* gene and were discarded. The remaining 25 events, containing both *Rpi-vnt1* and *Rpi-sto1*, were further tested using vector backbone gene-specific PCR analysis (Figure 3). We found that six events contained vector backbone sequences (Table 2). The remaining 19 events were vector backbone free and are therefore designated as cisgenic events. The 19 cisgenic events were transferred to the greenhouse for phenotypic characterisation. Three weeks after transfer to the greenhouse, five events displayed abnormal plant morphology that consisted of curly leaves and dwarfed growth (Additional file 1), a phenomenon that is commonly observed after regeneration [25]. The five events with these aberrant phenotypes were disregarded for further studies and the remaining 14 events were tested for their responsiveness to *Avrvnt1* and *Avrsto1* after agroinfiltration. Five events responded only to *Avrvnt1* and not to *Avrsto1*. Eight events responded to both *Avrvnt1* and *Avrsto1* infiltration, showing that both *Rpi-vnt1* and *Rpi-sto1* were functionally expressed (Table 4). The latter eight plants also displayed resistance in DLA to all *P. infestans*

Table 2 Marker-free transformation of two *R* genes (*Rpi-vnt1.1:Rpi-sto1*) to different potato varieties; Marker-free transformation frequencies

Variety	explants #	shoots #	PCR + #	frequency %	bbf #	bbf %
Atlantic	200	497	0/0/12	2.4	9	75
Bintje	200	590	2/0/6	1.0	5	83
Potae9	200	428	0/0/7	1.6	5	71
total	600	1515	2/0/25	1.7	19	76

#explants; number of explants; #shoots: number of shoots tested; #PCR+: the number of shoots containing *Rpi-sto1*, *Rpi-vnt1.1*, or both genes, respectively, as detected by PCR; % frequency: transformation frequency, as percentage of PCR + shoots, carrying both *Rpi-sto1* and *Rpi-vnt1.1*, over the number of tested shoots; #bbf: number of vector backbone free events; % bb: percentage of vector backbone containing PCR + shoots, carrying both *Rpi-sto1* and *Rpi-vnt1.1*.

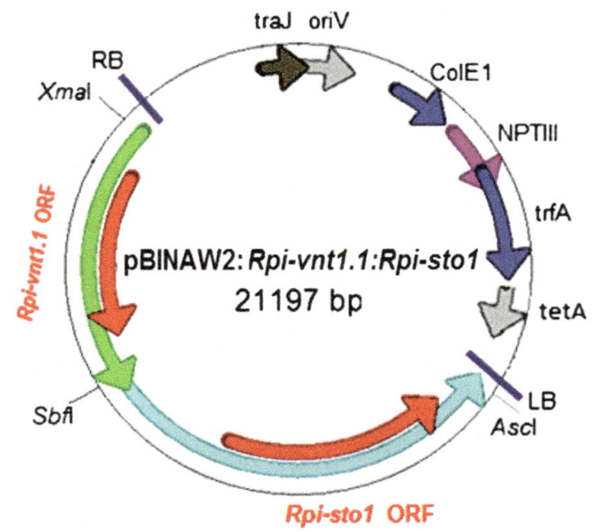

Figure 2 Schematic diagram of the marker-free double gene construct pBINAW2:*Rpi-vnt1.1:Rpi-sto1*. In light green and light blue arrows the *Rpi-vnt1.1* and *Rpi-sto1* genes are shown, respectively. The red arrows indicate the coding regions of *Rpi-vnt1.1* or *Rpi-sto1*. Unique restriction enzyme recognition sites *XmaI*, *SbfI* and *AscI* are shown. RB: right border of T-DNA, LB: left border of T-DNA, TetA, trfA, NPTIII, ColE1, oriV and traJ are vector backbone sequences for plasmid stability and replication in bacterial hosts *Agrobacterium tumefaciens* and *Escherichia coli*.

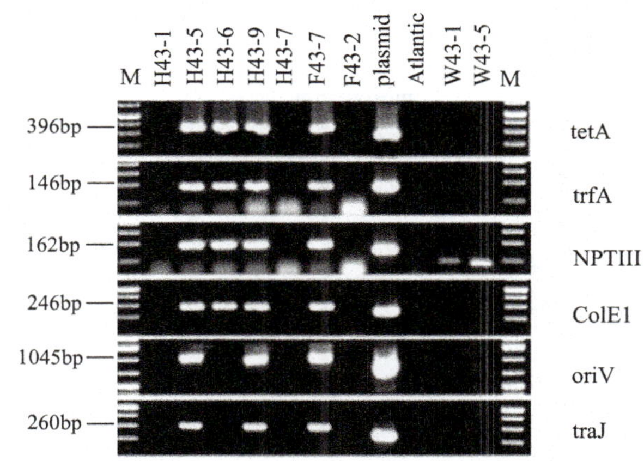

Figure 3 Vector backbone integration in marker-free transformation events. Atlantic (H), Bintje (F) and Potae9 (W), were transformed with construct pBINAW2: *Rpi-vnt1.1:Rpi-sto1*. PCR analysis was performed using primers specific for tetA, trfA, NPTIII, ColE1, oriV and traJ to detect vector backbone integration. The plasmid pBINAW2:*Rpi-vnt1.1:Rpi-sto1* was used as a positive control and the untransformed Atlantic as a negative control. Only the NPTIII primers amplified an a-specific fragment of similar size as shown here for the backbone free events W43-1 and W43-5 in untransformed Potae9. None of the other primers amplified an a-specific band in Potae9 or Bintje (data not shown) M: molecular weight marker.

isolates tested. Figure 4 shows an example of the validation of functional expression for both transferred *R* genes in event H43-7 (Atlantic background) by agroinfiltration and resistance assays in the DLA. Using the single gene-containing transgenic plants as reference it was demonstrated that stacking of *R* genes with different resistance spectra leads to complementary broad spectrum resistance (Table 1). Interestingly, the two introduced *R* genes are complementing the resistance spectrum that was already present in Potae9 plants. Using the pursued experimental setup we were able to select two cisgenic events in Atlantic, five cisgenic events in Bintje and one cisgenic event in Potae9 containing and functionally expressing a stack of two late blight *R* genes.

Table 3 Marker-free transformation of two *R* genes (*Rpi-vnt1.1:Rpi-sto1*) to different potato varieties; Identification of PCR-positive shoots in different time ranges after marker-free transformation

Variety	31-50 days	51-70 days	71-90 days	91-110 days	111-130 days	Total
Atlantic	4/197	4/174	4/111	0/15	0/0	12/497
Bintje	2/199	3/194	1/165	0/32	0/0	6/590
Potae9	4/183	2/143	1/78	0/24	0/0	7/428

Number of PCR-positive shoots carrying both *Rpi-sto1* and *Rpi-vnt1.1* over the number of tested shoots.

Comparison of marker-assisted- and marker-free transformation efficiencies

Kanamycin resistance assisted selection is routinely used for plant transformation. It is, therefore, interesting to compare the efficiency of marker-free transformation in the cisgenesis pipeline to marker-assisted transformation. Marker-assisted transformation efficiency was 100% when expressed as the percentage of rooting shoots being PCR positive for the gene of interest (Table 5). In this definition, marker-free transformation efficiency ranged from 1 to 2.4% over the three varieties.

For a better comparison of marker-assisted and marker-free transformation, it was essential to use a different definition for transformation efficiency that also takes shoot regeneration efficiency into account. We define marker-assisted transformation frequency as the percentage of PCR positive events among the number of explants used for transformation. Marker-free transformation frequency is defined as the percentage of shoots that is PCR positive. In variety Atlantic a high marker-assisted transformation frequency (71%) was observed whereas the other two varieties, Bintje and Potae9, had significantly lower marker-assisted transformation frequencies (10-13%) (Tables 5 and 6). In marker-free transformation, variety dependent differences in transformation frequencies were less dramatic (2.4, 1.0 and 1.6% in Atlantic, Bintje and Potae-9, respectively) and statistically insignificant (Table 6). Not only the frequency

Table 4 Phenotypic characterization of vector backbone free (cisgenic) events in different potato varieties carrying the *Rpi-vnt1 and Rpi-sto1* genes

Cisgenic event	Variety	Plant morphology	Agroinfiltration		DLA				
			Avrvnt1	*Avrsto1*	EC1	IPO-C	DHD11	90128	PIC99189
H43-1	Atlantic		+	–	S	R	R	R	R
H43-2	Atlantic	curly leaf	n	n	n	n	n	n	n
H43-3	Atlantic	curly leaf	n	n	n	n	n	n	n
H43-4	Atlantic		–	–	S	S	S	S	S
H43-7	Atlantic		+	+	R	R	R	R	R
H43-8	Atlantic		+	+	R	R	R	R	R
H43-10	Atlantic		+	–	S	R	R	R	R
H43-11	Atlantic	curly leaf	n	n	n	n	n	n	n
H43-12	Atlantic		+	–	S	S	S	S	S
F43-1	Bintje		+	–	S	R	R	R	R
F43-2	Bintje		+	+	R	R	R	R	R
F43-3	Bintje		+	+	R	R	R	R	R
F43-4	Bintje		+	+	R	R	R	R	R
F43-5	Bintje		+	+	R	R	R	R	R
W43-1	Potae9		+	+	R	R	R	R	R
W43-2	Potae9	curly leaf	n	n	n	n	n	n	n
W43-3	Potae9		+	–	S	R	R	R	R
W43-4	Potae9	dwarf	n	n	n	n	n	n	n
W43-5	Potae9		+	+	R	R	R	R	R

Cisgenic events functionally expressing both R genes were highlighted by bold font. n: no data, +: responsive to infiltration with the indicated Avr gene; -: not responsive to infiltration with the indicated Avr gene R: resistant to the indicated isolate in detached leaf assays (DLA). S: susceptible to the indicated isolate in DLA.

of transformation, also the timing of transformation was different between marker-free and marker-assisted transformation. In the marker-free transformation experiments, the majority of the PCR-positive shoots was obtained between 1 and 3 months after co-cultivation (Table 3). This was quicker than marker-assisted transformation of the *Rpi-vnt1* and *Rpi-sto1* genes individually, which took 2-4 months (Table 5). Finally, we compared vector backbone integration frequencies among the different marker-free and marker-assisted transformation experiments. We did not find significant differences in vector backbone integrations frequency when the different varieties or both of the marker-assisted transformation constructs were compared (Table 7). Only when vector backbone integration frequency was compared between the marker-free (24%; Table 2) and marker-assisted transformation experiments (57%; Table 5) we found that marker-free transformation was associated with less vector backbone integration.

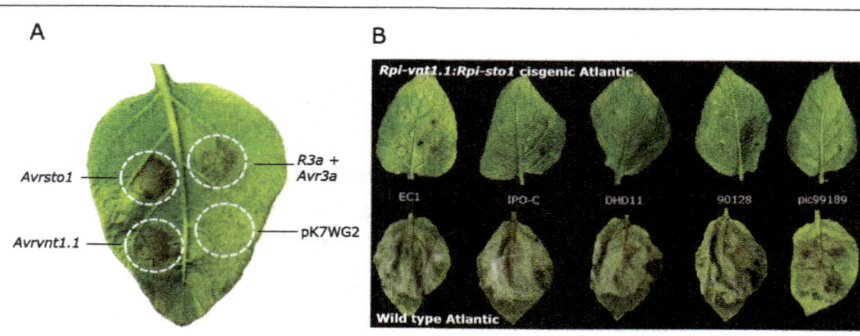

Figure 4 Functional validation of cisgenic transformants by agroinfiltration and resistance assays. A. *Avrvnt1*- and *Avrsto1*-induced hypersensitive responses in cisgenic transformant H43-7 (*Rpi-vnt1:Rpi-sto1* in Atlantic background). *Avrvnt1 and Avrsto1* were infiltrated in cisgenic plants. A 1:1 mixture of *R3a* and *Avr3a* and pK7WG2 were infiltrated as positive and negative controls, respectively. **B**. Detached leaf assays for cisgenic transformant H43-7. Different isolates are shown in the middle. Cisgenic transformant are shown on the top and the wild type Atlantic on the bottom of the panel.

Table 5 Marker-assisted transformations of single *R* genes to different varieties

Inserted *R* gene	Variety	Plant ID	Explants #	Regeneration time (days)	shoots %	shoots #	rooting %	PCR + %	frequency %	vbf %	vbf in DLA #	vbf and R in DLA #
Rpi-vnt1.1	Atlantic	H13	200	60-120	76	30[a]	100	100	76	40	12	12
Rpi-sto1		H09	200	60-120	66	30[a]	100	100	66	50	15	15
Rpi-vnt1.1	Bintje	F13	200	60-120	13	26[b]	77	100	10	40	8	8
Rpi-sto1		F09	200	60-120	10	20[b]	100	100	10	45	9	9
Rpi-vnt1.1	Potae9	W13	200	60-120	16	31[b]	84	100	13	47	12	12
Rpi-sto1		W09	200	60-120	19	37[b]	61	100	11	39	9	9

#exp; number of explants, % sht; percentage of number of shoots over number of explants, % rt; percentage of number of rooted shoots over the number of shoots, % PCR+; percentage of PCR positive shoots over the number of shoots, % freq; transformation frequency, calculated by % sht × % rt × % PCR+, a Among regenerated shoots, 30 plants were tested. b All regenerated shoots were tested. % vbf; percentage of backbone free plants out of plants tested. #vbf plants DLA: number of vector backbone free plants tested in detached leaf assays. #vbf R plants DLA: number of vector backbone free resistant plants in detached leaf assay.

Discussion

Cisgenesis, is a new approach for traditional plant breeding that uses genetic modification technology to introduce natural genes from within a plant species or from crossable plant species, into varieties [26]. Therefore, any gene "alien" to the breeder's gene pool can be avoided in the end product which is causal to many environmental and consumers' concerns about GM food crops [22]. Not only can widely used susceptible varieties, like Bintje and Atlantic, be converted into resistant varieties, also resistant varieties, like Potae9, can be complemented with additional resistance genes to avoid or delay future resistance breakdown. In order to complement existing varieties with stacks of cisgenic *R* genes, two choices must be made: 1. The method to introduce the *R* gene stack and 2. The method to exclude sequences of foreign origin from transformation events. With respect to the introduction method, in this study we chose transformation by marker-free binary vectors and subsequent regeneration in medium without selective antibiotics followed by PCR-based selection of transformation events [27]. Alternatives involving the removal of a selectable marker gene by site specific recombination pose disadvantages because of remnant sequences of foreign origin [28].

The average marker-free transformation frequency was 1.3% and seems to be genotype independent. In a previous marker-free transformation study in potato [27] a T-DNA of 6 kb was transformed with a frequency of 3.5% when *A. tumefaciens* strain AGL0 was used, and 0.4% when *A. tumefaciens* strain LB4404 was used. It can not be concluded that AGL1 + virG, which was used in this study, was less efficient in transferring the T-DNA than AGL0 in the study from de Vetten et al. [27]. From unpublished experiments in our laboratory it is known that regeneration time increases with the size of the T-DNA. We, therefore, assume that the lower transformation frequency in our study is rather related to the larger T-DNA size (11 kb) of the *Rpi-vnt1:Rpi-sto1* construct. Therefore, for stacking of more than two genes in cisgenic transformation, the effect of an increased insert size (e.g. >11 kb) on transformation frequency remains to be tested. It is known that marker-assisted transformation frequency is highly genotype dependent in potato [29]. Also here we found that transformation frequencies ranged from 10-71% in different varieties (Table 5). This variation was remarkably less (1-2.4%) in marker-free transformation experiments (Table 2). It must be noted that transformation frequencies can vary between different experiments and that we here only performed a limited number of experiments. However, the currently presented experiments show that marker-free transformation is less prone to varietal differences than marker assisted transformation. This could be caused by

Table 6 Pairwise comparisons of transformation frequencies in groups of transformation experiments

Group 1			Group 2			Pearson analysis*	
	Non transformed	Transformed		Non transformed	Transformed	Chi square	P (2-tailed)
Atlantic (MF)	497	12	Bintje (MF)	590	8	1.6	0.2
Atlantic (MF)	497	12	Potae9 (MF)	428	7	0.7	0.7
Potae9 (MF)	428	7	Bintje (MF)	590	8	0.1	0.7
Atlantic (MA)	15	35	Bintje (MA)	45	5	38	0
Atlantic (MA)	15	35	Potae9 (MA)	44	6	35	0
Potae 9 (MA)	44	6	Bintje (MA)	45	5	0.1	0.7

Transformation frequencies in marker-free transformations (MF) are derived from the number of transformed shoots (transformed) and the total number of shoots minus the number of transformed shoots (non-transformed). Transformation frequencies in marker-assisted transformations (MA) are derived from the number of rooting shoots (transformed) and the number of explants used minus the number of rooting shoots (non-transformed).
*null hypothesis: group 1 equals group 2.

Table 7 Pairwise comparisons of vector backbone integration frequencies in groups of transformation experiments

Group 1				Group 2				Pearson analysis*	
	Experiments	bb	bbf		Experiments	bb	bbf	Chi square	P (2-tailed)
Atlantic	H09 + H13	28	28	Potae9	W09 + W13	29	20	0.88	0.34
Atlantic	H09 + H13	28	28	Bintje	F09 + F13	22	18	0.23	0.6
Potae9	W09 + W13	29	20	Bintje	F09 + F13	22	18	0.16	0.69
Rpi-sto1	H09 + F09 + W09	42	36	*Rpi-vnt1*	H13 + F13 + W13	37	30	0.028	0.86
Markerfree	H43 + F43 + W43	6	21	Marker	H09	13	15	3.560	0.059
Markerfree	H43 + F43 + W43	6	21	Marker	H13	15	13	5.72	0.017

*null hypothesis: group 1 equals group 2; bb: vector backbone containing event; bbf: vector backbone free event. In the experiments columns, H, F, W represent the varieties Atlantic, Bintje and Potae9 respectively. Extensions 09, 13 and 43 represent the constructs pBINPLUS:*Rpi-sto1*, pBINPLUS:*Rpi-vnt1.1*, pBINAW2:*Rpi-vnt1.1*:*Rpi-sto1*, respectively.

differences in antibiotic tolerance between the varieties that provides transformed cells different abilities to develop into a shoot.

In terms of vector backbone integration, marker-free transformation apparently produces a lower percentage (24%) of vector backbone integrations compared to marker-assisted transformation (40-50%). Again, the number of experiments is limited and firm conclusions cannot be drawn. The vector backbone and border sequences in pBINPLUS and pBINAW2 are highly similar and we do not expect that these differences affect vector backbone integration. A potential explanation could be that the presence of the NPTII gene directly next to the left border of the T-DNA would stimulate selection of higher levels of backbone integration. As it is known that left border recognition is inaccurate in *Solanaceae*, [30], especially when agrobacterium strain AGL1 is used [31], positioning of NPTII near the left border would force the integration of the complete T-DNA. So, it might also lead to higher levels of vector backbone integration. In marker-free transformation, six plants out of 14 tested cisgenic plants did not appropriately express *Rpi-sto1* as observed using agroinfiltration of the corresponding *Avr* genes (like H43-1, -10, -12; Table 4). An obvious explanation could be that T-DNA insertion did not proceed all the way to the left border resulting in 3′ truncations of *Rpi-sto1*. These non-functional cisgenic events and the corresponding DNA samples were discarded in an early phase during the selection and, unfortunately, this hypothesis could not be confirmed. We observed and described some cisgenic plants differing morphologically from wild type varieties in the greenhouse (Table 4, Additional file 1). This is a generally observed phenomenon and in tissue culture-based breeding schemes it should be considered that aberrant plant phenotypes must be selected against [29].

According to the established experimental scheme, it takes less than one year to obtain potato plants with cisgenic *R* gene stacks from the *R* gene construct preparation to the functional validation of the resulting cisgenic plants by performing DLAs. As 2-3 shoots per explant can be collected and 30 independent transformed plants are required considering backbone integration and expression, it is recommended that between 1000-1500 explants are to be treated in a marker-free-transformation experiment of potato. The efficiency of PCR analysis can be improved by a factor 10 by pooling ten shoots, so that the labour intensity of the selection of marker-free transformation events is considered reasonable as compared to the marker-assisted transformations. Considering 2-3 years' field trials, it takes totally 3-4 years to produce late blight resistant cisgenic events in established potato varieties, which can be released for seed tuber multiplication. This time span is remarkably short compared to the conventional breeding scheme. The cisgenic potatoes selected in this study will be further tested for several years to evaluate whether the transferred *R* genes are stably expressed over many vegetative cycles. Chimeras and epigenetic silencing are issues that could affect stability of resistance. Also agronomic performance needs to be assessed and confirmed in multiple growing seasons.

Conclusions

We have set up and pursued an effective cisgenic marker-free transformation strategy for commercial potato varieties. It was found that marker-free transformation frequency was much less genotype dependent than marker-assisted transformation. Also the frequency of vector backbone integration tended to be lower in the marker-free transformations as compared to the marker-assisted transformations. The susceptibility or the narrow late blight resistance spectra of the selected varieties were upgraded to broad spectrum resistance after the successful introduction of two cisgenic late blight *R* genes. According to the recent conclusion of the European Food Safety Authority GMO Panel, cisgenic plants have a risk level similar to conventionally bred plants [32]. The cisgenic potatoes, generated in this study, will offer a safe, environmentally friendly, alternative to the current agricultural practice which is highly dependent on the use of chemical late blight control agents. For developing countries, where chemical control agents are unaffordable, cisgenic upgrades of local potato varieties might even ensure food security.

Methods

Plant material

The potato varieties Atlantic, Desiree, Bintje, and Potae9 were clonally maintained *in vitro* using Murashige and Skoog medium [33] supplemented with 3% (w/v) sucrose at 20°C at Wageningen UR Plant Breeding, Wageningen, The Netherlands. The varieties Potae9 from DPR Korea, which is resistant to late blight, was used for testing its reaction to certain late blight isolates and for transformation experiments to broaden its resistance spectrum.

Phytophthora infestans isolates and late blight resistance tests

Five *P. infestans* isolates (Additional file 2) were used in Detached Leaf Assays (DLAs); The European isolates IPO-C (race 1, 2, 3, 4, 5, 6, 7, 10, 11) and 90128 (race 1, 3, 4, 7, 8, 10, 11); the American isolates, EC1 (race 1, 3, 4, 7, 10, 11) and pic99189 (race 1, 2, 5, 7, 10, 11) and the Korean isolate DHD11 (race 1, 2, 3, 4, 6, 7, 10, 11). The DLAs were performed as described previously [34].

Vector construction

The single *R* gene constructs used in our study have been described before. Genomic DNA fragments from *S. venturii,* and *S. stoloniferum,* encompassing the *Rpi-vnt1.1* and *Rpi-sto1* genes, respectively, are cloned in the pBINPLUS binary vector [13,16]. The genomic fragments comprise the entire genes including their native promoters and terminators. In order to combine *Rpi-sto1* and *Rpi-vnt1.1* into one markerfree transformation vector, first the *Asc*I and *Sbf*I fragment from the pBINPLUS:*Rpi-sto1* vector, encompassing the *Rpi-sto1* gene, was ligated into the corresponding restriction sites of pBINAW2 [35]. pBINAW2 is a modified version of pBINPLUS where the entire T-DNA, including the NPTII gene, and the adjacent TetR gene from the vector backbone was removed and replaced by a minimal T-DNA containing only left and right border and a small multiple cloning site. To the pBINAW2: *Rpi-sto1* construct, the *Rpi-vnt1.1* gene was added using a *Sbf*I fragment, encompassing the *Rpi-vnt1.1* gene from the pBINPLUS:*Rpi-blb3*:*Rpi-vnt1.1*:*Rpi-sto1* described by Zhu et al. [23]. The clone with the desired *Rpi-vnt1.1* insert orientation, in tandem with *Rpi-sto1* (pBINAW2: *Rpi-vnt1.1*:*Rpi-sto1*, Figure 2; Additional file 3) was selected using restriction analysis. All ligation mixtures was transformed to ElectroMAX *E.coli* DH10b competent cells (Life technologies). Subsequently, the stability of the *R* gene constructs in *Agrobacterium* strain AGL-1 + VirG and functionality of the *R* genes in *N. benthamiana* were carried out using PCR and co-agroinfiltration with corresponding *Avr* genes, respectively. These tests confirmed the stability and activity of the constructs.

Potato transformation

Marker assisted transformation performed as described previously [36]. Marker-free transformations are derived from this protocol but kanamycin was omitted as a selection agent. Briefly, internodes of 2-5 mm in length were cut from thick stems of 4-week-old *in vitro*-grown plants and were used as explants in transformation experiments. After pre-culture on R3B medium (MS + 3% sucrose + 0.8% agar + 4 mg/ml NAA + 1 mg/ml BAP, pH5.8) supplemented with PACM (MS + 3% sucrose + 0.2% casein hydrolysate + 1 mg/ml 2,4-D + 1 mg/ml kinetin, pH6.5) for two days, explants were inoculated with *Agrobacterium* strain AGL1 + VirG + binary plasmid resuspended in LB medium to an OD_{600} of 0.2. After 2 days cocultivation, the explants were transferred to ZCVK medium (MS + 2% sucrose + 0.8% agar + 1 mg/ml zeatin + 200 mg/ml cefotaxim + 200 mg/ml vancomycin, pH5.8) for regeneration of shoots. Explants were transferred to fresh medium every two weeks. Shoots were transferred to CK medium (MS + 2% sucrose + 0.8% agar + 200 mg/ml cefotaxim + 200 mg/ml vancomycin, pH5.8) to induce root formation. To guarantee that regenerated plants were derived from independent transformation events, only shoots from physically separated positions on each explant were collected. Three weeks later, the rooted plantlets were analysed by PCR to determine the presence of the desired *R* genes. The transformation frequency was calculated as a percentage of the number of *R* gene-PCR positive shoots over the number of tested shoots.

For marker-assisted transformation, 100 mg/ml Kanamycin was added to ZCVK medium and CK medium for selection of transgenic shoots.

Functional tests of resistance (*R*) genes

Agroinfiltration was performed as previously described [37]. Two leaves per plant from three copies of each of the transformants were infiltrated with the following constructs: two effectors (*Avrvnt1* and *IpiO = Avrsto1*) [13,24], *R3a* [38] and *Avr3a* [39] as the positive control and empty pK7WG2 [40] as the negative control. *Agrobacterium tumefaciens* strain from glycerol stocks was grown in 3 ml of LB medium supplemented with appropriate antibiotics at 28°C overnight. The next day, the cultures were transferred to 15 ml of YEB medium (5 g beef extract, 5 g bacteriological peptone, 5 g sucrose, 1 g yeast extract, 2 ml 1 M $MgSO_4$ in 1 litre of milli-Q water) supplemented with antibiotics, 10 µl of 200 mM acetosyringone and 1000 µl of 1 M MES. On the third day, the cells were harvested and resuspended in MMA solution (20 g sucrose, 5 g MS salts and 1.95 g MES in 1 litre of distilled water, adjusted to pH5.6) supplemented with 1 ml of 200 mM acetosyringone to a final OD_{600} of 0.3. Leaves of 4- to 5-weeks old, greenhouse-grown, plants were infiltrated with this suspension. Responses were scored 3 to 4 days after infiltration.

DNA extraction and polymerase chain reaction (PCR)

Total genomic DNA was isolated from young leaves as described by Fulton et al. [41]. The Retsch machine (RETSCH Inc., Hannover, Germany) was used to grind young plant materials frozen in liquid nitrogen. Primers used for analysis of *R* genes, vector backbone integration are listed in Additional file 4. A pooled sampling method was exploited for PCR analysis of shoots in marker-free transformation. DNA extraction was carried out first by pooling one small leaf from each of ten shoots. If in this first round pools were found which were PCR-positive for both *R* genes and PCR-negative for backbone integration, a second round of PCR was carried out on genomic DNA of individual shoots within the pools. PCR reactions for *Rpi-sto1, Rpi-vnt1.1,* NPTIII, trfA, ColE1, oriV and traJ were performed using DreamTaq^{TM} polymerase (Fermentas) in a standard PCR program (94°C for 60 s followed by 30 cycles of 94°C for 30 s, 58°C for 60 s, 72°C for 90 s and a final extension time of 5 min at 72°C).

Statistical analysis

Transformation and vector backbone integrations frequencies are binary data and, therefore, the Pearson Chi-square test was chosen to compare the independent samplings of transformation events in the different transformation experiments. Calculations were performed using the IBM SPSS Statistics 22 software pack. Groupwise comparisons with one degree of freedom were applied.

Additional files

Additional file 1: Aberrant plant morphologies after transformation and regeneration.

Additional file 2: Characteristics of *P. infestans* isolates used in this study.

Additional file 3: Sequence of the pBINAW2:*Rpi-vnt:1.1:Rpi-sto1* construct.

Additional file 4: Primers used for PCR analysis of transformants.

Competing interests

The authors declare that they have no competing interests.

Authors' contributions

JKR wrote the manuscript and was involved in the experimental guidance of KCJ, KTY and MJ established the collaboration between the P-Y and the Wageningen institutes. KTY designed and cloned the marker-free transformation construct. MJ designed the integrated pest management strategy as pursued in the "EuropeAid" and "International Cooperation" projects and he, thereby, provided the scientific blueprint for this study. KCJ, MB and KSJ performed the transformations, PCRs, and late blight resistance assays. RV was involved in manuscript writing. EJ and JV provided the experimental design and supervised the manuscript writing. All authors read and approved the final manuscript.

Acknowledgements

pBINAW2 was kindly provided by Dr Annemarie Wolters, Wageningen UR plant Breeding. KRJ, TYK, CJK, SJK and MJ were financially supported by the European Commission (EuropeAid project DCI-FOOD/2009/218-671), and the Dutch Ministry of Agriculture, Nature and Fisheries (International Cooperation project BO-10-010-112 and BO-10-001-200). JHV was supported by the DuRPh program, financed by the Dutch Ministry of Economic affairs, formerly the Ministery of Agriculture, Nature and Fisheries.

Author details

[1]Wageningen UR Plant Breeding, Wageningen University & Research Centre, P.O. Box 386, 6700 AJ Wageningen, The Netherlands. [2]Graduate School Experimental Plant Sciences, Wageningen University, Wageningen, The Netherlands. [3]Research Institute of Agrobiology, Academy of Agricultural Sciences, Pyongyang, DPR Korea. [4]Plant Research International, Wageningen University and Research Centre, Wageningen, The Netherlands.

References

1. Muller KO, Black W: **Potato breeding for resistance to blight and virus diseases during the last hundred years.** *Z Pflanzenzuchtg* 1951, **31**:305–318.
2. Malcolmson JF, Black W: **New *R* genes in *Solanum demissum* Lindl. and their complementary races of *Phytophthora infestans* (Mont.) de Bary.** *Euphytica* 1966, **15**:199–203.
3. Malcolmson JF: **Races of *Phytophthora infestans* occurring in Great Britain.** *Trans Br Mycol Soc* 1969, **53**:417–423.
4. Bradshaw JE, Bryan GJ, Lees AK, McLean K, Solomon-Blackburn RM: **Mapping the *R10* and *R11* genes for resistance to late blight (*Phytophthora infestans*) present in the potato (*Solanum tuberosum*) *R* gene differentials of Black.** *Theor Appl Genet* 2006, **112**:744–751.
5. Wastie RL: **Breeding for resistance.** *Adv Plant Pathol* 1991, **7**:193–224.
6. Que Q, Chilton MDM, Fontes CM D, He C, Nuccio M, Zhu T, Wu Y, Chen JS, Shi L: **Trait stacking in transgenic crops: the challenges and opportunities.** *GM Crops* 2010, **1**:220–229.
7. Strange RN, Scott PR: **PLANT DISEASE: a threat to global food security.** *Annu Rev Phytopathol* 2005, **43**:83–116.
8. Tomczynska I, Stefanczyk E, Chmielarz M, Karasiewicz B, Kaminski P, Jones JD, Lees AK, Sliwka J: **A locus conferring effective late blight resistance in potato cultivar Sarpo Mira maps to chromosome XI.** *Theor Appl Genet* 2014, **127**(3):647–657.
9. Rietman H, Bijsterbosch G, Cano LM, Lee HR, Vossen JH, Jacobsen E, Visser RGF, Kamoun S, Vleeshouwers VGAA: **Qualitative and quantitative late blight resistance in the potato cultivar Sarpo Mira is determined by the perception of five distinct RXLR effectors.** *Mol Plant Microbe Interact* 2012, **25**(7):910–919.
10. Song J, Bradeen JM, Naess SK, Raasch JA, Wielgus SW, Haberlach GT, Liu J, Kuang H, Austin-Phillips S, Buell CR, Helgeson JP, Jiang J: **Gene *RB* cloned from *Solanum bulbocastanum* confers broad spectrum resistance to potato late blight.** *Proc Natl Acad Sci U S A* 2003, **100**:9128–9133.
11. van der Vossen EAG, Sikkema A, te Lintel HB, Gros J, Stevens P, Muskens M, Wouters D, Pereira A, Stiekema WJ, Allefs S: **An ancient *R* gene from the wild potato species *Solanum bulbocastanum* confers broad-spectrum resistance to *Phytophthora infestans* in cultivated potato and tomato.** *Plant J* 2003, **36**:867–882.
12. van der Vossen EAG, Gros JE, Sikkema A, Muskens M, Wouters D, Wolters P, Pereira A, Allefs S: **The *Rpi-blb2* gene from *Solanum bulbocastanum* is an *Mi-1* gene homolog conferring broad-spectrum late blight resistance in potato.** *Plant J* 2005, **44**:208–222.
13. Vleeshouwers VGAA, Rietman H, Krenek P, Champouret N, Young C, Oh SK, Wang M, Bouwmeester K, Vosman B, Visser RGF, Jacobsen E, Govers F, Kamoun S, van der Vossen EAG: **Effector genomics accelerates discovery and functional profiling of potato disease resistance and *Phytophthora infestans* avirulence genes.** *PLoS ONE* 2008, **3**:e2875.
14. Lokossou AA, Park TH, Van Arkel G, Arens M, Ruyter-Spira C, Morales J, Whisson SC, Birch PRJ, Visser RGF, Jacobsen E, van der Vossen EAG: **Exploiting knowledge of *R/Avr* genes to rapidly clone a new LZ-NBS-LRR family of late blight resistance genes from potato linkage group IV.** *Mol Plant Microbe Interact* 2009, **22**:630–641.

15. Vossen JH, Nijenhuis M, Arens de Reuver MJB, Van der Vossen EAG, Jacobsen E, Visser RGF: **Cloning and exploitation of a functional *R* gene from *Solanum chacoense*.** Patent application, published by: IPO: PCT/NL2010/050612.
16. Pel MA, Foster SJ, Park TH, Rietman H, Van Arkel G, Jones JDG, Eck HJ V, Jacobsen E, Visser RGF, Vossen EAG Van D: **Mapping and cloning of late blight resistance genes from *Solanum venturii* using an interspecific candidate gene approach.** *Mol Plant Microbe Interact* 2009, **22**:601–615.
17. Foster SJ, Park TH, Pel MA, Brigneti G, Sliwka J, Jagger L, van der Vossen EAG, Jones JDG: ***Rpi-vnt1.1*, a *Tm-2* homolog from *Solanum venturii*, confers resistance to potato late blight.** *Mol Plant Microbe Interact* 2009, **22**:589–600.
18. Mullins E, Milbourne D, Petti C, Doyle-Prestwich BM, Meade C: **Potato in the age of biotechnology.** *Trends Plant Sci* 2006, **11**(5):254–260.
19. Barrell PJ, Meiyalaghan S, Jacobs JM, Conner AJ: **Applications of biotechnology and genomics in potato improvement.** *Plant Biotechnol J* 2013, **11**(8):907–920.
20. Jacobsen E, Schouten HJ: **Cisgenesis strongly improves introgression breeding and induced translocation breeding of plants.** *Trends Biotechnol* 2007, **25**:219–223.
21. Jacobsen E: **Cisgenesis: a modern way of domesticating traits of the breeders' gene pool.** *CAB Reviews* 2013, **8**:56.
22. Eurobarometer: **Europeans and Biotechnology in 2010; Winds of change?** http://ec.europa.eu/research/science-society/document_library/pdf_06/europeans-biotechnology-in-2010_en.pdf accessed March 13, 2014.
23. Zhu SX, Li Y, Vossen JH, Visser RGF, Jacobsen E: **Functional stacking of three resistance genes against *Phytophthora infestans* in potato.** *Transgenic Res* 2012, **21**:89–99.
24. Pel MA: **Mapping, isolation and characterization of genes responsible for late blight resistance in potato.** In *PhD thesis.* Wageningen University; 2010.
25. Wheeler VA, Evans NE, Foulger D, Webb KJ, Karp A, Franklin J, Bright SWJ: **Shoot formation from explant cultures of fourteen potato cultivars and studies of the cytology and morphology of regenerated plants.** *Ann Bot* 1985, **55**(3):309–320.
26. Jacobsen E, Schouten HJ: **Cisgenesis, a new tool for traditional plant breeding, should be exempted from the regulation on genetically modified organisms in a step by step approach.** *Potato Res* 2008, **51**:75–88.
27. de Vetten NCM, Wolters AMA, Raemakers K, van der Meer I, ter Stege R, Heeres E, Heeres P, Visser RGF: **A transformation method for obtaining marker-free plants of a cross-pollinating and vegetatively propagated crop.** *Nat Biotechnol* 2003, **21**:439–442.
28. Joshi SG: **Towards durable resistance to apple scab using cisgenes.** In *PhD thesis.* Wageningen University; 2010.
29. Heeres P, Schippers-Rozenboom M, Jacobsen E, Visser RGF: **Transformation of a large number of potato varieties: genotype-dependent variation in efficiency and somaclonal variability.** *Euphytica* 2002, **124**:13–22.
30. Thomas CM, Jones JDG: **Molecular analysis of Agrobacterium T-DNA integration in tomato reveals a role for left border sequence homology in most integration events.** *Mol Genet Genomics* 2007, **278**(4):411–420.
31. Petti C, Wendt T, Meade C, Mullins E: **Evidence of genotype dependency within *Agrobacterium tumefaciens* in relation to the integration of vector backbone sequence in transgenic *Phytophthora infestans*-tolerant potato.** *J Biosci Bioeng* 2009, **107**(3):301–306.
32. EFSA: **Scientific opinion addressing the safety assessment of plants developed through cisgenesis and intragenesis.** *EFSA J* 2012, **10**:2561.
33. Murashige T, Skoog F: **A revised medium for rapid growth and bioassays with tobacco tissue cultures.** *Physiol Plant* 1962, **15**:473–497.
34. Vleeshouwers VGAA, Van DW, Keizer LC P, Sijpkes L, Govers F, Colon LT: **A laboratory assay for *Phytophthora infestans* resistance in various *Solanum* species reflects the field situation.** *Eur J Plant Pathol* 1999, **105**:241–250.
35. de Vetten NCM, Visser RGF, Jacobsen E, van der Vossen EAG, Wolters AMA: **Use of *R*-genes as a selection marker in plant transformation and use of cisgenes in plant transformation.** patent application, published by: IPO: WO 2008091154 A1.
36. Visser RGF: **Regeneration and transformation of potato by *Agrobacterium tumefaciens*.** In *Plant Tissue Culture Manual.* Edited by Lindsey K. Dordrecht, Boston, London: Kluwer Academic Publishers; 1991:1–9.
37. Vleeshouwers VGAA, Rietman H: **In planta expression systems.** In *Oomycete Genetics and Genomics.* Edited by Lamour KH, Kamoun S. Hoboken, New Jersey: John Wiley & Sons; 2009:455–475.
38. Huang S, van der Vossen EAG, Kuang H, Vleeshouwers VGAA, Zhang N, Borm TJA, van Eck HJ, Baker B, Jacobsen E, Visser RGF: **Comparative genomics enabled the isolation of the *R3a* late blight resistance gene in potato.** *Plant J* 2005, **42**:251–261.
39. Armstrong MR, Whisson SC, Pritchard L, Bos JIB, Venter E, Avrova AO, Rehmany AP, Bohme U, Brooks K, Cherevach I, Hamlin N, White B, Fraser A, Lord A, Quail MA, Churcher C, Hall N, Berriman M, Huang S, Kamoun S, Beynon JL, Birch PRJ: **An ancestral oomycete locus contains late blight avirulence gene *Avr3a*, encoding a protein that is recognized in the host cytoplasm.** *Proc Natl Acad Sci USA* 2005, **102**:7766–7777.
40. Karimi M, Inze D, Depicker A: **GATEWAY(TM) vectors for *Agrobacterium*-mediated plant transformation.** *Trends Plant Sci* 2002, **7**:193–195.
41. Fulton T, Chunwongse J, Tanksley S: **Microprep protocol for extraction of DNA from tomato and other herbaceous plants.** *Plant Mol Biol Rep* 1995, **13**:207–209.

17

Expression, purification, and characterization of rhTyrRS

Yongjiang Lang[1,2], Yanling Zhang[1,2], Ling Zhan[1,2], Zhe Feng[1,2], Xiushi Zhou[1,2], Min Yu[1,2*] and Wei Mo[1,2*]

Abstract

Background: Aminoacyl-tRNA synthetases (AARSs) catalyze the first step of protein synthesis. Emerging evidence indicates that AARSs may have additional functions, playing a role in signal transduction pathways regulating thrombopoiesis and inflammation. Recombinant human tyrosyl-tRNA synthetase (rhTyrRS) is engineered with a single amino acid substitution that unmasks its cytokine activity. An industrial production method that provides high yield as well as high purity, quality, and potency of this protein is required for preclinical research.

Results: We expressed codon-optimized rhTyrRS in *Escherichia coli* under fermentation conditions. Soluble protein was purified by a three-step purification method using cation exchange chromatography, gel filtration chromatography, and anion exchange chromatography. We also established a method to test the biological activity of rhTyrRS by measuring aminoacylation and IL-8 release in rhTyrRS-treated HL-60 cells.

Conclusions: The characterization of purified rhTyrRS indicated that this protein can be used in pharmacodynamic and pharmacokinetic studies.

Keywords: Recombinant human tyrosyl-tRNA synthetase, Expression of rhTyrRS, Protein purification, Biological activity assay

Background

Aminoacyl-tRNA synthetases (AARSs) are enzymes that catalyze the essential first step of protein synthesis by covalently attaching each of the 20 standard amino acids to their cognate tRNA. Because the reactions require the capacity to recognize tRNAs as well as small chemicals such as amino acids and ATP, the structures of AARSs are well equipped for interacting with diverse molecules that may be associated with their functional versatility [1,2].

AARSs, such as glutamyl-prolyl-tRNA synthetase, glutaminyl-tRNA synthetase, and lysyl-tRNA synthetase, are now known to be multifunctional, participating in a variety of functions including transcription, translation, splicing, inflammation, thrombopoiesis, and apoptosis [3]. Human tyrosyl-tRNA synthetase (TyrRS) was the first example of a tRNA synthetase that played a role in cell signaling and thrombopoiesis. It has been demonstrated that human full-length TyrRS has no known cytokine activity [4]. However, proteolytic removal of the C-domain from TyrRS activates its cytokine function [1,2,5]. The N-terminal fragment, named mini-TyrRS, specifically stimulates migration of endothelial cells and polymorphonuclear leukocytes [1,2,6,7]. Mini-TyrRS has also been shown to activate thrombocytopoiesis in a chick chorioallantoic membrane thrombopoiesis assay [8].

As in some CXC chemokines, such as IL-8, an Glu-Leu-Arg motif embedded in the sequence of mini-TyrRS was shown by mutational analysis to be critical for its cytokine activity [9-11]. Although there is no crystal structure reported for full-length TyrRS, high-resolution crystal structures of mini-TyrRS and of the C-domain enabled reconstruction of the full-length enzyme, and has provided the model and structural basis for its cytokine activity [8,12]. It is predicted that the C-terminus of the full-length protein is tethered to the N-terminal region, thus preventing ELR substrate recognition [13,14].

Recombinant human TyrRS (rhTyrRS) is a full-length human tyrosyl-tRNA synthetase with a mutation at Y341A that blocks the H-bond between the ELR motif and Y341. As a result, the ELR motif is exposed, and mini-rhTyrRS-like activity can then promote thrombocytopoiesis [15,16].

* Correspondence: minyu@shmu.edu.cn; weimo1025@shmu.edu.cn
[1]The Key Laboratory of Molecular Medicine, Ministry of Education, Fudan University, Shanghai, P.R. China
[2]The Department of Biochemistry and Molecular Biology, School of Basic Medical Sciences, Fudan University, Shanghai, P.R. China

As a potential thrombocytopoietic drug, large-scale expression and purification of rhTyrRS would be desirable. Preclinical testing involving *in vitro* and animal studies could then be carried out to evaluate its toxic and pharmacologic effects. In this study, rhTyrRS was expressed at a high level in *E. coli* and purified for future preclinical testing.

Methods

Cells and antibodies

The competent *E. coli* strain BL21 (F-ompT hsdS (rB-mB-) gal dcm; providded by aTyr Pharma) was used as the host for rhTyrRS expression. This strain was transformed with the pET24a inducible expression vector in which the His-tag sequence was deleted and the T7 promoter was replaced with a Tac promoter.

A mouse anti-human IL-8 monoclonal antibody (BC0 13615, Proteintech Group), rabbit anti-human IL-8 polyclonal antibody (BC013615, Proteintech Group), and goat anti-rabbit IgG-HRP SC-2004 antibody (D2111, Santa Cruz Biotechnology) were used.

Growth conditions

A clone with a high rhTyrRS expression level was cultivated in LB agar and M9CA medium (10 g glucose, 6 g Na_2HPO_4, 35 g KH_2PO_4, 2.93 g NaCl, 0.4 g NH_4Cl, 1.2 g $MgSO_4$, and 1 mL trace element solution per liter). Trace element solution (1 L) consisted of 2.8 g $FeSO_4 \cdot 7H_2O$, 2 g $MnCl_2 \cdot 4H_2O$, 2.8 g $CoSO_4 \cdot 7H_2O$, 1.5 g $CaCl_2 \cdot 2H_2O$, 0.2 g $CuCl_2 \cdot 2H_2O$, and 0.3 g $ZnSO_4 \cdot 7H_2O$. Feeding solutions were 50% glycerol, 250 g/L glucose, and 100 g/L yeast extract. The glucose and $MgSO_4$ solutions were sterilized separately. Kanamycin sulfate was added to a final concentration of 100 μg/mL in both the M9CA and feeding solutions.

Fermentation was performed in a 5-L Bioflo 3000 fermenter (New Brunswick Scientific, New Brunswick, NJ, USA) with automated control of: the pH at 7.0 by the addition of ammonium hydroxide, dissolved O_2 at 70% by providing pure oxygen, and agitation at 700 rpm. Air was provided at a flow rate of 4.0 L/min and the temperature was controlled at 30°C. Fermentation was conducted according to the process developed by Shiloach *et al.* (1996) [17]. The batch phase ends when cells have used up the available glucose. The best indications that the batch phase has ended include a sharp decrease in stirrer speed and an increase in pO_2. The bacterial concentration was measured off-line by the optical density at 600 nm and induced with 0.5 mM isopropylthio-β-galactoside (IPTG) once it reached an OD600 of 30 (~10 h). After 6 h of 0.5 mM IPTG induction, the cells were harvested by centrifugation at 6,000 × *g* for 30 min. The cell pellet was stored at −70°C.

Cell lysis

Harvested cell pellets were resuspended in 10 volumes of 20 mmol/L HAc-NaAc buffer (pH 6.0) and subjected to two cycles of microfluidization at 1000 bar. The crude extract was then centrifuged at 10,000 × *g* for 60 min.

Cation exchange chromatography

The clarified supernatant was loaded onto a SP Sepharose Fast Flow column (GE) that was pre-equilibrated with 20 mmol/L HAc-NaAc buffer (pH 6.0). The bound proteins were eluted with a linear NaCl gradient (0 to 1 mol/L). Fractions containing rhTyrRS were pooled and analyzed by SDS-PAGE.

Gel filtration

The pooled fractions were loaded onto a Sephadex-G50 column (GE) pre-equilibrated with 20 mmol/L phosphate buffer (PB; pH 7.0) and rhTyrRS was washed with 20 mmol/L PB at 5 mL/min. Fractions were collected from the column and analyzed by SDS-PAGE.

Anion exchange chromatography

The diluted product solution was loaded onto a 50-mL Q Sepharose Fast Flow column (GE) pre-equilibrated with 20 mmol/L PB (pH 7.0). RhTyrRS was eluted with a linear NaCl gradient (0 to 1 mol/L) and identified by SDS-PAGE via Coomassie blue staining.

SDS-PAGE and sequencing of the N-terminal amino acids

Electrophoresis was carried out in 1-mm-thick gels using BioRad MiniGel apparatus. Coomassie staining was performed as previously described, except that microwave heating was used at each staining step to reduce the total staining and destaining procedure time to 30 min. The protein concentration was measured using BSA as a standard.

The sequence of the N-terminal amino acids of purified rhTyrRS was determined using a protein sequencer (PPSQ-33A, USA).

Western blotting

Proteins resolved in a pre-cast Bis-Tris gel (BioRad) were electrotransferred to a PVDF membrane followed by blocking in 10% BSA solution prepared in TBST (Tris-buffered saline with 0.1% Tween-20). The membrane was then incubated with anti-rhTyrRS monoclonal antibody (1/5000) for 90 min at room temperature. After washing, the membrane was incubated with peroxidase-conjugated goat anti-mouse IgG (1/100,000) for 60 min at room temperature. All antibody incubations and washing steps were carried out in TBST. The immunoreactive bands were visualized with a Western Blot kit (Thermo 34077).

Reversed-phase HPLC

Reversed-phase HPLC was used to study the purity of rhTyrRS. A Higgins Proto 300 C_4 column (Waters 150 × 4.6 mm) was used at a flow rate of 1 mL/min and at a temperature of 40°C. The column was eluted with a linear 30-min gradient from 5% B to 95% B (A = 0.1% Trifluoroacetic Acid (TFA) in HPLC water; B = 0.1% TFA in acetonitrile) and monitored by absorbance at 215 nm with a total run time of 40 min.

Mass spectrometry analysis

The purified protein was analyzed by mass spectrometry. Molecular weight measurements were made by LC-MS with a quadrupole-time-of-flight mass spectrometer at the Fudan University Institute of Biomedical Science.

Aminoacylation assay

Aminoacylation activity was determined at ambient temperature in 150 mmol/L Tris–HCl (pH 7.5), 150 mmol/L KCl, 10 mmol/L $MgCl_2$, 10 mmol/L β-mercaptoethanol, 4 mmol/L ATP, and 10 mmol/L tyrosine (including 3 mmol/L [^{3}H]-tyrosine, GE Healthcare). Human tRNATyr was prepared as previously described [12]. Before each assay, 100 mmol/L human tRNATyr was annealed by heating at 65°C for 5 min and cooled to room temperature. Reactions were initiated by the addition of rhTyrRS (10 nmol/L) to the reaction mixture. Aliquots were taken at fixed intervals and spotted onto Whatman filter discs saturated with 5% trichloroacetic acid and dried. The filters were placed into ice-cold 5% trichloroacetic acid and washed three times with fresh 5% trichloroacetic acid at 4°C and once with 95% ethanol. The level of tRNA aminoacylation was then quantitated by liquid scintillation counting.

Biological activity assay

RhTyrRS activity was tested using ELISA to measure IL-8 secretion by rhTyrRS-treated HL-60 cells. Cells were thawed from vials and cultured according to the American Type Culture Collection instructions, except that RPMI-1640 (ATCC 30–2001) was supplemented with 10% heat-inactivated fetal bovine serum (Life Technologies, formerly Invitrogen 10082–147) and used as growth medium. Cells were plated in a 96-well plate at a density of 1×10^6 cells per 1 mL of medium on the day of the assay. The rhTyrRS was added immediately after cell plating. Cells were plated at 10^8 μl^{-1} per well. The protein was diluted in sterile phosphate-buffered saline (PBS; Life Technologies, formerly Invitrogen 10010–023) to a concentration of 10 μmol/L (10× protein stock). The 10× protein stock was serially diluted and 12 μL was added to each well such that the final protein concentrations were 500 nmol/L, 250 nmol/L, 125 nmol/L, 62.5 nmol/L, 31.25 nmol/L, and 15.625 nmol/L. PBS was used as a control. Cells were incubated at 37°C and 5% CO_2 for 24 h. The 96-well plate was then spun for 10 min at room temperature in a swinging bucket centrifuge at 1,000 × *g*. The plate was tilted at a 45° angle and 100 μL of media per well was transferred to a new 96-well plate. The amount of secreted IL-8 protein in the media was then determined by sandwich ELISA.

Results

rhTyrRS expression and purification

To evaluate the long-term stability of the plasmid/strain combination, bacteria were first grown in shake flasks for

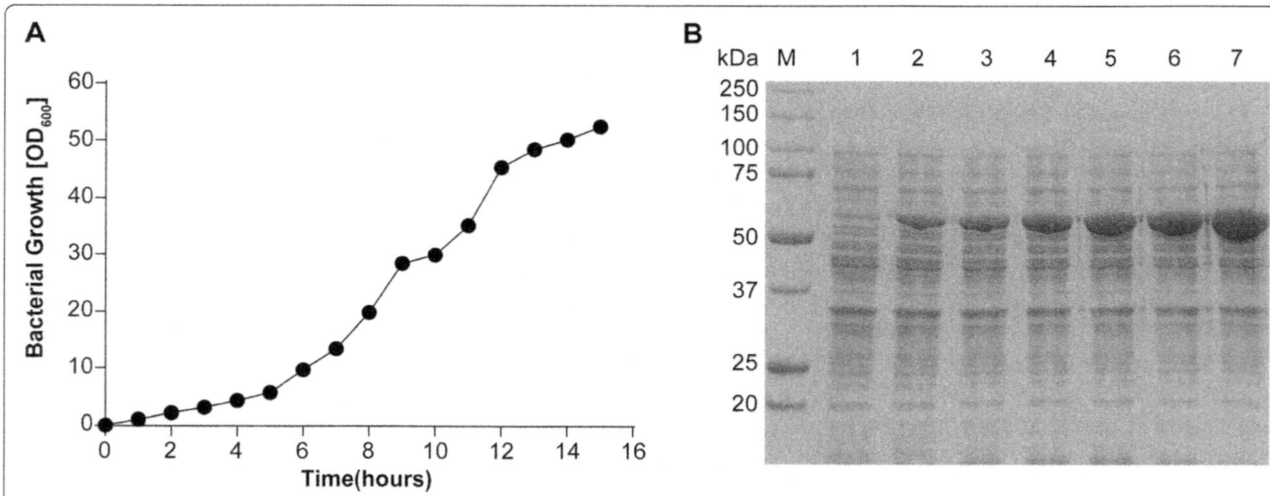

Figure 1 Expression and purification of rhTyrRS. A: Fed-batch growth curve of BL21 cells expressing rhTyrRS. A 5-L fermentor was inoculated with a 150-mL seed culture at time 0 and samples were drawn at regular intervals to monitor the optical density at 600 nm. rhTyrRS expression was induced with IPTG at approximately OD600 = 30. **B**: SDS-PAGE analysis of rhTyrRS expressed by fed-batch fermentation. Lane 1, standard protein marker (20–250 kD); Lane 2, uninduced BL21 cells; Lanes 3–8, BL21 cells induced with 0.5 mM IPTG for 1, 2, 3, 4, 5, and 6 h.

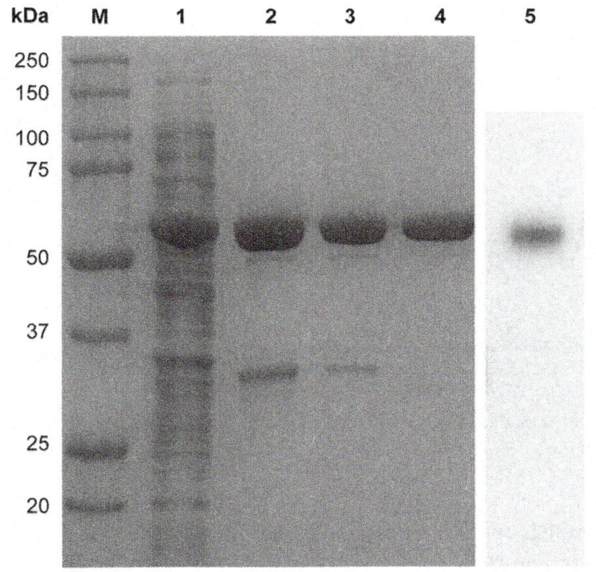

Figure 2 Purification of rhTyrRS using a three-step procedure. M, standard protein marker (20–250 kD). Lane 1, soluble lysate from induced cells; Lane 2, cation exchange chromatography wash fraction; Lane 3, gel filtration wash fraction; Lane 4, anion exchange chromatography wash fraction; Lane 5, western blot assay of rhTyrRS.

25 generations by repeated dilution before the analysis of rhTyrRS expression. A typical growth curve is shown in Figure 1A. SDS-PAGE analysis of expressed rhTyrRS is shown in Figure 1B. The expression level after 25 generations was comparable to that produced by the cells after five generations. Approximately 1.5–2.0 g/L of rhTyrRS was produced under these growth conditions. When rhTyrRS was expressed in *E. coli* strain BL21 using M9CA medium and fed-batch methodology, approximately 50% of the protein was soluble.

After 6 h of IPTG induction, cells were harvested by centrifugation and lysed by two cycles of microfluidization. The concentrated media were subjected to cation exchange, gel filtration, and anion exchange chromatography, yielding rhTyrRS of up to 97% purity. RhTyrRS was eluted with 0.25 mol/L NaCl-PB via cation exchange, then subjected to gel filtration to exchange the buffer to PB; all pooled fractions were then loaded onto a 50 mL Q Sepharose Fast Flow column pre-equilibrated with 20 mmol/L PB (pH 7.0). RhTyrRS was eluted with less than 0.1 mol/L NaCl-PB during anion exchange. Around 1.0 g of purified rhTyrRS was generated from 3 L of culture. The products from the optimized rhTyrRS purification process, consisting of cell lysis followed by cation exchange chromatography, gel filtration, and anion exchange chromatography, are shown in Figure 2.

rhTyrRS characterization

The identity of rhTyrRS was verified by Edman degradation to compare the N-terminal sequence of rhTyrRS to the expected mini-TyrRS sequence. Up to 15 amino acids from the N-terminus of rhTyrRS were consistently identical to the reference sequence: Met-Gly-Asp-Ala-Pro-Ser-Pro-Glu-Glu-Lys-Leu-His-Leu-Ile-Trp. We also sought to confirm the identity of rhTyrRS by resolving the sample (s) in parallel with a reference standard by SDS-PAGE followed by western blotting using a polyclonal antibody (Figure 2).

Reversed-phase HPLC

A reversed-phase HPLC was developed as a purity indication assay. Figure 3 shows the analysis of rhTyrRS purified using chromatography steps. This method will be transferred to the contract manufacturer organization or its approved vendor, as appropriate. The retention time of rhTyrRS in the C4 column was 19.5 min with purity up to 97%.

Mass spectrometry analysis

An accurate mass can be detected by intact mass analysis using LC-MS. Peaks corresponding to various protonated rhTyrRS species are marked. Mass analysis determined that rhTyrRS has a molecular weight of 59.18 kD (Figure 4). This is identical to that identified by SDS-PAGE.

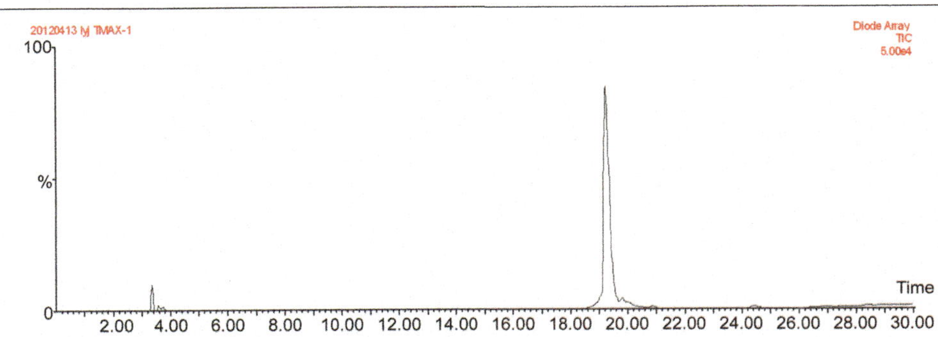

Figure 3 Purity analysis of rhTyrRS by HPLC. A single symmetrical peak with a 97% purity was indicated in the HPLC analytical chart.

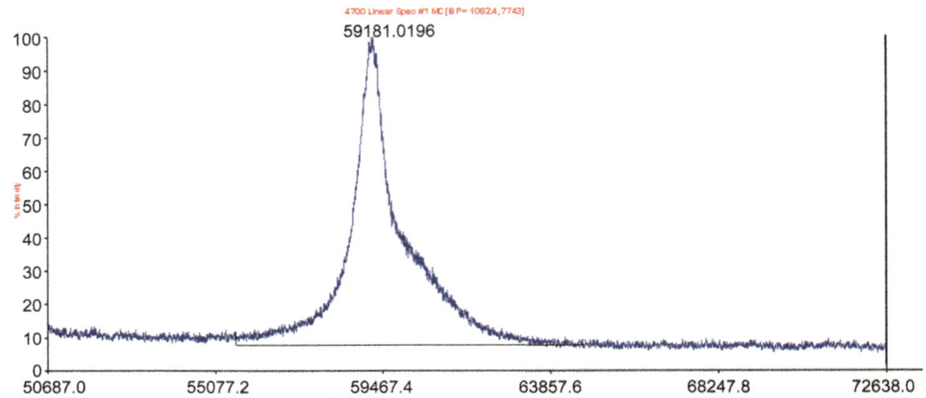

Figure 4 Molecular mass determination of purified rhTyrRS by LC-MS. Peaks corresponding to various protonated rhTyrRS species are marked. The mass analysis determined that rhTyrRS has a molecular weight of 59.18 kD.

Aminoacylation assay

To test whether rhTyrRS retained the ability to aminoacylate its cognate tRNA molecule, we took advantage of the inherent synthetase activity of rhTyrRS to verify proper protein folding and assess biological activity during the expression and purification process. The incorporation of ^{3}H tyrosine into tRNA is a well-established and reproducible assay [4,5] (Figure 5).

Biological activity assay

We used a sandwich ELISA to test IL-8 released by HL-60 cells in the presence of rhTyrRS. The standard IL-8 concentrations used for ELISA were 1200 ng/mL, 600 ng/mL, 300 ng/mL, 150 ng/mL, 75 ng/mL, and 32.5 ng/mL as shown in Figure 6A, and R^2 was 0.9924. Figure 6B shows the amounts of IL-8 released by HL-60 cells incubated in the presence of rhTyrRS at concentrations between 62.5 nmol/L and 1000 nmol/L, based on the standard IL-8 shown in Figure 6A.

Discussion and conclusions

To facilitate the use of rhTyrRS proteins in regulating thrombopoietic activity and to enable structural characterization of this protein, we established a system for the production of large amounts of rhTyrRS. A purity of approximately 97% was routinely achieved with <2% individual contaminants and little to no aggregation.

The final, optimized process consists of cell lysis followed by a sequence of three chromatography steps. The first step (cation exchange) captures rhTyrRS while contaminants flow through, the second step (gel filtration) exchanges the buffer to PB, and the last step (anion exchange) removes the remaining process-related impurities. The final concentration is 1 mg/ml in PBS (pH 7.4).

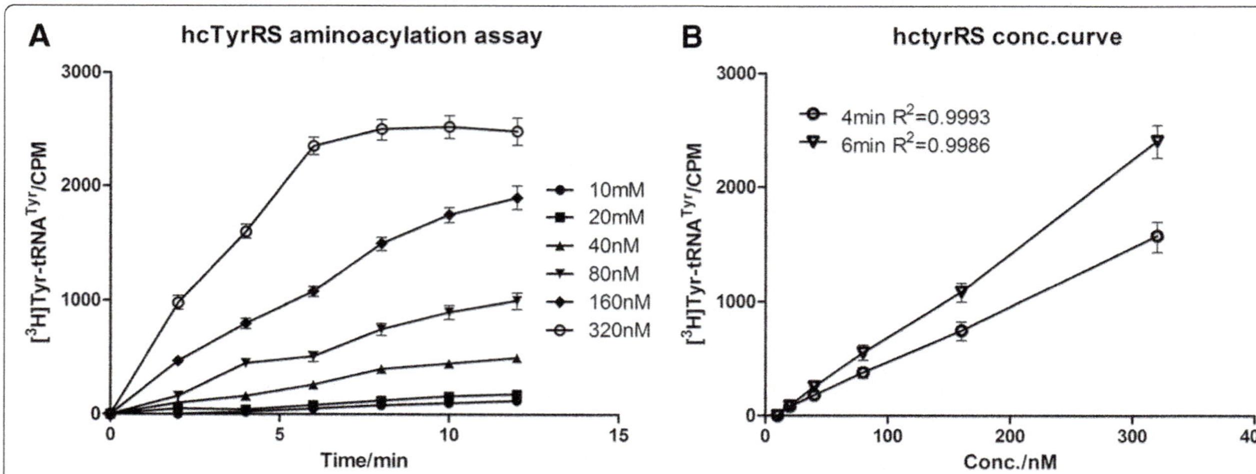

Figure 5 rhTyrRS is a functional aminoacyl-tRNA synthetase. A: The ability of rhTyrRS to charge tRNA with ^{3}H-labeled tyrosine was tested *in vitro* as a function of rhTyrRS concentration and time. The following concentrations of rhTyrRS were tested: 320 nM, 160 nM, 80 nM, 40 nM, 20 nM, and 10 nM. **B**: The ability of rhTyrRS to charge tRNA after 4 and 6 min.

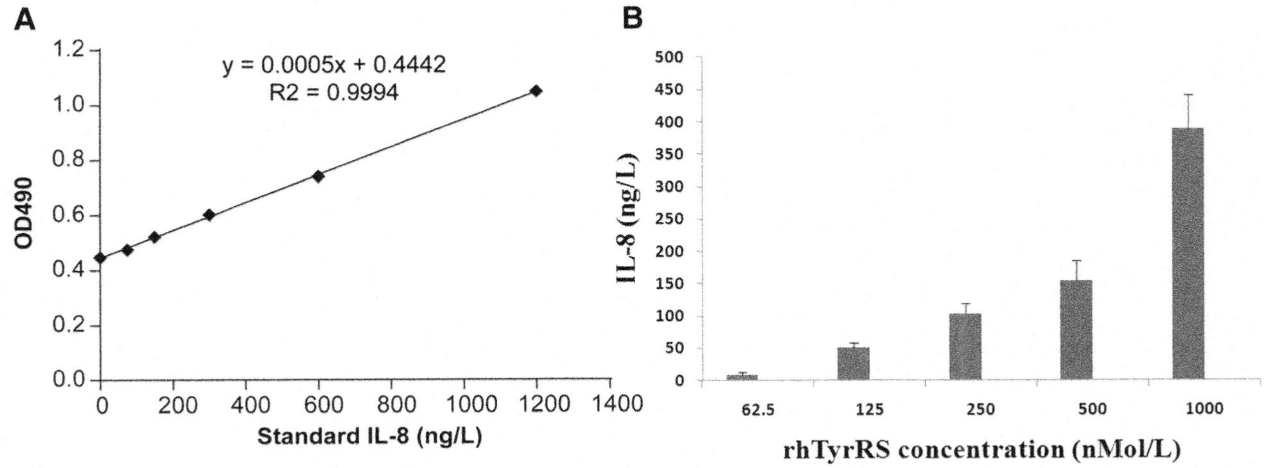

Figure 6 Dose-dependent induction of interleukin-8 (IL-8) production in HL-60 cells by rhTyrRS.

Preliminary characterization of rhTyrRS was performed using western blotting, HPLC, mass spectrometry, and N-terminal sequencing to verify the identity of the protein.

The ability of the purified rhTyrRS to aminoacylate its cognate tRNA molecule was also tested, and analysis of the inherent synthetase activity of rhTyrRS was performed to verify proper protein folding and assess biological activity during the expression and purification processes. Kinetic parameters from different preparations of rhTyrRS will be compared with a well-characterized reference standard. Purification lots that are within the established specifications will be deemed acceptable for release. This assay is still under development but a representative example of the linearity of product accumulation over time is shown below with a range of rhTyrRS concentrations.

The Y341A substitution activated full-length TyrRS for stimulation of cell proliferation. Similar results were observed in an endothelial cell migration assay. In these experiments, wounds were created in monolayers of human umbilical vein endothelial cells [9]. We predict that rhTyrRS not only has IL-8-like cytokine activity, but also can enhance IL-8 production from monocytes [16]. The biological assays of HL-60 stimulation with rhTyrRS presented here demonstrate that this may be the case. The findings presented here will be important for future preclinical studies.

Competing interests

The authors declare that they have no competing interests.

Authors' contributions

YL, LZ and XZ performed the protein expression and purification; YZ and ZF analyzed the protein activity; YL and WM designed the experiments, analyzed data, and wrote the paper; MY supervised the experiments and designed the study. All authors read and approved the final manuscript.

Authors' information

Yongjiang Lang and Yanling Zhang are first author.

Acknowledgments

We thank aTyr Pharma for providing the plasmid, *E. coli* strain, and the antibody for rhTyrRS. We really appreciate the advise of Dr. Xianglei Yang and Dr. Paul Schimmel from The Scripps Research Institute. This study was supported by the National Natural Science Foundation of China (NSFC 21242009, NSFCJ121041, JYH6273102/007/016/007), and Science and Technology Commission of Shanghai Municipality (STCSM 10431901000). This study was also supported by The State Key Laboratory of Biotherapy/Collaborative Innovation Center for Biotherapy, West China Hospital, Sichuan University.

References

1. Wakasugi K, Schimmel P: **Two distinct cytokines released from a human aminoacyl-tRNA synthetase.** *Science* 1999, **287:**147–151.
2. Wakasugi K, Schimmel P: **Highly differentiated motifs responsible for two cytokine activities of a split human tRNA synthetase.** *J Biol Chem* 1999, **274:**23155–23159.
3. Yang XL, Kapoor M, Otero FJ, Slike BM, Tsuruta H, Frausto R, Bates A, Ewalt KL, Cheresh DA, Schimmel P: **Gain-of-function mutational activation of human tRNA synthetase procytokine.** *Chem Biol* 2007, **14:**1323–1333.
4. Kapoor M, Otero FJ, Slike BM, Ewalt KL, Yang XL: **Mutational separation of aminoacylation and cytokine activities of human tyrosyl-tRNA synthetase.** *Chem Biol* 2009, **16:**531–539.
5. Jia J, Li B, Jin Y, Wang D: **Expression, purification and characterization of human tyrosyl-tRNA synthetase.** *Protein Expr Purif* 2003, **27:**104–108.
6. Kleeman TA, Wei D, Simpson KL, First EA: **Human tyrosyl-tRNA synthetase shares amino acid sequence homology with a putative cytokine.** *J Biol Chem* 1997, **272:**14420–14425.
7. Nangle LA, Zhang W, Xie W, Yang XL, Schimmel P: **Char-cot-Marie-Tooth disease-associated mutant tRNA synthetases linked to altered dimer interface and neurite distribution defect.** *Proc Natl Acad Sci U S A* 2007, **104:**11239–11244.
8. Park SG, Schimmel P, Kim S: **Aminoacyl tRNA synthetases and their connections to disease.** *Proc Natl Acad Sci U S A* 2008, **105:**11043–11049.
9. Yang XL, Guo M, Kapoor M, Ewalt KL, Otero FJ, Skene RJ, McRee DE, Schimmel P: **Functional and crystal structure analysis of active site adaptations of a potent anti-angiogenic human tRNA synthetase.** *Structure* 2007, **15:**793–805.
10. Wakasugi K, Slike BM, Hood J, Otani A, Ewalt KL, Friedlander M, Cheresh DA, Schimmel P: **A human aminoacyl-tRNA synthetase as a regulator of angiogenesis.** *Proc Natl Acad Sci U S A* 2002, **99:**173–177.
11. Howard OM, Dong HF, Yang D, Raben N, Nagaraju K, Rosen A, Casciola-Rosen L, Hartlein M, Kron M, Yang D, Yiadom K, Dwivedi S, Plotz PH, Oppenheim JJ: **Histidyl-tRNA synthetase and asparaginyl-tRNA synthetase, autoantigens in**

myositis, activate chemokine receptors on T lymphocytes and immature dendritic cells.J. *Exp Med* 2002, **196**:781–791.

12. Liu J, Yang XL, Ewalt KL, Schimmel P: **Mutational switchingof a yeast tRNA synthetase into a mammalian-like synthetase cytokine.** *Biochemistry* 2002, **41**:14232–14237.
13. Yang XL, Schimmel P, Ewalt KL: **Relationship of two human tRNA synthetases used in cell signaling.** *Trends Biochem Sci* 2004, **29**:250–256.
14. Yang XL, Liu J, Skene RJ, McRee DE, Schimmel P: **Crystal structure of an EMAP-II-like cytokine released froma human tRNA synthetase.** *Helv Chim Acta* 2003, **86**:1246–1257.
15. Yang XL, Otero FJ, Skene RJ, McRee DE, Schimmel P, Ribas de Pouplana L: **Crystal structures that suggest late development ofgenetic code components for differentiating aromatic side chains.** *Proc Natl Acad Sci U S A* 2003, **100**:15376–15380.
16. Curtis BM, Williams DE, Broxmeyer HE, Dunn J, Farrah T, Jeffery E, Clevenger W, deRoos P, Martin U, Friend D: **Enhanced hematopoietic activity of a human granulocyte/macrophage colony-stimulating factor-interleukin 3 fusion protein.** *Proc Natl Acad Sci U S A* 1991, **88**:5809–5813.
17. Shiloach J, Kaufman J, Guillard AS, Fass R: **Effect of glucose supply strategy on acetate accumulation, growth, and recombinant protein production by Escherichia coli BL21 (λDE3) and Escherichia coli JM109.** *Biotechnol Bioeng* 1996, **49**:421–428.

A new acidophilic thermostable endo-1,4-β-mannanase from *Penicillium oxalicum* GZ-2: cloning, characterization and functional expression in *Pichia pastoris*

Hanpeng Liao, Shuixian Li, Haiping Zheng, Zhong Wei, Dongyang Liu, Waseem Raza, Qirong Shen and Yangchun Xu*

Abstract

Background: Endo-1,4-β-mannanase is an enzyme that can catalyze the random hydrolysis of β-1, 4-mannosidic linkages in the main chain of mannans, glucomannans and galactomannans and has a number of applications in different biotechnology industries. *Penicillium oxalicum* is a powerful hemicellulase-producing fungus (Bioresour Technol 123:117-124, 2012); however, few previous studies have focused on the cloning and expression of the endo-1,4-β-mannanase gene from *Penicillium oxalicum*.

Results: A gene encoding an acidophilic thermostable endo-1,4-β-mannanase (E.C. 3.2.1.78) from *Penicillium oxalicum* GZ-2, which belongs to glycoside hydrolase family 5, was cloned and successfully expressed in *Pichia pastoris* GS115. A high enzyme activity (84.4 U mL^{-1}) was detected in the culture supernatant. The recombinant endo-1,4-β-mannanase (rPoMan5A) was tagged with 6 × His at its C-terminus and purified using a Ni-NTA Sepharose column to apparent homogeneity. The purified rPoMan5A showed a single band on SDS-PAGE with a molecular mass of approximately 61.6 kDa. The specific activity of the purified rPoMan5A was 420.9 U mg^{-1} using locust bean gum as substrate. The optimal catalytic temperature (10 min assay) and pH value for rPoMan5A are 80°C and pH 4.0, respectively. The rPoMan5A is highly thermostable with a half-life of approximately 58 h at 60°C at pH 4.0. The K_m and V_{max} values for locust bean gum, konjac mannan, and guar gum are 7.6 mg mL^{-1} and 1425.5 μmol min^{-1} mg^{-1}, 2.1 mg mL^{-1} and 154.8 μmol min^{-1} mg^{-1}, and 2.3 mg mL^{-1} and 18.9 μmol min^{-1} mg^{-1}, respectively. The enzymatic activity of rPoMan5A was not significantly affected by an array of metal ions, but was inhibited by Fe^{3+} and Hg^{2+}. Analytical results of hydrolytic products showed that rPoMan5A could hydrolyze various types of mannan polymers and released various mannose and manno-oligosaccharides, with the main products being mannobiose, mannotriose, and mannopentaose.

Conclusion: Our study demonstrated that the high-efficient expression and secretion of acid stable and thermostable recombinant endo-1, 4-β-mannanase in *Pichia pastoris* is suitable for various biotechnology applications.

Keywords: Endo-1,4-β-mannanase, Gene cloning, Expression system, *Pichia pastoris*, *Penicillium oxalicum*

* Correspondence: ycxu@njau.edu.cn
Jiangsu Collaborative Innovation Center for Solid Organic Waste Utilization, College of Resources and Environmental Science, Nanjing Agricultural University, Nanjing 210095, China

Background

Mannan polysaccharides are the major components of softwood hemicellulose and can be classified into linear mannan, glucomannan, galactomannan, and galactoglucomannan [1]. They consist of a backbone of β-1,4-linked mannose, which may be interrupted by D-glucose in glucomannans. Due to the complex nature of these polysaccharides, a combination of different enzymes such as endo-1,4-β-mannanase (EC 3.2.1.78), β-mannosidase (EC 3.2.1.25), β-glucosidase (EC 3.2.1.21), α-galactosidase (EC 3.2.1.22) and acetyl mannan esterase (EC 3.1.1.6) are required to complete the hydrolysis of heteromannans. Among them, the most important is endo-1,4-β-mannanase (EC 3.2.1.78), which can catalyze random internal hydrolysis of β-1,4-mannosidic linkages in the main structure of β-1,4-mannans, glucomannans and galactomannans by releasing small manno-oligosaccharides [2]. This type of enzyme has shown a wide range of potential uses in industrial applications, including food/feed, pharmaceutical, pulp/paper, gas well stimulation [3], as well as for second generation biofuels [4,5]. Mannans, the major components of hemicellulose, associate with lignin and cellulose in lignocellulosic biomass, physically hindering cellulase access to the fibers and reducing the efficiency of enzymatic hydrolysis [6]. In the bioenergy process, the breakdown of lignocellulose to fermentable sugars by enzymatic hydrolysis is one of the decisive bottlenecks, due to the recalcitrance of the plant cell wall and the high cost of enzymes. Based on this application, these enzymes have recently received much attention for industrial use, with an ever-increasing demand for renewable bioenergy utilization.

During the past few decades, many endo-1,4-β-mannanases have been purified and characterized from plants, bacteria, fungi and metazoans, among which the filamentous fungi are considered to be good candidates for the industrial production of endo-1,4-β-mannanases [7]. Many studies have been performed focusing on discovering new endo-1,4-β-mannanases with robust enzymatic activity by various methods, such as mutating enzyme-producing strains and optimizing fermentation conditions, and on producing endo-1,4-β-mannanases by fermentation on an industrial scale [8,9]. However, a suitable endo-1,4-β-mannanase for industrial use involves not only a high enzymatic hydrolysis activity; but thermo- and acidic-stability are also needed as many industrial bioprocesses require the use of enzymes at low pH and high temperature conditions to prevent contamination by other microbes [10-12].

P. oxalicum GZ-2, an agricultural biomass-degrading filamentous fungus, was isolated in our lab from decaying-wood acidic soil and showed favorable acidophilic hemicellulase-producing ability [13]. When *P. oxalicum* GZ-2 was incubated with agricultural waste (corn cob) as carbon source, an obvious mannanase activity could be determined in the culture supernatant (data not shown). In this study, a new acidic thermostable GH5 endo-1,4-β-mannanase gene (*poman5A*) from *P. oxalicum* GZ-2 was cloned and heterologously expressed in *Pichia pastoris* GS115. To our knowledge, this is the first example of cloning and expression of an acidic thermostable endo-1,4-β-mannanase gene from *P. oxalicum*. The biochemical characteristics of the recombinant enzyme were also evaluated. The superior biochemical and hydrolytic properties of rPoMan5A make it a highly useful candidate for various industrial applications.

Results

Cloning and sequence analysis of the endo-1,4-β-mannanase gene from *P. oxalicum* GZ-2

A 772-bp gene fragment was amplified from the gDNA of strain GZ-2 using the degenerate primers manA-df and manA-dr. Sequence analysis indicated that the fragment showed 76% identity with an endo-1,4-β-mannanase from *Aspergillus aculeatus* (GenBank: L35487.1). To amplify the full-length endo-1,4-β-mannanase gene the 5' and 3' flanking regions were obtained using self-formed adaptor PCR (SEFA-PCR). All fragments were assembled with core regions to create a 1671 bp sequence containing a complete chromosomal gene of 1380 bp predicted by FGENESH (http://linux1.softberry.com/berry.phtml). The full-length cDNA sequence of endo-1,4-β-mannanase (PoMan5A) was contained in an open reading frame of 1380 bp, which was cloned from the synthesized cDNA using the manA-f and manA-r primers. Three introns interrupted the PoMan5A coding sequence according to the alignment of the DNA and cDNA sequences. A putative 33-residue signal peptide at the N-terminus was predicted by the SignalP server system (http://www.cbs.dtu.dk/services/SignalP/). The deduced mature protein of PoMan5A contained of 426 residues with a calculated molecular mass of 45.8 kDa and a theoretical pI of 5.09 (http://web.expasy.org/compute_pi/). The deduced amino acid sequence of PoMan5A showed the highest identity of 100% with the putative endo-1,4-β-mannanase from *Penicillium oxalicum* 114-2 (EPS31069). PoMan5A contained an N-terminal fungal cellulose binding domain and a glycosyl hydrolase family 5 domain. After multiple sequence alignments with other GH5 endo-1,4-β-mannanase amino acid sequences, the two catalytic glutamates (Glu 320 and Glu 391) are located in highly conserved regions of the active site in PoMan5A. Several amino acid residues, Arg169, Trp279, Asn283, His357, and Tyr393 are conserved in PoMan5A and other fungal GH5 endo-β-1, 4-mannanases (Figure 1). The nucleotide sequence of *poMan5A* was deposited into GeneBank under accession number KF233753.

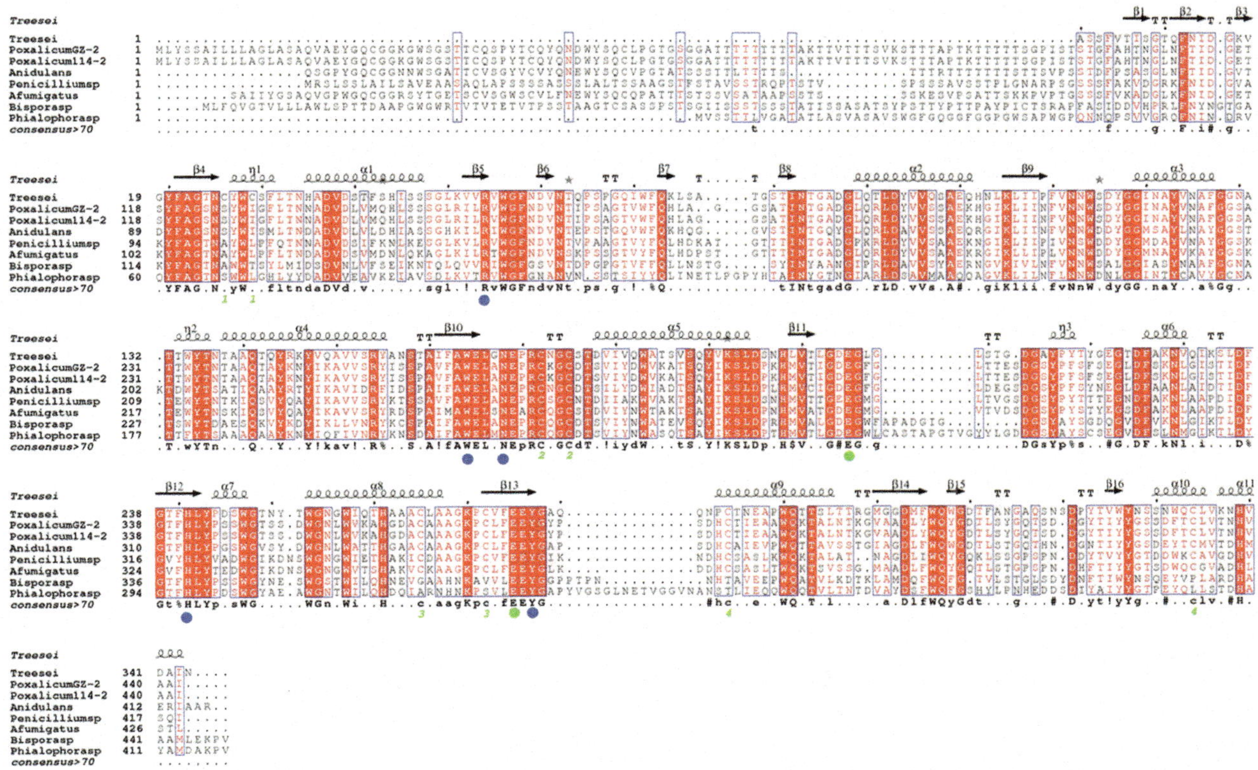

Figure 1 Amino acid sequence alignment of endo-β-1, 4-mannosidase from *P. oxalicum* GZ-2 with related fungi. The secondary structural elements (α-helices displayed as squiggles, β-strands rendered as arrows, and strict β-turns as TT letters) of rPoMan5A are using *Trichoderma reesei* β-mannanase as a template (pdb no. 1QNO_A). The alignment includes endo-β-1, 4-mannanases from *P. oxalicum* GZ-2 (GenBank: AGW24296.1), *Penicillium oxalicum* 114-2 (GenBank: EPS31069.1), *Penicillium* sp. (GenBank: AFC38441.1), *Bispora* sp. MEY-1 (GenBank: ACH56965.1), *Phialophora* sp. (GenBank: ADF28533.1), *Aspergillus fumigatus* (GenBank: ACH58410.1), *Aspergillus nidulans* (GenBank: AGG69666.1), and *Trichoderma reesei* QM6a (GenBank: XP_006962944). Strictly conserved residues are highlighted with a red background and conservatively substituted residues are boxed. Pairs of Cys residues that formed disulfide bonds are indicated by green-colored numbers. The conserved catalytic residues Glu320 and Glu391 are indicated by a green dot and other conserved amino acid residues are marked with a blue dot. The figure was made using ESPript 3.0.

Heterologous expression and purification of the *P. oxalicum* endo-1,4-β-mannanase

The gene expression vector *pPIC-man5A* was successfully constructed for use in this study. In this recombinant plasmid, cDNA fragments were fused with the α-factor signal sequence at its N-terminus allowing secretion of enzyme into the liquid culture that was expressed under the control of the AOX1 promoter. The resulting plasmid was then transformed into *Pichia pastoris* by electroporation. After transformation and screening on plates containing Zeocin in increasing concentrations, 20 transformants showing robust antibiotic-resistance were further evaluated for their ability to secrete endo-1,4-β-mannanase in a small-scale liquid fermentation. One transformant with the highest enzyme activity was used to produce endo-1,4-β-mannanase in a 1 L Erlenmeyer flask induced by 1% methanol at 28°C for 7 days. The highest extracellular endo-1,4-β-mannanase activity (84.4 U mL^{-1}) was obtained after a 6 day incubation at 28°C. SDS-PAGE analysis of the culture supernatants of the induction period from 1 to 7 day showed an apparent protein band (Figure 2A) corresponding to endo-1,4-β-mannanase activity in the zymogram analysis (Figure 2B). No endo-1,4-β-mannanase activity was detected in the culture medium of the control strain (transformed with pPICZαA) under the same culture conditions. The recombinant endo-1,4-β-mannanase was purified to electrophoretic homogeneity through Ni-NTA chromatography. After Ni-NTA purification, a 4.5-fold purification was achieved with 56% recovery. The purified rPoMan5A showed a single band with an apparent molecular mass of 61.6 kDa by SDS-PAGE (Figure 3).

Biochemical properties of the purified recombinant endo-1,4-β-mannanase

Effect of temperature and pH on rPoMan5A activity and stability

The effects of temperature and pH on rPoMan5A activity were investigated using locust bean gum as the substrate. The rPoMan5A showed optimal activity at the

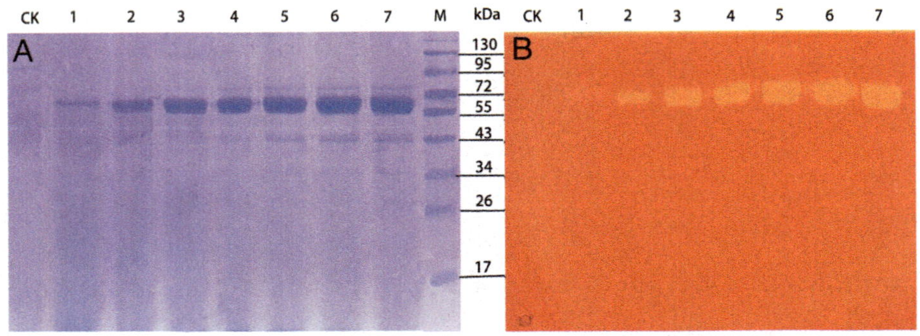

Figure 2 SDS-PAGE and zymography analysis of culture supernatants from recombinant *Pichia pastoris* during 7-day fermentation. SDS-PAGE **(A)** and zymography **(B)** of secreted proteins produced by recombinant *Pichia pastoris* fusing the *P. oxalicum* GZ-2 endo-β-1, 4-mannosidase gene. Lane CK: *Pichia pastoris* GS115 with the empty pPICZαA after induction for 4 days; Lane 1-7: culture supernatant after induction for 1, 2, 3, 4, 5, 6, and 7 days; Lane M: protein molecular mass makers. Twenty microliters of culture supernatants was added into each gel lane.

temperature of 80°C and pH 3.0-4.0 and exhibited over 50% activity in temperatures ranging from 40°C to 90°C (Figure 4A). The thermal stability profiles showed that rPoMan5A was stable at temperatures below 70°C (Figure 4B). Almost no enzyme activity (>98%) was lost after 1 h incubation at 60°C, and the half-life at this temperature was 58 h. At 70°C, the half-life of the enzyme was only 21 min. The rPoMan5A retained more than 96% of its maximal activity at pH 2.0 (Figure 5). The pH stability profiles showed that the enzyme was highly stable within a broad range of pH values, ranging from 3.0 to 7.0 (Figure 5).

Effect of metal ions and chemical reagents on endo-1, 4-β-mannanase activity

The presence of most metal ions at 1 mM or 10 mM (Ca^{2+}, Cd^{2+}, Cr^{3+}, Co^{2+}, Ba^{2+}, Mg^{2+}, Mn^{2+}, Ni^{2+}, Li^{+} and Fe^{2+}) had little or no effect on the activity of rPoMan5A. At the high concentrations of Cu^{2+} and Fe^{3+} (10 mM), the activity of rPoMan5A was reduced to almost 50%. The whole activity of rPoMan5A was lost in the presence of 10 mM Hg^{2+} (Table 1). β-Mercaptoethanol (1 mM, 76.2%), dithiothreitol (DTT) (1 mM, 81.2%) and 0.1% SDS (92.0%) partially influenced the rPoMan5A activity, while 0.1% Triton-X 100 strongly decreased the rPoMan5A activity (25%).

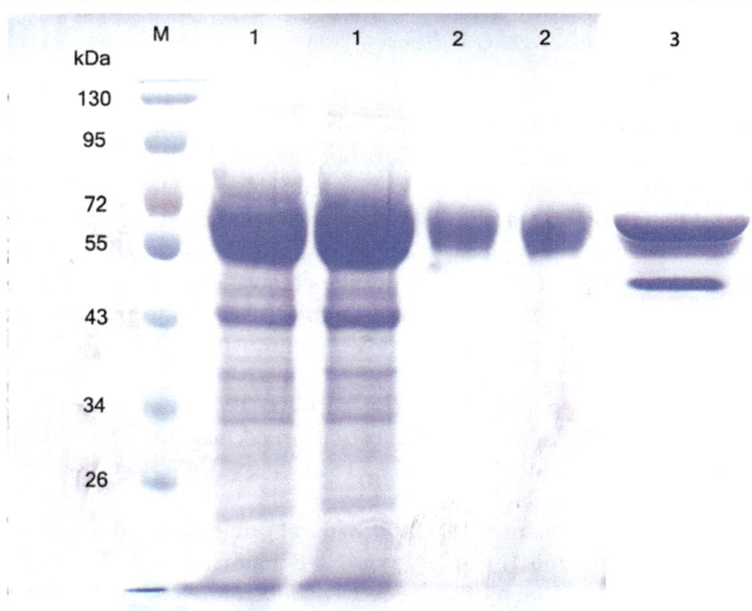

Figure 3 SDS-PAGE analysis of the deglycosylation of rPoMan5A with Endo H_f and purified rPoMan5A fractions. Lane 1, concentrated proteins from an ammonium sulfate precipitation; Lane 2, purified rPoMan5A; Lane 3, deglycosylated rPoMan5A after endo H_f treatment; Lane M, protein molecular mass makers.

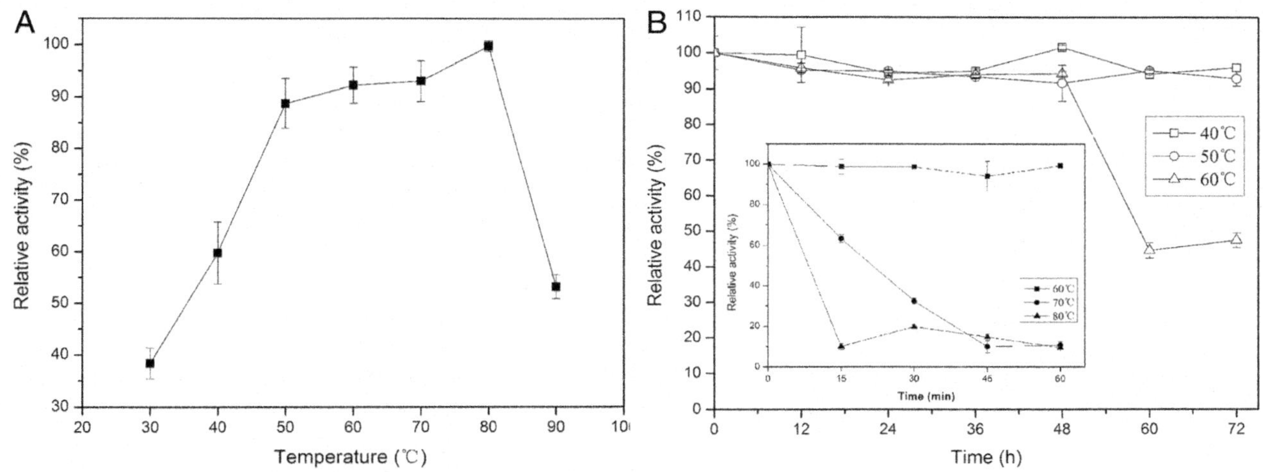

Figure 4 Optimal temperature (A) and thermostability (B) of purified rPoMan5A. The optimal temperature for endo-β-1, 4-mannosidase activity was measured at different temperatures using 0.5% locust bean gum as the substrate. The thermal stability was evaluated as the relative residual activity after incubation without the substrate at different temperatures at pH 4.0. Data are the mean of three replicates, and bars indicate the standard deviation of three replicates.

Substrate specificity and kinetics parameters

The rPoMan5A exhibited the highest activity with locust bean gum (100%) and lower activity for guar gum (30.0%). There was no enzyme activity with sodium carboxymethyl cellulose, filter paper, and beechwood xylan as substrate. The K_m and V_{max} values of purified rPoMan5A were 7.6 mg mL^{-1} and 1425.4 U mg^{-1} for locust bean gum, 2.1 mg mL^{-1} and 154.8 U mg^{-1} for konjac glucomannan, and 2.3 mg mL^{-1} 18.9 U mg^{-1} for guar gum, respectively (Table 2).

Analysis of the hydrolysis products

The hydrolysis products of locust bean gum, konjac glucomannan, and guar gum by rPoMan5A were analyzed by thin-layer chromatography (TLC). As shown in Figure 6, the rPoMan5A hydrolyzed various types of mannan polymers, such as galacto-and glucomannans and released various mannose and manno-oligosaccharides. The hydrolysis of locust bean gum yielded mannobiose (M2), mannotriose (M3), and mannopentaose (M5), and other unidentifiable oligosaccharides as the main products

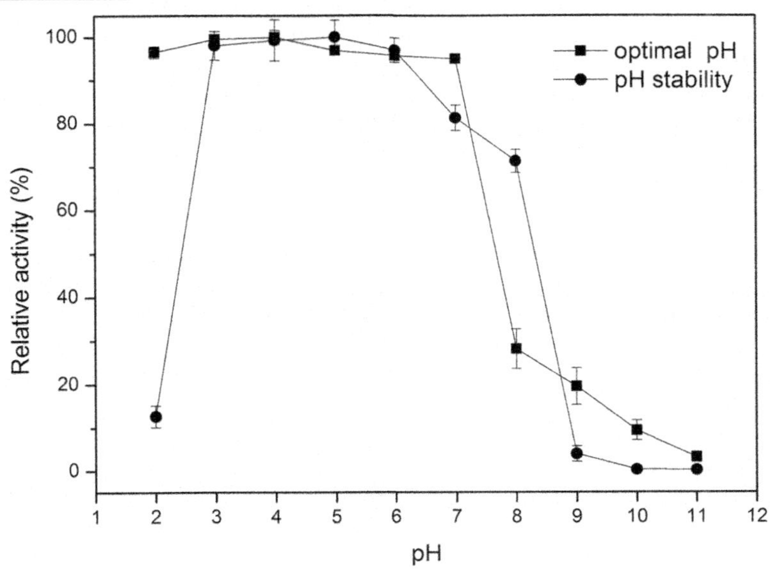

Figure 5 Optimal pH and pH stability of purified rPoMan5A. The influence of pH on enzyme activity was determined in different 50 mM buffers using 0.5% locust bean gum as the substrate. The pH stability was shown as the remaining activity after incubation for 30 min at 60°C in buffers at various pH without the substrate. Data are the mean of three replicates, and bars indicate standard deviation of three replicates.

Table 1 Effects of metal ions on the activity of the rPoMan5A

Metal ions	Relative activity (%)	
	1 mM	10 mM
Control	100.0 ± 2.7	
Ca^{2+}	95.2 ± 0.3	98.6 ± 0.9
Cd^{2+}	97.2 ± 0.7	105.1 ± 2.7
Cr^{3+}	100.9 ± 0.3	101.1 ± 0.3
Co^{2+}	100.0 ± 1.4	94.0 ± 0.2
Cu^{2+}	86.5 ± 3.0	57.1 ± 3.8
Ba^{2+}	98.6 ± 1.7	104.6 ± 0.9
Mg^{2+}	97.6 ± 0.2	103.6 ± 0.8
Mn^{2+}	85.4 ± 1.3	99.6 ± 2.4
Ni^{2+}	95.4 ± 1.5	98.7 ± 5.6
Li^{+}	97.1 ± 1.0	105.7 ± 3.2
Fe^{2+}	97.1 ± 1.0	103.5 ± 2.0
Fe^{3+}	64.9 ± 3.3	50.0 ± 2.8
Hg^{2+}	64.9 ± 3.3	0.0 ± 0.0

Data are the mean ± standard deviation of three repeats.

(Figure 6A). The degradation of konjac glucomannan resulted in the formation of a mixture of mannose and manno-oligosaccharides (Figure 6B). Mannotriose (M3), mannotetraose (M4), and mannopentaose (M5) were the major products of hydrolysis of guar gum by rPoMan5A (Figure 6C). The release of mannose required at least 4 hours when locust bean gum were used as the substrate, but for konjac glucomannan and guar gum, there is already slight band after 5 min.

Discussion

Our previous study demonstrated that the *P. oxalicum* GZ-2 is a robust hemicellulase-producing fungus during submerged fermentation used agricultural residues as carbon source [13]. In this study, a new endo-1,4-β-mannanase gene from *P. oxalicum* GZ-2 was cloned and successfully expressed in *Pichia pastoris*. The gene sequence of endo-1,4-β-mannosidase showed highest sequence similar (74%) to mannanase from *Aspergillus aculeatus* base on BLAST on NCBI (http://blast.ncbi.nlm.nih.gov/Blast.cgi). The amino acid sequence of endo-1,4-β-mannanase from *P. oxalicum* GZ-2 has 100% identity with the putative endo-1,4-β-mannanase from *Penicillium oxalicum* 114-2, followed by 71% with the likely endo-1,4-β-mannosidase from the *Penicillium roqueforti*. An abundance of endo-1,4-β-mannanases from other fungi genera such as *Aspergillus* and *Trichoderma* have previously been cloned and expressed in *Pichia pastoris* [5,7,14,15]. However, only few endo-1,4-β-mannanases from *Penicillium* genera have been cloned and expressed to date [16-18].

Based on amino acid sequence similarities, endo-1,4-β-mannanase has been classified into three families of glycoside hydrolases (family 5, 26, and 113, http://www.cazy.org). Most of fungal β-mannanases are predominantly belonging to family 5, and the bacterial β-mannanases belong to GH 5 or GH 26 [17]. Homology amino acid sequence comparison with other known fungal GH 5 β-mannanases showed that several catalytic active-sites are highly conserved. These features demonstrate that the rPoMan5A is a member of the GH5. A cellulose-binding domain (CBD) was identified in rPoMan5A at the N-terminus when aligned with available protein sequences from the NCBI databases. In GH5 fungal endo-1,4-β-mannanases, a CBD has been found at either the C-, or N-terminus of some enzymes that contained a catalytic domain (CD). For example, the CBD of *Trichoderma reesei* endo-1,4-β-mannanase (amino acid position 373-410) [19] and *Phanerochaete chrysosporium* (amino acid position 25-53) [20] are located at the C terminus and N terminus, respectively. The CBD play a very important role in the endo-1,4-β-mannanases activity. The CBD of *Trichoderma reesei* endo-1,4-β-mannanase has the ability to bind cellulose and increase the hydrolysis rate [21]. The endo-1,4-β-mannanase of *Aspergillus usamii* YL-01-78 consists of only a catalytic domain (CD), while a fusion of CBD from the *Trichoderma reesei* cellobiohydrolase into this enzyme could improve its catalytic activity and/or thermostability [22]. However, the mechanism of CBDs in catalysis still needs to be explored.

When the recombinant yeast was incubated at 28°C in a 1 L shake flask culture, a high level expression of endo-1,4-β-mannanase was obtained in *Pichia pastoris* (262 mg L^{-1}). In comparison with other fungal endo-1,4-β-mannanase expression levels in yeast, like *Bispora* sp. MEY-1 MAN5A expressed 148 mg L^{-1} [23], *A. niger* endo-1,4-β-mannanase expressed 243 mg mL^{-1} [5], and *Trichoderma reesei* endo-1,4-β-mannanase expressed 150 mg L^{-1} [19], the yield of rPoMan5A was superior. The crude supernatant of *Pichia pastoris* showed good activity (84.4 U mL^{-1}) and high specific activity (420.9 U mg^{-1}), as

Table 2 Kinetic parameters of rPoMan5A for the hydrolysis of various mannans

Substrate	V_{max} (U mg^{-1})	K_m (mg mL^{-1})	k_{cat} (s^{-1})	k_{cat}/K_m (mL mg^{-1} s^{-1})
Locust bean gum	1425.5	7.6	1463.5	192.6
Konjac glucomannan	154.8	2.1	158.9	75.7
Guar gum	18.9	2.3	19.4	8.4

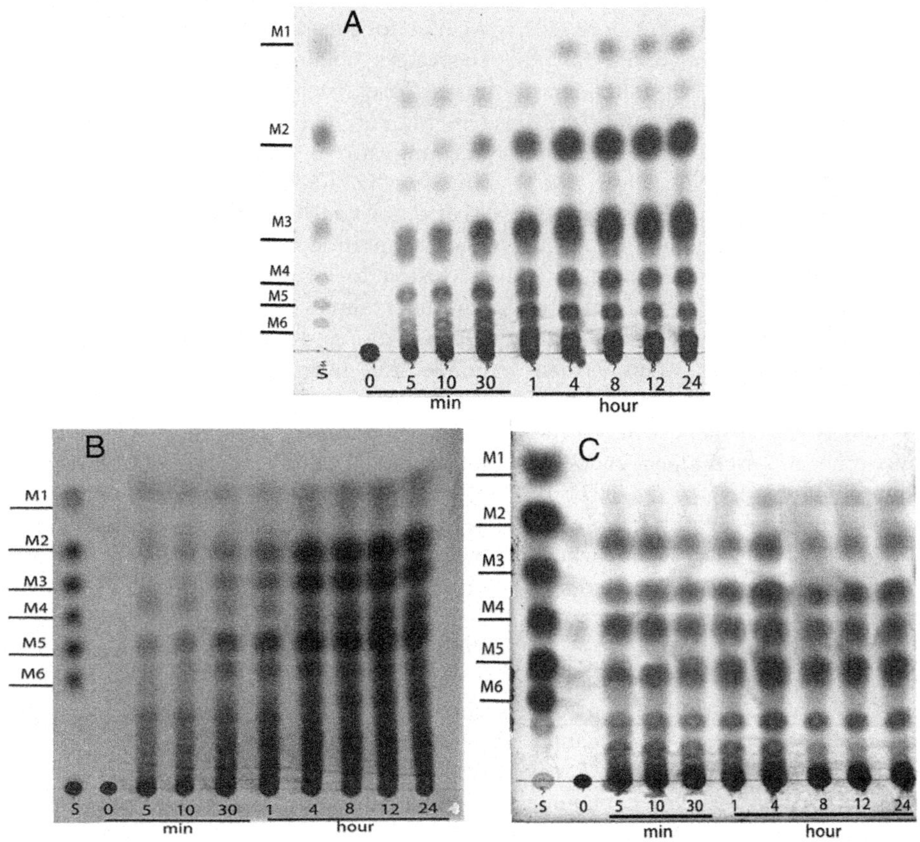

Figure 6 TLC analysis of the products after hydrolysis of various mannan polymers (0.5%) by rPoMan5A. The purified enzyme rPoMan5A and 0.5% various substrates (**A**: locust bean gum, **B**: konjac glucomannan, **C**: guar gum) were incubated in 50 mM acetate buffer (pH 4.0) at 60°C. Incubation times (hour or minute) are indicated. S, mannose and manno-oligosaccharides were used as standards.

well as high purity of the crude enzymes. Since cultivation experiments were performed in shake flasks which more or less limited the growth of yeast, it will result in considerably higher enzyme yields in a fermenter under controlled and optimized conditions [24]. Therefore, those features suggested the enzyme has great potential for large-scale production.

The purified rPoMan5A showed a single 61.6 kDa protein band on an SDS-PAGE (Figure 3), higher than that of the calculated molecular mass of the mature peptide (45.8 kDa). This may be due to the glycosylation of the expressed protein in *Pichia pastoris.* Two putative glycosylation sites in the PoMan5A amino acid sequence (^{127}NFTI and ^{229}NWSD) were predicted by the NetNGlyc 1.0 Server (http://www.cbs.dtu.dk/services/NetNGlyc/). The decrease in the molecular mass of rPoMan5A was verified by deglycosylation with Endo H_f on an SDS-PAGE (Figure 3). Many researchers had observed that the glycosylation of recombinant endo-1,4-β-mannanase in *Pichia pastoris* was able to obviously increase the molecular mass, such as endo-1,4-β-mannanase from *Aspergillus niger* BK01 [5], Man5C1 from *Penicillium pinophilum* C1 [16], and Anman5A from *Aspergillus niger* LW-1 [7].

The optimal temperature and pH for the rPoMan5A activity was 80°C and pH 3.0-4.0, respectively. For most fungal endo-1,4-β-mannanases the optimal temperature and pH are between 40-70°C and pH 4.0-6.0, respectively [25], such as endo-1,4-β-mannanase from *Rhizomucor miehei* (optimal temperature 55°C and pH 7.0) [26], MANI and MANII from *Aspergillus fumigatus* IMI 385708 (60°C and pH 4.5) [27], endo-1,4-β-mannanase from *P. oxalicum* SO (60°C and pH 5.0) [28], endo-1,4-β-mannanase from *A. niger* BK01 (80°C and pH 4.5) [5], and rMan5C1 from *P. pinophilum* C1 (70°C and pH 4.0) [16]. However, the rPoMan5A possesses some superior properties compared to these fungal endo-1,4-β-mannanases, such as a long half-life (58 h) at high temperatures (60°C) and a wide range of pH adaptability (pH 2.0-7.0) and stability (pH 3.0-7.0). Most fungal endo-1,4-β-mannanases were stable at temperatures below 50°C, such as rMan5F63 from *Penicillium freii* F63 (half-life of 30 min at 60°C) [18], recombinant man5XZ7 from thermophilic fungus

Thielavia arenaria XZ7 (half-life of 30 min at 60°C) [29], and recombinant Man5XZ3 from *Aspergillus nidulans* XZ3 (half-life of 240 min at 60°C). Although rPoMan5A was unstable (less than 13% residual activity after incubation 30 min at pH 2.0) at pH 2.0, it exhibited 96.3% activity at such low pH (10 min assay). In comparison, the endo-1,4-β-mannanase from *Aspergillus niger* BK01 showed only 20% activity at pH 2.5 [5], and rMan5C1 from *P. pinophilum* C1 likewise lost 80% activity at pH 2.0 [16]. Many industrial bioprocesses involving enzymes depend on the use of low pH conditions to prevent contamination by other microbes. Therefore, rPoMan5A has the potential to be used in such processes, where fermentation requires low pH conditions. At its optimal catalytic temperature (80°C), rPoMan5A was unstable. This characteristic is similar to many other reported fungal endo-1,4-β-mannanases [18]. The enzyme binding to the substrate could partially stabilize the enzyme conformation, and having no substrate to bind resulted in the loss of enzymatic activity during a thermostability assay [16]. Therefore, endo-1,4-β-mannanase showed acidophilic and thermostable characteristics, features that not only increase the rate of hydrolysis but also reduce cost of production in the food industry. Consistent with our results, Hg^{2+} has been shown to strongly inhibit other endo-1,4-β-mannanases [7,26]. Inhibition of rPoMan5A activity by Hg^{2+} may be due to the oxidation of specific residues containing sulfhydryl groups such as cysteine. The mature protein of rPoMan5A has 10 cysteine residues. The influence of Fe^{3+} on rPoMan5A is similar to other endo-1,4-β-mannanases [15,30]. The activity of rPoMan5A in 1 mM reducing agents such as β-Mercaptoethanol and DTT as well as 0.1% SDS remained above 75%. SDS has been shown to almost completely inhibit endo-1,4-β-mannanase activity [11,16,23].

The rPoMan5A exhibited different activity against various substrates, such as locust bean gum (100% activity), konjac glucomannan (73.6% activity), and guar gum (30.0% activity). RmMan5A from *Rhizomucor miehei* displayed a same activity order: locust bean gum > konjac glucomannan > guar gum, when used to hydrolyze various mannans [26]. Locust bean gum is a galactomannan with a high mannose/galactose ratio of 4:1; konjac glucomannan is a glucomannan containing a mannose/glucose ratio of 3:1 and guar gum is a galactomannan with mannose/galactose ratio of 2:1 [31]. The structural differences between these three mannan polymers may lead to various enzymatic activities, when they are used as substrates. There was no enzyme activity detected on (hemi) cellulosic substrates, such as beechwood xylan, CMC-Na, and filter paper. The cellulase-free characteristic of rPoMan5A is in agreement with the most of fungal endo-1,4-β-mannanases [5,26,29]. The cellulase-free characteristic of rPoMan5A is valuable for the applications involving bleaching pulp and paper [10]. The K_m value of rPoMan5A for locust bean gum (7.6 mg mL^{-1}) was close to the values obtained for some other fungal endo-1,4-β-mannanases, such as rMan5F63 from *Penicillium freii* F63 (7.8 mg mL^{-1}) [18], rMan5C1 from *P. pinophilum* C1 (5.6 mg mL^{-1}) [16], and Man5XZ7 from *Thielavia arenaria* XZ7 (5.3 mg mL^{-1}) [29]. The K_m value of rPoMan5A for konjac glucomannan and guar gum was 3-fold lower than for locust bean gum.

Most endo-1,4-β-mannanases degrade the backbone of various mannan polysaccharides to produce primarily mannotriose and mannobiose with no mannose. For example, the hydrolysis by endo-1,4-β-mannanase from *Aspergillus niger* BK01 showed that mannobiose was the main product and mannose was not liberated using locust bean gum as substrate [5]. However, some endo-1,4-β-mannanases also released mannose from mannan polymers [18,23]. In this study, the endo-1,4-β-mannanases whether liberate mannose or not, were dependent on different substrates and hydrolysis time. Similar enzymatic properties were reported for RmMan5A from *Rhizomucor miehei* [26]. The action of the rPoMan5A on different substrates indicated that its function was not only as endo-β-1,4-mannanase but also as a 1,4-β-mannosidase. An abundance of manno-oligosaccharides primarily containing various oligosaccharides (M2 to M6) was obtained from hydrolyzing the inexpensive and widely available locust bean gum and konjac glucomannan by rPoMan5A. Because monogastric animals cannot effectively digest the mannans and galactomannans into nutritional components, the incorporation of endo-1,4-β-mannanase into their diets would help them to more easily digest the nutritional elements. Thus, rPoMan5A has a potential application in the food/feed industries. Although the enzyme rPoMan5A exhibited attractive biochemical properties like acid-resistant and thermostability, but there is need to explain the mechanisms of its superior properties. The molecular enzyme engineering should be considered to improve the catalytic efficiency and characteristics in the further studies.

Conclusion

In this study, a new gene encoding a GH5 acidophilic thermostable endo-1,4-β-mannanase from *Penicillium oxalicum* was cloned and expressed in *Pichia pastoris.* High stability across a broad acidic pH range and thermostability make rPoMan5A better suited for many industrial applications in the food/feed and paper/pulp industries compared to most other fungal endo-1,4-β-mannanases.

Methods

Plasmids, strains and culture conditions

The *P. oxalicum* GZ-2 referred to in this study was isolated and identified as previously reported [13] and has been deposited at the China General Microbiological Culture Collection Center (CGMCC 7527). *Escherichia coli* strain TOP 10 (Invitrogen, San Diego, CA) grown in Luria Bertani (LB) medium at 37°C was used as the host cell for plasmid cloning and propagation of the recombinant expression pPICZαA vector. *Pichia pastoris* GS115 (Invitrogen, San Diego, CA) was used for heterologous protein expression. All media and protocols for *Pichia pastoris* are described in the *Pichia* expression manual (Invitrogen, San Diego, CA). Manno-oligosaccharides were purchased from Megazyme (Bray, Ireland). *P. oxalicum* GZ-2 was maintained and cultured according previously study [32]. *Pichia pastoris* GS115 was cultivated at 30°C in yeast extract peptone dextrose (YPD) medium (1% yeast extract, 2% tryptone, and 2% dextrose).

Total RNA isolation, cDNA synthesis and cloning of the endo-1,4-β-mannanase gene

The *P. oxalicum* GZ-2 strain was grown at 30°C in a 250 mL shaker flask containing 50 mL basal culture medium in a shaking incubator using 2% corncob powder as the sole carbon source. After a 72 h growth period, the mycelium was harvested for total RNA isolation. The genomic DNA of *P. oxalicum* GZ-2 was extracted as described by Möller et al. [33]. RNA isolation and cDNA synthesis was performed as previously reported [32]. The degenerate primers manA-df and manA-dr were used (Table 3), which were designed using the iCODEHOP (http://blocks.fhcrc.org/codehop.html) program based on the multiple sequence alignment of homologous amino acid sequences using the ClustalW2 program. The partial sequence of *P. oxalicum* GZ-2 endo-1,4-β-mannanase gene was amplified by polymerase chain reaction (PCR) using the degenerate primers. To obtain the full-length gene, self-formed adaptor PCR (SEFA-PCR) was performed to amplify the 5'-end and the 3'-end of *poman5A* according to the protocol developed by Wang et al. [34] with the primers listed in Table 3. By aligning the sequences of the 5'-end and 3'-end PCR products, the full-length cDNA sequence of *poman5A* was deduced and obtained through RT-PCR using the following specific primers: manA-f and manA-r. The PCR fragment was purified and ligated into the pMD-19T vector and was designated pMD-man5A. The cDNA encoding the mature *P. oxalicum* GZ-2 endo-1,4-β-mannanase gene was PCR amplified with PrimeSTAR™ HS DNA Polymerase (Takara, Dalian, China) using the recombinant plasmid pMD-man5A as a template, with specific primers manA-ef including an *Eco*RI site, and manA-er including an *Xba*I site. After digestion with *Eco*RI and *Xba*I, The DNA insert was ligated into the pPICZαA vector downstream of the α-factor signal peptide sequence. Proper construction was confirmed by restriction digestion and DNA sequencing and was designated as pPIC-man5A. The nucleotide sequence of *poMan5A* has been deposited in the GeneBank under accession number KF233753.

Expression of the endo-1,4-β-mannanase gene in *Pichia pastoris* in shaken flasks

The recombinant plasmid *pPIC-manA* was linearized with *Pme*I (New England BioLabs, Beverly, MA, USA), and then transformed into *Pichia pastoris* GS115 by electroporation (Gene Pulser Xcell™ Electroporation System, Bio-Rad, Hercules, CA, USA) according to the manufacturer's instructions. The control was transformed with empty pPICZαA at same condition. The transformants were plated on YPDS plates (1% yeast extract, 2% tryptone, 2% dextrose, 1 M sorbitol, 100 μg mL^{-1} Zeocin, and 2% agar)

Table 3 Primers used in this study

Primer name	Primer purpose	Primer sequence	Size (bases)
manA-df	Degenerate primers	CGGGTCTGGGGCTTYAAYGAYGT	23
manA-dr		GTG CCGTAGTAGATGGTRTTNCCRTCRT	28
manA-3-sp3	Amplify gene of the 3'-end by SEFA-PCR	ACTTTCCATCTGTACNNNNNNNNNTGTGAG	30
manA-3-sp2		TGACGACCGAGTCTGACGGAAGCTA	25
manA-3-sp1		CTGGGTCAAGGCGACTAGCCAATAT	25
manA-5-sp3	Amplify gene of the 5'-end by SEFA-PCR	GTGTTGATGGTGGCGNNNNNNNNNGCAAGA	30
manA-5-sp2		GGAGATGAGATGTACCGCGAAACCA	25
manA-5-sp1		CCATGTGCTTCGGGTCGAGTGATTT	25
manA-f	cDNA cloning primers	ATGACATTGGGATTGACTCAGACGA	25
manA-r		TCAGATCGCAGCGACATGATT	21
manA-ef	Specific expression primers	CGGAATTCCAGGTGGCGGAATATGGCCAGTGT	32
manA-er		GCTCTAGATAGATCGCAGCGACATGATTCGTCACCA	36
5'AOX	Confirmed primers	GACTGGTTCCAATTGACAAGC	21
3'AOX		GCAAATGGCATTCTGACATCC	21

R = A/G, W = A/T, Y = C/T, N = A/T/C/G.

and incubated at 30°C for 2-3 days until colonies appeared.

The colonies were first screened by selection on YPDZ (1% yeast extract, 2% tryptone, 2% dextrose, 2% agar) plates (Zeo1000) containing Zeocin at a final concentration of 1000 μg mL^{-1}, then on YPDZ plates containing Zeocin at final concentrations of 2000 μg mL^{-1} (Zeo2000) to screen for multiple copies of a cDNA gene targeted gene [35]. The Mut^{+} phenotype of *Pichia pastoris* recombinants was confirmed by PCR with 5′AOX and 3′AOX primers using DNA from the *Pichia pastoris* transformant as the template that was extracted using a Yeast Genomic DNA Extraction kit (Tiangen, Beijing, China). Twenty colonies from the Zeo2000 plates were inoculated into 10 mL of YPM medium (1% yeast extract and 2% tryptone) containing 1% methanol in a 50 mL flask and cultured at 28°C on a rotary incubator at 250 rpm to induce the expression of the endo-1,4-β-mannanase gene. The transformant with the highest endo-1,4-β-mannanase activity in a culture supernatant following a 7 day incubation was used for further fermentation in 1 L flasks following the method of Bai et al. [36].

Purification of recombinant endo-1,4-β-mannanase

All purification steps were performed at 4°C unless stated otherwise. The crude culture supernatant was harvested by centrifugation of the culture broth at 8000 rpm for 10 min at 4°C. The supernatant was subjected to precipitation at 80% ammonium sulfate saturation. After dialysis against 50 mM acetate buffer (pH 4.0), the proteins were collected and used for the next purification. The expressed 6 × His-tagged proteins were purified with Ni-NTA Sepharose (Qiagen, Valencia, CA) according to the manufacturer's instructions.

Enzyme activity and protein concentration assay

The enzyme activity was determined by the DNS method according to Miller [37] with some modifications. Briefly, 100 μL of enzyme diluted with 50 mM acetate buffer (pH 4.0) was added to 900 μL of 0.5% locust bean gum in 50 mM acetate buffer (pH 4.0) and incubated at 80°C for 10 min. The released reducing sugars were measured at 540 nm using D-mannose as the standard. One unit of endo-1,4-β-mannanase activity was defined as the amount of enzyme liberating 1 μmol of reducing sugars per minute under the conditions described above. Protein content was determined using a BCA protein assay kit (Dingguo Changsheng Biotechnology Co., Ltd, Beijing, China), and the standard curve was generated with bovine serum albumin (BSA) as the standard.

SDS-PAGE and zymography analysis

Sodium dodecyl sulfate polyacrylamide gel electrophoresis (SDS-PAGE) was performed using a 12% (w/v) polyacrylamide gel with a 5% stacking gel and the Mini-Protean II system (BioRad) according to the method described by Laemmli [38]. The proteins were stained with Coomassie Brilliant Blue R-250 (Sigma Chemical, St. Louis, USA). The zymogram analysis was performed as previously reported by co-polymerizing 0.05% (w/v) locust bean gum with 12% polyacrylamide gel [13]. After the separation of the enzyme samples by SDS-PAGE, the gel was soaked for 1 h in 2.5% (v/v) Triton X-100 to remove the SDS and refold the proteins in the gel. The gel was then thoroughly washed three times in MilliQ water and incubated at 50°C for 60 min in 50 mM acetate buffer (pH 4.0). The gel was stained with 0.1% (w/v) Congo red solution for 30 min and destained with 1 M NaCl until pale-red hydrolysis zones were observed against a red background.

Effect of temperature, pH and various reagents on enzyme activity

The optimal temperature for rPoMan5A activity was determined using the standard method as described above at various temperatures ranging from 30°C to 90°C. To estimate thermal stability, the diluted enzyme was pre-incubated in 50 mM acetate buffer (pH 4.0) at different temperatures (40-90°C) for various times in the absence of substrate and then assayed for remaining enzyme at 60°C using the standard DNS method.

For the estimation of optimal pH for rPoMan5A enzyme activity, 50 mM different buffer solutions like glycine-HCl buffer (pH 2.0-3.0), acetate buffer (pH 4.0-6.0), sodium phosphate buffer (pH 7.0-8.0), and glycine-NaOH buffer (pH 9.0-11.0) were used. The pH stability was determined at different pH-values for 30 min by incubation of diluted enzyme samples in the absence of substrate at 60°C. The residual enzyme activity was measured at 60°C for 10 min in the acetate buffer (pH 4.0) using the DNS method. The effect of metal ions on rPoMan5A activity was examined by incubating the enzyme with various metal ions ($BaCl_2$, $CuSO_4$, $FeCl_3$, $MnSO_4$, $CaCl_2$, $CdSO_4$, $FeSO_4$, $CrCl_3$, $CoCl_2$, LiCl, $NiSO_4$, $MgSO_4$ and $HgCl_2$) or reagents in 50 mM acetate buffer (pH 4.0) at 60°C for 10 min.

Substrate specificity and kinetic parameters

The specific activity of rPoMan5A was measured using different substrates (0.5% (w/v), pH 4.0) in 50 mM acetate buffer (locust bean gum, konjac mannan, guar gum, sodium carboxymethyl cellulose (CMC-Na), filter paper, and beechwood xylan). To determine the reaction rate of rPoMan5A, various mannans substrate concentrations were used to react 10 min at 80°C in acetate buffer (pH 4.0, 50 mM). The Michaelis-Menten constant (K_m) and the maximum velocity (V_{max}) were calculated using Lineweaver-Burk plots.

Thin-layer chromatography (TLC) analysis of the hydrolysis products

The analysis of hydrolysis products was performed by thin-layer chromatography using locust bean gum, konjac mannan and guar gum as substrates. The purified rPoMan5A (5 μg) was incubated in a reaction volume of 1 mL with 0.5% of substrate at 60°C in 50 mM acetate buffer (pH 4.0) for 24 h. Aliquots were collected at different time points and boiled for 10 min. All products were freeze-dried and dissolved in a methanol (70%). Then, 5 μL of each aliquot was spotted onto a silica plate (Merck, Germany, 10 × 20 cm), and the plates were developed twice with a solvent system consisted of chloroform-acetic acid-water (3:6:1, v/v) for 3-4 hours. Later, the plates were dried and hydrolysis products were detected by spraying with a 9:1 (v/v) mixture of methanol and sulfuric acid containing 0.2% orcinol and heating at 85°C for 5-10 min. Manno-oligosaccharides (mannose (M1), mannobiose (M2), mannotriose (M3), mannotetraose (M4), mannopentaose (M5), and mannohexaose (M6), Megazyme, Bray, Ireland) were used as standards.

Competing interests

The authors declare that they have no competing interests.

Authors' contributions

HPL performed the molecular genetic experiments, desired experiment and wrote the manuscript. SXL and HPZ did the enzyme production and purification, and the enzymatic activity assay. ZW and DYL directed the study and checked all the data. WR helped to revise the manuscript. YCX was the corresponding author, he and QRS supervised the work and contributed to the manuscript. All authors read the manuscript and approved to submit to this journal.

Authors' information

Jiangsu Collaborative Innovation Center for Solid Organic Waste Utilization, College of Resources and Environmental Science, Nanjing Agricultural University, Nanjing, 210095, China.

Acknowledgements

This research was financially supported by the Agricultural Ministry of China (2011-G27), Special Fund for Agro-scientific Research in the Public Interest (201203001), National key technology R&D program (L020130249).

References

1. De O, Petkowicz C, Reicher F, Chanzy H, Taravel F, Vuong R: **Linear mannan in the endosperm of *Schizolobium amazonicum*.** *Carbohydr Polym* 2001, **44**(2):107–112.
2. Songsiriritthigul C, Buranabanyat B, Haltrich D, Yamabhai M: **Efficient recombinant expression and secretion of a thermostable GH26 mannan endo-1, 4-β-mannosidase from *Bacillus licheniformis* in *Escherichia coli*.** *Microb Cell Fact* 2010, **9**(1):20.
3. Zhou H, Yang Y, Nie X, Yang W, Wu Y: **Comparison of expression systems for the extracellular production of mannanase Man23 originated from *Bacillus subtilis* B23.** *Microb Cell Fact* 2013, **12**(1):78.
4. Yoshida S, Sako Y, Uchida A: **Cloning, sequence analysis, and expression in *Escherichia coli* of a gene coding for an enzyme from *Bacillus circulans* K-1 that degrades guar gum.** *Biosci Biotechnol Biochem* 1998, **62**(3):514–520.
5. Bien-Cuong D, Thi-Thu D, Berrin J-G, Haltrich D, Kim-Anh T, Sigoillot J-C, Yamabhai M: **Cloning, expression in *Pichia pastoris*, and characterization of a thermostable GH5 mannan endo-1, 4-β-mannosidase from *Aspergillus niger* BK01.** *Microb Cell Fact* 2009, **8**(1):59.
6. Ishizawa CI, Jeoh T, Adney WS, Himmel ME, Johnson DK, Davis MF: **Can delignification decrease cellulose digestibility in acid pretreated corn stover?** *Cellulose* 2009, **16**(4):677–686.
7. Li J-F, Zhao S-G, Tang C-D, Wang J-Q, Wu M-C: **Cloning and functional expression of an acidophilic β-mannanase gene (Anman5A) from *Aspergillus niger* LW-1 in *Pichia pastoris*.** *J Agric Food Chem* 2012, **60**(3):765–773.
8. Wu M, Tang C, Li J, Zhang H, Guo J: **Bimutation breeding of *Aspergillus niger* strain for enhancing β-mannanase production by solid-state fermentation.** *Carbohydr Res* 2011, **346**(14):2149–2155.
9. Jiang Z, Wei Y, Li D, Li L, Chai P, Kusakabe I: **High-level production, purification and characterization of a thermostable β-mannanase from the newly isolated *Bacillus subtilis* WY34.** *Carbohydr Polym* 2006, **66**(1):88–96.
10. Dhawan S, Kaur J: **Microbial mannanases: an overview of production and applications.** *Crit Rev Biotechnol* 2007, **27**(4):197–216.
11. Zhao J, Shi P, Luo H, Yang P, Zhao H, Bai Y, Huang H, Wang H, Yao B: **An acidophilic and acid-stable β-mannanase from *Phialophora* sp. P13 with high mannan hydrolysis activity under simulated gastric conditions.** *J Agric Food Chem* 2010, **58**(5):3184–3190.
12. Lu H, Luo H, Shi P, Huang H, Meng K, Yang P, Yao B: **A novel thermophilic endo-beta-1,4-mannanase from *Aspergillus nidulans* XZ3: functional roles of carbohydrate-binding module and Thr/Ser-rich linker region.** *Appl Microbiol Biotechnol* 2014, **98:**2155–2163.
13. Liao H, Xu C, Tan S, Wei Z, Ling N, Yu G, Raza W, Zhang R, Shen Q, Xu Y: **Production and characterization of acidophilic xylanolytic enzymes from *Penicillium oxalicum* GZ-2.** *Bioresour Technol* 2012, **123**(2):117–124.
14. Lim J, Bakar F, Yusof H, Murad A: **Cloning and expression of a *Trichoderma longibrachiatum* β-mannanase gene in *Pichia pastoris*.** *Afr J Biotechnol* 2014, **11**(7):1705–1718.
15. Chen X, Cao Y, Ding Y, Lu W, Li D: **Cloning, functional expression and characterization of *Aspergillus sulphureus* β-mannanase in *Pichia pastoris*.** *J Biotechnol* 2007, **128**(3):452–461.
16. Cai H, Shi P, Luo H, Bai Y, Huang H, Yang P, Yao B: **Acidic β-mannanase from *Penicillium pinophilum* C1: cloning, characterization and assessment of its potential for animal feed application.** *J Biosci Bioeng* 2011, **112**(6):551–557.
17. Cai H, Shi P, Huang H, Luo H, Bai Y, Yang P, Meng K, Yao B: **An acidic β-mannanase from *Penicillium* sp. C6: gene cloning and over-expression in *Pichia pastoris*.** *World J Microbiol Biotechnol* 2011, **27**(12):2813–2819.
18. Wang Y, Shi P, Luo H, Bai Y, Huang H, Yang P, Xiong H, Yao B: **Cloning, over-expression and characterization of an alkali-tolerant endo-β-1, 4-mannanase from *Penicillium freii* F63.** *J Biosci Bioeng* 2012, **113**(6):710–714.
19. Stålbrand H, Saloheimo A, Vehmaanperä J, Henrissat B, Penttilä M: **Cloning and expression in *Saccharomyces cerevisiae* of a *Trichoderma reesei* beta-mannanase gene containing a cellulose binding domain.** *Appl Environ Microbiol* 1995, **61**(3):1090–1097.
20. Benech R-O, Li X, Patton D, Powlowski J, Storms R, Bourbonnais R, Paice M, Tsang A: **Recombinant expression, characterization, and pulp prebleaching property of a *Phanerochaete chrysosporium* endo-β-1,4-mannanase.** *Enzyme Microb Technol* 2007, **41**(6):740–747.
21. Hägglund P, Eriksson T, Collén A, Nerinckx W, Claeyssens M, Stålbrand H: **A cellulose-binding module of the *Trichoderma reesei* β-mannanase Man5A increases the mannan-hydrolysis of complex substrates.** *J Biotechnol* 2003, **101**(1):37–48.
22. Tang C-D, Li J-F, Wei X-H, Min R, Gao S-J, Wang J-Q, Yin X, Wu M-C: **Fusing a carbohydrate-binding module into the *Aspergillus usamii* β-mannanase to improve its thermostability and cellulose-binding capacity by in silico design.** *PLoS One* 2013, **8**(5):8.
23. Luo H, Wang Y, Wang H, Yang J, Yang Y, Huang H, Yang P, Bai Y, Shi P, Fan Y: **A novel highly acidic β-mannanase from the acidophilic fungus *Bispora* sp. MEY-1: gene cloning and overexpression in *Pichia pastoris*.** *Appl Microbiol Biotechnol* 2009, **82**(3):453–461.
24. Gasser B, Saloheimo M, Rinas U, Dragosits M, Rodríguez-Carmona E, Baumann K, Giuliani M, Parrilli E, Branduardi P, Lang C: **Protein folding and conformational stress in microbial cells producing recombinant proteins: a host comparative overview.** *Microb Cell Fact* 2008, **7**(1):11.
25. Chauhan PS, Puri N, Sharma P, Gupta N: **Mannanases: microbial sources, production, properties and potential biotechnological applications.** *Appl Microbiol Biotechnol* 2012, **93:**1817–1830.

26. Katrolia P, Yan Q, Zhang P, Zhou P, Yang S, Jiang Z: **Gene cloning and enzymatic characterization of an alkali-tolerant endo-1, 4-β-mannanase from *Rhizomucor miehei*.** *J Agric Food Chem* 2013, **61**(2):394–401.
27. Puchart V, Vršanská M, Svoboda P, Pohl J, Ögel ZB, Biely P: **Purification and characterization of two forms of endo-β-1, 4-mannanase from a thermotolerant fungus *Aspergillus fumigatus* IMI 385708 (formerly *Thermomyces lanuginosus* IMI 158749).** *Biochimica et Biophysica Acta* 2004, **1674**(3):239–250.
28. Kurakake M, Sumida T, Masuda D, Oonishi S, Komaki T: **Production of galacto-manno-oligosaccharides from guar gum by β-mannanase from *Penicillium oxalicum* SO.** *J Agric Food Chem* 2006, **54**(20):7885–7889.
29. Lu H, Zhang H, Shi P, Luo H, Wang Y, Yang P, Yao B: **A family 5 β-mannanase from the thermophilic fungus *Thielavia arenaria* XZ7 with typical thermophilic enzyme features.** *Appl Microbiol Biotechnol* 2013, **97:**8121–8127.
30. Tamaru Y, Araki T, Amagoi H, Mori H, Morishita T: **Purification and characterization of an extracellular beta-1, 4-mannanase from a marine bacterium, *Vibrio* sp. strain MA-138.** *Appl Environ Microbiol* 1995, **61**(12):4454–4458.
31. Van Zyl WH, Rose SH, Trollope K, Görgens JF: **Fungal β-mannanases: mannan hydrolysis, heterologous production and biotechnological applications.** *Process Biochem* 2010, **45**(8):1203–1213.
32. Liao H, Sun S, Wang P, Bi W, Tan S, Wei Z, Mei X, Liu D, Raza W, Shen Q, Xu Y: **A new acidophilic endo-beta-1,4-xylanase from *Penicillium oxalicum*: cloning, purification, and insights into the influence of metal ions on xylanase activity.** *J Ind Microbiol Biotechnol* 2014, **41:**1071–1083.
33. Möller E, Bahnweg G, Sandermann H, Geiger H: **A simple and efficient protocol for isolation of high molecular weight DNA from filamentous fungi, fruit bodies, and infected plant tissues.** *Nucleic Acids Res* 1992, **20**(22):6115–6116.
34. Wang S, He J, Cui Z, Li S: **Self-formed adaptor PCR: a simple and efficient method for chromosome walking.** *Appl Environ Microbiol* 2007, **73**(15):5048–5051.
35. Cheng Y-F, Yang C-H, Liu W-H: **Cloning and expression of *Thermobifida* xylanase gene in the methylotrophic yeast *Pichia pastoris*.** *Enzyme Microb Technol* 2005, **37**(5):541–546.
36. Bai Y, Wang J, Zhang Z, Shi P, Luo H, Huang H, Luo C, Yao B: **Research expression of an extremely acidic β-1, 4-glucanase from thermoacidophilic *Alicyclobacillus* sp. A4 in *Pichia pastoris* is improved by truncating the gene sequence.** *Microb Cell Fact* 2010, **33**(1):9.
37. Miller GL: **Use of dinitrosalicylic acid reagent for determination of reducing sugar.** *Anal Chem* 1959, **31**(3):426–428.
38. Laemmli UK: **Cleavage of structural proteins during the assembly of the head of *bacteriophage* T4.** *Nature* 1970, **227**(5259):680–685.

A simple method for semi-random DNA amplicon fragmentation using the methylation-dependent restriction enzyme MspJI

Hiroshi Shinozuka[1,2], Noel O I Cogan[1,2], Maiko Shinozuka[1,2], Alexis Marshall[1,2], Pippa Kay[1], Yi-Han Lin[1,2], German C Spangenberg[1,2,3] and John W Forster[1,2,3*]

Abstract

Background: Fragmentation at random nucleotide locations is an essential process for preparation of DNA libraries to be used on massively parallel short-read DNA sequencing platforms. Although instruments for physical shearing, such as the Covaris S2 focused-ultrasonicator system, and products for enzymatic shearing, such as the Nextera technology and NEBNext dsDNA Fragmentase kit, are commercially available, a simple and inexpensive method is desirable for high-throughput sequencing library preparation. MspJI is a recently characterised restriction enzyme which recognises the sequence motif CNNR (where R = G or A) when the first base is modified to 5-methylcytosine or 5-hydroxymethylcytosine.

Results: A semi-random enzymatic DNA amplicon fragmentation method was developed based on the unique cleavage properties of MspJI. In this method, random incorporation of 5-methyl-2′-deoxycytidine-5′-triphosphate is achieved through DNA amplification with DNA polymerase, followed by DNA digestion with MspJI. Due to the recognition sequence of the enzyme, DNA amplicons are fragmented in a relatively sequence-independent manner. The size range of the resulting fragments was capable of control through optimisation of 5-methyl-2′-deoxycytidine-5′-triphosphate concentration in the reaction mixture. A library suitable for sequencing using the Illumina MiSeq platform was prepared and processed using the proposed method. Alignment of generated short reads to a reference sequence demonstrated a relatively high level of random fragmentation.

Conclusions: The proposed method may be performed with standard laboratory equipment. Although the uniformity of coverage was slightly inferior to the Covaris physical shearing procedure, due to efficiencies of cost and labour, the method may be more suitable than existing approaches for implementation in large-scale sequencing activities, such as bacterial artificial chromosome (BAC)-based genome sequence assembly, pan-genomic studies and locus-targeted genotyping-by-sequencing.

Keywords: Second-generation sequencing technology, Genotyping-by-sequencing, 5-methylcytosine, Restriction endonuclease, DNA shearing

* Correspondence: john.forster@ecodev.vic.gov.au
[1]Department of Economic Development, Jobs, Transport and Resources, Biosciences Research Division, AgriBio, Centre for AgriBioscience, 5 Ring Road, La Trobe University Research and Development Park, Bundoora, Victoria 3083, Australia
[2]Dairy Futures Cooperative Research Centre, Bundoora, Australia
Full list of author information is available at the end of the article

Background

Massively parallel short-read sequencing technologies have become commonly used not only for *de novo* genome sequencing, but also for a wide range of biological purposes, such as resequencing and large-scale genotyping studies. Fragmentation at random nucleotide locations is an essential component of library construction for the various short-read sequencing instruments [1], through delivery of multiple read initiation points in template molecules. Sequence information may then be decoded through computational assembly of the short reads. Physical shearing is recommended by the manufacturers of all second-generation massively parallel DNA sequencing systems, due to the high reproducibility and randomness of fragmentation. However, the process is likely to require the use of dedicated instruments. The Nextera technology (Illumina, California, USA) and the NEBNext dsDNA Fragmentase kit (New England Biolabs, Massachusetts, USA) are alternative random DNA fragmentation methods which require only standard laboratory instruments [2-4]. The Nextera technology uses a transposon-transposase combination for random fragmentation of template DNA and attachment of transposon ends at the cleaved sites, permitting subsequent PCR amplification and sequencing. With the NEBNext dsDNA Fragmentase kit, double-stranded template DNA is fragmented in two sequential steps: nicks are enzymatically introduced into DNA, which is then cleaved at the nicked sites. These enzyme-based methods, however, require DNA sample preparation (buffer replacement and DNA concentration adjustment) for effective digestion, and the size of products is sensitive to both DNA sample quality and reaction duration, all of which require optimisation for each sample in order to achieve the desired outcome.

MspJI is a recently characterised modification-dependent endonuclease [5]. The enzyme was isolated from *Mycobacterium* sp. JLS and recognises sites containing the sequence motif CNNR (R = G or A nucleotides) when the first base is a 5-methylcytosine ($5^{m}C$) or 5-hydroxymethylcytosine, and cleaves DNA at N_{12}/N_{16} bases distant from the modified cytosine on the 3′-side. Enzyme activity is enhanced by short double-stranded DNA that includes the MspJI recognition site (and so acts as an enzyme activator). Digestion of genomic DNA with the MspJI enzyme generates fragments 32-34 bp in length, containing ^{m}CpG or ^{m}CNG sites central to the fragment. Methylation status of the human genome has been analysed through sequencing of such fragments [6]. Due to these unique features as a methylation-dependent restriction enzyme with adjacent non-specific cleavage activity, MspJI is expected to be highly useful for DNA modification and epigenomic studies [7].

FspEI and LpnPI are also recently characterised modification-dependent endonucleases, derived from *Frankia* sp. EAN1pec and *Legionella pneumophila* Philadelphia 1, respectively [5]. The sequence recognition sites for FspEI and LpnPI are 5'-CC-3' and 5'-CCDG-3' (D = A, G or T nucleotides), respectively, when the second base is a $5^{m}C$ or 5-hydroxymethylcytosine. Similar to MspJI, activity of the two enzymes is stimulated by short DNA templates containing the recognition site, and the enzymes produce DNA fragments with 5'- termini including cohesive ends. These two enzymes hence provide potential alternatives to MspJI as a tool for DNA fragmentation.

The present study describes a simple and inexpensive method for generation of semi-randomly fragmented DNA from amplicon templates. DNA amplicons with randomly-incorporated 5-methyl-2'-deoxycytidine-5'-triphosphate (5-methyl-dCTP) were synthesised with DNA polymerase, and then digested with the MspJI restriction enzyme. The size range of the MspJI-digested fragments was capable of control through optimisation of 5-methyl-dCTP concentration. A purification procedure is unnecessary for DNA digestion with MspJI, which permits high-throughput sequencing library preparation. Short DNA fragments were also generated from a range of templates with a whole genome amplification kit based on activity of the Φ29 DNA polymerase, using the same methodology. Illumina sequencing libraries with inserts of 200 or 550 bp in length were successfully prepared using the MspJI-digested DNA, and were processed on the Illumina MiSeq platform.

Results and discussion

DNA amplification with 5-methyl-dCTP and MspJI digestion

PCR amplification in the presence of 5-methyl-dCTP was performed with locus-specific primers and *Agrobacterium tumefaciens* genomic DNA (*Agro* gDNA) as template (Additional file 1). No significant difference in yield of PCR amplicon due to 5-methyl-dCTP concentration levels (final concentrations of 2, 4, or 8 μM in the PCR solution) was observed, based on the results of agarose gel electrophoresis, which was consistent with the previous studies (Figure 1a) [8,9]. The PCR amplicons of the *Agro*_gc40, *Agro*_gc50 and *Agro*_gc60 sequences, of which G/C content ratios were 40, 50 and 60%, respectively, were digested using the MspJI enzyme to identify the consequence of $5^{m}C$-containing DNA cleavage in the size-resolution range afforded by a 2.5% (w/v) agarose gel (Figure 1b). The size range of the digested DNA largely depended on concentration of 5-methyl-dCTP in the amplification solution, such that a lower range was obtained by a higher 5-methyl-dCTP concentration. Size range was also related to G/C content ratio, such that smaller DNA fragments were identified when amplicons with a higher ratio were digested (Figure 1b). This is probably due to the presence of at least one C-G base pair in

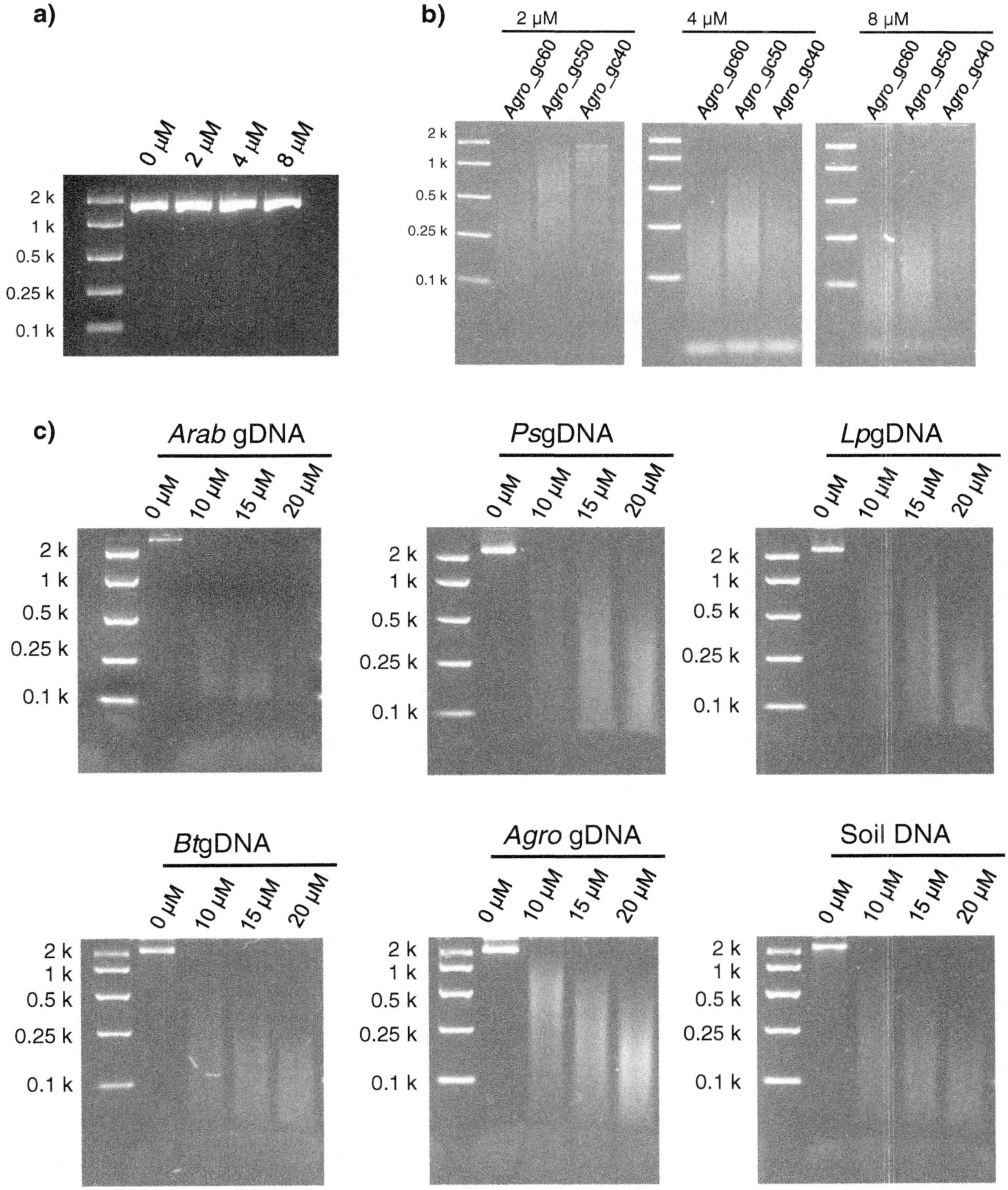

Figure 1 MspJI-enzymatic digestion of $5^{m}C$-containing PCR and Φ29 products. (**a**) DNA fragments amplified with the locus-specific PCR primers for the *Agro*_gc50 sequence under the presence of 5-methyl-dCTP (0, 2, 4 or 8 µM). (**b**) MspJI-digested DNA fragments derived from PCR products with each locus-specific primers and *Agro* gDNA as DNA template. Molar concentration denotes the 5-methyl-dCTP-concentration in PCR solution. (**c**) MspJI-enzymatic digestion of Φ29 enzyme-amplified DNA with randomly incorporated $5^{m}C$ from a range of DNA templates. 0, 10, 15 and 20 µM denote final concentrations of 5-methyl-dCTP in the REPLI-g WGA mixture.

the MspJI recognition site (5^{m}CNNR), which will produce a small bias in incidence toward regions of higher G/C content ratio.

The *Agro*_gc50 sequence was also amplified with four types of DNA polymerase. The sequence was successfully amplified with all polymerases in the presence of 5-methyl-dCTP, and the amplicons were digested with MspJI. No significant differences in size range were observed, suggesting that a variety of DNA polymerases may be used for the proposed DNA fragmentation method (Additional file 2). A further characterisation of MspJI enzymatic activity indicated that components of the PCR solution do not significantly affect activity of the MspJI enzyme when diluted in the reaction mixture; the MspJI-mediated digestion of amplicons is completed within 4 hours; and the digestion result is independent of input DNA amount, when performed in an appropriate volume of reaction mixture (Additional file 3).

Whole genome amplification (WGA) was performed using the QIAGEN REPLI-g mini kit in the presence of 5-methyl-dCTP (10 to 20 μM) with genomic DNA samples from *Arabidopsis thaliana* (L.) Heynh. ecotype Columbia (*Arab* gDNA), a field pea (*Pisum sativum* L. subsp. *sativum* var. *arvense* (L.) Poir.) genotype (*Ps*gDNA), a perennial ryegrass (*Lolium perenne* L.) genotype (*Lp*gDNA), a bovine (*Bos taurus* L.) genotype (*Bt*gDNA), *Agro* gDNA, and a DNA sample from soil harvested in South Australia (Soil DNA) as templates. Amplified product was visualised on an agarose gel, revealing no significant differences in DNA amplification due to variation of 5-methyl-dCTP concentration. The amplified DNA was digested with MspJI (Figure 1c). Similar size distribution patterns were detected across varying 5-methyl-dCTP concentrations: a majority of DNA fragments from the 20 μM 5-methyl-dCTP-containing solutions was shorter than 250 bp, and a proportion of DNA fragments from 10 μM 5-methyl-dCTP-containing solutions was close to 1,000 bp, or larger, in size.

Massively parallel sequencing of MspJI-digested templates

For exemplification of the effects on distribution of fragments generated with the present method, sequencing libraries were prepared from *Agro* gDNA and *Arab* gDNA, using the MspJI-based and standard physical fragmentation methods (Additional file 4). The libraries were sequenced on the Illumina MiSeq platform. Totals of 1,380,029 and 1,219,389 reads derived from Φ29 enzyme amplification were aligned with the *Agrobacterium* circular and linear chromosomes (Figure 2a, Table 1). From the physically sheared DNA, totals of 546,292 and 496,662 reads were aligned with the circular and linear chromosomes, respectively. When the short reads from the MspJI-digested and physically sheared DNA were aligned, 99.99% and 99.8% of the reference *Agrobacterium* genome sequence was covered, respectively. When the sequencing reads derived from *Arab* gDNA were aligned with the reference sequence, 3-5 million reads were assigned to each chromosome (Figure 2b, Table 1). Average depths of coverage from the enzymatically fragmented and physically sheared libraries were 43.3 and 53.3 times, respectively, which covered 97.9% and 99.3% of the reference *Arabidopsis* genome sequence. The results of resequencing activities suggested that MspJI successfully generates DNA templates with a moderate level of random fragmentation.

For a further exemplification of the effects on distribution of fragments generated with a methylation-dependent restriction enzyme, the *Bt*KIT1-10 and *Bt*KIT27-37 sequences were amplified from *Bt*gDNA, using the Roche Expand Long Range dNTPack kit in the presence of 7.5, 15 and 60 μM 5-methyl-dCTP (Additional file 1). Amplicons from the 7.5, 15 and 60 μM 5-methyl-dCTP reactions were digested with MspJI, FspEI or LpnPI (Additional file 5). Illumina sequencing libraries were prepared from the enzyme-digested DNA and sequenced on the MiSeq platform (Additional file 4). Totals of 3566-10,972 reads were aligned with the reference sequences, covering each nucleotide position of the reference. CVs for each data set were between 0.29 and 1.11 (Figure 3). The read alignment result indicated that the read distribution from the LpnPI-digested library was more skewed than those from the other libraries, and there was no large difference in the CVs between the MspJI- and FspEI-digested libraries. The average frequencies of potential MspJI-, FspEI- and LpnPI-recognition sites were once in every 4, 8 and 51.2 bp, respectively. This alignment result suggests that although the read distribution pattern depends on the frequency of the recognition sites, an incidence of once in every 8 bp is sufficient to generate reads with relatively even distribution.

Potential application 1: sequencing of BAC clones

Although high-throughput DNA sequencing technologies have delivered a cost-efficient whole-genome shotgun sequencing method for those species with large genome sizes, information from BAC-based genomic libraries is valuable for effective DNA sequence assembly [10]. Sequence information from BAC-ends is commonly used for *de novo* assembly of large genomes [11,12]. The BAC-end sequencing procedure, however, requires a large investment, as it depends on the Sanger sequencing method [13]. A simple sequencing method for BAC clones using high-throughput sequencing technologies is described here.

BAC DNA was amplified in the reaction mixture of the REPLI-g mini kit, in which 30 μM 5-methyl-dCTP were included (Figure 4a, Additional file 6). The amplicons were digested with MspJI. A sequencing library was

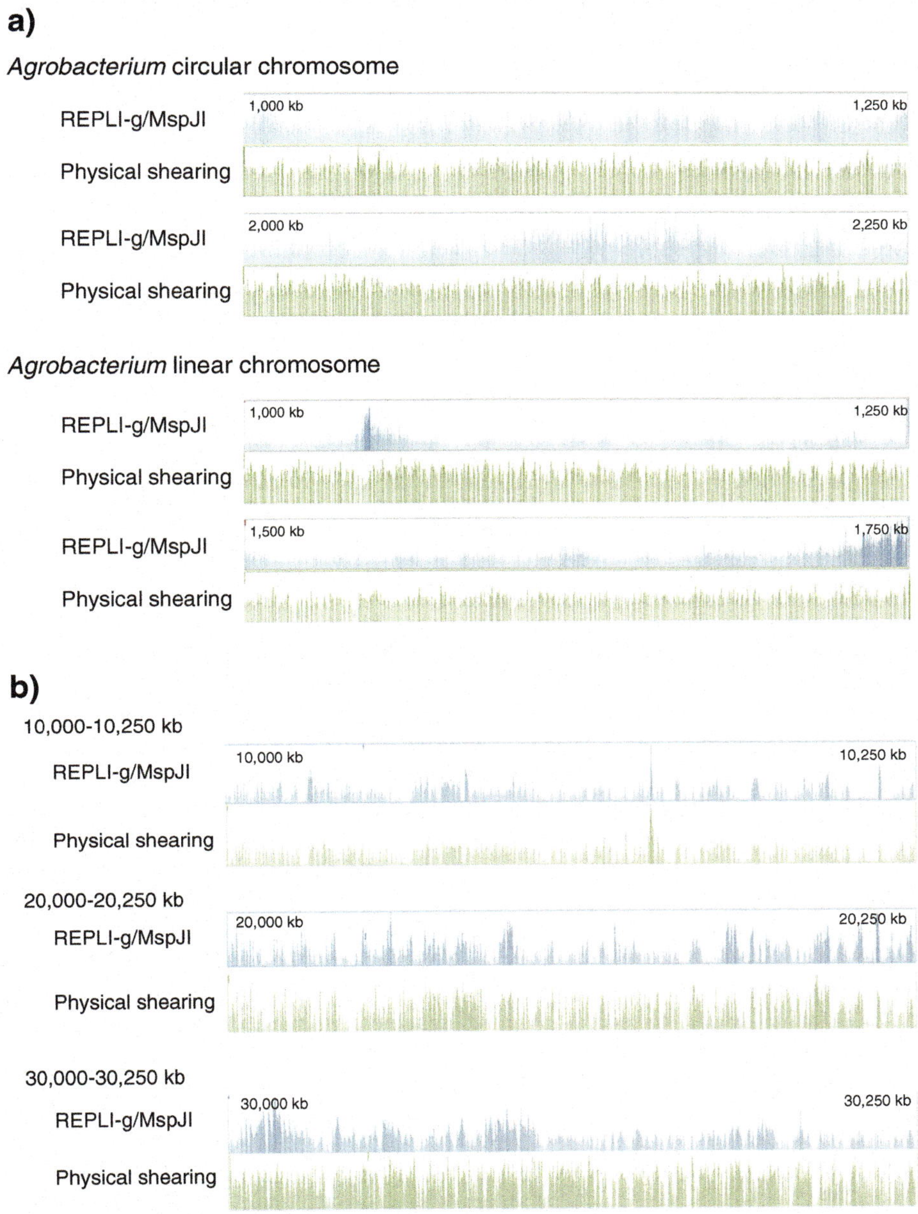

Figure 2 Illumina MiSeq short read-sequencing results of the libraries constructed from MspJI-digested and physically sheared DNA. The sorted alignment was visualised using the Tablet viewers. (**a**) Alignment results of Φ29 enzyme-amplified *Agro* gDNA-derived reads and physically sheared *Agro* gDNA-derived reads to the reference *Agrobacterium* genome sequences for the 1,000-1,250 and 2,000-2250 kb regions of the circular and linear chromosomes. (**b**) Alignment results of the *Arabidopsis* genome with the MspJI-enzymatic fragmentation and physical shearing methods. Read coverage depth for the 10,000-10,250, 20,000-20,250 and 30,000-30,250 kb regions of chromosome 1 was visualised.

prepared from the MspJI-digested DNA and processed on the Illumina MiSeq platform to generate 395,498-558,276 reads for each sample. Totals of 27,585-86,861 reads (7.0-15.6%) were aligned with the reference, which covered over 99.6% of the reference sequences (Figure 4b, Table 2). The alignment result indicated that 81%-86% of reads were derived from the *Escherichia coli* (*E. coli*) genome. When a subset of 100,000 reads was aligned, around 99% of the reference BAC insert sequences were covered an average redundancy of 21-27 times. A single run of the Illumina MiSeq platform may generate up to 50 million reads, permitting individual sequencing of a whole 384-well plate of BAC clones with moderate depth of read coverage. With the Illumina MiSeq platform, a single clone may be analysed at the cost of US$7.2 (Additional file 7).

Table 1 Result of *Agrobacterium* and *Arabidopsis* genome resequencing using the two different DNA fragmentation methods

Species	Fragmentation method	Chromosome	Number of reads	Ratio of coverage (%)	Average coverage depth (time)
Agrobacterium	REPLI-g/MspJI	Circular	1,380,029	99.9	64.3
		Linear	1,219,389	99.9	77.8
	Physical shearing	Circular	546,292	99.8	27.4
		Linear	496,662	99.9	34.1
Arabidopsis	REPLI-g/MspJI	1	4,611,463	97.3	36.4
		2	4,759,134	98	57.7
		3	4,702,657	98.3	47.9
		4	3,088,231	98	39.9
		5	4,222,266	98	37.8
	Physical shearing	1	5,116,095	98.9	41.5
		2	6,143,569	99.5	75
		3	6,083,236	99.5	62.4
		4	3,776,130	99.5	49.3
		5	4,980,370	99.4	45.1

Ratio of coverage denotes the value for specific nucleotides located on each chromosome.

Potential application 2: whole genome amplification and sequencing of bacterial and fungal genomes

High-throughput DNA sequencing technologies also provide an efficient method for pan-genome studies, especially for bacterial and fungal species [14,15]. Due to high levels of genomic diversity, a substantial number of bacteria or fungal strains must, however, be sequenced in order to define both core- and pan-genome constituents, and so a high-throughput library preparation method is required. Previously, direct WGA from fungal tissues was suggested as an efficient DNA sample preparation method [16]. A combination of the WGA and MspJI-based DNA fragmentation methods may permit high-throughput library processing.

Genomic DNA from the perennial ryegrass-associated endophyte (*Epichloë festucae* var. *lolii* syn. *Neotyphodium lolii*) was amplified from a section of fungal mycelium (Figure 5a). In the multiple displacement amplification (MDA) reaction mixture, 15 μM 5-methyl-dCTP was included, and the amplified DNA was digested with MspJI (Figure 5b). A sequencing library was prepared from the MspJI-digested DNA and processed on the Illumina MiSeq platform. Totals of 2.7 million reads were generated, and 81,237 and 59,445 reads were aligned with the reference contig, which represent 1.6 and 1.3 Mb sections,

Sequence	Enzyme	Average coverage depth (time)	CV
*Bt*KIT1-10	MspJI	207.2	0.29
	FspEI	213.9	0.33
	LpnPI	70.5	0.44
*Bt*KIT27-37	MspJI	120.9	0.44
	FspEI	187.0	0.38
	LpnPI	94.8	1.15

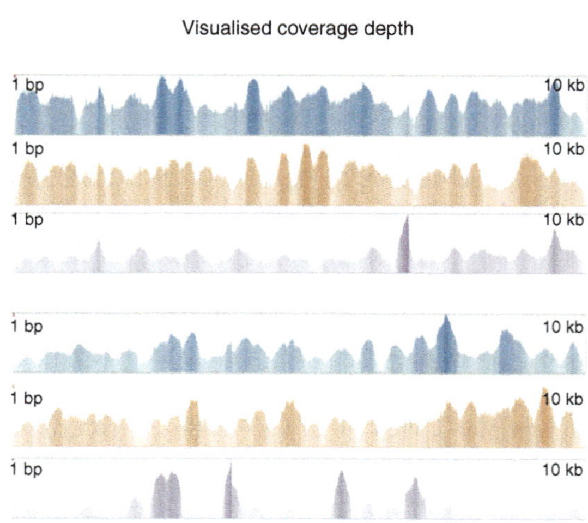

Figure 3 Short read-sequencing results of the libraries constructed from MspJI, FspEI and LpnPI-digested DNA. Read coverage depth was visualised using the Tablet viewer.

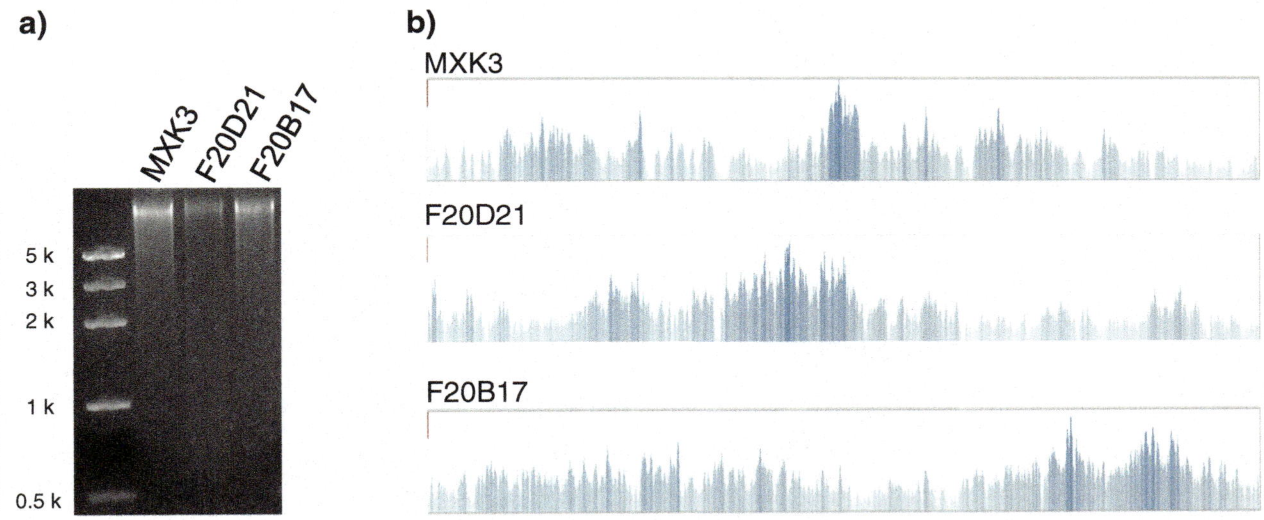

Figure 4 BAC clone sequencing based on the MspJI-based DNA fragmentation approach. (**a**) Amplified BAC DNA directly from glycerol stocks, using the Φ29 enzyme. In the amplification solution, 30 μM 5-methyl-dCTP were included. (**b**) Resequencing results of the BAC clones. Read coverage depth for each clone was visualised on the Tablet software.

respectively, of a genome 30 Mb in length (Table 3). A previous study reported that WGA was successfully achieved from as few as 24 fungal spores, which contribute to significantly reduced durations of DNA sample preparation [16]. DNA fragmentation with MspJI does not require several procedures subsequent to WGA, and so may enhance efficiency of the WGA-based sequencing library preparation (Figure 5c).

Potential application 3: sequencing of PCR amplicons

Massively parallel sequencing technologies have permitted whole genome re-sequencing in a cost-effective manner [17]. Subsequently, genome-wide association studies (GWASs) have identified DNA polymorphisms that are correlated with trait-specific variation [18]. The numbers of relevant DNA polymorphisms identified through GWASs have, however, been relatively small [17,18]. Identification of trait locus variation-related DNA polymorphisms could hence be usefully followed by conversion into specific PCR-based markers, permitting locus-targeted genotyping over larger numbers of individuals [19,20].

Through PCR with 5-methyl-dCTP, candidate sequences for genes involved in the perennial ryegrass flowering signaling pathway were amplified from the DNA samples of the p150/112 F_1 genetic mapping population (Figure 6a) [21]. The PCR amplicons were digested with MspJI, and sequencing library was prepared for the Illumina MiSeq platform (Figure 6b, Additional file 8). The sequencing reads were aligned against the references, and SNPs, which could be utilised for genetic linkage mapping studies, were identified in 7 of the sequences (Additional file 9). In the *Lp*CO and *Lp*FT sequence, no variation was identified between the two parental-derived sequence haplotypes. Due to the residual presence of heat-resistant DNA polymerase and dNTPs which could perform end-filling and adenine-tailing reactions during heat-inactivation of MspJI, the DNA fragments from restriction enzyme digestion could be directly used for DNA adapter ligation (Figure 6d). Cost assumption analysis suggested that the MspJI digestion method would be less expensive than the previously described cost-reduced physical shearing method (Additional file 10) [22]. Due to a lesser requirement for capital expenditure on

Table 2 Resequencing results of the BAC clones

		All reads				100,000 read subset		
BAC name	BAC length	Total reads	Aligned reads (%)	Ratio of coverage (%)	Average coverage depth (time)	Aligned reads	Ratio of coverage (%)	Average coverage depth (time)
MXK3	81 kb	395,498	27,585 (7.0%)	99.6%	82.3	7086	98.6%	21.3
F20D21	143 kb	558,276	86,861 (15.6%)	99.8%	149.4	15382	99.2%	26.5
F20B17	90 kb	478,300	46,409 (9.7%)	>99.9%	126.2	9173	99.9%	24.9

Ratio of coverage denotes the value for specific nucleotides located on each clone. A subset of 100,000 reads was obtained using the Seqtk software.

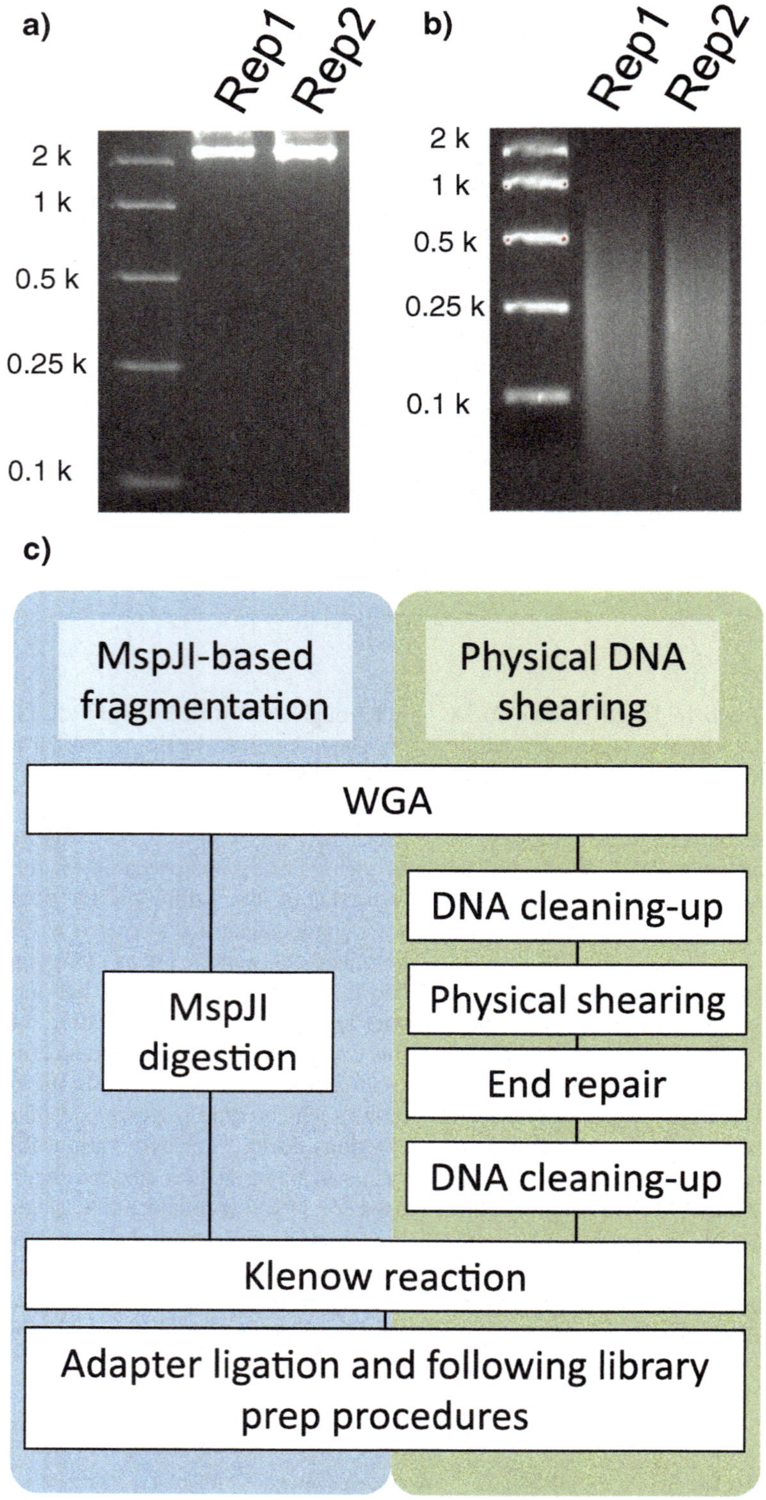

Figure 5 Sequencing of the fungal endophyte genome with the MspJI-based DNA fragmentation method. Two independent experiments (rep1 and rep2) were performed using the single endophyte strain. (**a**) WGA of the endophyte genome with the Φ29 enzyme in the presence of 5-methyl-dCTP (X μM). (**b**) MspJI digestion of the endophyte genome-derived amplicons. (**c**) Flowchart of the MspJI- and physical shearing-based library prep procedures from WGA products.

Table 3 Resequencing results of the ryegrass endophyte genome

Reference		Alignment				
UI	Length (bp)	Aligned reads	Ratio of coverage (%)	Average coverage depth (time)	Max coverage depth (time)	Mismatch (%)
gi|347366940	1,598,175	81,237	88.7	10.1	749	1.5
gi|347366939	1,323,136	59,445	86.7	8.8	304	1.4

UI denotes the unique identifier of the NCBI GenBank. Short read sequencing data from two experiments (rep1 and rep2) were combined for read alignment.

equipment, the MspJI digestion method would become further cost-competitive when sample number is less than 100,000 [22,23]. The operator-specific time for the MspJI digestion method was expected to be 40% less than the physical shearing method (Additional file 11).

Conclusions

The present study has reported a novel method for DNA fragmentation using the MspJI enzyme, which has been exemplified for a range of template types. A DNA sample preparation procedure, such as buffer replacement and DNA concentration adjustment, is not essential for MspJI digestion, which permits a simple DNA library preparation procedure from amplicons. A modified method involving combined use with other modification-dependent restriction enzymes may improve the random nature of the fragmentation. The size range of the resulting fragments was capable of control through adjustment of the 5-methyl-dCTP concentration in the amplification reaction solution, providing various fragment ranges from <100 bp to >2 kb. The method may hence be applicable for recombinant DNA purposes other than second-generation massively parallel short read sequencing technologies. Development of a computational methodology may improve sequencing efficiency with this method, through optimisation of 5-methyl-dCTP concentration and prediction of coverage for each nucleotide.

Methods

DNA sample preparation

DNA samples were prepared using the QIAGEN DNeasy kit (QIAGEN, Hilden, Germany) (*Arab* gDNA, *Ps*gDNA and *Lp*gDNA) and the Gentra PUREGENE® DNA Purification Kit (QIAGEN) (*Bt*gDNA), BioRad AquaPure Genomic DNA Kit (Bio-Rad Laboratories, CA, USA) (*Agro* gDNA). The Soil DNA sample was extracted with the MoBio Powersoil kit (MoBio, CA, USA), following a modified protocol [24]. DNA concentrations were adjusted to 5-30 ng/µl in the TE buffer using the NanoDrop system (Thermo Fisher Scientific, MA, USA).

DNA amplification with 5-methyl-dCTP

Locus-specific primers were prepared for amplification of the *Agro*_gc40, *Agro*_gc50 and *Agro*_gc60, *Bt*KIT1-10 and *Bt*KIT27-37 sequences (Additional file 1), and PCR was performed with Phusion Hot Start DNA polymerase (Thermo Fisher Scientific) and The Expand Long Range dNTPack (Roche Applied Science, Penzberg, Germany), following the manufacturer's protocol. 5-methyl-dCTP (TriLink, CA, USA) was added to the PCR mixture at final concentrations of from 0 to 60 µM. WGA was performed using the REPLI-g mini kit (QIAGEN). Following the manufacturer's protocol, 2.5 µl DNA (12.5-75 ng) was denatured with the D1 solution for three minutes, and then neutralised with the N1 solution. The amplification was performed in the reaction mixture with the presence of from 0 to 100 µM 5-methyl-dCTP (final concentrations) at 30°C for 16 hours. After incubation, the DNA polymerase was heat-inactivated, and the products were diluted with the same amount of water.

Restriction enzyme digestion

The amplified DNA (5 µl) was digested with 3 U of the MspJI, FspEI or LpnPI restriction enzyme (NEB) following manufacturer's protocol. After incubation at 37°C for 4-16 hours, the enzyme was heat-inactivated at 70°C for 20 minutes.

Illumina sequencing library construction

Sequencing libraries were constructed with the TruSeq DNA Sample Preparation kit (Illumina) or NEBNext® DNA Library Prep Master Mix Set for Illumina® (NEB) with modifications. End-filling and adenine-tailing reactions for the MspJI-digested REPLI-g products were performed with the A-Tailing Mix or Klenow Fragment (3' → 5' exo$^-$). For the PCR amplicons with the Expand Long Range dNTPack, the heat-inactivation procedure of the restriction enzyme also permitted end-filling of the restriction enzyme-digested fragments in the presence of the activated heat-resistant DNA polymerase and dNTPs. Blunt-ended DNA was purified with the AMPure XP bead kit (Life Technologies, CA, USA) and was used for the adenine-tailing reaction of the NEB kit. The 'Purify Ligation Products' process based on agarose gel electrophoresis (Illumina kit) was not performed. Following ligation of the DNA adapter index, the ligated DNA was purified with the AMPure XP bead kit and enriched through PCR. The size range of enriched DNA fragments was determined with the Agilent 2100

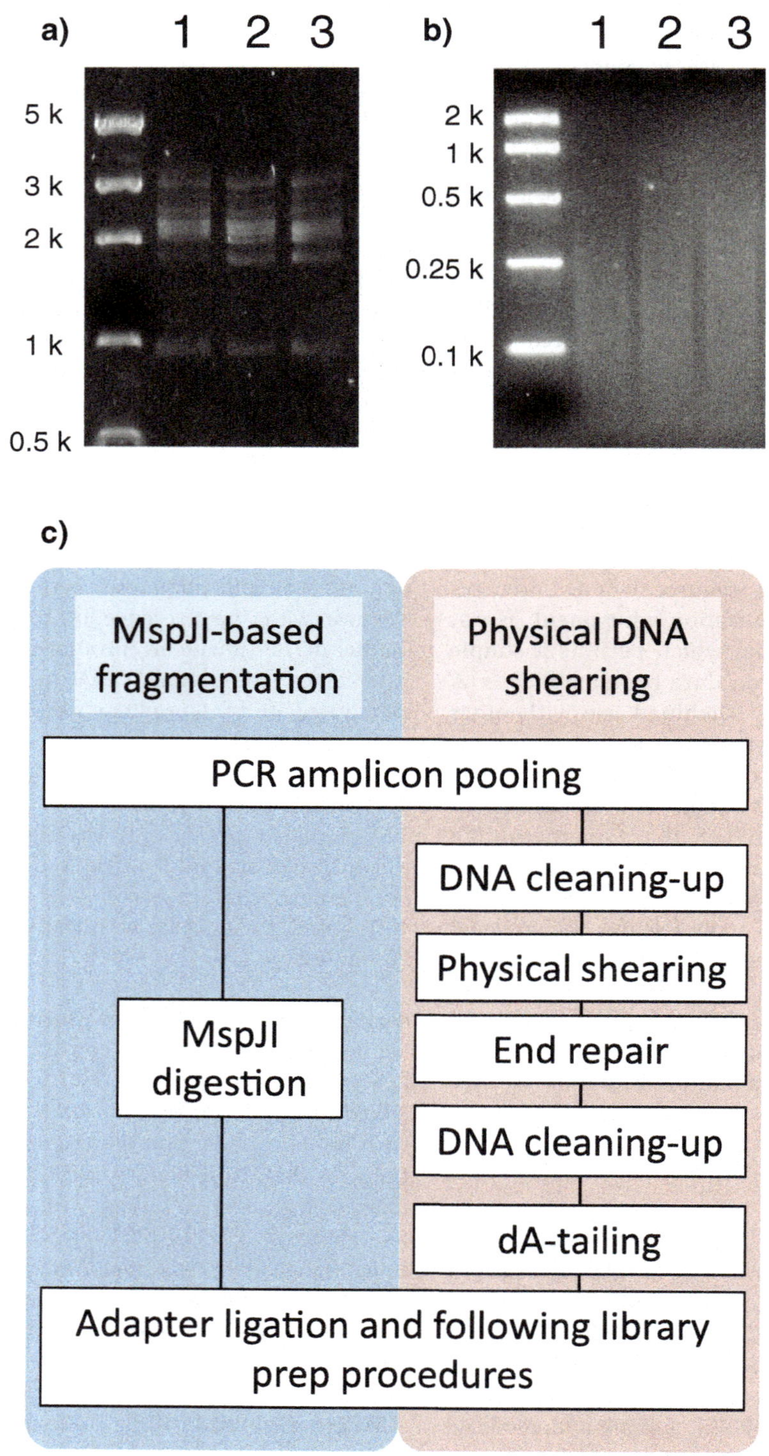

Figure 6 Resequencing of PCR amplicons with the MspJI-based DNA fragmentation method. (**a**) Pooled PCR amplicons containing $5^{m}C$. Locus-specific amplification was performed for each sequence independently. (**b**) MspJI digestion of the PCR amplicons. (**c**) Flowchart of the MspJI- and physical shearing-based library prep procedures from PCR amplicons.

Bioanalyzer and Agilent DNA 1000 Kit (Agilent Technologies, CA, USA). The sequencing library was quantified with the KAPA Library Quantification Kit (Kapa Biosystems, MA, USA), following the manufacturer's protocol.

Following the standard procedure, sequencing libraries from the *Agro* gDNA and *Arab* gDNA templates were prepared with the Illumina and NEB library prep kits, respectively. For these libraries, genomic DNA was fragmented with the S2 focused-ultrasonicator system (Covaris, MA, USA) following the manufacturer's protocol.

Sequencing library preparation from BAC-containing clone glycerol stocks

Arabidopsis BAC clones (MIXK3, F20D21 and F20B17) were amplified with the QIAGEN REPLI-g mini kit. Glycerol stock (4 μl) of a BAC-containing *E. coli* clone was mixed with the Buffer D1 (4 μl) and incubated on ice for 5 minutes. The Buffer N1 (8 μl) was added into the sample and mix by stirring with a tip. The sample was incubated at room temperature for 3 minutes. The reaction mixture, consists of 5.8 μl REPLI-g Reaction Buffer, 0.2 μl REPLI-g Mini DNA Polymerase and 0.4 μl 5-methyl-dCTP (750 μM), was added into 3.4 μl denatured sample, and the reaction mixture was incubated at 30°C for 16 hours. The amplicons (5 μl) were digested with MspJI and the end-filling reaction was performed with Klenow Fragment (3'→5' exo$^-$). Sequencing adopter ligation was performed with T4 ligase (NEB) and ligated DNA was cleaned with AMPure XP bead solution (x0.8) to exclude short DNA. DNA fragments were subsequently enriched through PCR with the phusion DNA polymerase Kit. Small fragments (<500 bp), in which fraction *E. coli* genome-derived fragments were highly prevalent, were removed through size-selection with AMPure XP bead solution (x0.6). The sequencing library was characterised with the Agilent 2100 Bioanalyzer, Agilent DNA 1000 Kit, and the Qubit® Fluorometer (Life Technologies), following the manufacturer's protocols.

Whole genome amplification and sequencing library preparation from perennial ryegrass-derived endophyte mycelium

Ryegrass endophyte genomic DNA was amplified with the QIAGEN REPLI-g mini kit. A section (2-3 mm^2) of endophyte mycelium was placed into 6 μl PBS solution. The Buffer D2 (7 μl) was added into the sample and incubated on ice for 10 minutes, following mixing by a vortex. The Stop Solution was, then, added and mixed by a vortex. The reaction mixture, consisting of 29 μl REPLI-g Reaction Buffer, 1 μl REPLI-g Mini DNA Polymerase and 1 μl 5-methyl-dCTP (750 μM), was added into 9 μl of denatured sample, and the sample was incubated at 30°C for 16 hours. The amplicons (5 μl) were digested with MspJI and the sequencing library preparation was performed following the BAC clone sequencing protocol. The PCR-enriched DNA was cleaned with AMPure XP bead solution (x0.8). The sequencing library was characterised with the Agilent 2100 Bioanalyzer and Qubit® Fluorometer.

Sequencing library preparation from PCR amplicon

Locus-specific primers for the *Lp*AP1, *Lp*CO, *Lp*CRY1, *Lp*FLD, *Lp*FT, *Lp*LHY, *Lp*PHYC, *Lp*TOC1 and *Lp*Vrn5 sequences, and the MyFi™ DNA Polymerase kit (BIOLINE), which contains DNA polymerase that lacks $3' \rightarrow 5'$ exonuclease activities, were used for PCR amplification (Additional file 1). The DNA samples of the p150/112 F_1 mapping population were used as DNA templates. In the PCR solution, 8 μM 5-methyl-dCTP was included [25]. The PCR products were pooled for MspJI digestion (37°C for 4 hours) and MspJI was inactivated through incubation at 70°C for 20 mins. The heat-inactivation procedure also permitted end-filling and adenine-tailing of the MspJI-digested fragments in the presence of the activated heat-resistant DNA polymerase and dNTPs. Sequencing adapter ligation was performed with T4 ligase and ligated DNA was cleaned with AMPure XP bead solution (x0.8) to remove short DNA. DNA fragments were, then, enriched through PCR and the product was cleaned with AMPure XP bead solution (x0.8). The sequencing library was characterised with the 2200 TapeStation system (Agilent) and Qubit® Fluorometer.

Massively parallel sequencing and read assembly

The Illumina MiSeq sequencing platform was used to generate sequence output for sequencing libraries with the Illumina MiSeq Reagent Kit v2 or v3. Reads were attributed by the use of sample-specific DNA bar codes. The generated sequence reads were then checked for quality and integrity using a custom PERL script. Any reads with more than 3 consecutive Ns or more than 3 nucleotides with PHRED score ≤ 20 or a median PHRED score < 20 or a read length <50 nucleotides were trimmed or removed. The specific DNA sequence reads were then reference-aligned to the respective amplicon, *Agrobacterium* C58 (NCBI accession numbers: AE007869 and AE007870) [26] or *Arabidopsis* Columbia sequence [27] (http://www.arabidopsis.org/index.jsp). Reference alignments were performed using the BWA software package and then converted to a sorted BAM file using the SAMtools software package (http://samtools.sourceforge.net/). The Seqtk software package was used for generation of a subset sequence data (https://github.com/lh3/seqtk). Alignment of the sequencing reads to the reference sequences was visualised using the Tablet software [28].

Availability of supporting datasets

The data sets supporting the results of this article are included within the article and its additional files.

Additional files

Additional file 1: DNA sequence name, sequence of locus-specific PCR primers, size and G/C content ratio of PCR amplicons, and type of DNA polymerase used for PCR amplification.

Additional file 2: MspJI-enzymatic digestion of $5^{m}C$-containing PCR amplicons. The *Agro*_gc50 sequence was amplified with the Immolase™, MangoTaq™ *Pfu* and Phusion Hot Start polymerases and digested with MspJI. Molar concentration denotes the 5-methyl-dCTP-concentration in the PCR solution.

Additional file 3: Characterisation of MspJI enzymatic activity. Through PCR, 8 μM $5^{m}C$-containing amplicons were generated, and the products were cleaned with the AMPure kit. A range of DNA amounts (100, 200, 500, and 1,000 ng) was used for MspJI digestion, of which 50 ng digested DNA was visualised on an agarose gel.

Additional file 4: Illumina MiSeq sequencing libraries constructed with the MspJI, FspEI or LpnPI-digested amplicons. Sharp peaks with red and blue triangles show size standards (15 bp and 1,500 bp, respectively) of the Agilent DNA 1000 Kit. A peak between 200 and 1,500 bp represents the size distribution of DNA fragments of the constructed DNA library.

Additional file 5: MspJI (a), FspEI (b) and LpnPI (c)-enzymatic digestion of $5^{m}C$-containing long PCR amplicons. Molar concentration denotes the 5-methyl-dCTP-concentration in PCR solution.

Additional file 6: Information on *Arabidopsis thaliana* genome-derived BAC clones used for short-read sequencing.

Additional file 7: Cost assumption (US$) for Sanger method-based BAC-end sequencing, and Φ29 and MspJI-based short-read sequencing. Cost for BAC-end sequencing was calculated based on previous publications [13].

Additional file 8: Illumina MiSeq sequencing library constructed with the MspJI-digested PCR amplicons. A peak between the lower and upper markers represents the size distribution of DNA fragments of the constructed DNA library.

Additional file 9: Genetic linkage mapping result of perennial ryegrass flowering signaling pathway-related sequences. The linkage maps were generated with the JoinMap® 3.0 program. The *Lp*FLD, *Lp*Vrn5, *Lp*TOC1, and *Lp*LHY loci were assigned to linkage groups (LGs) 2, 5, 6, and 7, respectively. Two loci (*Lp*AP1 and *Lp*PHY) were assigned to LG4. The loci mapped through the MspJI-based fragmentation method are indicated in red.

Additional file 10: Cost assumption for Illumina sequencing library preparation. Cost (US$) for major kit, reagent, consumable items and capital equipment was calculated based on previous publications [22,23].

Additional file 11: Time period requirement assumption for Illumina sequencing library preparation. Time periods required for library construction with the Covaris equipment-based and MspJI-based DNA fragmentation methods were estimated based on previous publications [22].

Competing interests

The authors declare that they have no competing interests.

Authors' contributions

HS performed the experimental work described in the study, co-conceptualised the study and prepared the manuscript. MS, AM, PK, and Y-HL assisted in performance of the experimental work. NOIC contributed to interpretation of data, co-conceptualised the study and contributed to preparation of the manuscript. GCS and JWF co-conceptualised the study and contributed to preparation of the manuscript. All authors read and approved the final manuscript.

Acknowledgments

This work was supported by funding from the Victorian Department of Economic Development, Jobs, Transport and Resources and the Dairy Futures Cooperative Research Centre (DFCRC). The authors would like to acknowledge Drs. Amanda Chamberlain, Tony Gendall, Helen Hayden Jatinder Kaur and Sukhjiwan Kaur, Ms. Joanne Rachel Ernest, Mr. Brett Mason and Mrs. Shimna Sudheesh for provision of the DNA samples, and plant and fungal materials. We would also like to acknowledge Mr. Sami Hakim for assistance during the sequencing process.

Author details

[1]Department of Economic Development, Jobs, Transport and Resources, Biosciences Research Division, AgriBio, Centre for AgriBioscience, 5 Ring Road, La Trobe University Research and Development Park, Bundoora, Victoria 3083, Australia. [2]Dairy Futures Cooperative Research Centre, Bundoora, Australia. [3]School of Applied Systems Biology, La Trobe University, Bundoora, Victoria 3086, Australia.

References

1. Poptsova MS, Il'icheva IA, Nechipurenko DY, Panchenko LA, Khodikov MV, Oparina NY, et al. Non-random DNA fragmentation in next-generation sequencing. Sci Rep. 2014;4:4532.
2. Knierim E, Lucke B, Schwarz JM, Schuelke M, Seelow D. Systematic comparison of three methods for fragmentation of long-range PCR products for next generation sequencing. PLoS One. 2011;6:e28240.
3. Syed F, Grunenwald H, Caruccio N. Optimized library preparation method for next-generation sequencing [advertising feature]. Nat Methods. 2009;6:i–ii.
4. Syed F, Haiying G, Nicholas C. Next-generation sequencing library preparation: simultaneous fragmentation and tagging using in vitro transposition [advertising feature]. Nat Methods. 2009;6:i–ii.
5. Zheng Y, Cohen-Karni D, Xu D, Chin HG, Wilson G, Pradhan S, et al. A unique family of Mrr-like modification-dependent restriction endonucleases. Nucleic Acids Res. 2010;38:5527–34.
6. Cohen-Karni D, Xu D, Apone L, Fomenkov A, Sun Z, Davis PJ, et al. The MspJI family of modification-dependent restriction endonucleases for epigenetic studies. Proc Natl Acad Sci. 2011;108:11040–5.
7. Chen W-H, Qin Z-J, Wang J, Zhao G-P. The MASTER (methylation-assisted tailorable ends rational) ligation method for seamless DNA assembly. Nucleic Acids Res. 2013;41:e93.
8. Wong KK, McClelland M. PCR with 5-methyl-dCTP replacing dCTP. Nucleic Acids Res. 1991;19:1081–5.
9. Wong KK, Markillie LM, Saffer JD. A novel method for producing partial restriction digestion of DNA fragments by PCR with 5-methyl-CTP. Nucleic Acids Res. 1997;25:4169–71.
10. Mascher M, Stein N. (2014) Genetic anchoring of whole-genome shotgun assemblies. Frontiers in Genetics 5. doi:10.3389/fgene.2014.00208
11. Lee KC, T-H CRO, Robertson JS, Pierce GJ, Paterson AH. A genome-wide BAC end-sequence survey of sugarcane elucidates genome composition, and identifies BACs covering much of the euchromatin. Plant Mol Biol. 2013;81:139–47. doi:10.1007/s11103-012-9987-x.
12. Breen J, Wicker T, Shatalina M, Frenkel Z, Bertin I, Philippe R, et al. A Physical Map of the Short Arm of Wheat Chromosome 1A. PLoS One. 2013;8:e80272. doi:10.1371/journal.pone.0080272.
13. Kim H, Miguel PS, Nelson W, Collura K, Wissotski M, Walling JG, et al. Comparative Physical Mapping Between Oryza sativa (AA Genome Type) and O. punctata (BB Genome Type). Genetics. 2007;176:379–90. doi:10.1534/genetics.106.068783.
14. Mira A, Martín-Cuadrado AB, D'Auria G, Rodríguez-Valera F. (2010) The bacterial pan-genome: a new paradigm in microbiology. International Microbiology 45–57. doi:10.2436/20.1501.01.110
15. Stajich JE, Harris T, Brunk BP, Brestelli J, Fischer S, Harb OS, et al. FungiDB: an integrated functional genomics database for fungi. Nucleic Acids Res. 2012;40:D675–81. doi:10.1093/nar/gkr918.
16. Lee J, Young JP. The mitochondrial genome sequence of the arbuscular mycorrhizal fungus Glomus intraradices isolate 494 and implications for the phylogenetic placement of Glomus. New Phytol. 2009;183:200–11.

17. Larkin DM, Daetwyler HD, Hernandez AG, Wright CL, Hetrick LA, Boucek L, et al. Whole-genome resequencing of two elite sires for the detection of haplotypes under selection in dairy cattle. Proc Natl Acad Sci. 2012;109:7693–8.
18. Huang X, Wei X, Sang T, Zhao Q, Feng Q, Zhao Y, et al. Genome-wide association studies of 14 agronomic traits in rice landraces. Nat Genet. 2010;42:961–7.
19. Mamanova L, Coffey AJ, Scott CE, Kozarewa I, Turner EH, Kumar A, et al. Target-enrichment strategies for next-generation sequencing. Nat Methods. 2010;7:111–8.
20. Fritsche S, Wang X, Li J, Stich B, Kopisch-Obuch F, Endrigkeit J, et al. A candidate gene-based association study of tocopherol content and composition in rapeseed (*Brassica napus*). Front Plant Sci. 2012;3:129.
21. Jones ES, Mahoney NL, Hayward MD, Armstead IP, Jones JG, Humphreys MO, et al. An enhanced molecular marker based genetic map of perennial ryegrass (*Lolium perenne*) reveals comparative relationships with other Poaceae genomes. Genome. 2002;45:282–95.
22. Rohland N, Reich D. Cost-effective, high-throughput DNA sequencing libraries for multiplexed target capture. Genome Res. 2012;22:939–46.
23. Lennon NJ, Lintner RE, Anderson S, Alvarez P, Barry A, Brockman W, et al. A scalable, fully automated process for construction of sequence-ready barcoded libraries for 454. Genome Biol. 2010;11:R15.
24. Hayden HL, Mele PM, Bougoure DS, Allan CY, Norng S, Piceno YM, et al. Changes in the microbial community structure of bacteria, archaea and fungi in response to elevated CO_2 and warming in an Australian native grassland soil. Environ Microbiol. 2012;14:3081–96.
25. Shinozuka H, Cogan NO, Spangenberg GC, Forster JW. Quantitative Trait Locus (QTL) meta-analysis and comparative genomics for candidate gene prediction in perennial ryegrass (*Lolium perenne* L.). BMC Genet. 2012;13:101.
26. Goodner B, Hinkle G, Gattung S, Miller N, Blanchard M, Qurollo B, et al. Genome sequence of the plant pathogen and biotechnology agent *Agrobacterium tumefaciens* C58. Sci (New York). 2001;294:2323–8.
27. Initiative AG. Analysis of the genome sequence of the flowering plant Arabidopsis thaliana. Nature. 2000;408:796–815.
28. Milne I, Stephen G, Bayer M, Cock PJA, Pritchard L, Cardle L, et al. Using Tablet for visual exploration of second-generation sequencing data. Brief Bioinform. 2012;14:193–202.

20

Isolation, production, purification and characterization of an organic-solvent-thermostable alkalophilic cellulase from *Bacillus vallismortis* RG-07

Rajeeva Gaur* and Soni Tiwari

Abstract

Background: The rising concerns about the scarcity of fossil fuels, the emission of green house gasses and air pollution by incomplete combustion of fossil fuel have also resulted in an increasing focus on the use of cellulases to perform enzymatic hydrolysis of the lignocellulosic materials for the generation of bioethanol. The aim of this study was to isolate a potential thermo-solvent tolerant cellulase producing bacterium from natural resources, and then applied for purification and characterization. The purified enzyme was to be accessible for the bioethanol production as well as industrial exploitation (discuss in our next study).

Results: It is the first instance when thermo-solvent tolerant cellulase producing bacterium was isolated from soil sample. The culture was identified as *Bacillus vallismortis* RG-07 by 16S rDNA sequence analysis. *Bacillus vallismortis* RG-07 reported maximum cellulase production from sugarcane baggase (4105 U ml^{-1}) used as agro-waste carbon source. The cellulase enzyme produced by the *Bacillus* sp. was purified by $(NH_4)_2SO_4$ precipitation, ion exchange and gel filtration chromatography, with overall recovery of 28.8%. The molecular weight of purified cellulase was 80 kDa as revealed by SDS-PAGE and activity gel analysis. The optimum temperature and pH for enzyme activity was determined as 65°C and 7.0 and it retained 95 and 75% of activity even at 95°C, and 9.0 respectively. The enzyme activity was enhanced in the presence of organic solvents (30%) n-dodecane, iso-octane, n-decane, xylene, toluene, n-haxane, n-butanol, and cyclohexane, after prolonged incubation (7 days). The enzyme activity was also stimulated by Ca^{2+}, mercaptoethanol, Tween-60, and Sodium hypochloride whereas strongly inhibited by Hg. Kinetic analysis of purified enzyme showed the K_m and V_{max} to be 1.923 mg ml^{-1} and 769.230 µg ml^{-1} min^{-1}, respectively.

Conclusion: The unique property of solvent-thermostable-alkalophilic, nature proves the potential candidature of this isolate for current mainstream biomass conversion into fuel and other industrial process.

Keywords: *Bacillus vallismortis*, 16S rDNA, Solvent tolerant, Heavy metal, Thermotolerant

Background

In present time, Cellulases have attracted much attention because of their application in various industrials processes, including food, textiles, laundry, pulp and paper as well as in agriculture [1]. Cellulases contribute to 8% of the worldwide industrial enzyme load and the demand is expected to increase by 100% within 2014 [2]. The increasing concern about the shortage of remnant energy, the release of green house gasses by incomplete burning of fossil fuel which create air pollution have resulted in rising center on the use of cellulases to carry out enzymatic hydrolysis of the lignocellulosic waste materials for the production of bioethanol [3,4]. Cellulases have a group of three enzymes namely endo-1,4-β-glucnase (Endogluconase), exo-1,4-β-glucnase (Exoglucanases) and β-glucosidase that synergistically hydrolyzed cellulose into soluble sugars and glucose [5]. Endogluconases attack the cellulose crystalline structure at random places, breaking the linear chains of glucose molecules to produce shorter chains. Each break generates two new chain ends. Exoglucanases act to these exposed ends of the chains and, working down the chains, liberate cellobiose and

* Correspondence: rajeevagaur@rediffmail.com
Department of Microbiology (Centre of Excellence), Dr. Ram Manohar Lohia Avadh University, Faizabad 224001, Uttar Pradesh, India

some glucose. Finally, β-glucosidases completes the saccharification by breaking cellobiose and small cello-oligosaccharides into glucose molecules [6].

Cellulases are inducible enzymes which are synthesized by microorganisms during their growth on cellulosic materials. Several microorganisms can produce cellulase enzyme including fungi, bacteria and actinomycetes. Presently, the majority of the commercial and laboratory cellulases are achieved by fungi due to their high enzyme activity, but several factors suggest that bacteria may have excellent potential [7]. Bacteria frequently have a higher growth rate than fungi allowing for higher rate of enzyme production. Most significantly, they show affinity to be more heat stable and are easier for genetic purpose. Various bacterial genera reported for cellulolytic activities include *Bacillus, Clostridium, cellulomonas, Rummminococcus, Alteromonas, Acetivibrio* etc. Among bacteria, *Bacillus* sp. including *B. brevis* [8], *B. pumilus* [9], *B. amyoliquefaciens* DL-3 [10], and *Bacillus subtilis* YJ1 [11] *Bacillus sp.* [12,13] are well recognized cellulase production under submerged condition [14,15]. The present study was to isolate a potential organic-solvent-thermostable alkalophilic cellulase producing bacteria from natural ecosystem. Afterthat, the cellulase enzyme was applied for purification and characterization of different parameters and purify enzyme was applied for the ethanol production and industrial exploitation which discuss in our next study.

Results and discussion

Isolation, screening and identification of thermotolerant cellulase producing bacterial cultures

Fifty (50) bacterial strains were isolated from the soil samples on CMC agar plates. For checking the cellulolytic activity of the isolates on plates, plates were stained with congo-red and NaOH solution. The zones of clearance by isolates reflect their extent of cellulolytic activity. Those having clearance zone greater than >1.0 cm were considered as significant. Among fifty (50) bacterial isolates, only twenty nine (29) bacterial isolates exhibited good cellulase activity which was reassessed by loading their culture broth in the wells on CMC agar plates which stained with congo-red and NaOH solution. The culture broth of good cellulase producers cleared more than >1.0 cm zone within 4–5 h of incubation at 65°C, thereby indicating an extra-cellular nature of the cellulase. The isolate RG-07, showing maximum clearance zone diameter was selected for further studies.

The efficient strain RG-07 was rod-shaped, gram-positive, aerobe and facultative, motile, with positive acetylmethylcarbinol, catalase and oxidase test. It grew over a wide range of pH (4.0-11), temperature (10-85°C), NaCl concentration (0.0-12%), and was able to hydrolyze gelatin, casein, starch, tween-20, tween-40 and tween-80 and produce acid from glucose, xylose, mannitol and arabinose. It gives positive test for citrate utilization and nitrate reduction. The strain was halotolerant as it grew in the presence of 0.0-12% NaCl. On account of morphological and biochemical characteristics, it was identified as *Bacillus sp.* by MTCC IMTECH, Chandigarh (India). Analysis of 16S rDNA sequence revealed its 99.8% homology with *Bacillus vallismortis* strains, and was designated as *Bacillus vallismortis* RG-07. The 16S rDNA sequence was submitted to Gene bank [JQ: 619483] which doi is http://www.ncbi.nlm.nih.gov/nuccore/JQ619483 (Additional file 1). The strain RG-07 was in the same cluster of phylogenetic tree (Figure 1) with different strains of *Bacillus vallismortis.* However, the 16S rDNA sequence analysis indicates that it is a different and novel strain of *Bacillus vallismortis.*

Effect of agro-waste materials on cellulase production

The effect of agro-waste materials as carbon source on cell growth and cellulase production by *Bacillus vallismortis* was investigated with sugarcane baggase, rice bran, wheat bran, rice husk and maize bran. From Figure 2 it revealed that 2% sugarcane baggase (4105 U ml^{-1}) was found as most suitable substrate for cellulase production followed by rice husk (3509 U ml^{-1}) and rice bran (3110 U ml^{-1}), within 48 h. Minimum cellulase production was reported from the wheat bran (2890 U ml^{-1}) and maize bran (2545 U ml^{-1}). Similarly, Sadhu et al. [16] also reported that sugarcane baggase was the best carbon sources for cellulase production by *Bacillus sp.* Annamalai et al. [17] and Jo et al. [14] were also reported that rice husk and rice bran were the best carbon source for cellulase production by *Bacillus halodurans* CAS 1 and *Bacillus amy-loliquefaciens* DL-3, respectively. The cellulosic waste materials such as rice bran, sugarcane baggase and rice hulls were used as best carbon sources for cellulase production due to their inducible nature [18]. The cellulase produced by the hydrolysis of cellulosic biomass by *Bacillus vallismortis* could be valuable for bio-ethanol production, single cell protein and other industrially required chemicals.

Purification of extracellular cellulase

The crude enzyme extract was first concentrated by ammonium sulphate precipitation. Maximum activity was observed in the fraction obtained by the addition of ammonium sulphate in 80% with protein content of 21.54 mg ml^{-1}. This fraction had 14864.94 U mg^{-1} of specific activity with recovery of 83.8% and with regard to purification it showed 4.7-fold purification (Table 1).

The active fraction of ammonium sulphate precipitation method was used for further purification by using ion exchange chromatography. Sample (1 ml) was loaded into the Q-Sepharose column pre-equilibrated with sodium

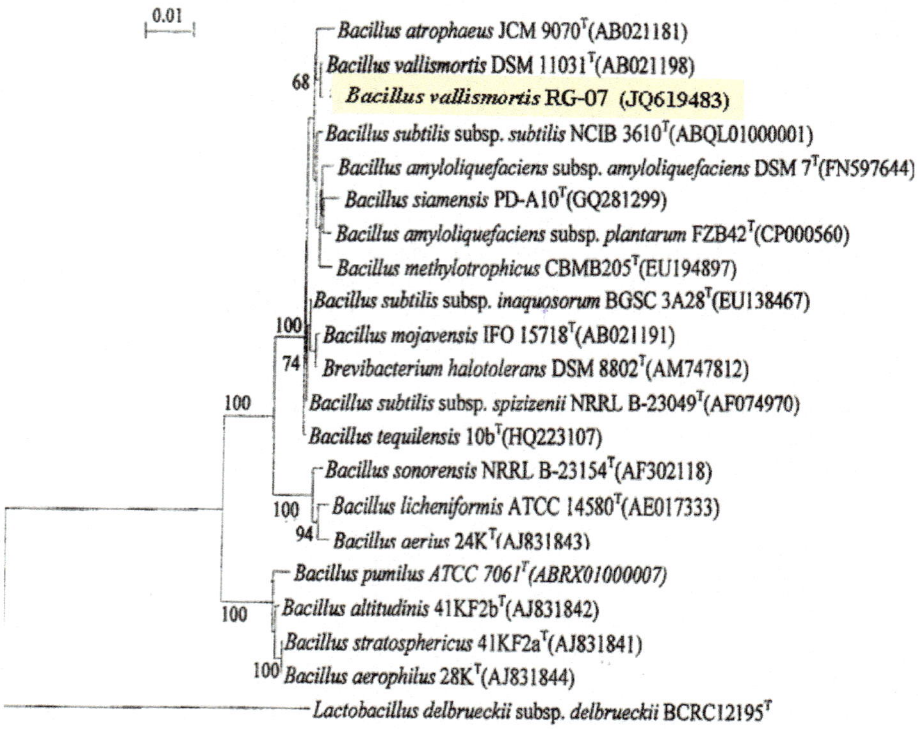

Figure 1 Phylogenetic tree showing relation between Strain RG-07 and other *Bacillus* strains. In phylogenetic tree analysis, the strain was in the same cluster with different strains of *Bacillus* and showed only 99.8% homology to other *Bacillus vallismortis* sp., so it could be stated that therefore it is different the from reported *Bacillus vallismortis*. The phylogenetic tree was drawn by MEGA 5 software using Neighbour-joining method and the significance of junctions was established using bootstrap method (1000 replicates).

phosphate buffer (100 mM, pH 7.0) and allowed to pass through the column. The un-bound fraction was collected and analyzed for cellulase activity and protein content. There was no cellulase activity in the un-bound fraction, while 1.8 mg ml^{-1} of protein was estimated. The absence of enzyme in un-bound fraction suggested that total cellulase was bound to matrix. The bound enzyme was eluted by sodium phosphate buffer (100 mM, pH 7.0) having NaCl with increasing concentration at gradient of 0.1 M. 10 ml solution of each concentration of NaCl was used to

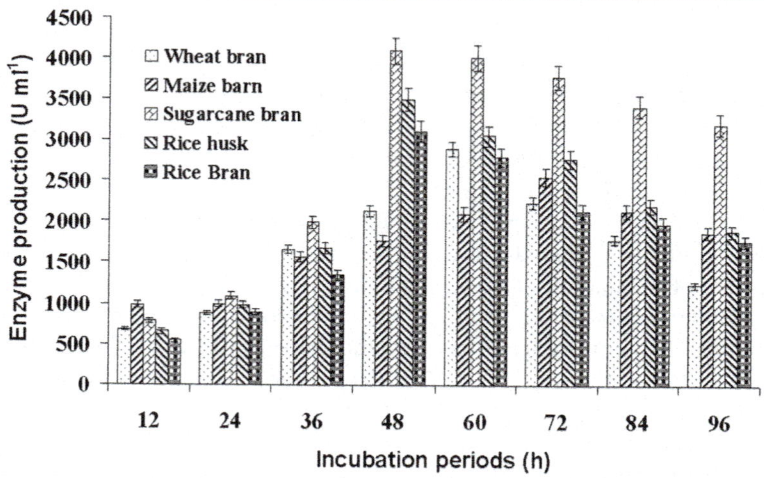

Figure 2 Effect of different agro-waste materials on cellulase production by *Bacillus vallismortis* RG-07. Five different agro-waste materials like sugarcane baggase, rice barn, rice husk, wheat bran and maize barn (2%) used as carbon sources for cellulose production, which inoculated with bacterial culture and incubated at 65°C, pH-7.0 for 12-96h.

Table 1 Purification of cellulase from *Bacillus vallismortis* RG-07

Purification steps	Total activity (U)	Total protein (mg)	Specific activity (U/mg)	Yield (%)	Purification fold
Crude	3,81,788	120.75	3161.81	100	1.0
Ammonium sulphate	3,20,190.9	21.54	14864.94	83.8	4.7
Q-Sepharose	2,88089.4	7.79	36981.95	75.4	11.6
Sephadex G-75	109,968.9	0.89	123560.56	28.8	39.1

evade the bound enzyme. The cellulase activity was detected in the fraction released by the addition of 0.5 M NaCl anion-exchange chromatography of cellulase on column resulted in one prominent peak at the 21th fraction (Figure 3a).

The active fraction was applied on Sephadex G-75 column. Figure 3b shows the fractionation pattern of cellulase on Sephadex G-75 column. One distinctive protein peak was appeared that overlapped with the cellulase activity. The purification process resulted in 39.1-fold purification factor and a final recovery of 28.8% of the enzyme with specific activity of 123560.56 U mg^{-1} (Table 1). Vijayaraghavan and Vincent, [12] reported 14.5-fold purified cellulase with 24% recovery after gel chromatography for purification of cellulase from *Bacillus* sp.

The purity of the enzyme was confirmed by the presence of a single band on SDS-PAGE and its molecular weight was approximately 80 kDa (Figure 4), which was similar to alkalophilic *Bacillus* sp. HSH-810 where it is 80 kDa [19] but different from *Bacillus sp.* cellulase (54 kDa) [10,20].

Characterization of purified enzyme

Effect of temperature on enzyme activity and stability

Effect of temperature on purified enzyme activity was recorded over a broad range of temperature (30-80°C) with the optimal activity at 65°C and declined thereafter (Figure 5). The cellulase of strain RG-07 was completely stable in the broad temperature range of 30-90°C during 1 h incubation. However, with further increase in every 5°C temperature, there was a gradual decrease in enzyme stability ranging between 15-20% upto 105°C. The enzyme retained 95, 90 and 73% activity even after treatment at 95, 100 and 105°C, respectively (Figure 5). Similarly, endoglucanase of *Bacillus licheniformis* C108 was highly stable up to 100°C [21]. The cellulase of strain RG-07 is more thermostable than cellulase studied by several other researchers. Most of workers have reported that thermostable cellulase stable up to 60-100°C but retained only ~50% activity at 100°C [22,21]. Most other thermotolerant *Bacillus* cellulase reported to so far, cellulases exhibited higher temperature optimum for activity and showed good thermal stability [23,24]. These are the properties considered to be very important for industrial cellulose saccharification. Hence it is evident that the cellulase of strain RG-07 is more thermostable, and may be applied to several biotechnological and industrial purposes.

Effect of pH on enzyme activity and stability

The pH activity and stability of the purified cellulase was determined by measuring the enzyme activity at varying pH values ranging from 4–10 using different suitable buffers. It observed that maximum cellulase activity was established at pH 7.0, however it was found to be most stable at pH 7.5 (Figure 6). Similar optimum enzyme activity was also found at pH 7.0 in *Pseudomonas fluorescence* [25], *Bacillus amyoliquefaciens* DL3 [10]. The relative activities at pH 4.0, 4.5, 5.0 5.5, 6.0, 6.5, 7.0, 7.5 and 8.0 were determined to be 53, 62, 75, 85, 95, 104, 135, 110 and 98%, respectively. At pH above 7.5, the cellulase activity decreased rapidly. The cellulase from *Bacillus vallismortis* RG-07 was stable in a range of pH 4.0-9.0 and at pH 10.0 approximately 80% of its activity was retained (Figure 6). Cellulases are generally stable over a wide range of pH from 5 to 10 [19,24].

Effect of metal ions on enzyme activity and stability

Bacillus vallismortis RG-07 cellulase was activated by 10 mM Ca^{2+}, Mg^{+2}, and Na^{+1} but inhibited by all other metal ions to a variable extent. Results suggest that cellulase showed maximum relative activity (184.4, 175.1 and 141.4%) and stability (160.3, 130.3, and 125.3%) in the presence of calcium, magnesium and sodium ions. Yoon et al. [26] and Bakare et al. [25] had also reported that Ca^{2+}, Mg^{+2}, and Na^{+1} ions strongly stimulated cellulase activity. Similarly Wang et al. [27] also reported that *Paenibacillus* sp. strain B39 showed maximum enzyme activity in the presence of 1 mM Ca^{2+}. Cellulase activity was slightly inhibited was by Ni^{2+}, Cu^{2+} and Zn^{2+} (Table 2). Cellulase was strongly inhibited in the presence of Mn^{2+}, and Hg^{2+}. Similar results were reported for *Bacillus* strain [23,16] and *Bacillus amyoliquefaciens* DL-3 [10]. It has been reported that the inhibition of cellulase activity by Hg^{+} ion might be related to its binding with thiol groups, tryptophan residue, or the carboxyl group of amino acid residues in the enzyme [28]. The inhibition of cellulase by Co^{2+}, and Cu^{2+} ions could be due to competition between the exogenous cations and the protein-associated cations, resulting in decreased metallo-enzyme activity.

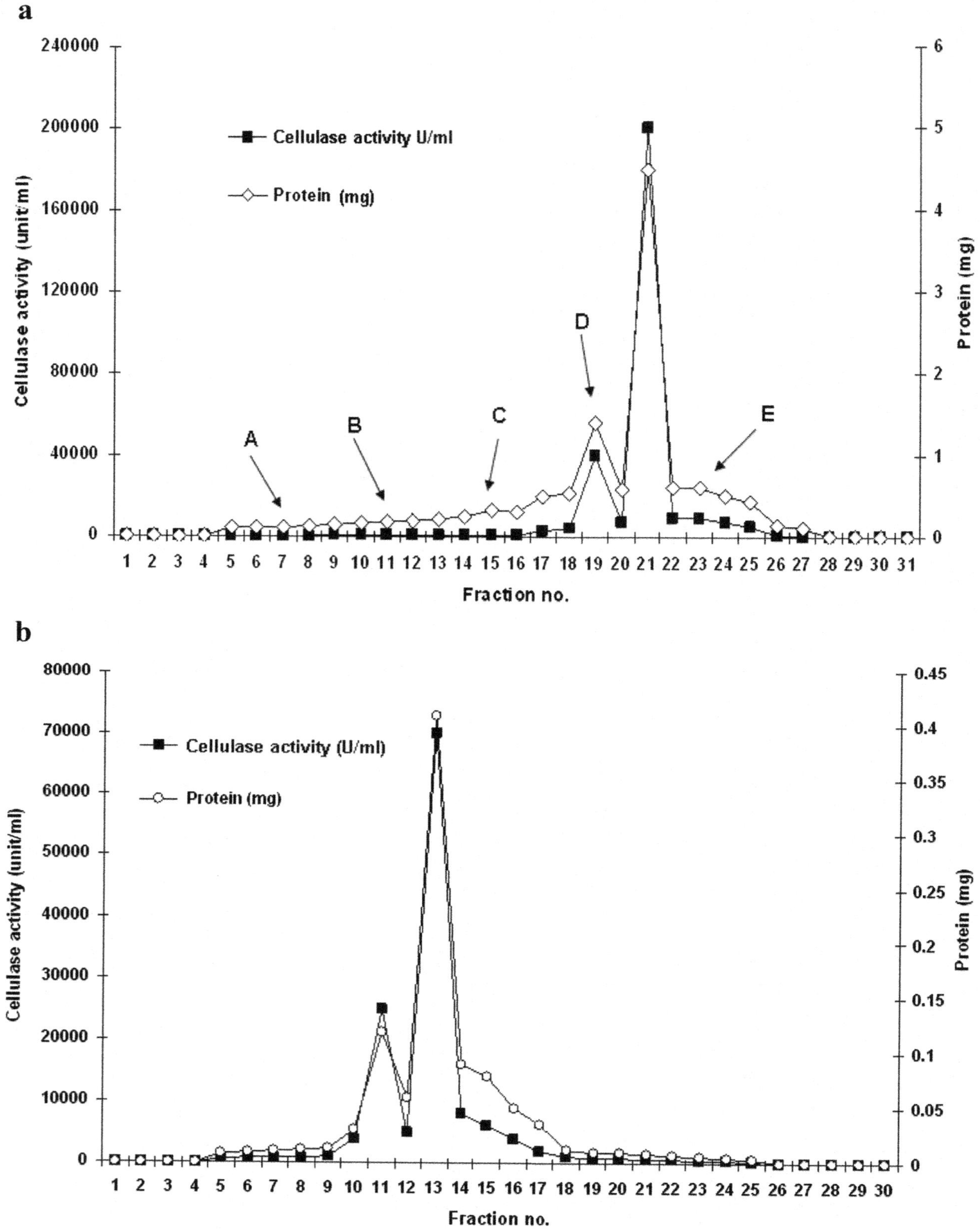

Figure 3 (See legend on next page.)

(See figure on previous page.)
Figure 3 Chromatographic purification profile of extracellular cellulase. a Chromatographic purification profile of extracellular cellulase from *Bacillus vallismortis* RG-07 on anion-exchange column. Column was equilibrated with sodium phosphate buffer (100 mM, pH 7.0). The cellulase was eluted with a gradient of sodium chloride (0.1 M-0.5 M) in sodium phosphate buffer (100 mM, pH 7.0). A-0.1 M, B-0.2 M, C-0.3 M, D-0.4 M, E-0.5 M. **b** Chromatographic purification profile of extracellular cellulase on Sephadex G-75 gel filtration chromatography. Column was equilibrated with sodium phosphate buffer (100 mM, pH 7.0). The sample was loaded and eluted with the same buffer.

Effect of organic solvents on cellulase stability

In another approach, the effect of various organic solvents (30%, v/v) on cellulase stability was also investigated for one week, and the results are depicted in Table 3. The cellulase of *Bacillus vallismortis* RG-07 is extraordinarily stable in the presence of all organic solvents under study. It was observed that except benzene, propanol and ethanol, presence of other solvents enhanced the cellulase activity. After incubation with n-dodecane, n-decane, xylene, iso-octane, toluene, n-haxane, n-butanol, acetone, methanol, and cyclohexane the cellulase activity increased to 242.5 (48 h), 178.7 (48 h), 117 (48 h), 170.3 (24 h), 180.2 (24 h), 143 (24 h), 160.3(24 h), 115.9 (24 h), 131.9 (48 h), and 133% (48 h), respectively. The presence of benzene, ethanol and propanol marginally reduced the cellulase with residual activities of 79.9, 88.2 and 84.2%, respectively (Table 3). Annamalai et al. [17] also reported an organic solvent stable alkaline cellulase of *Bacillus halodurans* CAS 1 strain rand with enhanced activity in the presence of organic solvents (25%, v/v). Zaks and Klibanov, [29] suggested that stimulation of enzyme activity by organic solvents might be due to the residues of carried-over non-polar hydrophobic solvent providing an interface, thereby keeping the enzyme in an open conformation which resulting stimulated activation. It is therefore; evident from our study that cellulase of *Bacillus vallismortis* RG-07 is remarkably stable in the presence of broad range hydrophilic as well as hydrophobic organic solvents employed in this study.

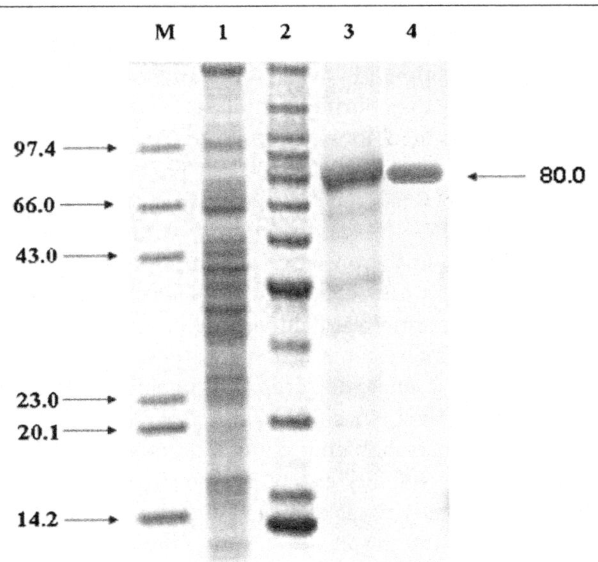

Figure 4 SDS-PAGE of different fractions of cellulase of *Bacillus vallismortis* RG-07 obtained during purification steps. Lane 1: Marker proteins; lane 2: Crude supernatant; lane 3: Concentrated enzyme after ammonium sulphate precipitation; lane 4: Purified enzyme obtained after anion-exchange chromatography; lane 5: Purified enzyme obtained after gel-filtration chromatography. Molecular weights were presented in the form of kDa.

Effect of inhibitors on enzyme stability

When the *Bacillus vallismortis* RG-07 cellulase enzyme was incubated with EDTA, Iodo-acetic acid (IAA), *p*-chloromercuribenzoate (*p*-CMB), Dithiothreitol (DTT), Urea and β-mercaptoethanol, the enzyme activity was retained at 110%, 70%, 50%, 134%, 114%, and 167% of the original activity at 10 mM (Table 4). Similarly, Yin et al. [11] reported that *Bacillus subtilis* YJ1 retained full activity in the presence of 10 mM EDTA and β-mercaptoethanol, while Wang et al. [30] reported that cellulase of *Paenibacillus* sp. showed 22 and 29% activity with 10 mM. IAA and PCMB inhibits cellulase activity because they can bind with the SH group with different degree interaction and subsequently inhibit the activity. However, the β-mercaptoethanol and DTT can reduce the disulfide bonds and re-nature their activity, if the oxidation or aggregation of these enzyme proteins occurs during purification and storage. These phenomenons suggested that the active site of the enzyme contains -SH group [31].

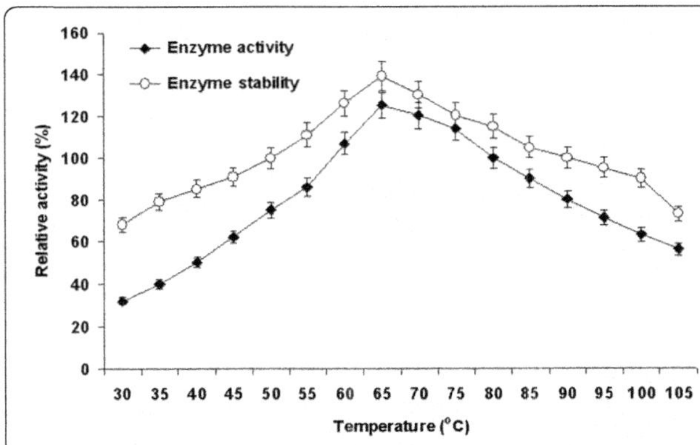

Figure 5 Effect of temperature on enzyme activity and stability. For enzyme activity reaction mixture was incubated at different temperatures (30-105°C) and for stability enzyme was pre-incubated at respective temperatures for 1h and reaction was conducted as standard assay method.

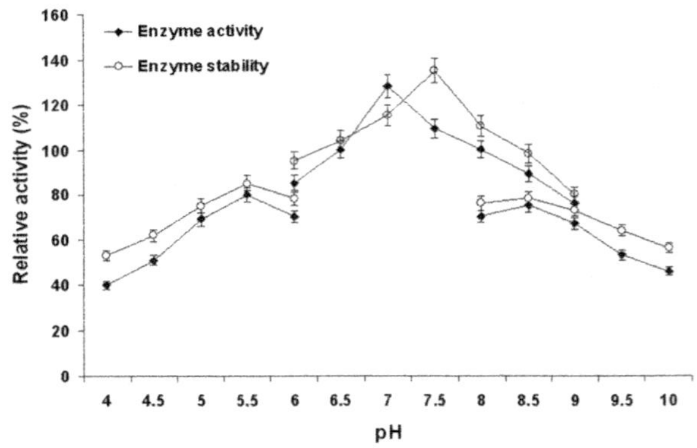

Figure 6 Effect of pH on enzyme activity and stability. For enzyme activity the reaction was assayed at respective pH and for stability enzyme was pre-incubated with buffers (100 mM, in ratio 1:1) of different pH (4–10) at 65°C for 1 h and assayed by standard assay method.

Table 2 Effect of metal ions on enzyme activity and stability

Metal ions	Concentration (mM)	Residual activity (%)	
		Activity	Stability
Control		100.0	100.0
$CaCl_2$	5	156.4	125.4
	10	184.4	160.3
$NiCl_2$	5	121.4	93.1
	10	90.4	87.7
$FeSO_4$	5	112.9	100.8
	10	102	86.5
$MgCl_2$	5	145.9	115.6
	10	175.1	130.3
$CuSO_4$	5	87.4	82.5
	10	56.7	49.5
$HgCl_2$	5	46	52
	10	25	32
$MnCl_2$	5	76.9	83.9
	10	61.9	50.2
KCl	5	100.2	106.4
	10	92.1	90.1
NaCl	5	127.8	101.4
	10	141.4	125.3
$ZnSO_4$	5	95	76
	10	63	59

Enzyme activity was determined at 65°C in the presence of metal ions in the reaction mixture directly and for stability enzyme was pre-incubated with different metal ions at 65°C for 1 h and assayed as standard assay method. The enzyme activity without incubation with metal ions was taken as 100%. Mean standard deviation for all the values is $< \pm$ 5.0%.

Effect of surfactant, detergent and oxidizing agents on enzyme stability

In order to have applications in detergent industries, cellulase must be stable to various detergent ingredients, such as surfactants. As shown in Table 5 the enzyme was appreciably stable in the presence of non-ionic surfactants like tween-40, tween-60, tween-80 and tritone-100 and detergent SDS. However, these compounds slightly inhibited the cellulase activity with 89.6, 93, 87, 92.5, 95.7% of residual activity at concentration 1.0% (v/v). Though, cellulase from a halotolerant isolate, *Bacillus* sp. L1 retained 91% activity towards triton X-100, tween-20, and tween-80 and lost about 41.1% activity in the presence of SDS [32]. Yin et al. [11] reported that, a highly thermo-stable and alkaline cellulase enzyme, there was 95% stability after 1 h incubation with SDS (10 mM). In another study, Sadhu et al. [16] and Wang et al. [30] have reported that SDS and tween-80 moderately inhibited cellulase activity 59-71% and 50-59%, respectively. Thus from the above study it clear that cellulase of *Bacillus vallismortis* RG-07 highly stable with 1% SDS (95%). This resistance, which is essential requirements, suggests that the enzyme may be used as an effective additive in detergents.

The cellulase was substantially stable with commercial detergents at lower concentration (0.1%, w/v). However, higher concentration (1.0%, w/v) led to decrease the enzyme activity. The enzyme showed maximum stability in the presence of ariel, having residual activity of 89% after incubation at 65°C for 1 h (Table 5). The enzyme had 80%, 75.5, 68.4 and 55.9% residual activities in the presence of surf excel, tide, fena and henko, respectively under similar condition (Table 5). Similarly Annamalai et al. [17] reported that purified cellulase from *B. halodurans* CAS 1 retained its activity in the presence of some commercial detergents such as rin (85.33%), ariel (76.67%), henko (64.67%) and tide (80.33%). Sadhu et al. [16] also reported 57-72% residual activity with cellulase from *B*acillus strain in the presence of commercial detergents. Similar results are reported from *Thermonospora* sp. with different commercial detergent [33].

Among the oxidizing agents tested the cellulase activity enhanced in presence of sodium hypochlorite and H_2O_2 with residual activities 127.2 and 118.7% at concentration 0.1%, whereas, higher concentrations (0.5 and 1.0%) decreased the stability except sodium hypochlorite (115.4% residual activity at concentration 0.5%, w/v) (Table 5). Likewise, Wang et al. [27] also reported that lipase from *B. cepacia* was highly stable in the presence of hydrogen peroxide, sodium hypochlorite and sodium perborate after 1 h. The stability profile of the cellulase in the presence of detergents and oxidizing agents prove its potential application in the detergent formulations as these agents are the active components of house hold detergents.

Table 3 Stability of cellulase in presence of various organic solvents

Organic solvents (30%)	log *P*	Residual activity (%)							
		1 h	24 h	48 h	72 h	96 h	120 h	144 h	168 h
Methanol	−0.76	100.5	124.8	131.9	125.3	119.9	112.4	106.5	97.1
Iso-propanol	−0.28	90.7	104.3	99.5	92.5	91.8	91.0	88.3	84.2
Ethanol	−0.24	98	106.9	105.1	102.9	98.8	95.1	91.6	88.2
Benzene	2.13	99.7	112	120	109	96	89	85	79.9
Cyclohexane	3.3	90	118	133	123.9	112.9	109	93	89.9
Acetone	−0.23	105.2	115.9	110.9	104.1	101	96	92	90.3
Butanol	−0.80	120.7	160.3	145.2	130.4	122.6	112.3	107	100
Toluene	2.5	120.9	180.2	150.4	128.5	120.9	112.8	105.2	102.4
Iso-octane	2.9	120.5	170.3	135.6	130.2	123.2	120.4	115.6	100.2
Xylene	3.1	90	112	117	113	110	102	96	90
Hexane	3.6	128	143	124	128	113	114	104	100
n-decane	5.6	135.6	161.8	178.7	158.6	136.5	111.9	107	97.0
n-dodecane	6.0	141.6	170.7	242.5	168.4	179.8	130.5	113.0	100.0

Enzyme was pre-incubated with different organic solvents at a concentration of 30% (v/v) at 65°C for different time period and assayed as standard assay method. The enzyme activity without incubation with organic solvent was taken as 100%. Mean standard deviation for all the values is < ± 5.0%.

Kinetic analysis

Kinetic analysis with CMC revealed the K_m and V_{max} to be 1.923 mg ml^{-1} and 769.230 μg ml^{-1} min^{-1}, respectively by Linweaver-Burk plot (Figure 7). Mawadza et al. [23] also reported similar value of K_m (1.5 and 1.7 mg ml^{-1}) and V_{max} (1.5 and 0.9 mmol min^{-1} mg^{-1}) from *Bacillus* sp. using CMC as substrate. If the K_m value is low, it means that enzyme had a stronger affinity with substrate.

Conclusion

It is the first instance when a thermo-tolerant cellulase being reported from a thermo-tolerant solvent tolerant *Bacillus vallismortis* RG-07 isolate. The strain is unique with respect to increased cellulase activity in the presence of various solvents of greater hydrophobicity (log P ≥ 4.5). The cellulase activity in broad pH and temperature range of 4.0-10.0 and 30-105°C clearly indicate the thermo-alkaline nature of this enzyme. Considering its stability under high temperature as well as mild alkaline condition, cellulase may be exploited for industrial usage.

Table 4 Effect of different inhibitors on enzyme stability

Reagents	Concentration	Residual activity (%)
Control		100.0
β-mercaptoethanol	0.1 (%)	179.6
	1.0 (%)	167.0
EDTA	5 mM	120.9
	10 mM	110.5
Urea	5 mM	119.0
	10 mM	114.0
IAA	5 mM	105.4
	10 mM	70.0
p-CMB	5 mM	90.0
	10 mM	50.0
DTT	5 mM	115.0
	10 mM	134.0

Enzyme was pre-incubated with different inhibitors at 65°C for 1 h and assayed as standard assay method. The enzyme activity without incubation with inhibitors was taken as 100%. Mean standard deviation for all the values is < ± 5.0%.

Methods

Isolation, screening and identification of thermo-tolerant cellulase producing bacteria

The soil samples were collected aseptically from different garden site of University campus (Dr. Ram Manohar Lohia Avadh University Faizabad) to isolate cellulase producing bacteria. 1 g soil was suspended in 9.0 ml sterile distilled water, agitated for a min and 0.1 ml suspension was spread over CMC agar plates (pH 7.0) containing, 2.0%, CMC; 0.5%, Ammoium sulphate; 2%, agar and incubated at 65°C, till sufficient growth appeared. After sufficient growth incubated plates were overlaid with congo-red solution (0.1%) for 10 min and then washed with 1 N NaOH solution for de-staining. If a strain was cellulolytic then it started hydrolyzing the cellulose present in the surrounding and in the zone degradation there was no red color formation. Selection was done as per colonies with and without clear and transparent zone as cellulase producing and cellulase non-producing strain, respectively. Bacterial colonies showing clear zones were selected, streaked twice on CMC agar plates for purification and maintained as pure culture over CMC agar slants (pH 7.0, 4°C). The isolate

Table 5 Effect of surfactants, commercial detergents and oxidizing agents on cellulase stability

Surfactants	Concentration (%)	Residual activity (%)
Control		100.0
Tween-40	0.1	119.7
	1.0	89.6
Tween-60	0.1	124.9
	1.0	93.0
Tween-80	0.1	116.4
	1.0	87.0
Triton-X-100	0.1	105
	1.0	92.5
SDS	0.1	130
	1.0	95.7
Tide	0.1	90.7
	1.0	75.5
Ariel	0.1	99.7
	1.0	89.0
Surf Excel	0.1	94.3
	1.0	80
Fena	0.1	89.4
	1.0	68.4
Henko	0.1	88.2
	1.0	55.9
Sodium perborate	0.1 (%)	87.9
	0.5 (%)	68.2
	1.0 (%)	57.7
Sodium hypochlorite	0.1 (%)	127.2
	0.5 (%)	115.4
	1.0 (%)	94.6
H_2O_2	0.1 (%)	118.7
	1.0 (%)	96.9

Enzyme was pre-incubated with different surfactants, commercial detergents and oxidizing agents at 65°C for 1 h and assayed as standard assay method. The enzyme activity without incubation with surfactants, commercial detergents and oxidizing agents was taken as 100%. Mean standard deviation for all the values is < ± 5.0%.

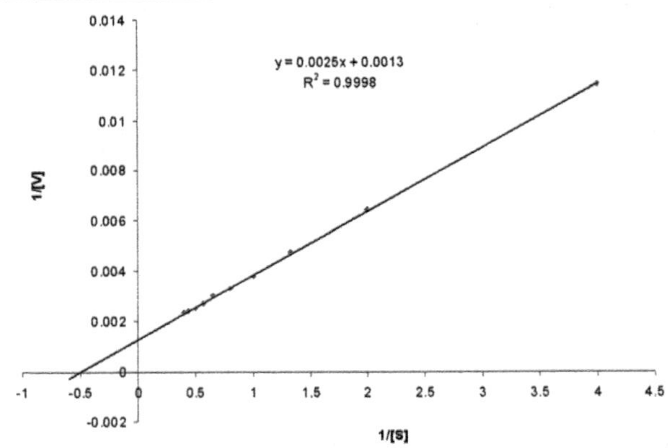

Figure 7 Lineweaver-Burke plot of initial velocity data of *Bacillus vallismortis* RG-07 cellulase on CMC. CMC concentration is 0.25, 0.5, 0.75, 1.0, 1.25, 1.5, 1.75, 2.0, 2.25, 2.5, 2.75, 3.0, 3.25, 3.5, 3.75, 4.0 mg ml^{-1}, respectively.

having maximum clearance zone was selected for further studies. The selected bacterial isolate RG-07 was identified by morphological and biochemical characterization as per the Bergey's Manual of Systematic Bacteriology [34]. The identity of RG-07 was authenticated from Institute of Microbial Technology (IMTECH), Chandigarh, India based on the phenotypic (16S r DNA) and biochemical tests.

Crude enzyme preparation and enzyme assay

To obtain crude enzyme 24 h old cultures were transferred to micro-centrifuge tubes and centrifuged at 12000 rpm for 10 min. Cells were discarded and resultant supernatant was used as the crude enzyme for various enzyme assay. Cellulase was assayed by measuring the reducing sugar released by reaction on CMC. Cellulase assay was done by Nelsom [35] and Somogy [36] methods, in this method a reaction was performed in which a reaction mixture contain 500 μl of substrate (1.0% soluble CMC in 1.0 M phosphate buffer pH 7.0.), 100 μl of the enzyme and 400 μl distilled water. The reaction mixture was incubated for 10 min at 65°C. Reaction was stopped by adding 1 ml of alkaline copper tartrate solution and incubated in boiling water bath for 10 min and cooled, then added arsenomolybdate solution for color stabilization. Optical density of each sample with reaction mixture was taken at 620 nm in a spectrophotometer (Shimadzu, Japan). One unit of enzyme activity was defined as the amount of enzyme that liberates 1.0 μg of glucose min/ml.

Effect of agro-waste materials on cellulase production

To study the effect of agro-waste materials for cellulase production, growth was carried out in 150 ml Erlenmeyer flasks in 50 ml of basal medium contained; 0.05% ammonium sulphate, 0.5% $MgSO_4 \cdot 7H_2O$ (pH 7.0) supplemented with 2% (w/v) agro-waste materials to be investigated such as sugarcane baggase, rice bran, rice husk, wheat bran, and maize bran at 65°C for 12–96 h. The enzyme activity was estimated by Nelsom [35] and Somogy [36] methods as discuss in enzyme essay section.

Cellulase purification

The crude culture supernatant obtained from 24 h old cultures of *Bacillus vallismortis* RG-07 grown under optimal conditions was subjected to purification. All the purification steps were carried out at 4°C. The crude culture filtrate was subjected to a cooling centrifugation at 10,000 rpm to remove the cells and the residual medium. Supernatant was precipitated overnight with $(NH_4)_2SO_4$ (80% saturation) and the pellets was recovered by centrifugation at 12000 rpm for 10 min. The pellet was re-suspended in a small amount 100 mM phosphate buffer, pH 7.0, and dialyzed overnight against the same buffer. The corresponding precipitates were recovered, dissolved individually in fresh buffer and assayed for both total protein content and cellulase activity. The dialyzed sample was applied to a Q-Sepharose (Sigma-Aldrich, USA) column equilibrated with sodium phosphate buffer (100 mM, pH 7.0). The desired enzyme fraction was allowed to bind with matrix for 2 h at 4°C. The unbound fraction was collected and analyzed for enzyme activity and for protein content. The bound fractions were eluted with a linear gradient of NaCl (0.1-0.5 M, 10 ml each) in the same buffer at a flow rate of 1.0 ml min^{-1}.

The active fractions were collected and dialyzed against sodium phosphate buffer (pH 7.0). The dialyzed sample was further purified by gel-filtration chromatography for purification up to homogeneity. The Sephadex-75 column (Sigma Aldrich Pvt. Ltd., USA, 1.5 × 40 cm) was equilibrated with sodium phosphate buffer (100 mM, pH 7.0) and 1 ml of concentrated sample was applied to the column. The flow rate was adjusted to 5–6 ml h^{-1} and fraction of 2 ml each was collected. Cellulase activity and estimation of protein content were determined for each individual fraction.

Determination of protein concentrations

The protein content of individual fraction obtained after different steps of chromatography was monitored by measuring the extinction at 280 nm. Quantified protein for each fraction was done by the method of Lowry et al. [37] using Bovine serum albumin (BSA) as standard and expressed as mg ml^{-1}.

Polyacrylamide gel electrophoresis

The active fraction, with maximum specific activity, obtained after gel filtration chromatography along with crude, ammonium sulphate and anion-exchange chromatography was electrophorezed by Sodium Dodecyl Sulphate-Poly Acrylamide Gel Electrophoresis in a 12.5% polyacrylamide gel according to the method of Laemmli [38]. Approximate molecular weight of the cellulase was estimated by SDS-PAGE against the molecular mass markers i.e. lysozyme (14.3 kDa), β-lactoglobulin (20 kDa), Carbonic anhydrase (29 kDa), ovalbumin (43 kDa), bovine serum albumin (66 kDa) and phosphorylase B (97.4 kDa) (Sigma-Aldrich Pvt Ltd., USA) run with the samples.

Characterization of purified enzyme

Effect of temperature on enzyme activity and stability

The influence of temperature on activity of cellulase was studied by incubating the reaction mixture at different temperatures (30, 35, 40, 45, 50, 55, 60, 65, 70, 75, 80, 85, 90, 95, 100, and 105°C). The enzyme was incubated at different temperatures (30, 35, 40, 45, 50, 55, 60, 65, 70, 75, 80, 85, 90, 95, 100, and 105°C) for 1 h to study the stability of the enzyme. The residual cellulase activity was measured by conducting the reaction at temperature 65°C and pH 7.0 [35,36]. The activity of the enzyme was considered as 100% under standard assay conditions.

Effect of pH on cellulase activity and stability

The effect of pH on cellulase activity was measured in the pH range of 4 to 10, using the appropriate buffers at concentration of 100 mM (4.0-6.0, sodium acetate; 6.0-8.0, sodium phosphate; 8.0-10.0, Tris–HCl) under standard assay conditions [35,36]. To study stability as a function of pH, 100 μl of the purified enzyme was mixed with 100 μl of the buffer solutions and incubated at 65°C for 1 h then aliquots of the mixture were taken to measure the residual cellulase activity (%) under standard assay conditions [35,36].

Effect of metal ions on enzyme activity and stability

The effect of various metal ions (5 mM and 10 mM) on enzyme activity was investigated using $FeSO_4$, $CaCl_2$, KCl, NaCl, $MgCl_2$, $MnCl_2$, $ZnSO_4$, $CuSO_4$, $HgCl_2$ and $NiCl_2$. The enzyme was incubated with different metals at 65°C for 1 h to study metal ion stability and assayed under standard assay conditions [35,36].

Effect of organic solvent on cellulase stability

Purified enzyme having maximum cellulase activity was incubated with 30% (v/v) of different organic solvent viz., n-dodecane, n-decane, iso-octane, xylene, n-hexane, n-butanol, cyclohexane, acetone, toluene, benzene, ethanol, methanol and propanol for one week in screw crapped tubes at 65°C and 120 rpm. The residual cellulase activity was estimated against the control, in which solvent was not present [35,36].

Effect of inhibitors on cellulase activity

The effects ethylene diamine tetra acetic acid (EDTA), *β*-mercaptoethanol, Phenyl methyl sulphonyl flouride (PMSF) and urea as inhibitors on cellulase activity were investigated at a concentration of 5 mM and 10 mM in order to characterize enzyme. Crude enzyme was pre-incubated with the above mentioned reagents for 1 h at 65°C and residual activity (%) was determined under standard assay conditions [35,36].

Effect of surfactants, commercial detergents and oxidizing agents on enzyme stability

The cellulase sample was incubated with surfactants *viz.*, Triton-X-100, Tween-40, Tween-60, Tween-80, SDS (0.1 and 1.0%, v/v), commercial detergents *viz.*, surf, aerial, ghari, henko and fena (0.1 and 1.0%, w/v), and oxidizing agents *viz.*, H_2O_2 (0.1 and 1.0%, v/v), sodium perborate and sodium hypochlorite (0.1, 0.5 and 1.0%, v/v) for 1 h at 65°C and then the residual activity (%) was tested under standard assay conditions [35,36].

Kinetic analysis

The influence of substrate concentration on the reaction velocity of the purified cellulase was studied with CMC [39]. The purified cellulase was incubated with various concentration of CMC ranged from 0.25-4.0 mg ml^{-1}. In all cases, the enzymatic activity was assayed under standard conditions. The Michaelis constant (K_m) and maximum velocity (V_{max}) was determined from Lineweaver-Burk plots of Michaelis-Menten equation [40].

Statistical analysis

Each experiment was performed twice, each in triplicate and standard deviation for each experimental results were calculated using the Microsoft Excel.

Availability of Supporting Data' (ASD)

Name of the repository is NCBI (National Center for Biotechnology Information) where our data is deposited and a link to the dataset DOI is http://www.ncbi.nlm.nih.gov/nuccore/JQ619483. The supporting data also include in this manuscript as a supporting file [41].

Additional file

Additional file 1: Bacillus vallismortis strain RG-07 16S ribosomal RNA gene, partial sequence.

Competing interests

The author(s) declare that they have no competing interests.

Authors' contributions

ST carried out the research work and drafted the manuscript. RG has designed the experiment, contributed substantially to analysis and interpretation of data and have given final approval of the version to be published. Both authors read and approved the final manuscript.

Acknowledgements

Financial assistance by Council of science and technology, U.P., is greatly acknowledged by Rajeeva Gaur and Soni Tiwari.

References

1. Bhat MK, Bhat S. Cellulose degrading enzymes and their potential industrial applications. Biotechnol Adva. 1997;15:583–620.
2. Costa RB, Silva MVA, Freitas FC, Leitao VSF, Lacerda PSB, Ferrara MA, et al. Mercado e Perspectivas de Uso de Enzimas Industriais e Especiais no Brasil. In: Bon EPS, Ferrara MA, Corvo ML, Vermelho AB, Paiva CLA, Alencastro RB, Coelho RRR, editors. Enzimas em Biotecnologia, Producao, Aplicac oes e Mercados. 1st ed. Rio de Janeiro: Intercie^ncia; 2008. p. 463–88.
3. Zaldivar J, Nielsen J, Olsson L. Fuel ethanol production from lignocellulose: a challenge for metabolic engineering and process integration. Appl Microbiol Biotechnol. 2001;56:17–34.
4. Sun Y, Cheng J. Hydrolysis of lignocellulosic materials for ethanol production: a review. Bioresour Technol. 2002;83:1–11.
5. Lynd LR, Weimer PJ, Zyl WH, Isak S. Microbial cellulose utilization: fundamentals and biotechnology microbiology. Mole Biol Reviews. 2002;66:506.
6. Harrison MJ, Nouwens AS, Jardine DR, Zachara NE, Gooley AA, Nevalainen H, et al. Modified glycosylation of cellobiohydrolase I from a high cellulase-producing mutant strain of *Trichoderma reesei.* European J Biochem. 1998;256:119–27.
7. Nagendran S, Hallen-Adams HE, Paper JM, Aslam N, Walton JD. Reduced genomic potential for secreted plant cell wall degrading enzymes in the ectomycorrhizal fungus *Amanita bisporigera*, based on the secretome of *Trichoderma reesei*. Fungal Gene Biol. 2009;46:427–35.
8. Singh VK, Kumar A. Production and purification of an extracellular cellulase from *Bacillus brevis* VS-1. Biochem Molecul Biol Inter. 1998;45:443–52.
9. Gachomo WE. *Bacillus pumilus* BPCRI 6, a promising candidate for cellulase production under conditions catabolite repression. Afri J Biotechnol. 2003;2:140–6.
10. Lee YJ, Kim BK, Lee BH, Jo KI, Lee NK, Chung CH, et al. Purification and characterization of cellulase produced by *Bacillus amyoliquefaciens* DL-3 utilizing rice hull. Bioresour Technol. 2008;99:378–86.
11. Yin LJ, Lin HH, Xiao ZR. Purification and characterization of a cellulase from *Bacillus subtilis* YJ1. J Marine Sci Technol. 2010;18:466–71.
12. Vijayaraghavan P, Vincet SGP. Purification and characterization of carboxymethyl cellulase from *Bacillus* sp. Isolated from a paddy field. Polish J Microbiol. 2012;61:51–5.
13. Rawat R, Tewari L. Purification and characterization of an acidothermophilic cellulase enzyme produced by *Bacillus subtilis* strain LFS3. Extremophiles. 2012;16:637–44.
14. Jo KI, Lee YJ, Kim BK, Lee BH, Jung CH, Nam SW. Pilot-scale production of carboxymethylcellulase from rice hull by *Bacillus amyloliquefaciens* L-3. Biotechnol Biopro Engineer. 2008;13:182–8.
15. Mayende L, Wilhelmi BS, Pletschke BI. Cellulases (CMCases) and polyphenol oxidases from thermophilic *Bacillus* sp. isolated from compost. Soil Biol Biochem. 2006;38:2963–6.
16. Sadhu S, Saha P, Sen SK, Mayilraj S, Maiti TK. Production, purification and characterization of a novel thermotolerant endoglucanase (CMCase) from *Bacillus* strain isolated from cow dung. Springer Plus, 2013, 2:10. http://www.springerplus.com/content/2/1/10 (25)
17. Annamalai N, Rajeswari MV, Elayaraja S, Balasubramanian T. Thermostable, haloalkaline cellulase from *Bacillus halodurans* CAS 1 by conversion of lignocellulosic wastes. Carbohy Poly. 2013;94:409–15 [27].
18. Lee BH, Kim BK, Lee YJ, Chung CH, Lee JW. Industrial scale of optimization for the production of carboxymethylcellulase from rice bran by a marine bacterium, *Bacillus subtilis subsp. subtilis* A-53. Enzy Microb Technol. 2010;46:38–42.
19. Kim JY, Hur SH, Hong JH. Purification and characterization of an alkaline cellulase from a newly isolated alkalophilic *Bacillus* sp. HSH-810. Biotechnol Lett. 2005;27:313–6.
20. Bajaj BK, Pangotra H, Masood A, Wani Sharma P, Sharma A. Partial purification and characterization of a highly thermostable and pH stable endogluconase from a newly isolated *Bacillus* strain M-9. Indian J Chem Technol. 2009;16:382–7.
21. Aygan A, Karcioglu L, Arikan B. Alkaline thermostable and halophilic endoglucanase from *Bacillus licheniformis* C108. Afri J Biotechnol. 2011;10:789–96.
22. Liang Y, Feng Z, Yesuf J, Blackburn JW. Optimization of growth medium and enzyme assay conditions for crude cellulases produced by a novel thermophilic and cellulolytic bacterium, *Anoxybacillus* sp. Appl Biochem Biotechnol 2009, doi:10.1007/s12010-009-8677-x.
23. Mawadza C, Hatti-Kaul R, Zvauya R, Mattiasson B. Purification and characterization of cellulases produced by two *Bacillus* strains. J Biotechnol. 2000;83:177–87.
24. Tahir SR, Bakhsh A, Rao AQ, Naz M, Saleem M. Isolation, purification and characterization of extracellular β-glucosidase from *Bacillus* sp. Adva Environ Biol Report. 2009;3:269.

25. Bakare MK, Adewale IO, Ajai A, Shonukan OO. Purification and characterization of cellulase from the wild-type and two improved mutants of *Pseudomonas fluorescens*. Afri J Biotechnol. 2005;4:898–904.
26. Yoon S, Kim MK, Hong JS, Kim MS. Production of polygalacturonase from *Ganoderma lucidum*. Korean J Mycol. 1994;22:286–97.
27. Wang X, Yu X, Xu Y. Homologous expression, purification and characterization of a novel high-alkaline and thermal stable lipase from *Burkholderia cepacia* ATCC 25416. Enzy Microb Technol. 2009;45:94–102.
28. Lusterio DD, Suizo FG, Labunos NM, Valledor MN, Ueda S, Kawai S, et al. Alkali-resistant, alkaline endo-1,4-β-glucanase produced by *Bacillus* sp. PKM-5430. Biosci Biotechnol Biochem. 1992;56:1671–2.
29. Zaks A, Klibanov AM. Enzymatic catalysis in nonaqueous solvents. J Biol Chem. 1988;263:3194–201.
30. Wang C-M, Shyu C-L, Ho S-P, Chiou S-H. Characterization of a novel thermophilic, cellulosedegrading bacterium *Paenibacillus* sp. strain B39. Lett Appl Microbiol. 2008;47:46–53.
31. Singh A, Agrawal AK, Abidi AB, Darmwal NS. Properties of cellobiase from *Aspergillus niger*. Appl Microbiol Biotechnol. 1990;34:356–8.
32. Li X, Yu HY. Purification and characterization of an organic-solvent-tolerant cellulase from a halotolerant isolate, *Bacillus* sp. L1. J Indust Microbiol Biotechnol. 2012;39:1117–24.
33. George PS, Ahmad A, Rao MB. Studies on carboxymethyl cellulose produced by an alkalothermophilic actinomycete. Bioresour Technol. 2001;77:171–5.
34. Creig RN, Holt GJ. Bergey's manual of systematic bacteriology. London: Williams and Willkins; 1984.
35. Nelson N. A photometric adaptation of the Somogyi method for the determination of glucose. J Biol Chem. 1944;153:375–80.
36. Somogyi M. Notes on sugar determination. J Biol Chem. 1952;195:19–23.
37. Lowry OH, Rosenbrough MJ, Farr AL, Randell RJ. Protein measurement with folin phenol reagent. J Biol Chem. 1951;193:265–75.
38. Laemmli UK. Cleavage of structural proteins during the assembly of the head of *bacteriophage* T4. Nature. 1970;227:680–5.
39. Robyt JF, White BJ. Biochemical techniques: Theory and practical. Inc Illinois, U.S.A.: Waveland Press; 1990.
40. Lineweaver H, Burk D. The determination of enzyme dissociating constants. J Am Chem Soc USA. 1934;56:658–66.
41. Rai P, Tiwari S, Gaur R. *Bacillus vallismortis* strain RG-07 16S ribosomal RNA gene, partial sequence. National Center for Biotechnology Information. http://www.ncbi.nlm.nih.gov/nuccore/JQ619483.

DNA polymerase preference determines PCR priming efficiency

Wenjing Pan[1], Miranda Byrne-Steele[2], Chunlin Wang[3], Stanley Lu[4], Scott Clemmons[4], Robert J Zahorchak[2] and Jian Han[2*]

Abstract

Background: Polymerase chain reaction (PCR) is one of the most important developments in modern biotechnology. However, PCR is known to introduce biases, especially during multiplex reactions. Recent studies have implicated the DNA polymerase as the primary source of bias, particularly initiation of polymerization on the template strand. In our study, amplification from a synthetic library containing a 12 nucleotide random portion was used to provide an in-depth characterization of DNA polymerase priming bias. The synthetic library was amplified with three commercially available DNA polymerases using an anchored primer with a random 3′ hexamer end. After normalization, the next generation sequencing (NGS) results of the amplified libraries were directly compared to the unamplified synthetic library.

Results: Here, high throughput sequencing was used to systematically demonstrate and characterize DNA polymerase priming bias. We demonstrate that certain sequence motifs are preferred over others as primers where the six nucleotide sequences at the 3′ end of the primer, as well as the sequences four base pairs downstream of the priming site, may influence priming efficiencies. DNA polymerases in the same family from two different commercial vendors prefer similar motifs, while another commercially available enzyme from a different DNA polymerase family prefers different motifs. Furthermore, the preferred priming motifs are GC-rich. The DNA polymerase preference for certain sequence motifs was verified by amplification from single-primer templates. We incorporated the observed DNA polymerase preference into a primer-design program that guides the placement of the primer to an optimal location on the template.

Conclusions: DNA polymerase priming bias was characterized using a synthetic library amplification system and NGS. The characterization of DNA polymerase priming bias was then utilized to guide the primer-design process and demonstrate varying amplification efficiencies among three commercially available DNA polymerases. The results suggest that the interaction of the DNA polymerase with the primer:template junction during the initiation of DNA polymerization is very important in terms of overall amplification bias and has broader implications for both the primer design process and multiplex PCR.

Keywords: PCR, DNA polymerase, Priming bias, Next generation sequencing, PPI, Polymerase preference index, iC-Architect

* Correspondence: jhan@hudsonalpha.org
[2]HudsonAlpha Institute for Biotechnology, Huntsville, AL 35806, USA
Full list of author information is available at the end of the article

Background

The polymerase chain reaction (PCR) is one of the most important developments in modern biotechnology. However, PCR is known to introduce biases during amplification, particularly during multiplex PCR when several templates are amplified simultaneously [1,2]. Extreme base compositions (sequences with mostly G/C or A/T composition) are recognized to be problematic for both traditional Sanger sequencing and next generation sequencing platforms [3]. However, recent evaluations of biases generated in high throughput sequencing data have pinpointed the amplification step as the primary cause [4,5]. Factors such as thermocycler make, model, and ramping speed were demonstrated to affect the uniformity of the amplified library [4]. However, the DNA polymerase was identified as the primary source of bias with a variety of commercially available DNA polymerases skewing the amplification profile of the Neandertal genome with regard to GC content and template length [5].

While most applications focus on or even require the removal of the polymerase-dependent bias, we have taken an alternative approach in which we systematically define the amplification bias and utilize it to improve PCR success rates. The initiation of DNA polymerization on the template strand is a critical step in the polymerization process and is likely affected by differences among DNA polymerases and their interaction with the primer:template junction. In support of this notion, a study by Hansen et al. demonstrates that the use of random hexamer priming induces biases in the nucleotide composition at the beginning of transcriptome sequencing reads [6]. In our study, we used high throughput sequencing to test the hypothesis that DNA polymerases have a bias for different oligonucleotides used as primers to initiate DNA synthesis. In order to define this source of DNA polymerase bias, we carefully examined the contribution of the DNA sequence from a ten base pair (bp) window surrounding the primer:template junction including six base pairs (bps) of the primer:template duplex, which rests in the palm of the polymerase prior to nucleotide addition [7,8], and the four bps of single-stranded DNA template immediately following the 90° kink at the junction, which we termed the "runway".

First, we created a synthetic sequencing library containing a twelve nucleotide random insertion (12 N) and flanking sequences without the use of amplification. This synthetic library (termed SL) provides a library of random sequences that reflect the complete pool of template sequences available prior to amplification. The SL was utilized as the template for several amplification experiments. The sequencing results of the amplification experiments were compared to the SL, which was sequenced directly (no amplification). After identifying sequence motifs that were preferentially amplified, single-primer templates were created in order to verify the DNA polymerase bias. We then developed a primer-design program, iC-Architect, which uses the observed bias in order to improve primer design.

Methods

Synthetic library and barcode production

The synthetic library (SL) was produced by the ligation of a barcode segment and a synthesized oligonucleotide (Figure 1). The synthesized oligonucleotide contains a 12 N random region, which was converted to a double stranded template as described below. The barcode portion of the SL design allows a unique barcode to be ligated to each of the different sample amplification test sets so that they can be pooled for high throughput sequencing. The barcode oligo consists of the Illumina adaptor B sequence, a filler region (200 bp long), and a 4 bp barcode followed by a SfiI site that matches the synthesized oligonucleotide's 5' end. The barcode portion was created by PCR amplification using a 200 bp portion of the human IgG C-kappa domain as the template to serve as the filler sequence. The filler sequence is used to make the end-product length optimal for bridge PCR during high throughput sequencing. Forward and reverse primers were designed so that they included the Illumina adaptor B sequence and the barcode (XXXX) with the SfiI ligation site, respectively. The forward primer utilized is as follows: 5'-CAAGCAGAAGACGGCATACGAGATCGGTCTCGG CATTCCTGCTGAACCGCTCTTCCGATCTCCAGAG AGGCCAAAGTACAG-3', while the reverse primer is 5'-CGTAGGCCACTGAGGCCXXXXTCGCCCCGGTTGA AGCT. Each unique barcode was produced by amplification with Qiagen's TopTaq Master Mix Kit in a Bio-Rad C1000 thermocycler as follows: initial denaturation at 94°C for 3 minutes; 35 cycles at 94°C for 30 s, 65°C for 30 s, and 72°C for 40 s; a final extension at 72°C for 10 minutes and then hold at 4°C. All amplified products were run on a 1.5% agarose gel, and the gel band of correct size was extracted and purified using Qiagen's Qiaquick Gel Extraction Kit as per the manufacturer's instructions.

The 99 bp synthesized oligo was purchased from IDT with standard desalting. The design includes a 17 bp SfiI ligation site, a 7 bp spacer, a 12 N random site, an additional 2 bp spacer followed by a 3 N random site to facilitate sequencing, and the Illumina sequencing adaptor A (Figure 1). Both the 7 bp and 2 bp spacer regions adjacent to the 12 N random site were required so that non-specific cleavage of the 12 N region did not occur during digestion with SfiI. The additional 3 N random site after the 2 bp spacer is required during sequencing so that the different clusters can be distinguished during first few cycles of sequencing. The synthesized oligo was converted to a double stranded template using a single reverse primer (5'-AATGATACGGCGACCACCGAGATCT-3') and

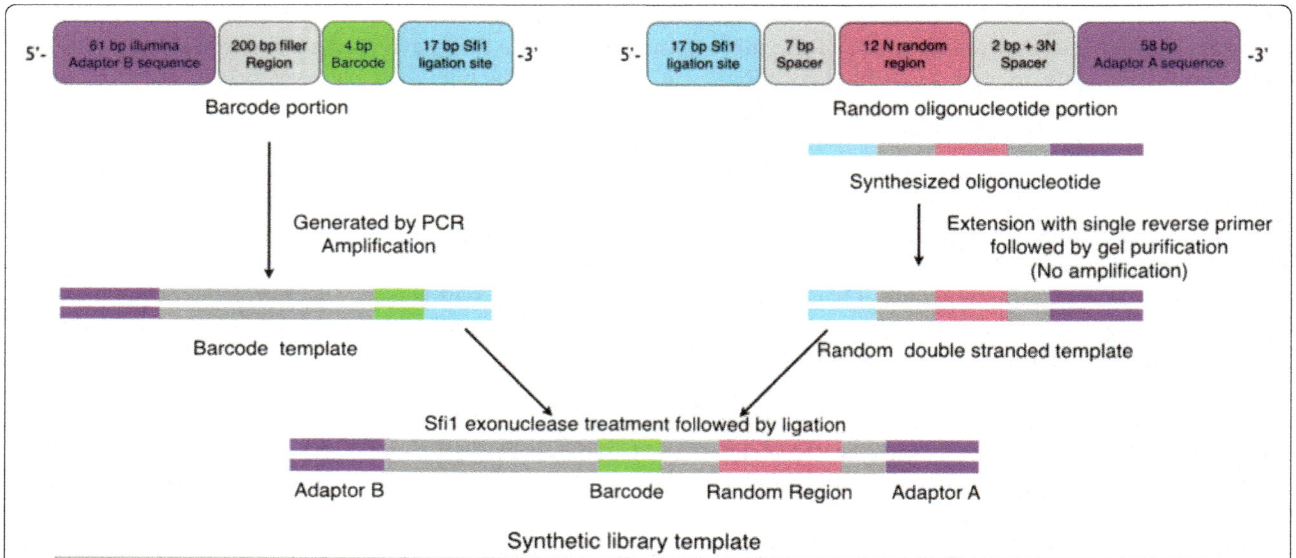

Synthetic library template sequence information

Sequence information	Adaptor B sequence	200 bp insert sequence	Barcode	Ligation Site with spacer	12N	Spacer with 3N	Adaptor A sequence	Sequence information
5' end	CAAGCAGA AGACGGCA TACGAGAT CGGTCTCG GCATTCCT GCTGAACC GCTCTTCC GATCT	CCAGAGAGGCCAAAGTACAGTGGAAGGT GGATAACGCCCTCCAATCGGGTAACTCC CAGGAGAGTGTCACAGAGCAGGACAGC AAGGACAGCACCTACAGCCTCAGCAGCA CCCTGACGCTGAGCAAAGCAGACTACGA GAAACACAAAGTCTACGCCTGCGAAGTC ACCCATCAGGGCCTGAGCTCGCCCGTC ACAAAGAGCTTCAACCGGGGCGAG	XXXX	GGCCTCAGTGGC CTCTGCAT	NNNNNN NNNNNN	ATNNN	AGATCGGA AGAGCGTC GTGTAGGG AAAGAGTG TAGATCTCG GTGGTCGC CGTATCATT	3' end

Figure 1 Design of the barcode and synthetic random oligonucleotide used to create the SL. The barcode portion was created through PCR amplification, while the synthesized random oligonucleotide was purchased and converted to a double stranded template. Both portions were digested with SfiI endonuclease, ligated, gel purified, and sequenced to produce the SL. Exact sequences used to generate the final synthetic library are provided.

35 cycles of 2 steps of annealing and extension including 55°C for 30 s and 72°C for 30 s with Qiagen's HotStar HiFidelity DNA polymerase. An excess of cycles was used during this step to ensure that all of the single-stranded oligo was converted to double stranded template. The newly generated double stranded synthetic templates were gel purified, and the product was then directly ligated to the barcode portion after SfiI digestion. After ligation, the SL was gel purified after extraction from a 3% agarose gel and sequenced on the Illumina Hiseq 2000 platform using single end reads from adaptor A through SeqWright's sequencing service. For all subsequent amplification experiments, a unique barcode portion was ligated to the amplified product, gel purified, and sequenced as described. The barcodes associated with a given amplification experiment are summarized in Additional file 1: Table S1.

Amplification experiments

For all amplification experiments, 2 μL of SL at 50 ng/μL served as the template for amplification, and the same reverse primer as the SL production was used for all amplification experiments. All PCRs were performed as a 50 μL total reaction with 2 μL each of forward and reverse primer at 10 pmol/μL. All reactions were performed in the buffer system supplied by the respective DNA polymerase vendor without modification. For the amplified background experiment, the forward primer (5'-GCATG GCCTCAGTGGCCTCT-3') was placed 4 bp upstream of the 12 N random site. Amplification was performed in a Bio-Rad C1000 thermocycler as follows: initial denaturation at 94°C for 3 minutes; 35 cycles at 94°C for 30 s, 65°C for 30 s, and 72°C for 30 s; a final extension at 72°C for 10 minutes and with a hold at 4°C.

For both the Promega and Qiagen's Taq DNA polymerases, two repeat amplification experiments were performed for both types of polymerase, while one amplification experiment was performed with Qiagen's HotStar HiFidelity DNA Polymerase. The analysis for only one commercial family B DNA polymerase was performed because we needed to ensure that we allotted enough sequencing reads for sufficient coverage of the

12 N random region of all of the currently described amplification experiments in one Illumina HiSeq sequencing lane. Since the synthetic library is considered a low diversity library, we expected the number of sequencing reads output to be at the lower end of the spectrum. Since we needed to include the different annealing temperatures and repeat experiments for the two Taq DNA polymerases, we limited the number of commercial DNA polymerases in the experiment to three. All polymerases utilized the same forward primer (5'-GCATGGCCTCAGTGGCCTCTGCATNNNNNN-3') covering 6 nucleotides of the 12 N random portion of the template. All polymerases were subjected to the same cycling conditions, and amplification was performed as follows: initial denaturation at 94°C for 150 s; 35 cycles at 94°C for 30 s, 70–60°C touchdown for 120 s, and 72°C for 30 s; a final extension at 72°C for 10 minutes and then hold at 4°C. For the experiment testing different annealing temperatures, all PCR parameters remained the same with the exception of the annealing temperatures: 63, 65, and 67°C for 100 s.

Statistical analysis of the observed bias value

To assure that the proportion of each unique sequence in the SL was represented in the amplification experiments, we counted the reads for each sequence combination for four observation windows (4 bp-runway, 6 bp-primer:template interaction only, 8 bp- primer:template and runway, and 10 bp-primer:template and runway; Figure 2) for barcode 1 (the SL) and designated that result as the background for the generated template for all further analyses.

Since the number of the sequence reads differs between barcoded samples, normalization of the data was required prior to comparing samples. To do so, we adjusted the reads from data sets of differing scales to a notionally common scale. A z-score was calculated that adjusted a given sample size to that of the designated background sample (SL), and this standard score procedure was applied to all observation regions for all barcodes.

To ascertain the degree of bias in a given sample, a relative value for each specific sequence from the observation region was calculated. The normalized reads from the PCR amplified sample (NR_{PCR}) were divided by the reads for the same sequence from the synthetic library (SR_{SL}). For example, for a 6 bp observation window, assuming the specific DNA sequence "ATCGAT" results in 10 reads from the SL, and the normalized reads from the Taq polymerase PCR sample is 20, the "ATCGAT" at the end of primer has an observed bias value of 2. Theoretically, if there is no priming bias for the DNA polymerase, the observed bias value for all sequences in different observation regions should be similar to each other and close to a value of "1". The observed bias value (OBV) formula follows:

$$OBV = \frac{NR_{PCR}}{SR_{SL}}$$

The statistics for all data comparisons are summarized in Additional file 2: Table S2.

Single-primer test

Single-primer templates are templates with the same primer binding site on both the sense and anti-sense strands. In total, 24 unique single-primer templates were created and utilized for additional amplification experiments. The single-primer templates were produced by first designing primers that contained the primer site of interest and a portion of the human IgG C-kappa domain. Therefore, the forward primer consisted of a 19 nucleotide filler region, the 8 nucleotide primer site to be tested, and 19 nucleotides specific for the sense strand of the kappa domain, while the reverse primer included the identical filler and primer site with nucleotides specific for the anti-sense strand of the kappa domain (Additional file 3: Figure S1). The kappa domain was amplified as follows: initial denaturation at 94°C for 3 minutes; 35 cycles at 94°C for 30 s, 56°C for 30 s, and 72°C for 40 s; a final extension at 72°C for 10 minutes and then hold at 4°C. Twenty-four generated templates were separated on 1.5% agarose gel, and the band of the correct size was gel purified. All single-primer templates were adjusted to the same concentration of 0.001 ng/μL prior to the single-primer amplification test.

For each of the 24 templates, a unique single primer for each template was designed that includes the 19 nucleotide filler region and 6 nucleotides of the 8 nucleotide primer site. In each reaction, 2 μL of generated template and 2 μL of single primer at 10 pmol/μL were used with either Qiagen TopTaq or Qiagen HotStar HiFidelity DNA polymerase. End-point PCR was performed using two annealing temperatures, 57°C and 59°C for 30 s, with either 20 or 25 cycles. An internal control for each reaction was also performed with primers specific for an internal portion of the C-kappa region, which was the same for all 24 generated templates (forward-5'-TCTGTCTTCATCTTCCCGCCA-3'; reverse-5'AAGCTC TTTGTGACGGGCGAG-3'). Negative control amplification experiments, in which no template was added to the reaction, were performed, and no amplification was observed (data not shown).

Results

Amplification experiments and barcode analysis

In order to define the DNA polymerase bias, we created a synthetic library (SL) of sequences that are identical in

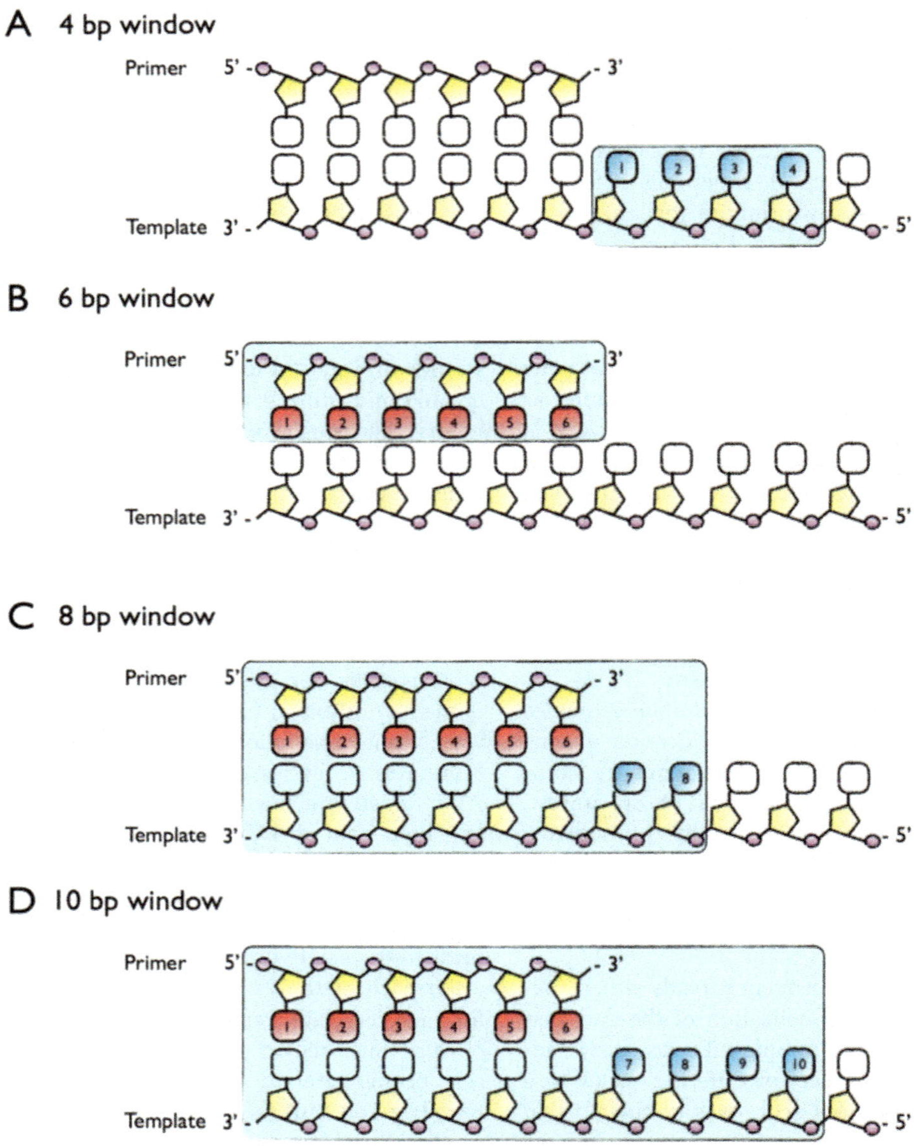

Figure 2 Observation windows for the 12 N portion of the SL. (A) The 4 bp window includes the nucleotides 4 bp ahead of the primer: template junction; **(B)** the 6 bp window includes the primer:template interaction; **(C)** the 8 bp window includes the primer:template interaction and 2 bp ahead of the primer:template junction; and **(D)** the 10 bp window includes the primer:template interaction and 4 bp ahead of the junction.

sequence with the exception of a 12 nucleotide random insertion (Figure 3A). The SL was utilized as a template for a variety of amplification experiments, and all amplified libraries, including the non-amplified SL, were pooled and extensively sequenced using the Illumina HiSeq platform. Two primer sets were designed to amplify the SL. The forward and reverse primers of primer set I are located outside the random-insertion region (Figure 3B). The forward primer of primer set II is located at the boundary of the random-insertion region with six random nucleotides at the 3' end of each primer (Figure 3C). The reverse primer of primer set II is identical to the reverse primer of primer set I. The SL was amplified with both primer sets I and II using three commercially available DNA polymerases: two family A DNA polymerases, Qiagen TopTaq (QTT-A) and Promega GoTaq (PGT-A); and one family B DNA polymerase Qiagen HotStar HighFidelity (QHH-B). Repeat amplification experiments were performed at several annealing temperatures (touchdown 70–60°C, 63°C, 65°C, and 67°C) as described in the Methods.

Prior to data analysis, all high throughput sequencing data were first separated into individual data sets by utilizing the barcodes associated with each experiment, and

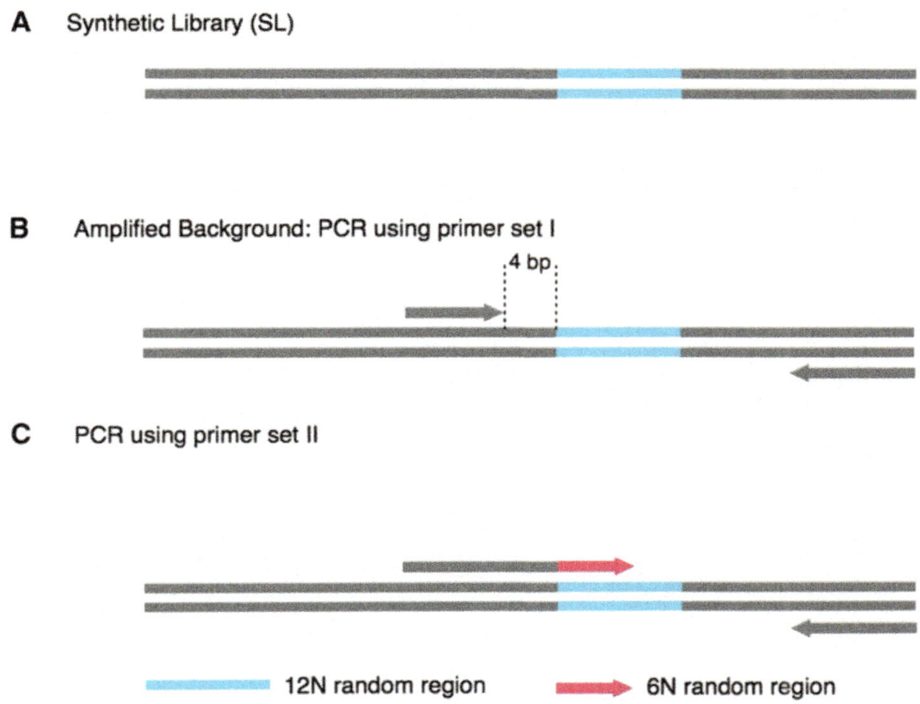

Figure 3 Experimental set-up for the amplification of the SL with two different primer sets. (A) The SL that contains a 12 N random region (blue) serves as the template for all additional amplification experiments; **(B)** in the amplified background experiment, the forward primer is placed 4 bp upstream of the 12 N region; **(C)** in the polymerase amplification tests, the forward primer (grey and pink) covers 6 N of the 12 N region.

the data were normalized as described in the Methods. High throughput sequencing of the data sets returned 91,434,220 sequence reads, after filtering by the barcode sequences. The barcode sequences, number of associated reads, and the experimental description matching each barcode are shown in Additional file 1: Table S1. Sequencing of the unamplified SL confirmed that all of the possible combinations of template random nucleotide sequences are represented.

In order to analyze the data relevant to the random region, we filtered the data based on a 4 bp window (runway), 6 bp window (primer:template interaction), 8 bp window (primer:template interaction and 2 bps of runway), and 10 bp window (primer:template interaction and 4 bps of runway) as demonstrated in Figure 2. An observed bias value (OBV) was calculated and assigned to each sequence motif in the respective observation window (Methods). Since the ten bp observation window lacks sufficient read coverage (Additional file 1: Table S1), it was not utilized for further analysis due to lack of statistical significance.

Observation of bias

When primer set I is used to amplify the SL (Figure 3B), the random sequences are faithfully amplified (Figure 4A). However, when a primer with six random bases (6 N) at the 3' end is used to amplify the library (Figure 3C), priming bias is observed (Figure 4B-D). Furthermore, the observed bias is polymerase specific with the two commercially available family A DNA polymerases (QTT-A and PGT-A) preferring a set of sequence motifs different from those preferred by the family B DNA polymerase (QHH-B; Figure 5A-B). For instance, the sequence motif "GGGGGCGG" is the top-ranked one among 65,536 possible motifs for all the QTT-A DNA polymerase amplification experiments, including the touchdown, 63°C, 65°C, and 67°C annealing temperatures (Table 1). The same sequence motif is ranked second and third, respectively, for PGT-A polymerase. In contrast, this sequence motif is ranked 4,180 for the touchdown experiment of QHH-B. When comparing the top 30 ranked sequence motifs for the QTT-A touchdown PCR experiment across all family A DNA polymerase amplification experiments (including PGT-A), the same 30 sequence motifs appear within the top 87 of 65,536 possibilities. In contrast for the QHH-B touchdown experiment, these 30 sequence motifs are ranked from a range of 220 to 29,598 of 65,536 sequences (Table 1).

There is positive amplification bias towards increasing GC content for the six base pairs of primer:template interaction for all of the commercial DNA polymerases tested. This is evident when the OBV for each sequence

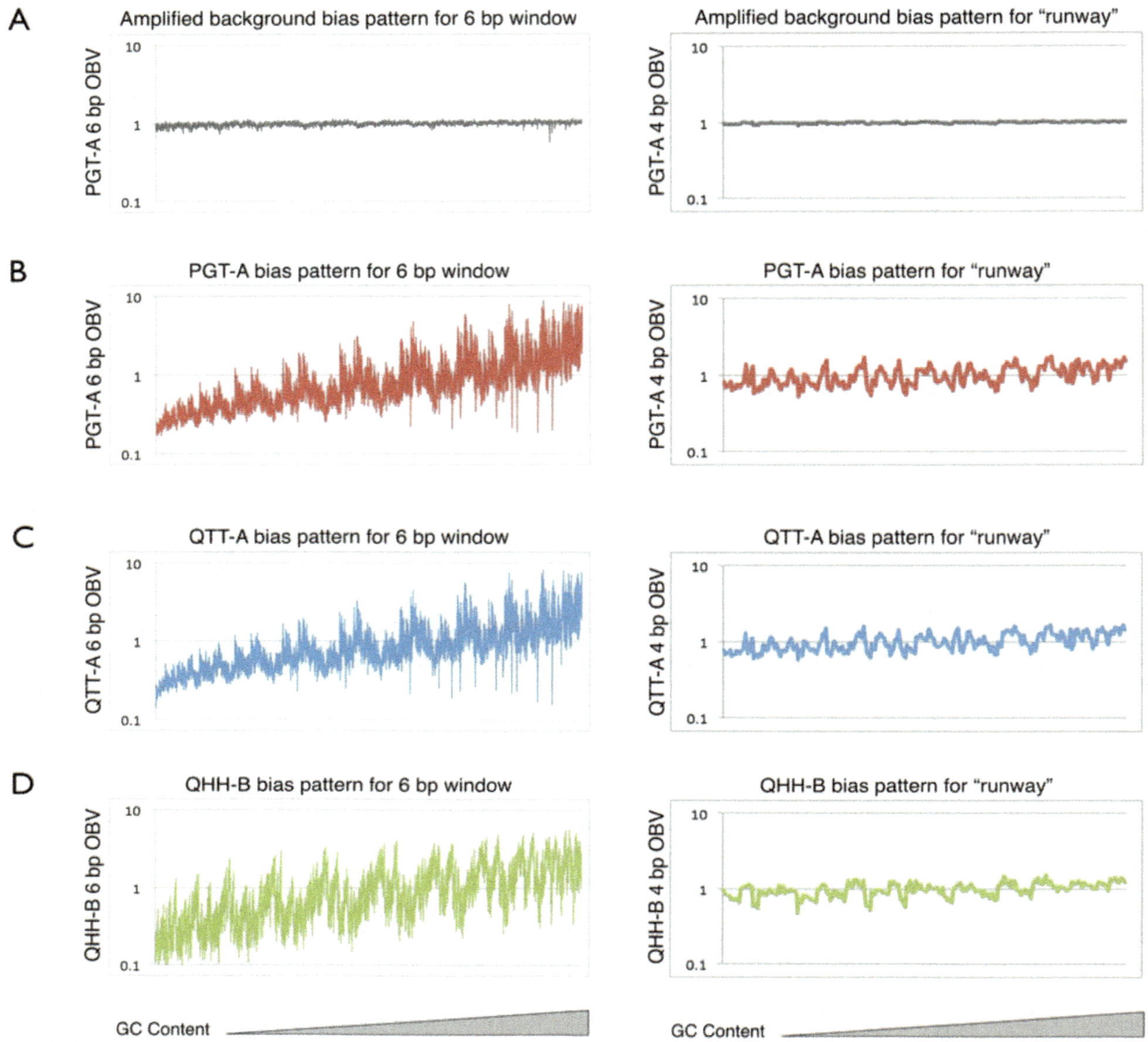

Figure 4 Bias profiles for the SL and amplification experiments. All bias profiles were created by plotting the observed bias value (OBV) for the 4 bp (n = 256) and 6 bp (n = 4,096) windows of observation on a logarithmic scale versus the amplified sequences, which were arranged from low to high GC content in the order of ATCG; **(A)** the bias observed after amplification with primer set I; **(B)** after amplification using primer set II with PGT-A DNA polymerase; **(C)** using primer set II with QTT-A DNA polymerase; **(D)** using primer set II with QHH-B DNA polymerase.

is plotted with the sequences arranged in order of increasing GC content (Figure 4B-D). For the top 1% of ranked sequences, the average GC content is 79% for the family A DNA polymerases and 83% for the QHH-B DNA polymerase. We also observed that the two bps located closest to the primer:template junction contained high GC content in sequences that were preferentially amplified. For instance, for the top 1% of ranked sequences of the 8 bp window, "GC" or "CG" is located at the 3' end of the primer for 88-91% of the sequences across all of the family A DNA polymerase amplification experiments (Table 2). In contrast, 25% of the top-ranked sequences of QHH-B polymerase contain these sequence motifs at the 3' end of the primer. The trend extends to the last three nucleotides on the 3' end of the primer. The sequence motifs "CGC", "CCG", "GCG", "GGC", "TGC", and "TCG" occur at the 3' end of the primer for 87-90% of the top 1% of ranked sequences for the family A DNA polymerases and 22% for the family B DNA polymerase (Table 2). In contrast, QHH-B has a preference for "GC" or "CG" at the 5' end of the 6 N portion of the primer with 89% of the top 1% of ranked

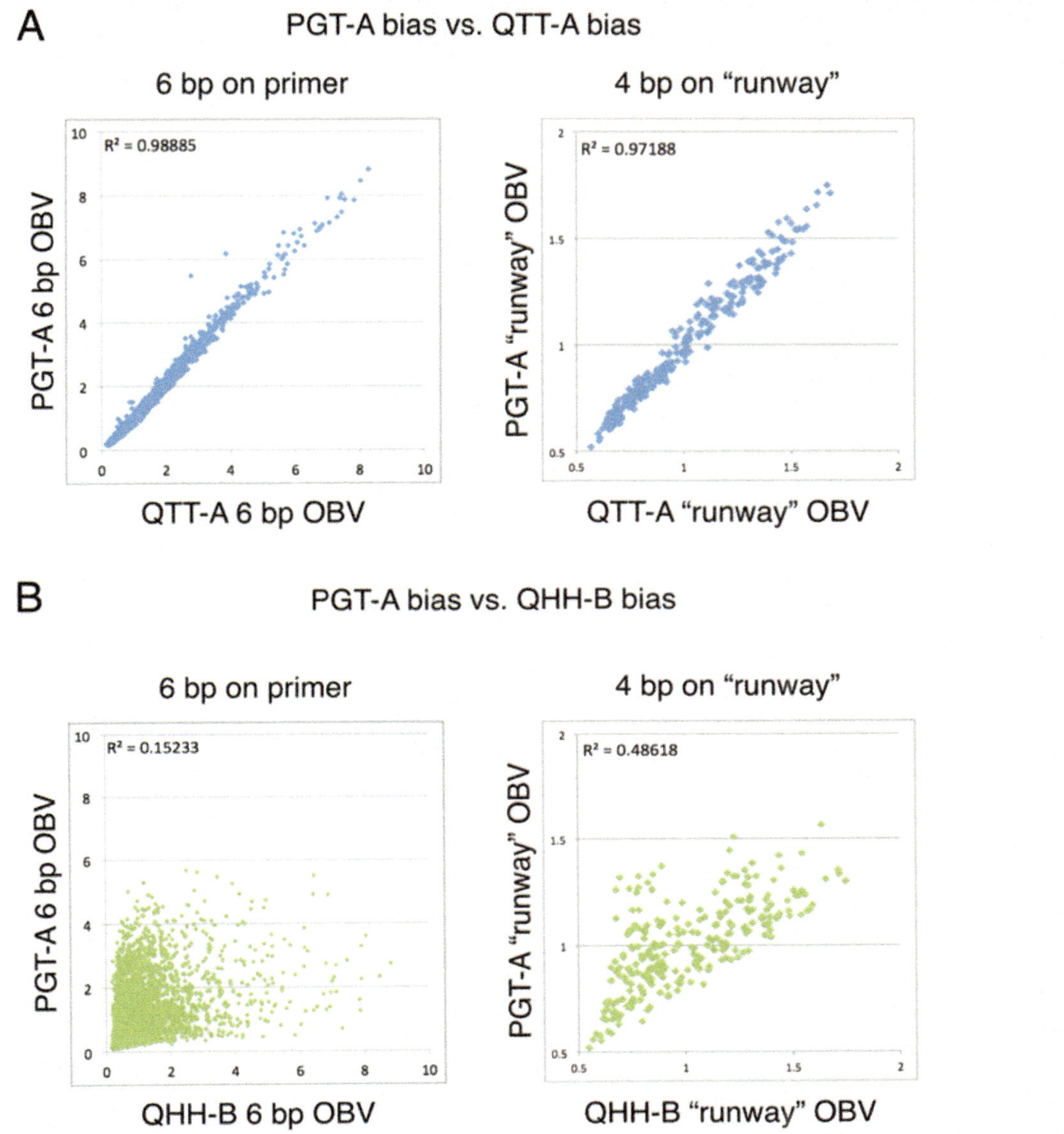

Figure 5 Correlation of amplified libraries. (A) A plot of the OBV of two family A polymerases (PGT-A versus QTT-A) demonstrates a high degree of correlation throughout the primer:template (6 bp) and runway (4 bp) regions; **(B)** a plot of the OBV of PGT-A versus OBV of QHH-B indicates that resulting amplification profile is weakly correlated for both the primer:template interaction (6 bp) and the runway (4 bp) for two differing families of polymerases.

sequences containing this motif, while only 6-12% of top ranked sequences from the family A experiments contain these sequences at their 5' end. Therefore, the specific location of the GC content contributes to the amplification bias throughout the primer:template interaction. Interestingly, two GC-rich motifs at the 3' end of the primer, "AGC" and "ACG", had consistently poor amplification results across all polymerases with only 1% of top ranked QTT-A and PGT-A results and 4% of top ranked QHH-B sequences containing these motifs.

When the OBV for each runway motif is plotted with the sequences arranged in order of increasing GC content, the bias profile does not demonstrate a similar pattern to the primer:template interaction (Figure 4B-D). The sequence motif 3'-"TC"-5' is preferred by QHH-B DNA polymerase in the first two positions after the primer:template junction (on the template sequence) with 19% of the top 1% of ranked sequences sharing this motif. In contrast, across all family A polymerase experiments, only 3-4% of top ranked sequences contain this motif at the beginning of the runway (Table 2). All of the DNA

Table 1 A comparison of the top thirty ranked QTT-A sequences across all amplification experiments

8 bp window sequence	Top 30 ranked QTT-A sequences: comparison of rank in other amplification experiments							
	QTT-A	QTT-AR	PGT-A	PGT-AR	QTT-A63	QTT-A65	QTT-A67	QHH-B
GGGGGC GG	1	3	2	3	1	1	1	4180
TTGGGC GG	2	1	3	9	8	5	2	20382
GGGCCG GG	3	12	5	2	5	3	3	606
TGGCCG GG	4	5	4	4	6	6	6	3065
GGTCCG GG	5	6	1	1	3	2	4	1666
GGGTGC GG	6	9	6	13	16	12	9	2377
GTGGGC GG	7	11	10	5	9	7	11	1464
TTGCCG GG	8	8	8	7	22	21	20	13440
GTGCCG GG	9	13	7	8	7	13	17	220
GGGGGC CG	10	19	21	26	2	4	5	3950
TGGTGC GG	11	18	9	14	37	28	19	8530
TTGGGC CG	12	10	19	24	21	27	14	26404
TGCCCG GG	13	14	13	6	13	15	21	3370
GGTCCG GC	14	17	11	12	18	18	23	3802
GGGGGC CA	15	31	24	29	12	14	15	8559
GGGCCG GC	16	32	30	15	10	9	28	2255
TGGCCG GC	17	15	15	16	26	23	22	5312
TTGTGC GG	18	27	23	40	51	69	44	17564
TGTCCG GG	19	21	12	11	32	33	33	9598
TTGGGC CA	20	16	34	35	58	51	47	29598
GGTGGC GG	21	22	16	17	19	22	24	1890
GGGGGC CC	22	36	47	45	15	17	13	10961
TTTGGC GG	23	33	32	22	57	71	39	24345
GGCCCG GG	24	26	14	10	17	16	18	508
TGGGCG GG	25	29	29	23	29	47	40	5616
GGGCGC GG	26	87	33	38	25	34	42	1337
TTGCCG GC	27	23	31	32	46	52	65	18324
TTGGGC GC	28	24	51	57	92	62	60	25137
GGGGGC GC	29	47	58	64	27	24	27	9051
GGGGGC TG	30	35	55	80	4	8	10	8558

The top thirty sequences of the QTT-A touchdown PCR experiments were ranked and compared to the QTT-A repeat experiment (QTT-AR), the PGT-A and PGT-A repeat experiments (PGT-AR), the QTT-A experiments performed at 63°C, 65°C, and 67°C (QTT-A63, QTT-A65, and QTT-A67), and the QHH-B touchdown PCR amplification experiment.

polymerases tested demonstrate a preference for 3'-"CC"-5' in this position with 23-30% of family A and 30% of the family B top ranked sequences sharing this motif.

In order to examine the effects of annealing temperature, we repeated amplification experiments at three additional annealing temperatures and compared the results to touchdown PCR experiments (Methods). All results were highly correlated with $R^2 > 0.98$ ($n = 4{,}096$, $p = 0.0000$; Figure 6), indicating that the annealing temperature over the tested range did not have a significant effect on the bias profile for the six bp observation window.

Overall, QTT-A and PGT-A DNA polymerases have a similar amplification profile with an R^2 of 0.98 ($n = 4{,}096$; $p = 0.0000$) for the primer:template interaction despite the fact that they were ordered from two separate vendors and likely work in different buffer conditions (Figure 5A). The similarity between the amplification profiles for both the primer:template interaction and the runway sequences of these two DNA polymerases is remarkable. However, we observed a different bias profile when the QTT-A polymerase was compared to the QHH-B DNA polymerase, a family B enzyme (Figure 5B). The bias profiles for the two different families of polymerases have an R^2 of

Table 2 The 8 bp window motif analyses for several DNA polymerase experiments

Top 1% (655 in 65,536)	8 bp window motif analyses for several DNA polymerase experiments							
	QTT-A	QTT-AR	PGT-A	PGT-AR	QTT-A63	QTT-A65	QTT-A67	QHH-B
(GC,CG) motif at 3′ end of primer	88%	89%	90%	90%	90%	91%	88%	25%
(GC,CG) motif at 5′ end of 6 N portion of the primer	8%	6%	8%	11%	12%	10%	8%	89%
(TC) motif at first 2 bp of runway	3%	4%	3%	3%	4%	4%	4%	19%
(CC) motif at first 2 bp of runway	28%	26%	28%	30%	23%	23%	25%	30%
(CGC,CCG,GCG, GGC,TGC,TCG) motif at 3′ end of primer	87%	88%	88%	89%	90%	90%	87%	22%
(AGC,ACG) motif at 3′ end of primer	1%	1%	1%	1%	1%	1%	1%	4%
(ACG) motif at 3′ end of primer	1%	1%	1%	1%	1%	1%	1%	2%
(AGC) motif at 3′ end of primer	0	0	0	0	0	0	0	2%

The motif percentage breakdown for the top 1% of ranked sequences are shown for several PCR experiments: QTT-A touchdown, QTT-AR, PGT-A, PGT-AR, QTT-A63, QTT-A65, QTT-A67, and the QHH-B touchdown PCR amplification experiment.

0.15 (n = 4,096, p = 0.0000) for the primer:template interaction and an R^2 of 0.49 (n = 4,096, p = 0.0000) for the runway sequences. An interaction between the primer and template DNA sequences independent of the polymerase cannot be the only factor leading to the observed bias; otherwise, all of the DNA polymerases would have amplified the SL similarly. The observed differences in amplification profiles are likely due to variation in the proprietary buffer conditions and structural differences between the two DNA polymerase families.

Consideration of DNA polymerase bias during primer design

Based on the observed frequencies of DNA sequence motifs from the high throughput sequencing experiments, we developed a primer-design program called iC-Architect to aid PCR primer design by displaying the observed bias alongside target sequences (Additional file 4: Supplementary Methods). The software models the observed DNA polymerase preference by calculating an index value for a sliding eight base pair window, termed the polymerase preference index or PPI, and assigning the PPI value to the template strand (Additional file 5: Figure S2). An example of the iC-Architect display for a sample target sequence is provided in Figure 7. The software aids in positioning the 3' end of the primer in places where the PPI value of the template is at a maximum. As demonstrated by the correlation coefficients in Figure 8A-B and Additional file 2: Table S2, the PPI accurately models the observed polymerase bias from high throughput sequencing experiments. However, there are a few notable outliers between the predicted and observed bias values. After examining the full-length sequence of these particular primers, significant hairpin secondary structure was predicted through IDT's OligoAnalyzer Tool (http://www.idtdna.com/analyzer/Applications/OligoAnalyzer/), and poor amplification efficiency was observed during the amplification experiments. This underscores the importance of

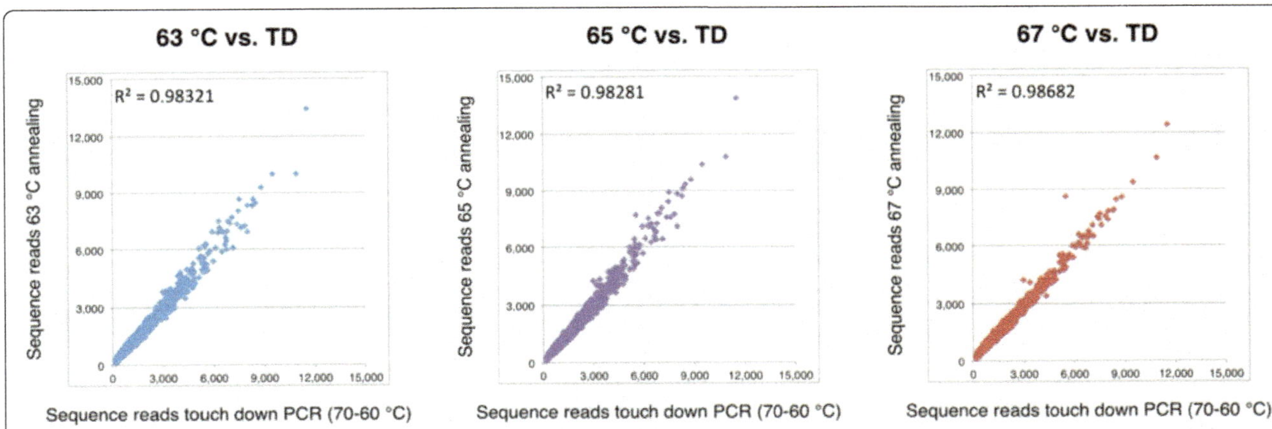

Figure 6 Affect of annealing temperature on the amplified library. Amplification was performed at 63, 65, and 67°C, and the sequenced reads were compared to touch down PCR experiments (70-60°C) for the 6 bp observation window.

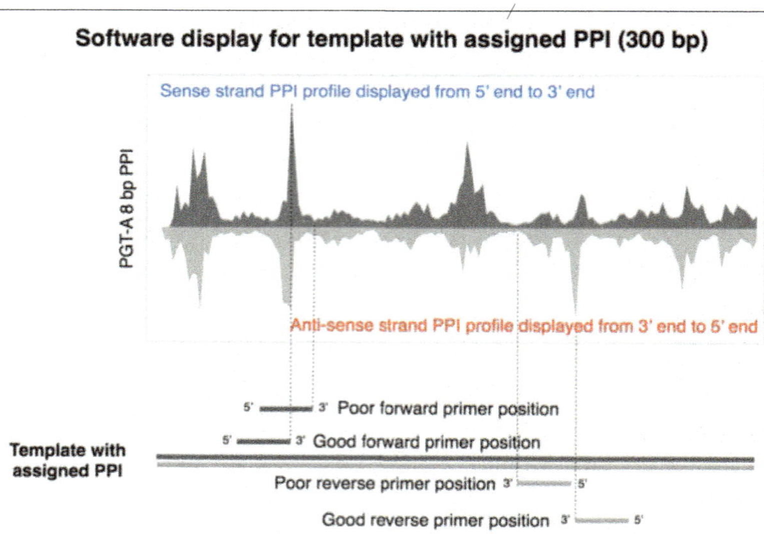

Figure 7 Output of the iC-Architect software. A graphical display of the output of the iC-Architect online software for a sample template strand. Maximum positions indicate the optimal placement for the 3′ end of the primer, while minimum positions should be avoided.

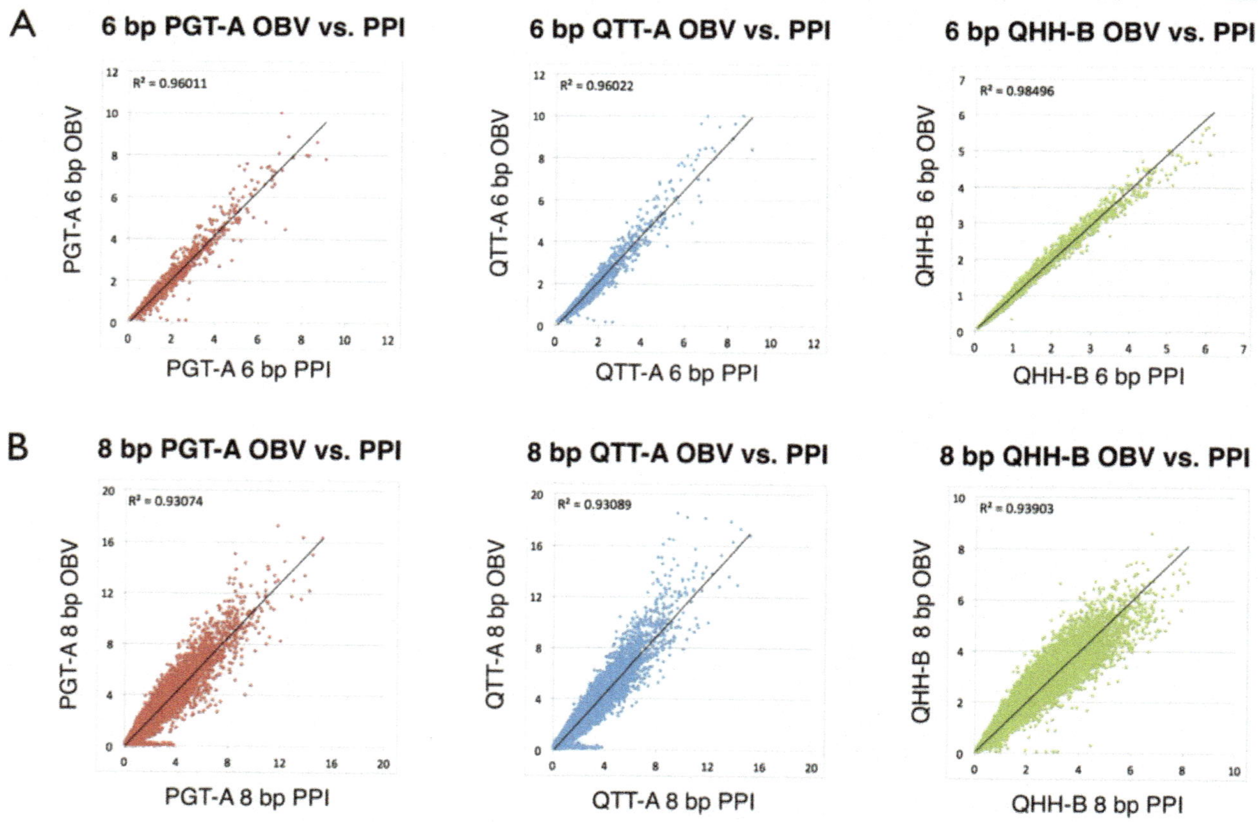

Figure 8 Agreement between the modeled and observed bias. The polymerase preference index (PPI) is highly correlated with the observed bias value (OBV) for both the **(A)** 6 bp and **(B)** 8 bp windows of observation for each of the DNA polymerase tested.

avoiding secondary structural elements such as hairpins when designing primers, and this additional step should be taken after considering the position of the primer on the template strand.

We tested 24 primers that were predicted to have a range of polymerase preference from poor to good as indicated by the directly observed bias (Methods). In all cases, the observed bias was in agreement with the optimal primer position provided by the iC-Architect software and therefore further served to validate our software. Since we wanted the amplification to be attributed to only one primer at a time, we generated single-primer templates. These templates include the same single-primer binding site on both the forward and reverse strands. We designed the single primer to examine an eight bp window, which includes six bps of the primer:template and two nucleotides of the runway. The amplification profiles demonstrate that the bias both observed and predicted through the software can be replicated experimentally (Figure 9).

We also selected a subset of primer sequences that showed positive amplification bias with the family A DNA polymerases and negative bias with the family B DNA polymerase, and vice versa. We performed the single-primer amplification tests with these two opposing subsets and were able to demonstrate that the two families of polymerases preferentially amplify certain sequences, yet fail to amplify others (Figure 10). Even though the two families of DNA polymerases have different bias profiles, the iC-Architect software is capable of predicting the best locations for the primer as long as the underlying data used to calculate the index refers to the specific commercially available DNA polymerase utilized in the experiment.

The preferential amplification of certain sequences based on the type of DNA polymerase utilized indicates that GC and T_m calculations should not be the only considerations when designing primers. For example, in the single-primer template amplification experiments, the same single primer was used with two different types of DNA polymerases to amplify an identical template under the same reaction conditions. Although the primers have the same T_m and GC content, differences in amplification efficiency were readily evident (Figure 10). The differences in amplification efficiency may be related to both polymerase type and varying buffer conditions among commercial vendors. In other words, we cannot rule out that differing buffer conditions between the family A and family B DNA polymerase contribute to the observed difference or if it is actual structural differences between the two families of DNA polymerases (or a combination of both). Furthermore, the arrangement of nucleotides in the primer itself has a large impact on PCR success. There are many examples from the current high throughput sequencing experiment of sequences that have the same nucleotide content (and therefore, the same GC content), but vastly different outcomes in terms of amplification success due to the actual arrangement of the nucleotides in the sequence (Figure 11 and Additional file 6: Table S3). For example, the sequence CCG TTT has an OBV of 0.48, while the sequence TTT CCG has an OBV of 4.55. This is not to say that T_m and GC content should be ignored during primer design. They should be considered in addition to the DNA polymerase preference for specific sequence motifs.

Discussion

Through the use of the next generation sequencing technology, the current study evaluated the polymerase preference by directly observing the priming efficiency of all possible hexamer primers. The crystal structure of catalytically active *Bacillus stearothermophilus* DNA polymerase crystals indicates that approximately ten bps of primer:template duplex DNA and four nucleotides of template in front of the primer template junction occupy the DNA binding site of the polymerase during synthesis [7,9]. In addition, the structure reveals a transition from A to B form DNA in the six bps adjacent to the primer: template junction. The tendency of a given nucleotide sequence toward A or B conformation is a key factor in many protein-DNA interactions [10,11]. For instance, when cyclic AMP receptor protein binds, the DNA undergoes a B to A like transition, and the binding is stronger when the central part of the DNA is in the A form [12]. Analogous to DNA polymerases, the RNA-DNA heteroduplex transiently formed during transcription [13] and the DNA in the active centers of HIV-1 reverse transcriptase [14] are also in the A form. It has been suggested that the stabilization of the A form may be critical for increasing the fidelity of DNA and RNA synthesis [15]. In this study, we observed that the two base pairs located at the primer:template junction contained a high GC content in sequences that were preferentially amplified. Several authors have demonstrated that poly (dG)-poly (dC) undergoes the transition from B form to A from more easily than poly (dA)-poly (dT), which tends to resist the B to A transition [12,16]. Indeed, the Gibb's free energy for the B to A transition for GC rich trimeric motifs is lower than the transition of similar AT rich sequence motifs [16]. It is reasonable to consider that a primer:template interaction that has a tendency to be of the A form or a primer:template interaction that can be more easily conformed to the A form is preferentially bound by the DNA polymerase, ultimately resulting in biased amplification.

The purpose of the current study was to demonstrate and define DNA polymerase-dependent priming bias. Positive amplification bias towards increasing GC content

8 bp window sequence	Target No.	QTT-A PPI	QTT-A OBV	Single-primer TM
CCGCCA GC	1	0.90	0.94	66.0 °C
GCCCCA CA	2	0.95	0.84	66.2 °C
GCCCCA GC	3	1.05	0.77	66.2 °C
GCCGCA AC	4	1.12	0.95	66.5 °C
GCCCGA GG	5	1.32	1.36	66.0 °C
CCCGCC TG	6	1.58	1.63	66.6 °C
GCCCAC CG	7	1.63	1.20	65.0 °C
CGCCTG CA	8	1.95	2.06	64.7 °C
GCAGCC GC	9	2.13	2.15	65.2 °C
CCGCGG CA	10	2.25	2.15	66.6 °C
GGCACC GG	11	2.33	2.59	65.0 °C
CGCTGC AA	12	2.34	2.09	65.0 °C
TGCGCC AG	13	3.40	3.48	65.5 °C
CTCGGC CG	14	4.36	4.34	64.5 °C
GGTCGC TG	15	4.48	4.98	64.8 °C
CTGGGC GA	16	4.52	4.36	64.6 °C
TCCCCG CT	17	4.65	5.17	64.6 °C
TGACCG GT	18	4.86	4.79	62.8 °C
TTCGGC GA	19	5.14	5.57	63.3 °C
GGTGGC TA	20	5.23	6.51	65.0 °C
TGTTGC GG	21	5.24	7.36	61.7 °C
TTTCCG CG	22	6.06	6.08	61.3 °C
TTTGGC CC	23	6.84	8.08	61.8 °C
TTGGGC CT	24	7.13	9.02	63.4 °C

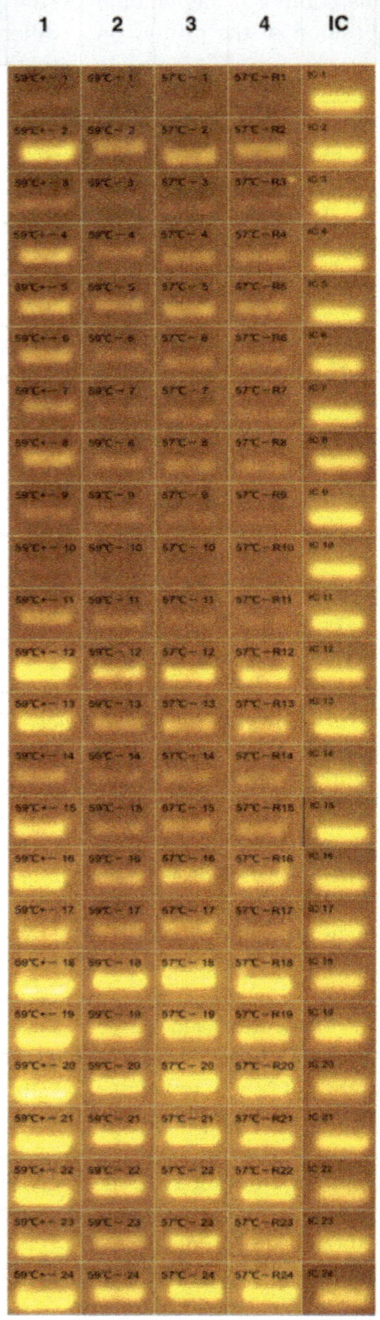

Figure 9 Single primer experiments validating suggested primer-design locations from the iC-Architect software. For each of the 24 single-primer templates, the agarose gel results are shown with calculated PPI and OBV values arranged from poor to good. Two repeat amplifications were performed at two temperatures: 59°C (lanes 1 and 2) and 57°C (lanes 3 and 4). All amplifications were stopped after 20 cycles, except lane 1, which was stopped at 25 cycles to demonstrate that 20 cycles was sufficient to see the difference in amplification. An internal control amplification was also performed on a portion of the sequence consistent with all templates. T_m values for the single-primers were calculated using the IDT OligoAnalyzer Tool.

was observed for the six base pairs of primer:template interaction for all of the commercial DNA polymerases tested, and several preferentially amplified sequence motifs were identified. Interestingly, Dabney et al. found that some commonly used polymerases strongly bias against amplification of endogenous DNA in favor of GC-

8 bp window sequence	Target No.	QTT-A PPI	QTT-A OBV	QHH-B PPI	QHH-B OBV	1	2	IC	Single-primer TM
GCCCCA GC	3	1.05	0.77	6.88	5.96				66.2 °C
GCCGCA AC	4	1.12	0.95	5.69	5.85				66.5 °C
CCCGCC TG	6	1.58	1.63	5.18	5.08				66.6 °C
GCCCAC CG	7	1.63	1.20	5.21	5.21				65.0 °C
CGCTGC AA	12	2.34	2.09	4.64	4.96				65.0 °C
TCCCCG CT	17	4.65	5.17	1.70	1.62				64.6 °C
TGTTGC GG	21	5.24	7.36	1.18	1.23				61.7 °C
TTTCCG CG	22	6.06	6.08	0.50	0.48				61.3 °C
TTTGGC CC	23	6.84	8.08	0.47	0.50				61.8 °C
TTGGGC CT	24	7.13	9.02	0.55	0.56				63.4 °C

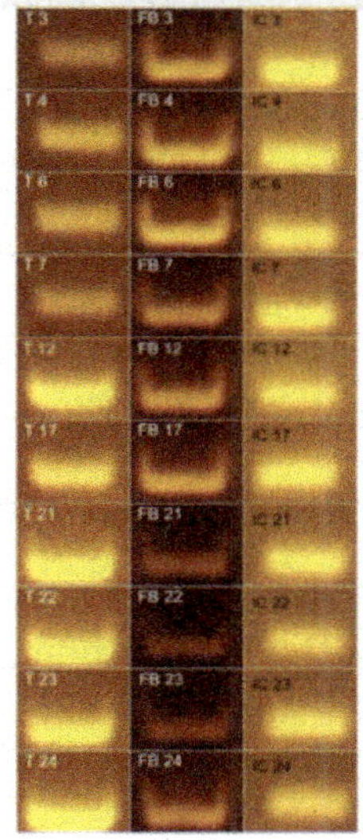

Figure 10 Single primer experiments demonstrating the difference in amplification bias between different polymerase families. For each of the 24 single-primer templates, a subset of sequences for the QHH-B DNA polymerase that have a PPI of the opposite trend with QTT-A DNA polymerase were selected. For agarose gel results, lane 1 shows the QTT-A results while lane 2 shows the QHH-B results, which has PPI values in reverse order. An internal control amplification was also performed on a portion of the sequence consistent with all templates.

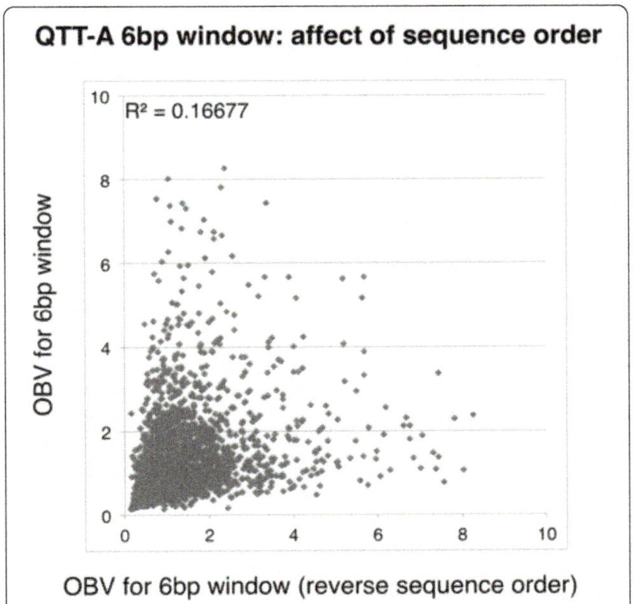

Figure 11 Importance of the 3′ end of the primer sequence on amplification bias. The OBV for the 6 bp window of (n = 4,096) was plotted against the OBV of the sequence of the reverse trimer order. For instance the OBV for the 6-mer "CCG AGG" is plotted against the OBV of "AGG CCG". Although the sequences have the same GC content, the amplification results are not correlated.

rich microbial contamination [5]. In their study, Phusion HF and AmpliTaq Gold showed a very pronounced bias towards molecules with >50% GC. Another study by Hansen et al. demonstrated that random hexamer priming during the generation of cDNA induces biases in the beginning of nucleotide sequencing reads. In their study, they attempted to discern whether the high throughput sequencing reads originated from the sense strand by second-strand synthesis (from the DNA polymerase) or the antisense strand by first-strand synthesis (from the reverse transcriptase) [6]. Since both strands displayed a similar bias pattern, they concluded that the second strand DNA synthesis is likely being primed by remaining random hexamers in the solution. They also noted slight differences in the patterns in the sense and antisense strands which they attributed to different sequence specificities of the reverse transcriptase and DNA polymerase or the effect of nick priming [6]. Our data is in agreement with their conclusion and suggests that a majority of the bias is likely due to priming by the random hexamers and the preference of the DNA polymerase for certain primer: template junctions. The reverse transcriptase may also have preferential sequence motifs as the RNA-DNA heteroduplex in the active center of HIV-1 reverse transcriptase [14] is of the A form; however, reverse transcriptase bias was not addressed in the currently described experiments.

In our study, the four base pairs of single-stranded DNA template immediately following the junction, which we termed the "runway", were also of interest because the polymerase can directly interact with the exposed bases of the single stranded template. This interaction can affect amplification. For instance, in archaeal family B DNA polymerases, the DNA polymerase possesses a read-ahead function in which polymerization will stall if an uracil is encountered 4 base pairs ahead of the primer:template junction [17,18]. This is not observed with family A DNA polymerases. The runway sequences in the described experiments did not have a clearly defined trend with regards to GC content like the primer:template interaction. However, some degree of bias was evident, and we were able to identify sequence motifs that were preferentially amplified. This suggests that when amplifying targets with universal primers, as in the case of certain multiplex PCR reactions, care should be taken to make sure that the primer-template junction is similar for all targets. If the primer:template junction varies ahead of the junction, this may result in the DNA polymerase preferentially amplifying "favorable" primer:template junctions over others.

In this work, all experiments were performed using the manufacturers' polymerase-buffer systems directly. Modifying buffer conditions may influence enzyme and/or DNA configuration and achieve a successful amplification; however, this post-primer design effort may be avoided by adjusting the primer position slightly during the initial design phase. This is particularly useful when designing primers for multiplex reactions where inter-loci balancing is necessary to reduce conflicts among primers and targets. During the design of primers for multiplex reactions, the iC-Architect software can be used to select forward and reverse primers of relatively high and similar PPI values for multiple targets or to avoid regions with very low PPI values. Ultimately, the software identifies primers with preferred motifs on the 3' end of the primer, which may be used to increase PCR success rates, especially when used in conjunction with T_m measurements.

Conclusions

Random hexamer priming bias was analyzed by comparing the amplified products of a synthetic library to the unamplified synthetic library. Three commercially available DNA polymerases were utilized to amplify the library, and two of the DNA polymerases (both Taq) demonstrated remarkably similar amplification profiles, especially when compared to the third DNA polymerase of a different family. Preferentially amplified sequence motifs at the 3' end of the primer were identified. These motifs demonstrated a marked GC-rich bias pattern. The identified bias patterns were used to guide primer design. Current primer design methods assume that all sequences can be used equally as priming sites, and melting temperature (T_m) and GC measurements are the most important predictors of priming efficiency. However, our study demonstrates that the template sequence ahead of the primer:template junction and the 3' end of the primer can have an effect on the amplification efficiency and should be considered in addition to T_m and GC measurements. In the future, we plan to extend the analysis to different commercially available DNA polymerases, particularly family B type DNA polymerases. In addition, it would be interesting to examine the effect of common PCR additives on the amplification bias profile and identify additives that may reduce bias. Use of the PPI software is available through the iC-Architect site at http://ic-architect.com/. After logging in, users can submit a template sequence. A downloadable csv file is generated, which provides the PPI profile for the template, and optimal primer positions are suggested. These results can be utilized together with T_m and GC calculations to guide the placement of a primer to preferential locations on the template.

Additional files

Additional file 1: Table S1. Barcodes and read coverage are presented. The barcode sequence, the number of total associated reads, the average number of reads per unique sequence in the 4 bp, 6 bp, 8 bp, and 10 bp windows, and the DNA polymerase used are provided for each of the pooled amplification experiments. During pooling, more of the SL and amplified background were included in the pool (as evidenced by the read distribution) to ensure that both of these libraries had an ample number of reads for downstream analysis.

Additional file 2: Table S2. Statistical analysis of compared data sets. A statistical comparison of several data sets is provided with Pearson R, R^2, p-value, and n.

Additional file 3: Figure S1. Design for generating single-primer templates and their amplification. **(a)** The forward primer consists of a filler sequence, an 8 bp primer testing window and 18 nucleotides specific for a portion of the sense strand of the human IgG C-kappa domain. The reverse primer includes the same filler and 8 bp priming testing site, but includes 18 bp specific for the antisense strand of the kappa domain. After amplification, both the sense and antisense strand contain the same 8 bp sequence for single-primer testing. **(b)** After the templates are generated, gel purified, and the concentration normalized over all samples, a single primer, which serves as both forward and reverse, consisting of 19 bp of the filler region and 6 bp of the 8 bp priming site, is used in the amplification experiment.

Additional file 4: Supplementary Methods.

Additional file 5: Figure S2. An example demonstrating the PPI algorithm. First, the 8 bp window is divided into four sections: 1 dimeric scale (DiSc) and three trimeric scales (TrSc). The PPI value for each of the scales is based on the nucleotide sequence of the dimer or trimer and its relative position in the 8 bp window. The product of the DiSc and three TrScs is calculated and assigned to the 6th position of the 8 bp window. The entire 8 bp window is then slid 1 bp in the 3' direction, and the process is repeated until the end of the template is reached. A PPI profile of both the sense and antisense strand can be generated in order to guide the placement of the 3' end of the primer into a favorable position.

Additional file 6: Table S3. The importance of the 3′ end of the primer sequence on amplification bias. A detailed demonstration of sequences that have the same GC content but very different amplification outcomes when amplified with QTT-A DNA polymerase.

Abbreviations

PCR: Polymerase chain reaction; bp: base pair; bps: Base pairs; SL: Synthetic library; OBV: Observed bias value; QTT-A: Qiagen TopTaq Family A; PGT-A: Promega GoTaq Family A; QHH-B: Qiagen HotStar HighFidelity Family B; PPI: Polymerase preference index; IC: Internal control; NGS: Next generation sequencing; T_m: Melting temperature.

Competing interests

Funding for project materials was provided by iCubate Inc. The authors declare financial competing interests for several of the authors listed. Jian Han is the chief scientific officer (CSO) and a shareholder for iCubate, Inc., while Stanley Lu and Scott Clemmons perform contract work for iCubate Inc. iCubate Inc. uses the iC-Architect software to design multiplex PCR panels for its open-platform diagnostic panels.

Authors' contributions

WP devised, executed all described experiments, performed data analysis, and contributed significantly to the writing of the manuscript. MBS aided with experimental design, data analysis and provided significant contribution to the writing of the manuscript. CW helped with data analysis and manuscript preparation. SL and SC developed the PPI software. RZ aided with experimental design and editing the manuscript. JH supervised the experimental design and execution and contributed to the writing of the manuscript. All authors read and approved the final manuscript.

Acknowledgements

We would like to thank Dr. Preti Jain for her initial input into this project and Jessica Alleyne and Dr. Chris Gunter for reading the original manuscript and providing insightful suggestions for improvement. We would also like to thank Dr. Ritalinda Lee for her help with the statistical analysis and manuscript editing.

Author details

[1]Biotechnology Science and Engineering Program, University of Alabama in Huntsville, Huntsville, AL 35899, USA. [2]HudsonAlpha Institute for Biotechnology, Huntsville, AL 35806, USA. [3]Stanford Genome Technology Center, Stanford University, Palo Alto, CA 94304, USA. [4]Diatherix Laboratories, Huntsville, AL 35806, USA.

References

1. Benjamini Y, Speed TP: **Summarizing and correcting the GC content bias in high-throughput sequencing.** *Nucleic Acids Res* 2012, **40**(10):e72.
2. Kanagawa T: **Bias and artifacts in multitemplate polymerase chain reactions (PCR).** *J Biosci Bioeng* 2003, **96**(4):317–323.
3. Dohm JC, Lottaz C, Borodina T, Himmelbauer H: **Substantial biases in ultra-short read data sets from high-throughput DNA sequencing.** *Nucleic Acids Res* 2008, **36**(16):e105.
4. Aird D, Ross MG, Chen WS, Danielsson M, Fennell T, Russ C, Jaffe DB, Nusbaum C, Gnirke A: **Analyzing and minimizing PCR amplification bias in Illumina sequencing libraries.** *Genome Biol* 2011, **12**(2):R18.
5. Dabney J, Meyer M: **Length and GC-biases during sequencing library amplification: a comparison of various polymerase-buffer systems with ancient and modern DNA sequencing libraries.** *Biotechniques* 2012, **52**(2):87–94.
6. Hansen KD, Brenner SE, Dudoit S: **Biases in Illumina transcriptome sequencing caused by random hexamer priming.** *Nucleic Acids Res* 2010, **38**(12):e131.
7. Johnson SJ, Taylor JS, Beese LS: **Processive DNA synthesis observed in a polymerase crystal suggests a mechanism for the prevention of frameshift mutations.** *Proc Natl Acad Sci USA* 2003, **100**(7):3895–3900.
8. Kiefer JR, Mao C, Braman JC, Beese LS: **Visualizing DNA replication in a catalytically active Bacillus DNA polymerase crystal.** *Nature* 1998, **391**(6664):304–307.
9. Kiefer JR, Mao C, Hansen CJ, Basehore SL, Hogrefe HH, Braman JC, Beese LS: **Crystal structure of a thermostable Bacillus DNA polymerase I large fragment at 2.1 A resolution.** *Structure* 1997, **5**(1):95–108.
10. Lu XJ, Shakked Z, Olson WK: **A-form conformational motifs in ligand-bound DNA structures.** *J Mol Biol* 2000, **300**(4):819–840.
11. Olson WK, Gorin AA, Lu XJ, Hock LM, Zhurkin VB: **DNA sequence-dependent deformability deduced from protein-DNA crystal complexes.** *Proc Natl Acad Sci USA* 1998, **95**(19):11163–11168.
12. Ivanov VI, Minchenkova LE, Chernov BK, McPhie P, Ryu S, Garges S, Barber AM, Zhurkin VB, Adhya S: **CRP-DNA complexes: inducing the A-like form in the binding sites with an extended central spacer.** *J Mol Biol* 1995, **245**(3):228–240.
13. Cheetham GM, Steitz TA: **Structure of a transcribing T7 RNA polymerase initiation complex.** *Science* 1999, **286**(5448):2305–2309.
14. Ding J, Das K, Hsiou Y, Sarafianos SG, Clark AD Jr, Jacobo-Molina A, Tantillo C, Hughes SH, Arnold E: **Structure and functional implications of the polymerase active site region in a complex of HIV-1 RT with a double-stranded DNA template-primer and an antibody Fab fragment at 2.8 A resolution.** *J Mol Biol* 1998, **284**(4):1095–1111.
15. Timsit Y: **DNA structure and polymerase fidelity.** *J Mol Biol* 1999, **293**(4):835–853.
16. Tolstorukov MY, Ivanov VI, Malenkov GG, Jernigan RL, Zhurkin VB: **Sequence-dependent B < −− > A transition in DNA evaluated with dimeric and trimeric scales.** *Biophys J* 2001, **81**(6):3409–3421.
17. Firbank SJ, Wardle J, Heslop P, Lewis RJ, Connolly BA: **Uracil recognition in archaeal DNA polymerases captured by X-ray crystallography.** *J Mol Biol* 2008, **381**(3):529–539.
18. Fogg MJ, Pearl LH, Connolly BA: **Structural basis for uracil recognition by archaeal family B DNA polymerases.** *Nat Struct Biol* 2002, **9**(12):922–927.

Development of genetic tools for *Myceliophthora thermophila*

Jing Xu[1,2†], Jingen Li[2†], Liangcai Lin[2], Qian Liu[2], Wenliang Sun[2], Bangquan Huang[1] and Chaoguang Tian[2*]

Abstract

Background: The thermophilic filamentous fungus *Myceliophthora thermophila* has many suitable characteristics for industrial biotechnology and could be a promising new chassis system for synthetic biology, particularly the ATCC 42464 strain, whose genome was sequenced in 2011. However, metabolic engineering of this strain using genetic approaches has not been reported owing to a lack of genetic tools for this organism.

Results: In the present study, we developed a high efficiency *Agrobacterium tumefaciens* mediated transformation system for *M. thermophila*, including an approach for targeted gene deletion using green fluorescence protein (GFP) as a marker for selection. Up to 145 transformants per 10^5 conidia were obtained in one transformation plate. Moreover, a *ku70* deletion mutant was constructed in the ATCC 42464 background using the tools developed in present study and subsequently characterized. The *ku70* deletion construct was designed using resistance to phosphinothricin as the selection marker. Additionally, a GFP-encoding cassette was incorporated that allowed for the selection of site-specific (no fluorescence) or ectopic (fluorescence) integration of the *ku70* construct. Transformants with ectopically integrated *ku70* deletion constructs were therefore identified using the fluorescent signal of GFP. PCR and Southern blotting analyses of non-fluorescent putative *ku70* deletion transformants revealed all 11 tested transformants to be correct deletions. The deletion frequency in a pool of 116 transformants analyzed was 58 %. Moreover, the homologous rate improved about 3 folds under *ku70* mutant using the *pyrG* as a test gene to disrupt in *M. thermophila*.

Conclusions: We successfully developed an efficient transformation and target gene disruption approach for *M. thermophila* ATCC 42464 mediated by *A. tumefaciens*. The tools and the *ku70* deletion strain developed here should advance the development of *M. thermophila* as an industrial host through metabolic engineering and accelerate the elucidation of the mechanism of rapid cellulose degradation in this thermophilic fungus.

Keywords: *Myceliophthora thermophila*, *Agrobacterium tumefaciens*, genetic transformation, *ku70*

Background

The thermophilic filamentous fungus *Myceliophthora thermophila* exhibits a large capacity for biomass degradation and represents a potential reservoir of novel enzymes for industrial applications, including abundant thermostable enzymes for biomass degradation [1–5]. *M. thermophila* also has the potential to be a cell factory to produce chemicals and biofuels from renewable ligno-cellulose [5]. Therefore, metabolic engineering of *M. thermophila* is attractive. To use *M. thermophila* in industrial biotechnology, the development of versatile genetic tools is urgently needed.

PEG-mediated protoplast transformation was developed to knock out target genes and for hetero-expression of interesting genes in several filamentous fungi, such as *Aspergillus* and *Trichoderma* [6–8]. It was previously reported this approach can be used to transform the industrial strain C1, previously known as *Chrysosporium lucknowense*, and recently renamed as an isolate of *Myceliophthora thermophila* [9]. However, preparation of fungal protoplasts was laborious, and heterokaryotic transformants might have genetic stability concerns [10, 11]. The C1 strain is a patented strain, therefore the ATCC 42464 strain is predominantly used for public research as the general wild-type strain of *M. thermophila*.

* Correspondence: tian_cg@tib.cas.cn
†Equal contributors
[2]Key Laboratory of Systems Microbial Biotechnology, Tianjin Institute of Industrial Biotechnology, Chinese Academy of Sciences, Tianjin 300308, China
Full list of author information is available at the end of the article

The whole genome sequence of ATCC 42464 has been completed and is open to the public [12]. Various studies have been performed with *M. thermophila* ATCC 42464, such as biochemical characterization of enzymes [13–15] and investigating cellulose degradation on a cellular and genomic level [4, 12, 16, 17]. However, molecular engineering using a genetic approach has not been reported in *M. thermophila* ATCC 42464 because of a lack of mature genetic tools. *Agrobacterium tumefaciens*-mediated transformation has become a common technique for filamentous fungi [18–22]. This technique allows transfer of a segment from its Ti plasmid into plant or fungi cells, and the T-DNA integrates into the nuclear chromosome either randomly or specifically. The direct transformation of conidia makes *A. tumefaciens* T-DNA transfer a powerful and easy to use tool for genetic transformation of filamentous fungi. A variety of fungi have been successfully transformed by this approach [18–22]. However, the application of this approach in thermophilic filamentous fungi, including *M. thermophila*, has not been tested so far.

In present study, we developed a high efficiency transformation and targeted gene disruption system mediated by *A. tumefaciens* for *M. thermophila* ATCC 42464. Using this technique, up to 145 transformants were obtained per transformation plate, and a *ku70* deletion strain was constructed in the ATCC 42464 background and characterized. All of the tools and strains developed here should facilitate the development of *M. thermophila* metabolic engineering and accelerate the elucidation of the fascinating cellulose degradation mechanism in this thermophilic fungus.

Results and discussion

Marker selection and genetic transformation in the ATCC 42464 strain

To perform *M. thermophila* transformation mediated by *A. tumefaciens*, the sensitivity of this fungus to hygromycin B (Amresco, Solon, OH, USA) and phosphinothricin (Sigma-Aldrich, St. Louis, MO, USA) was tested. Five concentrations of two antibiotics were used. The mature conidia were harvested in 0.05 % Tween 80, and 10^3 conidia were spread on MM (Vogel's minimal medium supplemented with 2 % sucrose) agar plates supplemented with various concentrations of antibiotic (Hygromycin B: 25, 50, 75, 100, 150 μg/mL; phosphinothricin: 25, 50, 100, 150, 200 μg/mL). After incubation at 45 °C for 5 days, Hygromycin B showed little inhibition for *M. thermophila* at more than 100 ug/mL, whereas phosphinothricin concentration completely inhibited the growth of this fungus at 100 ug/mL. Thus, the *bar* gene conferring fungal resistance to phosphinothricin was employed as the selection marker for *M. thermophila* transformation.

The Ti vector pPK2BarGFPD containing the *trpC* promoter from *Aspergillus nidulans* and the *tef* promoter from *Aureobasidium pullulans* (PtrpC and Ptef, respectively) developed in a previous study [23] was chosen as the test plasmid for genetic transformation in *M. thermophila* ATCC 42464. The vector contains a GFP reporter gene as well as a *bar* selection marker gene, which confers resistance to phosphinothricin. PCR analysis of the transformants showed the T-DNA was integrated in the chromosomes. High levels of green fluorescent protein (GFP) signal were clearly detected in the conidia and mycelia of transformants (Fig. 1), whereas no autofluorescent signal was observed in the parental strain. This suggested pPK2BarGFPD and its elements, the promoters and the GFP gene, were functioning and could be used for gene over-expression and protein localization analysis in *M. thermophila* ATCC 42464.

Optimization of *M. thermophila* transformation mediated by *A. tumefaciens*

In an attempt to develop a simple, high efficiency transformation system mediated by *A. tumefaciens* in *M. thermophila* ATCC 42464, we performed transformation optimization as described previously [21]. Acetosyringone (AS) serves as an inducer of *virulence (vir)* genes, whose expression is a prerequisite for *A. tumefaciens* T-DNA transfer. AS concentration influences the frequency of *A. tumefaciens*-mediated transformation for *Saccharomyces cerevisiae* and several filamentous fungi [11, 20, 24, 25]. Therefore, the impact of AS on this approach for *M. thermophila* was first investigated. As shown in Fig. 2a, using the binary plasmid pPK2BarGFPD, the maximum transformation efficiency (145 ± 10 transformants per plate) was obtained with AS at a concentration of 200 μM, and the number of transformants was reduced when the concentration was increased or decreased. The observation that the higher concentration of AS resulted in a reduction in transformation efficiency is similar to that for *Fusarium avenaceum*. However, it is in contrast to that for *Beauveria bassiana*, where an increase in the concentration from 100 μM to 800 μM resulted in increasing numbers of transformants [21, 26]. Based on these observation, the different fungus seems has different tolerance to Acetosyringone (AS), the low frequency of transformation induced by the increased AS might be attributed to the toxic effect of AS on *M. thermophila*, and the deficient level of *vir* gene expression at low AS concentrations might be responsible for poor efficiency of *A. tumefaciens* T-DNA transfer.

The transformation efficiency is also dependent on co-culture time of *M. thermophila* and *A. tumefaciens*. The largest number of transformants was observed with a

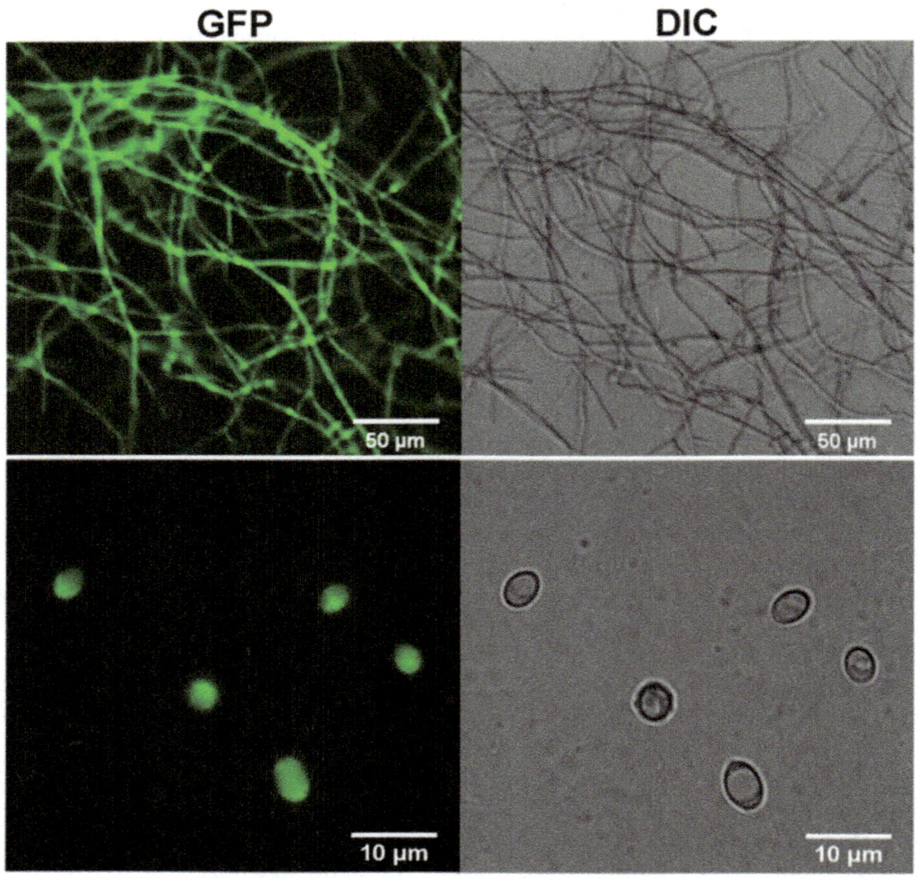

Fig. 1 Confocal fluorescence imaging of mycelia (Top) and conidia (Bottom) of *M. thermophila* pPK2BarGFPD transformants

co-culture time of 2 days, and alterative co-culture times impaired the transformation efficiency (Fig. 2b). Similar results were obtained in *Metarhizium anisopliae*, *Aspergillus terreus* and *Beauveria bassiana* transformation mediated by *A. tumefaciens* [2, 18, 27], whereas the co-culture time had no influence on a *Ganoderma lucidum* transformation system [28]. In conclusion, with the optimized approach, a high transformation efficiency of 145 transformants per 10^5 conidia was obtained. This was higher compared with the amount of transformants obtained in fungi using other approaches, typically less than 100 transformants per 10^5 conidia [18, 29–31].

Mitotic stability analysis of transformants

The mitotic stability of the T-DNA in *M. thermophila* transformants was examined next. Twenty randomly selected transformants containing pPK2BarGFPD were cultured on media without phosphinothricin for five generations. After five generations, 19 transformants (95 %) of 20 tested strains retained the resistance to phosphinothricin. Satisfyingly, PCR analysis showed the presence of *bar* in the 19 transformants, and GFP signal was observed under the fluorescence microscope.

ku70 disruption of *M. thermophila* ATCC 42464

The high transformation frequency, together with the precision and simplicity of T-DNA, make *A. tumefaciens*-mediated transformation a suitable genome mutagenesis approach in filamentous fungi, such as *Trichoderma*, *Aspergillus*, *Beauveria* and *Metarhizium* [22, 27, 32, 33]. Ku70 and Ku80 make up the Ku heterodimer, which binds to the ends of double-stranded DNA breaks and is required for the non-homologous end joining (NHEJ) pathway of DNA repair [34]. In some fungi, including the *M. thermophila* C1 strain, *Trichoderma reesei* and *Neurospora crassa*, mutation of *ku70* results in a dramatically increased homologous integration frequency with short homologous flanks [9, 35, 36]. Therefore, disruption of *ku70* of the ATCC 42464 strain was performed to test the efficiency of this approach for knocking out of a specific gene in wild-type *M. thermophila*.

The binary vector pPK2BarGFPD was employed for DNA manipulation and the *ku70* knockout cassette

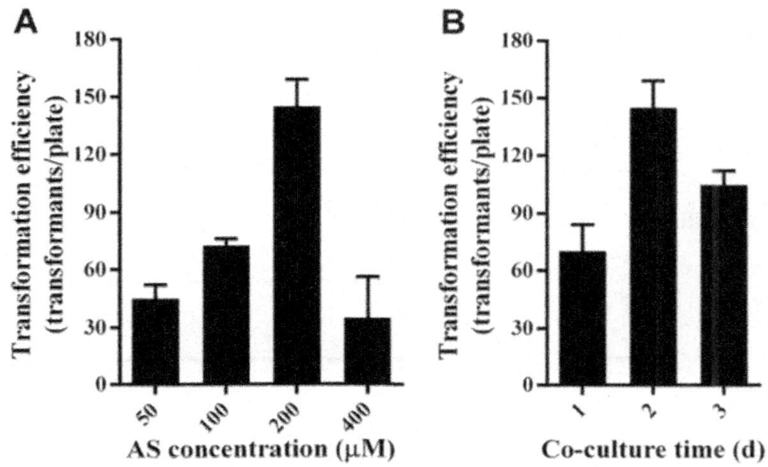

Fig. 2 Optimization of *A. tumefaciens*-mediated *M. thermophila* transformation. **a** The impact of acetosyringone (AS) on *A. tumefaciens*-mediated transformation efficiency. Co-cultivation was conducted in induction medium containing AS on indicated concentrations for 2 days. **b** The effect of co-culture time on transformation efficiency. Co-cultivation was conducted in induction medium containing 200 μM AS for various times indicated. Error bars indicate the standard deviation of three independent experiments

(Fig. 3) was constructed as described in the Methods section. The gene encoding green fluorescence protein on the binary vector served as the marker to differentiate between ectopic insertion and correct gene replacement. Fluorescence should not be detected in the transformants where the target gene (*ku70*) was properly replaced by the *bar* cassette (Fig. 4a). Conversely, if the fluorescent signal is detected in a transformant, this suggests the *bar-gfp* segment of the T-DNA was ectopically integrated elsewhere in the genome (random insertion) rather than at the correct gene locus. Using this strategy, we performed the *ku70* disruption experiment. One hundred sixteen primary transformants were obtained. Forty-six (39.7 %) transformants were filtered out by an obvious fluorescent signal, indicating they were ectopic insertion mutants. The remaining 70 transformants (60.3 %) with no detectable fluorescent signal under the microscope were likely free of ectopic insertions and passed the first round screening. Subsequent PCR analysis showed the *ku70* gene was properly replaced by the *bar* cassette in 68 of these 70 transformants. To further confirm this observation, we randomly selected eleven PCR-confirmed transformants to perform Southern blotting analysis, and the result showed all 11 tested mutants were correct (Fig. 4b and c). Taken together, these results indicate as many as 58 % (68 of 116) of the transformants could be correct *ku70* disruption mutants, indicating a very high rate of homologous recombination. Previous studies reported that the efficiencies of genetic targeted disruption via transformation mediated by *A. tumefaciens* varied greatly in fungal species, such as 0.04 % for *Blastomyces dermatitidis*, 29 % for *Aspergillus awamori*, 74 % for *Fusarium avenaceum* and 85 % for *Fusarium graminearum* [37, 38]. Compared with transformation by direct electroporation transformation of protoplasts or conidia, the transformation mediated by *A. tumefaciens* seems more likely to achieve a higher successful gene disruption rate, at least for some of the species. For example, in *Neurospora crassa* the homologous recombination rate was usually low (<10 %) without NHEJ disruption using electroporation [35]. Although future studies with more precise statistical analyses are

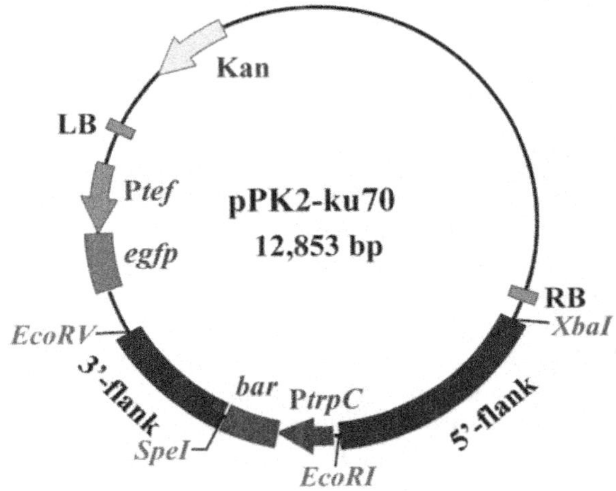

Fig. 3 Vector map of the pPK2-ku70 vector constructed based on binary vector pPK2BarGFPD. Kan, kanamycin resistance gene; *Ptef*, promoter of translation elongation factor gene from *Aureobasidium pullulans*; *egfp*, enhanced green fluorescence protein; *PtrpC*, promoter of tryptophan synthetase gene from *Aspergillus nidulans*; *bar*, phosphinothricin resistance gene; 3'-flank and 5'-flank, 3' and 5' flanking fragments of *ku70*, respectively; RB and LB, right and left border of T-DNA, respectively

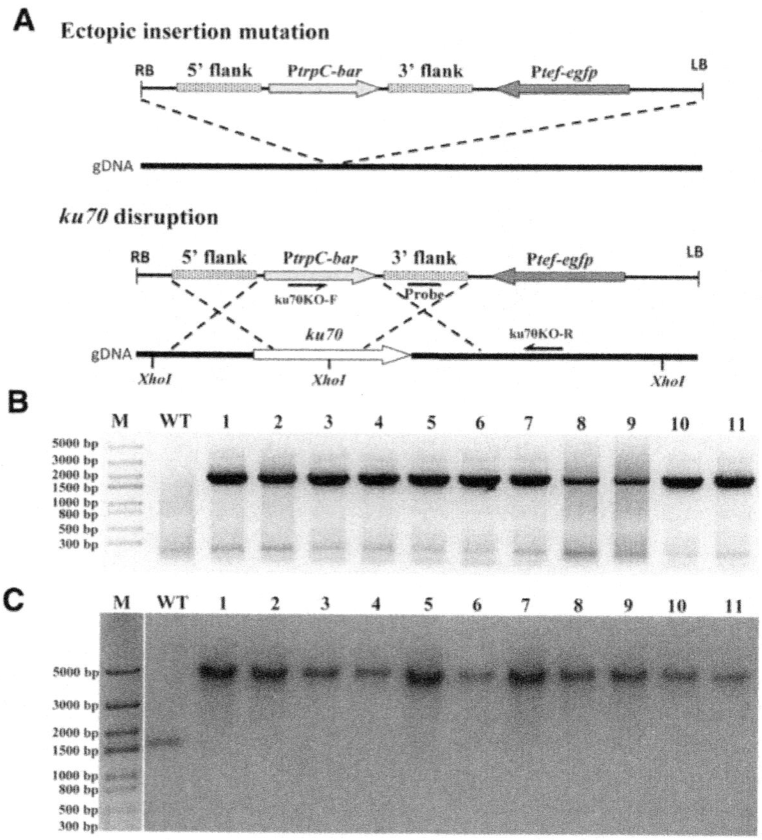

Fig. 4 *ku70* disruption of wild-type *M. thermophila* ATCC 42464. **a** Pattern of the *ku70* knockout cassette integrating into the chromosomes of *M. thermophila* ATCC 42464 via ectopic insertion or homologous recombination. **b** PCR analysis of *ku70* deletion transformants with one primer (ku70KO-F) located in the *bar* gene cassette and the other (ku70KO-R) located in the downstream of 3′ flank in the genomic DNA (expected product of 1872 bp) and **c** Southern blotting analysis of *ku70* mutants with *Xho*I digested genomic DNA and the probe amplified from the 3′ flank. Lines 1–11 in (**b**) and (**c**), genomic DNA from *ku70* mutants (expected product of 5056 bp). WT, genomic DNA from wild-type as negative control (expected product of 1703 bp). Primers are represented by solid black arrows. Abbreviations: gDNA, genomic DNA; RB and LB, right and left border of T-DNA, respectively

needed to support the improved homologous recombination rate, if validated, it will be a boost for genetic manipulation in this strain, and will speed up strain engineering for industrial applications.

There was no noticeable difference between *ku70* disruption mutants and the wild-type strain during our analysis (Fig. 5). After five generations of sub-culturing on MM agar medium, 10 randomly selected *ku70* mutants were shown to be mitotically stable. Taken together, these results indicate the approach we developed is an efficient system for gene disruption in *M. thermophila* ATCC 42464.

In order to check whether the gene homologous recombination rate was improved in *ku70* mutant compared to that in wild type strain, we chose the potential marker gene *pyrG* (MYCTH_2311494) as a test to disrupt in both wild-type and *ku70* deletion strain. As expected, 97 % (29/30) of transformants were correct *pyrG* deletion mutants under *ku70* background, whereas only 30 % (12/40) of transformants were correct in wild type background (Fig. 6). This result clearly shown the deletion of KU70 gene can improve the gene homologous recombination in this strain, similar as reported before [9].

Taken together, these results indicate the approach we developed is an efficient system for gene disruption in *M. thermophila* ATCC 42464 and will speed up its strain engineering for industrial applications.

Conclusions

We successfully developed a simple and highly efficient transformation approach mediated by *A. tumefaciens* for the thermophilic filamentous fungus *M. thermophila*. This study established a targeted gene deletion approach using green fluorescence protein as a selection helper to filter the ectopic insertion. The tools and *ku70* mutant

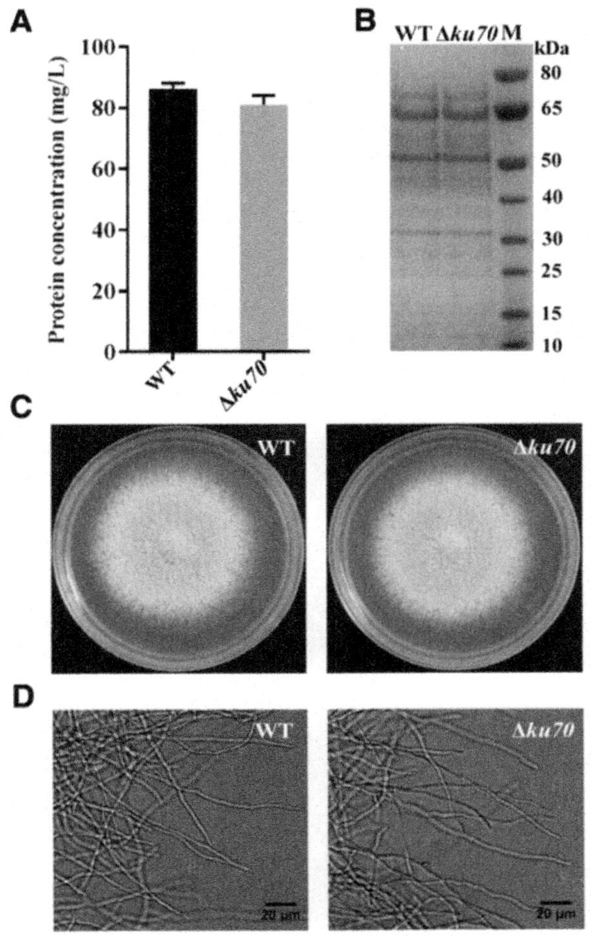

Fig. 5 Phenotypic characterization of *ku70* mutant. **a** Protein concentration in supernatants from wild-type (WT) and ku70 mutant cultures grown on 1 % Avicel for 3 d at 45 °C. **b** SDS-PAGE gel of wild-type and Δ*ku70* strains grown at 45 °C for 3 d on 1 % Avicel. **c** Conidiation of wild-type and Δ*ku70* strains growth at 45 °C for 9 d on agar plate without selective agent. **d** The hyphae of wild-type and Δ*ku70* strains growth at 45 °C for 24 h on liquid MM without selective agent

developed by the present study should facilitate *M. thermophila* strain development for industrial applications in the future.

Methods

Strains and culture conditions

Myceliophthora thermophila ATCC 42464 obtained from the American Type Culture Collection (ATCC) was grown on Vogel's minimal medium supplemented with 2 % sucrose (MM medium) at 45 °C for 15 d to obtain conidia. Sucrose was purchased from Amresco (Solon, OH, USA).

Escherichia coli DH5α was used for plasmid amplification and vector construction and cultivated in Luria–Bertani medium with antibiotics according to conditions for plasmid selection.

Agrobacterium tumefaciens AGL-1 [23] used for *M. thermophila* transformation, was grown in Luria–Bertani medium at 28 °C and kanamycin was added when needed to maintain binary plasmids [39].

Preparation and transformation of *M. thermophila* ATCC 42464 mediated by *Agrobacterium*

M. thermophila transformation mediated by *A. tumefaciens* was performed using a modification of the method described previously for *Fusarium circinatum* [21]. A single colony of *A. tumefaciens* harboring binary vector was inoculated into LB medium supplemented with 50 μg/mL kanamycin and cultivated with shaking of 250 rpm at 28 °C for 12–16 h to an optical density of 0.5–1.0 at 600 nm. Subsequently, cells were harvested, washed twice with induction medium (IM) containing AS and then diluted to an OD600 of 0.15. The culture was grown at 28 °C with shaking at 250 rpm for 6–8 h to an OD600 of 0.5–0.8. Conidia were harvested into 0.05 % Tween 80, filtered through glass wool to remove residual mycelia and adjusted to 10^6 conidia per milliliter calculated using a hemocytometer. An aliquot of 100 μL *A. tumefaciens* culture was mixed with an equal volume of *M. thermophila* conidia and spread on an IM agar plate (9.0 cm) containing AS and was covered with cellophane. After co-cultivation at 28 °C, the cellophane was transferred onto MM medium plate supplemented with 100 μg/mL phosphinothricin (Sigma-Aldrich, St. Louis, USA) and 300 μg/mL cefotaxime (Sigma-Aldrich), overlaid with Vogel's salt agar medium containing 2 % sucrose and 100 μg/mL phosphinothricin and then cultivated at 45 °C. Putative transformants were visible after approximately 3 days.

Mitotic stability analysis

Enhanced GFP (EGFP) was used as the reporter gene to elucidate mitotic stability of transformants. Twenty randomly selected successful transformants with pPK2BarGFPD were consecutively sub-cultured on agar plates without the selective agent. When the agar plates were full of mycelia and conidia, conidia were used to inoculate the next generation culture. After five generations, the GFP-signal was detected under fluorescence microscopy and PCR analysis targeting the *bar* gene with the specific primer sets, bar-F/bar-R (Table 1), was performed. Triplicate reactions were performed for each transformant.

Binary vector construction for *ku70* disruption

The master Ti vector pPK2BarGFPD [23] containing the phosphinothricin resistant *bar* marker, was employed to test transformation efficiency of *M. thermophila*

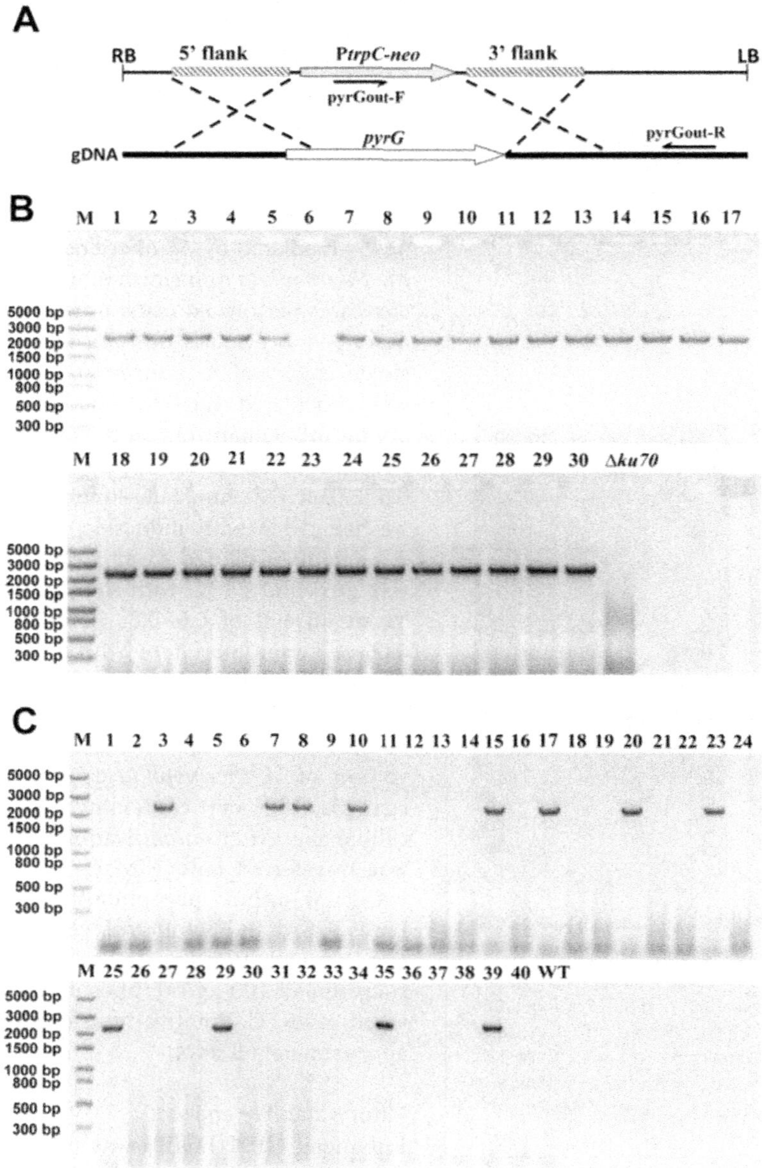

Fig. 6 pyrG gene deletion in wild type and Δku70 mutant of *M. thermophila* ATCC 42464. The experiment design (**a**), PCR analysis of transformants under wild-type background (**b**) and Δku70 mutant (**c**) with one primer (pyrGout-F) located in the neo gene cassette and the other (pyrGout-R) located in the downstream of 3′ flank in the genomic DNA. The genomic DNA from the wild-type (WT) and Δku70 strains was used as the negative control

mediated by *A. tumefaciens*. To disrupt the *ku70* gene involved in the nonhomologous end-joining (NHEJ) pathway, the plasmid pPK2-ku70 was constructed (Fig. 3). The 5′ and 3′ flanking fragments (2095 bp and 1246 bp) were amplified from *M. thermophila* genomic DNA with paired primers (Table 1) designed using the annotated genome of *M. thermophila* ATCC 42464 [12]. The two flanks were inserted between the *Xba*I/*Eco*RI and *Spe*I/*Eco*RV sites of pPK2BarGFPD, respectively, to result in pPK2-ku70. The inserted fragments were sequenced to confirm the authenticity of the modified plasmids. All restriction enzymes were obtained from Thermo Scientific (Waltham, MA, USA).

Identification of transformants by PCR and Southern blotting

Transformants harboring pPK2BarGFPD were verified by PCR analysis with paired primers, bar-F/bar-R (Table 1), to detect *bar* after fluorescence detection under the microscope.

The *ku70* deletion mutant transformants with ectopic plasmid insertions were recognized by checking

Table 1 Primers used for the manipulation of *ku70* and the identification of its mutants

Names	Primer sequences (5'-3')	Notes
bar-F	CTGGAGCTAGTGGAGGTCAAC	PCR detection of *bar*
bar-R	TTCAATCTTAAGAAACTTTATTG	PCR detection of *bar*
ku70-5 F	GCTCTAGAGCGCTACCACGGACGGATGATAC	Cloning *ku70* 5'flank
ku70-5R	CCGGAATTCGGCTTTGAAGGTGCAGGTGCGAC	Cloning *ku70* 5'flank
ku70-3 F	GGACTAGTGTTGGGATCGCATGGTTCGTTG	Cloning *ku70* 3'flank
ku70-3R	CCGATATCGGCTTCAGAATGCAGAGGTCAGAG	Cloning *ku70* 3'flank
Probe-F	GCGACGATCCCGAAATAC	Southern blotting of *ku70*
Probe-R	CCCTGACGAAGTAGCATCAT	Southern blotting of *ku70*
ku70KO-F	CTACACCCACCTGCTGAAGTCC	PCR detection of *ku70* mutation
ku70KO-R	GTCCCGCACTACCGTTGATGG	PCR detection of *ku70* mutation
ku70-F	CGTGGGAAGGTGACGATGATAGAC	PCR detection of *ku70* ORF
ku70-R	CATAATGCTCCTCGCTCGGGAAG	PCR detection of *ku70* ORF
PtrpC-F	AAAAAGCTTGGTACCGAGCTCCGACGTTAACTGATATTGAAGGAGC	Cloning *PtrpC*
PtrpC-R	CGTGCAATCCATCTTGTTCAATCATTTGGATGCTTGGGTAGAATAGGTAA	Cloning *PtrpC*
neo-F	TTACCTATTCTACCCAAGCATCCAAATGATTGAACAAGATGGATTGCACG	Cloning *neo*
neo-R	AAATTAATTAAGTTTAAACCTCGAGTCTAGAAGATCTTCAGAAGAACTCGTCAAGAAGGCGA	Cloning *neo*
pyrG-5 F	AAAGAATTCCGATGCAGATGCAACTCCGCTCCCT	Cloning *pyrG* 5'flank
pyrG-5R	AAAGGATCCAGGTTCGAGGCCTTGAGGTCCATGA	Cloning *pyrG* 5'flank
pyrG-3 F	CCGCTCGAGATCTGGCGCGTCTGGCGTGATTTGG	Cloning *pyrG* 3'flank
pyrG-3R	GGAAGATCTCGGGGTGAGTGTTGGGGTGTTGTTT	Cloning *pyrG* 3'flank
pyrGout-F	GGGCTGACCGCTTCCTCGTGCTTTA	PCR detection of *pyrG*
pyrGout-R	TGGCGTAGTGCGTGTTGAACTCGGC	PCR detection of *pyrG*

Underlined regions indicate restriction enzyme sites

fluorescence under the microscope. The green fluorescent signal suggested it was an ectopic insertion mutant rather than homologous recombination replacement. Proper gene replacement was confirmed by amplifying specific fragments with one primer located in the *bar* gene and the other located in the downstream of 3′ flank in the genomic DNA (Table 1). Mutants with correct placement of the *bar* gene underwent further homokaryon analysis. The knockout mutants containing no wild-type open reading fragment of *ku70* were determined by PCR analysis using paired primers, ku70-F/ku70-R (Table 1).

To verify the authenticity of *ku70* mutants, Southern blotting was performed with 20 μg genomic DNA digested by *Xho*I. Genomic DNA was extracted as described previously [40]. The digested DNA was separated by agarose gel electrophoresis and DNA transfer was performed as described previously [41]. A 441 bp PCR-amplified product used as the probe was generated with paired primers, Probe-F/Probe-R (Table 1). Probe preparation, membrane hybridization and visualization were performed according to the manufacturer's instructions (DIG High Prime DNA Labeling and Detection Starter Kit II, Roche, Mannheim, Germany).

pyrG disription in *ku70* mutant

The pyrG gene (MYCTH_2311494) encoding orotidine-5'-phosphate decarboxylase was disrupted through homologous replacement in the wild-type strain (WT) and *ku70* mutant. The P*trpC* and neomycin resistant elements were amplified from pCB1003 and pEGFP-N2 using paired primers PtrpC-F/R and neo-F/R, respectively, and fused using the primers PtrpC-F/neo-R with several restriction sites introduced to both ends. The PtrpC-neo cassette was cloned into the *Hind*III-*Bgl*II sites of pCAMBIA-0380 (CAMBIA, Canberra, Australia; p0380 herein) to generate p0380-neo. The 5′ and 3′ fragments of *pyrG* (2071 and 1972 bp) were separately amplified from *M. thernophila* genomic DNA via PCR with paired primers (Table 1). The amplified fragments were inserted into *Eco*RI/*Bam*HI and *Xho*I/*Bgl*II sites in p0380-neo plasmid, forming the disruption plasmid p0380-pyrGup-neo-pyrGdn. The constructed plasmids were transformed into WT and *ku70* deletion strains via *Agrobacterium*-mediated transformation. Colonies grown for 5 or 6 days on a selective medium at 45 °C were screened in terms of the neomycin resistance to G418 (40 μg mL/L), followed by identification via PCR with paired primers (Table 1).

Microscopy

Confocal microscopy was performed on a Leica TCS SP5 II microscope using a 100× 1.4 NA oil or 20× immersion objective, 30 % Laser (Argon, visible) Power, PMT Trans of 180 HV, Emission bandwidth of PMT was 500 nm to 600 nm. The images were processed and analyzed by ImageJ (version 1.47).

Phenotype characterization of the *ku70* mutant

Mature conidia from wild-type and *ku70* mutant strains were inoculated into 100 mL 1 × Vogel's salt supplemented with 1 % Avicel (PH101, Sigma-Aldrich, St. Louis, MO, USA) at 2.5×10^5 conidia per milliliter in a 250-mL Erlenmeyer flask. After growth for 16 h at 45 °C with shaking at 150 rpm under constant light, total secreted protein in the supernatant was determined using a Bio-Rad DC protein assay kit (Bio-Rad, Hercules, CA, USA) and 25 μl of supernatant was used for SDS-PAGE analysis. The experiments were performed in biological triplicates.

Abbreviations

AS: Acetosyringone; EGFP: Enhanced green fluorescent protein; DIC: Differential interference contrast; gDNA: Genomic DNA; IM: Induction medium; NHEJ: Nonhomologous end-joining; PCR: Polymerase chain reaction; RB and LB: Right and left border of T-DNA, respectively.

Competing interests

The authors declare that they have no competing interests.

Authors' contributions

JX, JL and CT designed the project. JX, LL, QL and WS carried out the experiments. JL, BH and CT wrote the manuscript. All the authors have read the article and approved the final manuscript.

Acknowledgements

We thank all members of the Tian laboratory for critical comments and technical assistance, and we thank Lixian Wang from Technology Support Center of Tianjin Institute of Industrial Biotechnology of CAS for her help with the confocal microscopy. This work was supported by the 973 Program of China (2011CBA00800), the 863 Program of China (SS2014AA021300) and Tianjin Zhuanxiang (13ZCDZSY05000).

Author details

[1]College of Life Sciences, Hubei University, Wuhan 430062, China. [2]Key Laboratory of Systems Microbial Biotechnology, Tianjin Institute of Industrial Biotechnology, Chinese Academy of Sciences, Tianjin 300308, China.

References

1. Roy SK, Dey SK, Raha SK, Chakrabarty SL. Purification and properties of an extracellular endoglucanase from *Myceliophthora thermophila* D-14 (ATCC 48104). J Gen Microbiol. 1990;136(10):1967–71.
2. Kaur G, Kumar S, Satyanarayana T. Production, characterization and application of a thermostable polygalacturonase of a thermophilic mould *Sporotrichum thermophile* Apinis. Bioresour Technol. 2004;94(3):239–43.
3. Vafiadi C, Topakas E, Biely P, Christakopoulos P. Purification, characterization and mass spectrometric sequencing of a thermophilic glucuronoyl esterase from *Sporotrichum thermophile*. FEMS Microbiol Lett. 2009;296(2):178–84.
4. Karnaouri A, Topakas E, Antonopoulou I, Christakopoulos P. Genomic insights into the fungal lignocellulolytic system of *Myceliophthora thermophila*. Front Microbiol. 2014;5:281.
5. Singh B. *Myceliophthora thermophila* syn. *Sporotrichum thermophile*: a thermophilic mould of biotechnological potential. Crit Rev Biotechnol 2014; 15: 1-11.
6. Gruber F, Visser J, Kubicek CP, Degraaff LH. The development of a heterologous transformation system for the cellulolytic fungus *Trichoderma reesei* based on a pyrg-negative mutant strain. Curr Genet. 1990;18(1):71–6.
7. Punt PJ, van den Hondel CA. Transformation of filamentous fungi based on hygromycin B and phleomycin resistance markers. Methods Enzymol. 1992;216:447–57.
8. Ruiz-Diez B. Strategies for the transformation of filamentous fungi. J Appl Microbiol. 2002;92(2):189–95.
9. Visser H, Joosten V, Punt PJ, Gusakov AV, Olson PT, Joosten R, et al. Development of a mature fungal technology and production platform for industrial enzymes based on a *Myceliophthora thermophila* isolate, previously known as *Chrysosporium lucknowense* C1. Ind Biotechnol. 2011;7(3):10.
10. Fungaro MHP, Rech E, Muhlen GS, Vainstein MH, Pascon RC, Dequeiroz MV, et al. Transformation of *Aspergillus nidulans* by microprojectile bombardment on intact conidia. Fems Microbiol Lett. 1995;125(2-3):293–7.
11. de Groot MJA, Bundock P, Hooykaas PJJ, Beijersbergen AGM. *Agrobacterium tumefaciens*-mediated transformation of filamentous fungi. Nat Biotechnol. 1998;16(9):839–42.
12. Berka RM, Grigoriev IV, Otillar R, Salamov A, Grimwood J, Reid I, et al. Comparative genomic analysis of the thermophilic biomass-degrading fungi *Myceliophthora thermophila* and *Thielavia terrestris*. Nat Biotechnol. 2011;29(10):922–7.
13. Karnaouri AC, Topakas E, Christakopoulos P. Cloning, expression, and characterization of a thermostable GH7 endoglucanase from *Myceliophthora thermophila* capable of high-consistency enzymatic liquefaction. Appl Microbiol Biotechnol. 2014;98(1):231–42.
14. Volkov PV, Rozhkova AM, Gusakov AV, Zorov IN, Sinitsyn AP. Glucoamylases from *Penicillium verruculosum* and *Myceliophthora thermophila*: Analysis of differences in activity against polymeric substrates based on 3D model structures of the intact enzymes. Biochimie 2015, 110:45-51.
15. Topakas E, Moukouli M, Dimarogona M, Christakopoulos P. Expression, characterization and structural modelling of a feruloyl esterase from the thermophilic fungus *Myceliophthora thermophila*. Appl Microbiol Biotechnol. 2012;94(2):399–411.
16. Kolbusz MA, Di Falco M, Ishmael N, Marqueteau S, Moisan MC, Baptista Cda S, et al. Transcriptome and exoproteome analysis of utilization of plant-derived biomass by *Myceliophthora thermophila*. Fungal Genet Biol. 2014;72:10–20.
17. van den Brink J, van Muiswinkel GC, Theelen B, Hinz SW, de Vries RP. Efficient plant biomass degradation by thermophilic fungus *Myceliophthora heterothallica*. Appl Environ Microbiol. 2013;79(4):1316–24.
18. Fang WG, Pei Y, Bidochka MJ. Transformation of *Metarhizium anisopliae* mediated by *Agrobacterium tumefaciens*. Can J Microbiol. 2006;52(7):623–6.
19. Duarte RTD, Staats CC, Fungaro MHP, Schrank A, Vainsten MH, Furlaneto-Maia L, et al. Development of a simple and rapid *Agrobacterium tumefaciens*-mediated transformation system for the entomopathogenic fungus *Metarhizium anisopliae* var. acridum. Lett Appl Microbiol. 2007;44(3):248–54.
20. Gouka RJ, Gerk C, Hooykaas PJJ, Bundock P, Musters W, Verrips CT, et al. Transformation of *Aspergillus awamori* by *Agrobacterium tumefaciens*-mediated homologous recombination. Nat Biotechnol. 1999;17(6):598–601.
21. Covert SF, Kapoor P, Lee M-h, Briley A, Nairn CJ. *Agrobacterium tumefaciens*-mediated transformation of *Fusarium circinatum*. Mycol Res. 2001;105(3):259–64.
22. Lin L, Wang F, Wei D. Chlorimuron ethyl as a new selectable marker for disrupting genes in the insect-pathogenic fungus *Metarhizium robertsii*. J Microbiol Methods. 2011;87(2):241–3.
23. Lin L, Fang W, Liao X, Wang F, Wei D, St Leger RJ. The MrCYP52 cytochrome P450 monoxygenase gene of *Metarhizium robertsii* is important for utilizing insect epicuticular hydrocarbons. PLoS One. 2011;6(12):e28984.
24. Winans SC. Two-way chemical signaling in *Agrobacterium*-plant interactions. Microbiol Rev. 1992;56(1):12–31.
25. Bundock P, Dendulkras A, Beijersbergen A, Hooykaas PJJ. Transkingdom T-DNA transfer from *Agrobacterium tumefaciens* to *Saccharomyces cerevisiae*. Embo J. 1995;14(13):3206–14.
26. Fang W, Zhang Y, Yang X, Zheng X, Duan H, Li Y, et al. *Agrobacterium tumefaciens*-mediated transformation of *Beauveria bassiana* using an

herbicide resistance gene as a selection marker. J Invertebr Pathol. 2004;85(1):18–24.

27. Wang D, He D, Li G, Gao S, Lv H, Shan Q, et al. An efficient tool for random insertional mutagenesis: *Agrobacterium tumefaciens*-mediated transformation of the filamentous fungus *Aspergillus terreus*. J Microbiol Methods. 2014;98:114–8.
28. Shi L, Fang X, Li MJ, Mu DS, Ren A, Tan Q, et al. Development of a simple and efficient transformation system for the basidiomycetous medicinal fungus *Ganoderma lucidum*. World J Microbiol Biotechnol. 2012;28(1):283–91.
29. Li M, Zhou L, Liu M, Huang Y, Sun X, Lu F. Construction of an engineering strain producing high yields of alpha-transglucosidase via *Agrobacterium tumefaciens*-mediated transformation of *Asperillus niger*. Biosci Biotechnol Biochem. 2013;77(9):1860–6.
30. Covert SF, Kapoor P, Lee MH, Briley A, Nairn CJ Agrobacterium tumefaciens-mediated transformation of *Fusarium circinatum*. Mycol Res. 2001, 105:259-264
31. Zhang YJ, Zhao JJ, Xie M, Peng DL. Agrobacterium tumefaciens-mediated transformation in the entomopathogenic fungus Lecanicillium lecanii and development of benzimidazole fungicide resistant strains. J Microbiol Methods. 2014;105:168–73.
32. Liu Q, Ying SH, Li JG, Tian CG, Feng MG. Insight into the transcriptional regulation of Msn2 required for conidiation, multi-stress responses and virulence of two entomopathogenic fungi. Fungal Genet Biol. 2013;54:42–51.
33. Yao HZ, Xiao LW, Tian HW, Qiao J. *Agrobacterium*-mediated transformation (AMT) of *Trichoderma reesei* as an efficient tool for random insertional mutagenesis. Appl Microbiol Biotechnol. 2007;73(6):1348–54.
34. Koh CM, Liu Y, Moehninsi, Du M, Ji L. Molecular characterization of *KU70* and *KU80* homologues and exploitation of a *KU70*-deficient mutant for improving gene deletion frequency in *Rhodosporidium toruloides*. BMC Microbiol. 2014;14:50.
35. Colot HV, Park G, Turner GE, Ringelberg C, Crew CM, Litvinkova L, et al. A high-throughput gene knockout procedure for *Neurospora* reveals functions for multiple transcription factors. Proc Natl Acad Sci U S A. 2006;103(27):10352–7.
36. Guangtao Z, Hartl L, Schuster A, Polak S, Schmoll M, Wang T, et al. Gene targeting in a nonhomologous end joining deficient Hypocrea jecorina. J Biotechnol. 2009;139(2):146–51.
37. Sorensen LQ, Lysoe E, Larsen JE, Khorsand-Jamal P, Nielsen KF, Frandsen RJ. Genetic transformation of *Fusarium avenaceum* by *Agrobacterium tumefaciens* mediated transformation and the development of a USER-Brick vector construction system. BMC Mol Biol. 2014;15:15.
38. Gauthier GM, Sullivan TD, Gallardo SS, Brandhorst TT, Vanden Wymelenberg AJ, Cuomo CA, et al. SREB, a GATA transcription factor that directs disparate fates in blastomyces dermatitidis including morphogenesis and siderophore biosynthesis. PLoS Pathog. 2010;6(4):e1000846.
39. Lazo GR, Stein PA, Ludwig RA. A DNA transformation-competent *Arabidopsis* genomic library in *Agrobacterium*. Nat biotechnol. 1991;9(10):963–7.
40. Raeder U, Broda P. Rapid preparation of DNA from filamentous fungi. Lett Appl Microbiol. 1985;1(1):17–20.
41. Sambrook J, Russell DW. Molecular cloning - A laboratory manual, 3rd edition. Cold Spring Harbor Laboratory Press 2001.

Permissions

All chapters in this book were first published in BMCB, by BioMed Central; hereby published with permission under the Creative Commons Attribution License or equivalent. Every chapter published in this book has been scrutinized by our experts. Their significance has been extensively debated. The topics covered herein carry significant findings which will fuel the growth of the discipline. They may even be implemented as practical applications or may be referred to as a beginning point for another development.

The contributors of this book come from diverse backgrounds, making this book a truly international effort. This book will bring forth new frontiers with its revolutionizing research information and detailed analysis of the nascent developments around the world.

We would like to thank all the contributing authors for lending their expertise to make the book truly unique. They have played a crucial role in the development of this book. Without their invaluable contributions this book wouldn't have been possible. They have made vital efforts to compile up to date information on the varied aspects of this subject to make this book a valuable addition to the collection of many professionals and students.

This book was conceptualized with the vision of imparting up-to-date information and advanced data in this field. To ensure the same, a matchless editorial board was set up. Every individual on the board went through rigorous rounds of assessment to prove their worth. After which they invested a large part of their time researching and compiling the most relevant data for our readers.

The editorial board has been involved in producing this book since its inception. They have spent rigorous hours researching and exploring the diverse topics which have resulted in the successful publishing of this book. They have passed on their knowledge of decades through this book. To expedite this challenging task, the publisher supported the team at every step. A small team of assistant editors was also appointed to further simplify the editing procedure and attain best results for the readers.

Apart from the editorial board, the designing team has also invested a significant amount of their time in understanding the subject and creating the most relevant covers. They scrutinized every image to scout for the most suitable representation of the subject and create an appropriate cover for the book.

The publishing team has been an ardent support to the editorial, designing and production team. Their endless efforts to recruit the best for this project, has resulted in the accomplishment of this book. They are a veteran in the field of academics and their pool of knowledge is as vast as their experience in printing. Their expertise and guidance has proved useful at every step. Their uncompromising quality standards have made this book an exceptional effort. Their encouragement from time to time has been an inspiration for everyone.

The publisher and the editorial board hope that this book will prove to be a valuable piece of knowledge for researchers, students, practitioners and scholars across the globe.

List of Contributors

Kai Lerche
Department of Cellular and Developmental Biology of Plants, University of Bielefeld, Universitätsstr. 25, D-33615 Bielefeld, Germany

Armin Hallmann
Department of Cellular and Developmental Biology of Plants, University of Bielefeld, Universitätsstr. 25, D-33615 Bielefeld, Germany

Ulrich M Tillich
Molecular Biotechnology and Functional Genomics, Technical University of Applied Sciences Wildau, Bahnhofstraße 1, 16-2001, D-15745 Wildau, Germany
Institute of Biology, Humboldt-University Berlin, Berlin, Germany

Nick Wolter
Molecular Biotechnology and Functional Genomics, Technical University of Applied Sciences Wildau, Bahnhofstraße 1, 16-2001, D-15745 Wildau, Germany

Philipp Franke
Molecular Biotechnology and Functional Genomics, Technical University of Applied Sciences Wildau, Bahnhofstraße 1, 16-2001, D-15745 Wildau, Germany

Ulf Dühring
Algenol Biofuels Germany GmbH, Berlin, Germany

Marcus Frohme
Molecular Biotechnology and Functional Genomics, Technical University of Applied Sciences Wildau, Bahnhofstraße 1, 16-2001, D-15745 Wildau, Germany

Tsevelmaa Nanjidsuren
Animal Biotechnology, Graduate School of Bio and Information Technology, Institute of Genetic Engineering, Hankyong National University, Ansung 456-749, Republic of Korea

Kwan-Sik Min
Animal Biotechnology, Graduate School of Bio and Information Technology, Institute of Genetic Engineering, Hankyong National University, Ansung 456-749, Republic of Korea

Maria N Somleva
Metabolix, Inc., 21 Erie St., Cambridge, MA 02139, USA

Chang Ai Xu
Metabolix, Inc., 21 Erie St., Cambridge, MA 02139, USA

Kieran P Ryan
Metabolix, Inc., 21 Erie St., Cambridge, MA 02139, USA

Roger Thilmony
USDA-ARS-CIU, 800 Buchanan St., Albany, CA 94710, USA

Oliver Peoples
Metabolix, Inc., 21 Erie St., Cambridge, MA 02139, USA

Kristi D Snell
Metabolix, Inc., 21 Erie St., Cambridge, MA 02139, USA

James Thomson
USDA-ARS-CIU, 800 Buchanan St., Albany, CA 94710, USA

Catherine A Lichten
SynthSys, University of Edinburgh, Mayfield Road, Edinburgh, UK
Department of Physiology, McGill University, Promenade Sir William Osler, Montreal, Canada

Rachel White
Biological Sciences, University of Edinburgh, Mayfield Road, Edinburgh, UK

Ivan BN Clark
SynthSys, University of Edinburgh, Mayfield Road, Edinburgh, UK

Peter S Swain
SynthSys, University of Edinburgh, Mayfield Road, Edinburgh, UK

Dongmei Han
Department of Biological Sciences, Bowling Green State University, 43403 Bowling Green, OH, USA

Hui Xu
Department of Biological Sciences, Bowling Green State University, 43403 Bowling Green, OH, USA

Rutika Puranik
Department of Biological Sciences, Bowling Green State University, 43403 Bowling Green, OH, USA

Zhaohui Xu
Department of Biological Sciences, Bowling Green State University, 43403 Bowling Green, OH, USA

David R Micklem
Department of Biomedicine, University of Bergen, N-5009 Bergen, Norway

BerGenBio A/S, N-5009, Bergen, Norway

Magnus Blø
Department of Biomedicine, University of Bergen, N-5009 Bergen, Norway
BerGenBio A/S, N-5009, Bergen, Norway

Petra Bergström
Department of Clinical Chemistry and Transfusion Medicine, Sahlgrenska University Hospital, S-413 45 Göteborg, Sweden

Erlend Hodneland
Department of Biomedicine, University of Bergen, N-5009 Bergen, Norway

Crina Tiron
Department of Biomedicine, University of Bergen, N-5009 Bergen, Norway

Torill Høiby
Department of Biomedicine, University of Bergen, N-5009 Bergen, Norway

Christine Gjerdrum
Department of Biomedicine, University of Bergen, N-5009 Bergen, Norway

Ola Hammarsten
Department of Clinical Chemistry and Transfusion Medicine, Sahlgrenska University Hospital, S-413 45 Göteborg, Sweden

James B Lorens
Department of Biomedicine, University of Bergen, N-5009 Bergen, Norway

Esther Yip-Mei Liu
Institute of Clinical Medicine, School of Medicine, National Yang-Ming University, Taipei 11217, Taiwan

Feng-Yee Chang
Division of Infectious Diseases, Department of Internal Medicine, Tri-Service General Hospital, National Defense Medical Center, Taipei 11490, Taiwan

Jen-Chang Chang
National Institute of Infectious Diseases and Vaccinology, National Health Research Institutes, Zhunan, Miaoli County 35053, Taiwan

Chang-Phone Fung
Institute of Clinical Medicine, School of Medicine, National Yang-Ming University, Taipei 11217, Taiwan
Division of Infectious Diseases, Department of Medicine, Taipei Veterans General Hospital and National Yang-Ming University, Taipei 11217, Taiwan

Akram Ghantous
Epigenetics Group, International Agency for Research on Cancer (IARC), 150 rue Albert-Thomas, 69008 Lyon, France

Richard Saffery
Cancer and Disease Epigenetics, Murdoch Childrens Research Institute Royal Children's Hospital, Flemington Road Parkville, Victoria 3052, Australia

Marie-Pierre Cros
Epigenetics Group, International Agency for Research on Cancer (IARC), 150 rue Albert-Thomas, 69008 Lyon, France

Anne-Louise Ponsonby
Environmental & Genetic Epidemiology Research, Murdoch Children's Research Institute Royal Children's Hospital, Flemington Road Parkville, Victoria 3052, Australia
Menzies Research Institute, University of Tasmania, Hobart TAS 7000, Australia

Steven Hirschfeld
Department of Health and Human Services, National Children's Study, Eunice Kennedy Shriver National Institute of Child Health and Human Development, National Institutes of Health U.S., 6100 Executive Boulevard, Room 3A01, Bethesda, MD 20892, USA

Carol Kasten
Department of Health and Human Services, National Children's Study, Eunice Kennedy Shriver National Institute of Child Health and Human Development, National Institutes of Health U.S., 6100 Executive Boulevard, Room 3A01, Bethesda, MD 20892, USA

Terence Dwyer
Environmental & Genetic Epidemiology Research, Murdoch Children's Research Institute Royal Children's Hospital, Flemington Road Parkville, Victoria 3052, Australia
Menzies Research Institute, University of Tasmania, Hobart TAS 7000, Australia
Chair, Steering Committtee, International Childhood Cancer Cohort Consortium (I4C)

Zdenko Herceg
Epigenetics Group, International Agency for Research on Cancer (IARC), 150 rue Albert-Thomas, 69008 Lyon, France

Hector Hernandez-Vargas
Epigenetics Group, International Agency for Research on Cancer (IARC), 150 rue Albert-Thomas, 69008 Lyon, France

Takayuki Sakurai
Department of Cardiovascular Research, Graduate school of Medicine, Shinshu University, 3-1-1 Asahi, Matsumoto, Nagano 390-8621, Japan

Satoshi Watanabe
Animal Genome Research Unit, Division of Animal Science, National Institute of Agrobiological Sciences, 2-1-2 Kannondai, Tsukuba, Ibaraki 305-8602, Japan

Akiko Kamiyoshi
Department of Cardiovascular Research, Graduate school of Medicine, Shinshu University, 3-1-1 Asahi, Matsumoto, Nagano 390-8621, Japan

Masahiro Sato
Section of Gene Expression Regulation, Frontier Science Research Center, Kagoshima University, 8-35-1 Sakuragaoka, Kagoshima, Kagoshima 890-8544, Japan

Takayuki Shindo
Department of Cardiovascular Research, Graduate school of Medicine, Shinshu University, 3-1-1 Asahi, Matsumoto, Nagano 390-8621, Japan

Rajiv Kumar Yadav
Ecologie Microbienne, UMR CNRS 5557, USC INRA 1364, Université Lyon 1, Université de Lyon, Villeurbanne, France

Florian Barbi
Ecologie Microbienne, UMR CNRS 5557, USC INRA 1364, Université Lyon 1, Université de Lyon, Villeurbanne, France

Antoine Ziller
Ecologie Microbienne, UMR CNRS 5557, USC INRA 1364, Université Lyon 1, Université de Lyon, Villeurbanne, France

Patricia Luis
Ecologie Microbienne, UMR CNRS 5557, USC INRA 1364, Université Lyon 1, Université de Lyon, Villeurbanne, France

Roland Marmeisse
Ecologie Microbienne, UMR CNRS 5557, USC INRA 1364, Université Lyon 1, Université de Lyon, Villeurbanne, France

M Sudhakara Reddy
Department of Biotechnology, Thapar University, Patiala 147004, Punjab, India

Laurence Fraissinet-Tachet
Ecologie Microbienne, UMR CNRS 5557, USC INRA 1364, Université Lyon 1, Université de Lyon, Villeurbanne, France

Chizue Hiruta
Okazaki Institute for Integrative Bioscience, National Institute for Basic Biology, National Institutes of Natural Sciences, Higashiyama 5-1, Myodaiji, Okazaki, Aichi 444-8787, Japan

Yukiko Ogino
Okazaki Institute for Integrative Bioscience, National Institute for Basic Biology, National Institutes of Natural Sciences, Higashiyama 5-1, Myodaiji, Okazaki, Aichi 444-8787, Japan Faculty of Life Science, Graduate University for Advanced Studies (SOKENDAI), 5-1 Higashiyama, Myodaiji, Okazaki, Aichi 444-8787, Japan

Tetsushi Sakuma
Department of Mathematical and Life Sciences, Graduate School of Science, Hiroshima University, Higashi-Hiroshima 739-8526, Japan

Kenji Toyota
Okazaki Institute for Integrative Bioscience, National Institute for Basic Biology, National Institutes of Natural Sciences, Higashiyama 5-1, Myodaiji, Okazaki, Aichi 444-8787, Japan Faculty of Life Science, Graduate University for Advanced Studies (SOKENDAI), 5-1 Higashiyama, Myodaiji, Okazaki, Aichi 444-8787, Japan

Shinichi Miyagawa
Okazaki Institute for Integrative Bioscience, National Institute for Basic Biology, National Institutes of Natural Sciences, Higashiyama 5-1, Myodaiji, Okazaki, Aichi 444-8787, Japan Faculty of Life Science, Graduate University for Advanced Studies (SOKENDAI), 5-1 Higashiyama, Myodaiji, Okazaki, Aichi 444-8787, Japan

Takashi Yamamoto
Department of Mathematical and Life Sciences, Graduate School of Science, Hiroshima University, Higashi-Hiroshima 739-8526, Japan

Taisen Iguchi
Okazaki Institute for Integrative Bioscience, National Institute for Basic Biology, National Institutes of Natural Sciences, Higashiyama 5-1, Myodaiji, Okazaki, Aichi 444-8787, Japan Faculty of Life Science, Graduate University for Advanced Studies (SOKENDAI), 5-1 Higashiyama, Myodaiji, Okazaki, Aichi 444-8787, Japan

Thomas B Jacobs
Institute for Plant Breeding, Genetics and Genomics, University of Georgia, Athens, Georgia 30602, USA
Center for Applied Genetic Technologies, University of Georgia, Athens, GA 30602, USA
Present address: Boyce Thompson Institute for Plant Research, Ithaca, NY 14853, USA

Peter R LaFayette
Department of Crop and Soil Sciences, University of Georgia, Athens, GA 30602, USA
Center for Applied Genetic Technologies, University of Georgia, Athens, GA 30602, USA

Robert J Schmitz
Department of Genetics, University of Georgia, Athens, GA 30602, USA

Wayne A Parrott
Institute for Plant Breeding, Genetics and Genomics, University of Georgia, Athens, Georgia 30602, USA
Department of Crop and Soil Sciences, University of Georgia, Athens, GA 30602, USA
Center for Applied Genetic Technologies, University of Georgia, Athens, GA 30602, USA

Yoshiko Nakagawa
Center for Animal Resources and Development, Kumamoto University, Kumamoto 860-0811, Japan

Tetsushi Sakuma
Department of Mathematical and Life Sciences, Graduate School of Science, Hiroshima University, Hiroshima 739-8526, Japan

Takuya Sakamoto
Department of Mathematical and Life Sciences, Graduate School of Science, Hiroshima University, Hiroshima 739-8526, Japan

Masaki Ohmuraya
Center for Animal Resources and Development, Kumamoto University, Kumamoto 860-0811, Japan

Naomi Nakagata
Center for Animal Resources and Development, Kumamoto University, Kumamoto 860-0811, Japan

Takashi Yamamoto
Department of Mathematical and Life Sciences, Graduate School of Science, Hiroshima University, Hiroshima 739-8526, Japan

Mihaela Škulj
Sandoz Biopharmaceuticals, Mengeš, Lek Pharmacetucals d.d, Kolodvorska 27, 1234 Mengeš, Slovenia

Dejan Pezdirec
Sandoz Biopharmaceuticals, Mengeš, Lek Pharmacetucals d.d, Kolodvorska 27, 1234 Mengeš, Slovenia

Dominik Gaser
Sandoz Biopharmaceuticals, Mengeš, Lek Pharmacetucals d.d, Kolodvorska 27, 1234 Mengeš, Slovenia

Marko Kreft
Celica Biomedical Centre, Tehnološki Park 24, 1000 Ljubljana, Slovenia
Laboratory of Neuroendocrinology - Molecular Cell Physiology, Institute of Pathophysiology, Faculty of Medicine, University of Ljubljana, Zaloška 4, 1000 Ljubljana, Slovenia
Biotechnical Faculty, University of Ljubljana, Večna Pot 111, 1000 Ljubljana, Slovenia

Robert Zorec
Celica Biomedical Centre, Tehnološki Park 24, 1000 Ljubljana, Slovenia
Laboratory of Neuroendocrinology - Molecular Cell Physiology, Institute of Pathophysiology, Faculty of Medicine, University of Ljubljana, Zaloška 4, 1000 Ljubljana, Slovenia

Kwang-Ryong Jo
Wageningen UR Plant Breeding, Wageningen University & Research Centre, P.O. Box 386, 6700 AJ Wageningen, The Netherlands
Graduate School Experimental Plant Sciences, Wageningen University, Wageningen, The Netherlands
Research Institute of Agrobiology, Academy of Agricultural Sciences, Pyongyang, DPR Korea

Chol-Jun Kim
Research Institute of Agrobiology, Academy of Agricultural Sciences, Pyongyang, DPR Korea

Sung-Jin Kim
Research Institute of Agrobiology, Academy of Agricultural Sciences, Pyongyang, DPR Korea

Tok-Yong Kim
Research Institute of Agrobiology, Academy of Agricultural Sciences, Pyongyang, DPR Korea

Marjan Bergervoet
Wageningen UR Plant Breeding, Wageningen University & Research Centre, P.O. Box 386, 6700 AJ Wageningen, The Netherlands

Maarten A Jongsma
Plant Research International, Wageningen University and Research Centre, Wageningen, The Netherlands

Richard GF Visser
Wageningen UR Plant Breeding, Wageningen University & Research Centre, P.O. Box 386, 6700 AJ Wageningen, The Netherlands

Evert Jacobsen
Wageningen UR Plant Breeding, Wageningen University & Research Centre, P.O. Box 386, 6700 AJ Wageningen, The Netherlands

Jack H Vossen
Wageningen UR Plant Breeding, Wageningen University & Research Centre, P.O. Box 386, 6700 AJ Wageningen, The Netherlands

Yongjiang Lang
The Key Laboratory of Molecular Medicine, Ministry of Education, Fudan University, Shanghai, P.R. China
The Department of Biochemistry and Molecular Biology, School of Basic Medical Sciences, Fudan University, Shanghai, P.R. China

Yanling Zhang
The Key Laboratory of Molecular Medicine, Ministry of Education, Fudan University, Shanghai, P.R. China
The Department of Biochemistry and Molecular Biology, School of Basic Medical Sciences, Fudan University, Shanghai, P.R. China

Ling Zhan
The Key Laboratory of Molecular Medicine, Ministry of Education, Fudan University, Shanghai, P.R. China
The Department of Biochemistry and Molecular Biology, School of Basic Medical Sciences, Fudan University, Shanghai, P.R. China

Zhe Feng
The Key Laboratory of Molecular Medicine, Ministry of Education, Fudan University, Shanghai, P.R. China
The Department of Biochemistry and Molecular Biology, School of Basic Medical Sciences, Fudan University, Shanghai, P.R. China

Xiushi Zhou
The Key Laboratory of Molecular Medicine, Ministry of Education, Fudan University, Shanghai, P.R. China
The Department of Biochemistry and Molecular Biology, School of Basic Medical Sciences, Fudan University, Shanghai, P.R. China

Min Yu
The Key Laboratory of Molecular Medicine, Ministry of Education, Fudan University, Shanghai, P.R. China
The Department of Biochemistry and Molecular Biology, School of Basic Medical Sciences, Fudan University, Shanghai, P.R. China

Wei Mo
The Key Laboratory of Molecular Medicine, Ministry of Education, Fudan University, Shanghai, P.R. China
The Department of Biochemistry and Molecular Biology, School of Basic Medical Sciences, Fudan University, Shanghai, P.R. China

Hanpeng Liao
Jiangsu Collaborative Innovation Center for Solid Organic Waste Utilization, College of Resources and Environmental Science, Nanjing Agricultural University, Nanjing 210095, China

Shuixian Li
Jiangsu Collaborative Innovation Center for Solid Organic Waste Utilization, College of Resources and Environmental Science, Nanjing Agricultural University, Nanjing 210095, China

Haiping Zheng
Jiangsu Collaborative Innovation Center for Solid Organic Waste Utilization, College of Resources and Environmental Science, Nanjing Agricultural University, Nanjing 210095, China

Zhong Wei
Jiangsu Collaborative Innovation Center for Solid Organic Waste Utilization, College of Resources and Environmental Science, Nanjing Agricultural University, Nanjing 210095, China

Dongyang Liu
Jiangsu Collaborative Innovation Center for Solid Organic Waste Utilization, College of Resources and Environmental Science, Nanjing Agricultural University, Nanjing 210095, China

Waseem Raza
Jiangsu Collaborative Innovation Center for Solid Organic Waste Utilization, College of Resources and Environmental Science, Nanjing Agricultural University, Nanjing 210095, China

Qirong Shen
Jiangsu Collaborative Innovation Center for Solid Organic Waste Utilization, College of Resources and Environmental Science, Nanjing Agricultural University, Nanjing 210095, China

Yangchun Xu
Jiangsu Collaborative Innovation Center for Solid Organic Waste Utilization, College of Resources and Environmental Science, Nanjing Agricultural University, Nanjing 210095, China

Hiroshi Shinozuka
Department of Economic Development, Jobs, Transport and Resources, Biosciences Research Division, AgriBio, Centre for AgriBioscience, 5 Ring Road, La Trobe University Research and Development Park, Bundoora, Victoria 3083, Australia
Dairy Futures Cooperative Research Centre, Bundoora, Australia

Noel O I Cogan
Department of Economic Development, Jobs, Transport and Resources, Biosciences Research Division, AgriBio, Centre for AgriBioscience, 5 Ring Road, La Trobe University Research and Development Park, Bundoora, Victoria 3083, Australia
Dairy Futures Cooperative Research Centre, Bundoora, Australia

Maiko Shinozuka
Department of Economic Development, Jobs, Transport and Resources, Biosciences Research Division, AgriBio, Centre for AgriBioscience, 5 Ring Road, La Trobe University Research and Development Park, Bundoora, Victoria 3083, Australia
Dairy Futures Cooperative Research Centre, Bundoora, Australia

Alexis Marshall
Department of Economic Development, Jobs, Transport and Resources, Biosciences Research Division, AgriBio, Centre for AgriBioscience, 5 Ring Road, La Trobe University Research and Development Park, Bundoora, Victoria 3083, Australia
Dairy Futures Cooperative Research Centre, Bundoora, Australia

Pippa Kay
Department of Economic Development, Jobs, Transport and Resources, Biosciences Research Division, AgriBio, Centre for AgriBioscience, 5 Ring Road, La Trobe University Research and Development Park, Bundoora, Victoria 3083, Australia

Yi-Han Lin
Department of Economic Development, Jobs, Transport and Resources, Biosciences Research Division, AgriBio, Centre for AgriBioscience, 5 Ring Road, La Trobe University Research and Development Park, Bundoora, Victoria 3083, Australia
Dairy Futures Cooperative Research Centre, Bundoora, Australia

German C Spangenberg
Department of Economic Development, Jobs, Transport and Resources, Biosciences Research Division, AgriBio, Centre for AgriBioscience, 5 Ring Road, La Trobe University Research and Development Park, Bundoora, Victoria 3083, Australia
Dairy Futures Cooperative Research Centre, Bundoora, Australia
School of Applied Systems Biology, La Trobe University, Bundoora, Victoria 3086, Australia

John W Forster
Department of Economic Development, Jobs, Transport and Resources, Biosciences Research Division, AgriBio, Centre for AgriBioscience, 5 Ring Road, La Trobe University Research and Development Park, Bundoora, Victoria 3083, Australia
Dairy Futures Cooperative Research Centre, Bundoora, Australia
School of Applied Systems Biology, La Trobe University, Bundoora, Victoria 3086, Australia

Rajeeva Gaur
Department of Microbiology (Centre of Excellence), Dr. Ram Manohar Lohia Avadh University, Faizabad 224001, Uttar Pradesh, India

Soni Tiwari
Department of Microbiology (Centre of Excellence), Dr. Ram Manohar Lohia Avadh University, Faizabad 224001, Uttar Pradesh, India

Wenjing Pan
Biotechnology Science and Engineering Program, University of Alabama in Huntsville, Huntsville, AL 35899, USA

Miranda Byrne-Steele
HudsonAlpha Institute for Biotechnology, Huntsville, AL 35806, USA

Chunlin Wang
Stanford Genome Technology Center, Stanford University, Palo Alto, CA 94304, USA

Stanley Lu
Diatherix Laboratories, Huntsville, AL 35806, USA

Scott Clemmons
Diatherix Laboratories, Huntsville, AL 35806, USA

Robert J Zahorchak
HudsonAlpha Institute for Biotechnology, Huntsville, AL 35806, USA

Jian Han
HudsonAlpha Institute for Biotechnology, Huntsville, AL 35806, USA

Jing Xu
College of Life Sciences, Hubei University, Wuhan 430062, China
Key Laboratory of Systems Microbial Biotechnology, Tianjin Institute of Industrial Biotechnology, Chinese Academy of Sciences, Tianjin 300308, China

Jingen Li
Key Laboratory of Systems Microbial Biotechnology, Tianjin Institute of Industrial Biotechnology, Chinese Academy of Sciences, Tianjin 300308, China

Liangcai Lin
Key Laboratory of Systems Microbial Biotechnology, Tianjin Institute of Industrial Biotechnology, Chinese Academy of Sciences, Tianjin 300308, China

Qian Liu
Key Laboratory of Systems Microbial Biotechnology, Tianjin Institute of Industrial Biotechnology, Chinese Academy of Sciences, Tianjin 300308, China

Wenliang Sun
Key Laboratory of Systems Microbial Biotechnology, Tianjin Institute of Industrial Biotechnology, Chinese Academy of Sciences, Tianjin 300308, China

Bangquan Huang
College of Life Sciences, Hubei University, Wuhan 430062, China

Chaoguang Tian
Key Laboratory of Systems Microbial Biotechnology, Tianjin Institute of Industrial Biotechnology, Chinese Academy of Sciences, Tianjin 300308, China

CPSIA information can be obtained
at www.ICGtesting.com
Printed in the USA
BVOW10*0652140716
455452BV00038B/24/P

9 781682 862438